Flow-Induced Vibration

Flow-Induced Vibration

Second Edition

Robert D. Blevins

KRIEGER PUBLISHING COMPANY
MALABAR, FLORIDA

Second Edition 1990
Reprint Edition 1994 w/updating and new preface
Reprint Edition 2001

Printed and Published by
KRIEGER PUBLISHING COMPANY
KRIEGER DRIVE
MALABAR, FLORIDA 32950

**FROM A DECLARATION OF PRINCIPLES JOINTLY ADOPTED BY A
COMMITTEE OF THE AMERICAN BAR ASSOCIATION AND A COM-
MITTEE OF PUBLISHERS:**
This publication is designed to provide accurate and authoritative information
in regard to the subject matter covered. It is sold with the understanding that the
publisher is not engaged in rendering legal, accounting, or other
professional service. If legal advice or other expert assistance is required, the
services of a competent professional person should be sought.

Library of Congress Cataloging-In-Publication Data

Blevins, Robert D.
   Flow-induced vibration / Robert D. Blevins. — 2nd ed.
      p. cm.
   Originally published: New York : Van Nostrand Reinhold.
c1990.
   Includes bibliographical references and indexes.
   ISBN 1-57524-183-8 (alk. paper)
      1. Vibration. 2. Fluid dynamics. I. Title.
[TA355.B52   1993]
620. 3—dc20   92-40221
                                                         CIP

10   9   8   7   6   5

Contents

Preface

The purpose of this book is to provide mechanical engineers, aeronautical engineers, civil engineers, marine engineers, and students with analytical tools for the analysis of the vibrations of structures exposed to fluid flow. The book assumes some knowledge of structural vibrations and fluid mechanics. Since the book is basically introductory, a review of vibrations of structures is provided in the appendices, and some explanatory material on the theory of random vibrations is also presented. The text is organized according to the fluid dynamic mechanism for exciting vibration rather than by type of structure, since any given structure may be subject to several mechanisms for flow-induced vibration.

This book represents the sum of twenty-five years of research at General Atomics, San Diego, California; the California Institute of Technology, Pasadena, California; David Taylor Model Basin, Carderock, Maryland; and Rohr Industries, Chula Vista, California. The second edition is revised and expanded from the first edition. The chapter on ideal fluid models is entirely new. All other chapters have been updated and expanded. There is new material on aeronautical problems of flutter and sonic fatigue in addition to the heat-exchanger, offshore, and wind-engineering topics treated in the first edition. Many exercises and examples have been added to make the book more suitable for use as a text book.

In this Krieger reprint edition, all known errors have been corrected and updates have been made to reflect current research. Since many systems that are prone to flow-induced vibration are also governed by the ASME code, new information on vibration-related code rules has been added.

Appendix E contains procedures for flow-induced vibration analysis given in Appendix N, Code Section 3, of the American Society of Mechanical Engineers Boiler and Pressure Vessel Code, 1992. Table 3-2 gives Code recommendations for calculating vortex-induced vibration of cylinders, Figure 5-6(a) gives Code recommendations for calculation of instability in tube arrays, and Sections 7.3.2 and 7.3.3 present methods consistent with the Code for turbulence-induced vibration.

vii

Computer programs that implement many of the solutions in this book are available from

Avian Software
P.O. Box 80374
San Diego, CA 92138

The programs include discrete vortex motion, vortex-induced vibration, flutter, heat-exchanger vibration analysis, and sonic fatigue of plates. As of this writing, the price is $179 for individual copies or class license. The programs are provided on disks suitable for IBM PC or compatible.

The author is indebted to many individuals for their help with the second edition. Colleagues, especially H. Aref, M. K. Au-Yang, B. Bickers, M. Gharib, I. Holehouse, T. M. Mulcahy, M. P. Paidoussis, M. W. Parkin, M. J. Pettigrew, J. Sandifer, T. Sarpkaya, and D. S. Weaver, provided insight, inspiration, and data. Kathleen Cudahy withstood the paper blizzard that covered our house and supported the author with kindness during endless rewrites. Boecky Yalof edited every word in the manuscript and the book is much better for her skill.

This book is dedicated to people who will never know my name or yours. They are the people who travel in the airplanes, work in the buildings, and are warmed by the power plants that this book endeavors to analyze and protect. They motivate our work and they are our first responsibility.

Nomenclature

A	Cross-sectional area or amplitude of vibration (one-half peak-to-peak displacement)
C_D	Drag coefficient
C_a	Added mass coefficient
C_L	Lift coefficient
C_M	Moment coefficient
C_m	Inertia coefficient
C_y	Vertical force coefficient
D	Width or characteristic length used in forming aerodynamic coefficients
F	Force per unit length or moment per unit length
G	Gust factor
I	Area bending moment of inertia
J_θ	Mass polar moment of inertia
Kc	Keulegan–Carpenter number
L	Spanwise structural length
M	Mass or added mass
Re	Reynolds number
S	Strouhal number
S_p	Pressure spectral density
SPL	Sound pressure level
U	Fluid velocity in free stream or at minimum area
V	Velocity of structure
$\underline{V}$	Volume
X	Position parallel to free stream
Y	Position normal to free stream
c	Speed of sound in fluid or chord of airfoil
f	Frequency (Hz)
g	Acceleration of gravity
$\bar{g}$	Peak-to-rms ratio
i	Integer or imaginary constant
k	Spring constant per unit length
l_c	Correlation length
m	Mass per unit length
p	Pressure or probability

q	Dynamic pressure, $\rho U^2/2$
r	Correlation function
t	Time
v	Transverse fluid velocity
w	Fluid displacement parameter
x	Displacement of coordinate parallel to free stream
y	Displacement of coordinate normal to free stream
z	Spanwise coordinate or complex number
α	Angle of attack
δ	Logrithmic decrement of damping
δ_r	Reduced damping, $2m(2\pi\zeta)/(\rho D^2)$
ζ	Damping factor
θ	Rotation angle
λ	Eigenvalue or wavelength
ν	Kinematic viscosity
ψ	Stream function
ϕ	Phase angle or potential function
τ	Time lag
ρ	Fluid density
ω	Circular frequency (radians/sec) or vorticity

Subscripts and Superscripts

a	Acoustic
D	Drag
L	Lift
i	Integer or imaginary constant $(-1)^{1/2}$
j	Integer
m	Maximum value
rms	Square root of mean of square
x	Parallel to free stream
y	Normal to free stream
θ	Torsion
$(\dot{\ })$	Differentiation with respect to time
$(\bar{\ })$	Average (mean) with respect to time
$\mathbf{B}$	Vector quantity (boldface)
$(\tilde{\ })$	Mode shape, a function of x, y, or z

Introduction

The flow of fluids around structures, ranging from clarinet reeds to skyscrapers, can cause destructive vibrations as well as useful motions. Wind rustles leaves, rings wind chimes, uproots crops, and destroys whole communities. The sea currents that fan leaves of kelp also bring offshore platforms crashing into the ocean. The liquid sodium that cools the core of a breeder reactor can break shielding and precipitate melting of the reactor core. Such flow-induced vibrations have become increasingly important in recent years because designers are using materials to their limits, causing structures to become progressively lighter, more flexible, and more prone to vibration.

A new jargon has developed to describe these flow-induced vibrations. Ice-coated transmission lines in Canada "gallop" in a steady wind; marine tow cables suspending hydrophones "strum" at predictable frequencies; slender wings of aircraft "flutter" above a critical speed; heat-exchanger tube arrays experience "fluid elastic instability." Each of these vibrations arises from distinct fluid-dynamic phenomena that can be classified by the nature of the flow and the structure, as shown in Figure A.

This book explores the vibrations of structures induced by a subsonic flow. Many of the structures considered are bluff. A *bluff* structure is one in which the flow separates from a large section of the structure's surface. Most civil engineering structures, such as bridges or heat-exchanger tubes, are bluff structures. The primary purpose of these structures is not to gain lift or minimize drag as it is with aircraft components, but rather to bear loads, contain flow, or provide heat transfer surface. These structures are not aerodynamically optimized. Flow-induced vibrations are usually regarded as a secondary design consideration, at least until a failure occurs.

The fluid flow and the structure are coupled through the force exerted on the structure by the fluid, as shown in Figure B. The fluid force causes the structure to deform. As the structure deforms, its orientation to the flow changes, and the fluid force may change. In some cases, such as the excitation of ship hull plates by turbulence of rushing water, the fluid force is independent of small changes in the position of the structure. In

1

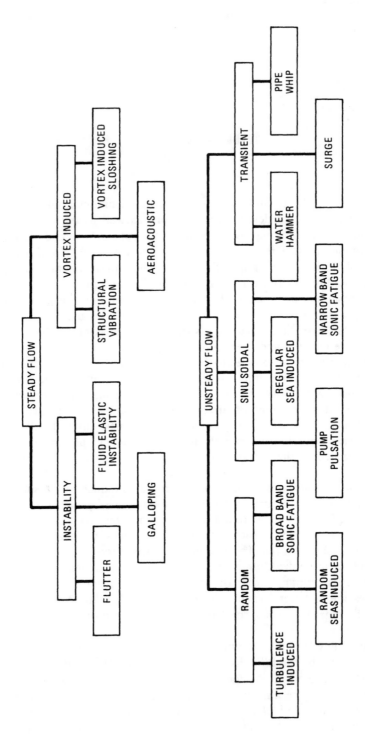

Fig. A A classification of flow-induced vibrations.

2

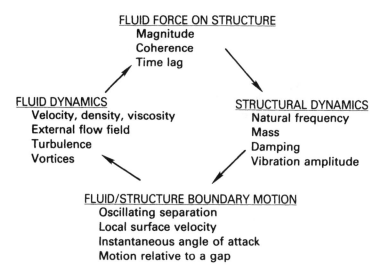

FLUID FORCE ON STRUCTURE
Magnitude
Coherence
Time lag

FLUID DYNAMICS
Velocity, density, viscosity
External flow field
Turbulence
Vortices

STRUCTURAL DYNAMICS
Natural frequency
Mass
Damping
Vibration amplitude

FLUID/STRUCTURE BOUNDARY MOTION
Oscillating separation
Local surface velocity
Instantaneous angle of attack
Motion relative to a gap

Fig. B Feedback between fluid and structure.

other cases, such as ice-coated power lines in a steady wind, the fluid force is completely determined by the orientation and velocity of the structure relative to the fluid flow. Finally, just as the fluid exerts a force on the structure, the structure exerts an equal but opposite force on the fluid. The structure force on the fluid can synchronize vortices in the wake and produce large-amplitude vibration.

In the analysis of flow-induced vibration, models are generated for both the structure and the fluid. Since most structures are near-linear in deformation with increasing load, the structures are modeled as linear oscillators. If only a single structural degree of freedom is excited, the structural motion is described by a single linear equation with fluid forcing. If more than one degree of structural freedom is present, such as translation and torsion, linear systems of equations are required. Fluid models are more difficult. Fluid mechanics is an inherently nonlinear, multi-degree-of-freedom phenomenon. Fluid models are generally formed from a combination of basic principles and experimental data.

From the database of physical laws and experimental data, model solutions are generated for the response of structure to flow. The solutions lead to a prediction that can be compared with new experimental data. The error between the new data and the prediction reflects the limitation of the models. Unfortunately, only a few general solutions are available. Most results rely on experimental data. There are sets of parameters that have proven useful in a wide variety of problems. These nondimensional parameters are discussed in Chapter 1.

In Chapter 2, ideal fluid models are considered. In Chapters 3 through 5, analysis is developed for the response of structures in a steady flow.

The effects of oscillatory flow, turbulence, and sound on a structure are examined in Chapters 6 and 7. In Chapter 8, damping of structures, which is the primary mechanism for limiting flow-induced vibration, is explored analytically and experimentally. Flow-induced sound (aero-acoustics) is investigated in Chapter 9, while the effect of internal flow on a pipe is presented in Chapter 10. The appendices present mathematical fundamentals of structural dynamics, aeroacoustics, and spectral analysis.

Chapter 1

Dimensional Analysis

Figure 1-1 shows a two-dimensional, spring-supported, damped building model exposed to a steady flow. This model is prototypical of elastic structures exposed to a fluid flow. The vibrations of this model can be described in terms of nondimensional parameters governing the fluid flow, the model, and the fluid–structure interaction. These parameters are useful for scaling flow-induced vibration and estimating the importance of different fluid phenomena.

1.1. NONDIMENSIONAL PARAMETERS

1.1.1. Geometry

Geometry is the most important parameter in determining the fluid force on a structure. The geometry of the building model can be specified by its *fineness ratio*:

$$\frac{l}{D} = \frac{\text{length}}{\text{width}} = \text{fineness ratio.} \tag{1-1}$$

Specification of geometry ordinarily includes the ratio of length in the third dimension to width (this is called *aspect* ratio) and the ratio of surface roughness to width.

1.1.2. Reduced Velocity, Dimensionless Amplitude

As the model vibrates in the flow, it traces out the path shown in Figure 1-2. For steady vibration, the length of the path for one cycle is U/f, where U is the free stream velocity and f is the frequency of vibration. The width of the path is $2A_y$, where A_y is the amplitude of the vibration. These path dimensions can be related to the structural dimension:

$$\frac{U}{fD} = \frac{\text{path length per cycle}}{\text{model width}} = \text{reduced velocity,} \tag{1-2}$$

$$\frac{A_y}{D} = \frac{\text{vibration amplitude}}{\text{model width}} = \text{dimensionless amplitude.} \tag{1-3}$$

5

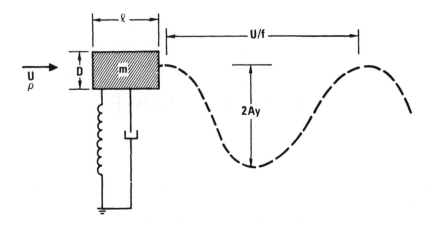

Fig. 1-1 Vibration of a two-dimensional, spring-supported, damped model.

The first of these parameters is commonly called *reduced velocity* or *nondimensional velocity*. Its inverse is called *nondimensional frequency*. The maximum model width (D) is ordinarily used in forming these parameters, because this width tends to govern the width of the wake. If reduced velocity is between 2 and 8, then the model often interacts strongly with vortex shedding in its own wake.

Two nondimensional parameters that are closely related to reduced velocity are Keulegan–Carpenter number and Strouhal number. The

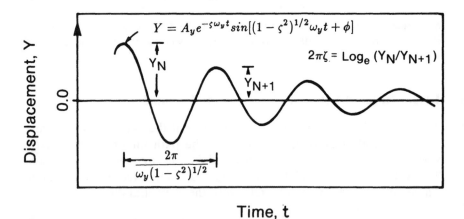

Fig. 1-2 Free vibration of a viscously damped structure; ω_y is the natural frequency of vibration in radians per second.

Keulegan–Carpenter number is used for oscillating flows, such as ocean waves; see Chapter 6. It is identical in form to the reduced velocity but with U defined as the amplitude of velocity of a flow that oscillates with frequency f about a structure of diameter D. The Strouhal number is $S = f_s D / U$, where f_s is the frequency of periodic vortex shedding from a structure of diameter D in a steady flow of velocity U; see Chapter 3.

1.1.3. Mass Ratio

The ratio of the model mass to the mass of fluid it displaces is proportional to

$$\frac{m}{\rho D^2} = \frac{\text{mass per unit length of model}}{\text{fluid density} \times \text{model width}^2} = \text{mass ratio}, \qquad (1\text{-}4)$$

where m ordinarily includes structural mass and the "added mass" of fluid entrained by the moving model (see Chapter 2). The mass ratio is a measure of the relative importance of buoyancy and added mass effects on the model. It is often used to measure the susceptibility of lightweight structures to flow-induced vibration. As the ratio of fluid mass to structural mass increases, so does the propensity for flow-induced vibration.

1.1.4. Reynolds Number

Boundary layer growth and flow separation are determined by fluid forces on a microscopic level. The boundary layer is impelled about the model by the inertia of the flow. Viscous friction at the model surface retards the boundary layer. It can be shown that the ratio of inertial force to viscous force in the boundary layer is (Schlichting, 1968)

$$\frac{UD}{\nu} = \frac{\text{inertial force}}{\text{viscous force}} = \text{Reynolds number}, \qquad (1\text{-}5)$$

where ν is the kinematic viscosity of the fluid and is equal to the absolute viscosity divided by density. The Reynolds number, abbreviated Re, scales the boundary layer thickness and transition from laminar to turbulent flow. Flow separates from the back of bluff bodies at Reynolds numbers, based on the width of the structure, greater than about 50.

1.1.5. Mach Number, Turbulence Intensity

Mach number is equal to

$$\frac{U}{c} = \frac{\text{fluid velocity}}{\text{speed of sound}} = \text{Mach number,} \qquad (1\text{-}6)$$

where c is the speed of sound in the fluid. Mach number measures the tendency of fluid to compress as it encounters a structure. Most of the analysis in this book is limited to Mach numbers less than 0.3, at which compressibility does not ordinarily influence the vibration.

Turbulence intensity, like Mach number, is a measurement relative to the free stream fluid velocity. The turbulence intensity,

$$\frac{u'_{rms}}{U} = \frac{\text{root mean square turbulence}}{\text{free stream fluid velocity}}, \qquad (1\text{-}7)$$

measures the turbulence in the flow. Usually the turbulence is generated upstream of the model. The model will respond to this random excitation. Low-turbulence wind tunnels typically have turbulence levels equal to 0.1% of the free stream velocity. Turbulence induced by wind is a factor of 100 greater (Section 7.4.1).

1.1.6. Damping Factor, Reduced Damping

The energy dissipated by a structure as it vibrates is characterized by its damping factor,

$$\zeta = \frac{\text{energy dissipated per cycle}}{4\pi \times \text{total energy of structure}} = \text{damping factor.} \qquad (1\text{-}8)$$

ζ is called the *damping factor* or *damping ratio*. It is often expressed as a fraction of 1, the critical damping factor. For linear, viscously damped structures, $2\pi\zeta$ is equal to the natural logarithm of the ratio of the amplitudes of any two successive cycles of a lightly damped structure in free decay, as shown in Figure 1-2. If the energy input to a structure by the flow is less than the energy expended in damping, then the flow-induced vibrations will diminish. Many real structures have damping factors on the order of 0.01 (i.e., 1% of critical); see Chapter 8.

A very useful parameter, variously called *mass damping, reduced damping,* or *Scruton number* (see Walshe, 1983), can be formed by the product of mass ratio and the damping factor:

$$\frac{2m(2\pi\zeta)}{\rho D^2} = \text{reduced damping.} \qquad (1\text{-}9)$$

Increasing reduced damping ordinarily reduces the amplitude of
flow-induced vibrations.

1.2. APPLICATION

The nondimensional parameters that have been found most useful in
describing the vibrations of an elastic structure in a subsonic (Mach
number less than 0.3) steady flow are:

1. Geometry (l/D).
2. Reduced velocity (U/fD).
3. Dimensionless amplitude (A_y/D).
4. Mass ratio ($m/\rho D^2$).
5. Reynolds number (UD/ν).
6. Damping factor (ζ).
7. Turbulence intensity (u'/U).

Some of these nondimensional parameters will appear in every
flow-induced vibration analysis. Other parameters may appear as well.

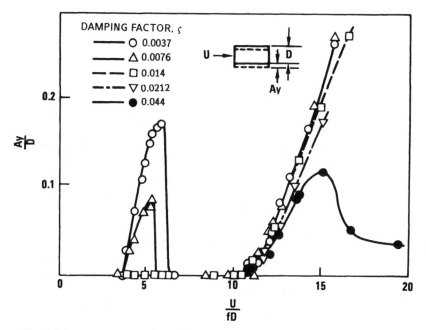

Fig. 1-3 Lateral response of a spring-supported, damped building model with a side
ratio of 2. The natural frequency of the model is 6 Hz, $D = 3.35$ in, and $m/(\rho D^2) = 129.5$.
(Novak, 1971.)

The objective of the majority of the analysis presented in this book is to predict the dimensionless amplitude of flow-induced vibration as a function of the remaining parameters,

$$\frac{A_y}{D} = F\left[\frac{l}{D}, \frac{UD}{v}, \frac{U}{fD}, \frac{m}{\rho D^2}, \zeta, \frac{u'}{U}\right]. \tag{1-10}$$

For example, Figure 1-3 shows the dimensionless amplitude of flow-induced vibration of a building model as a function of reduced velocity and damping factor. The vibrations occur at the natural frequency of the model. The two peaks in vibration amplitude are caused by different phenomena. Periodic vortex shedding occurs at a frequency given by $f_s = SU/D$, where S is approximately 0.2. The natural frequency is the characteristic frequency f and the reduced velocity at which the vortex shedding frequency equals the structural natural frequency is $U/(fD) \simeq 1/S = 5$. Thus, the first peak in the vibrations is caused by vortex shedding. The second peak, beginning at a reduced velocity of about 11, is associated with an instability called galloping, which is similar to aircraft flutter. Note that both sets of vibration decrease with increasing damping.

Exercise

1. Estimate each of the nondimensional parameters in this chapter for wind over your library building. D is the width of the building on the narrow side and L is the building height. Estimate the maximum wind velocity as 150 ft/sec (45 m/sec), or alternately, use the extreme wind data of Figs. 7-17 and 7-18 of Chapter 7. The room-temperature properties of air are: $\rho = 0.075 \text{ lb/ft}^3$ (1.2 kg/m^3), $v = 0.00016 \text{ ft}^2/\text{sec}$ (0.000015 m^2/sec), $c = 1100 \text{ ft/sec}$ (343 m/sec). Assume a typical building average density of 10 lb/ft^3 (150 kg/m^3). Calculate library mass per unit height. Estimate the natural frequency of the fundamental mode as $f = 10/N$ Hz, where N is the number of stories, or use Eq. 7-85. Estimate the damping factor as $\zeta = 0.0076$ or use the data presented in Chapter 8, Section 8.4.3. Estimate the rms turbulence as $u'_{rms} = 0.2U$ or use the data in Section 7.4.1, Eq. 7-64. If you don't feel your building moves significantly during the wind, then $A_y \approx 0$. Plot your library on Figure 1-3. How many stories high must your library be for $U/(fD) = 5$?

REFERENCES

Novak, M. (1971) "Galloping and Vortex Induced Oscillation of Structures," *Proceedings of the Conference on Wind Effects on Buildings and Structure,* held in Tokyo, Japan.

Schlichting, H. (1968) *Boundary Layer Theory,* McGraw-Hill, New York.

Walshe, D. E. (1983) "Scruton Number," *Jounal of Wind Engineering and Industrial Aerodynamics,* **12,** 99.

Chapter 2

Ideal Fluid Models

An ideal fluid is a fluid without viscosity. Not surprisingly, ideal fluid flows are modeled far more readily than viscous fluid flows. Ideal fluid flow modeling is developed in this chapter and applied to determine the added mass of accelerating bodies and fluid coupling between adjacent structures. Fundamentals of the discrete vortex dynamics are also presented.

2.1. FUNDAMENTALS OF POTENTIAL FLOW

Consider a reservoir of incompressible, inviscid (i.e., zero viscosity) fluid, shown in Figure 2-1(a). The reservoir contains a structural body that has a surface S. The fluid can move about the body and the body can move relative to the fluid. For two-dimensional motion, the fluid velocity vector is $\mathbf{V}$. The horizontal and vertical components of velocity are u and v, respectively:

$$\mathbf{V}(x, y, t) = u(x, y, t)\mathbf{i} + v(x, y, t)\mathbf{j}. \tag{2-1}$$

The two-dimensional rectangular coordinates are x and y. Time is represented by t, and $\mathbf{i}$ and $\mathbf{j}$ are the unit normal vectors in the x and y directions, respectively.

The motion of ideal fluid in a reservoir is governed by three sets of equations: (1) equation of continuity (conservation of mass); (2) Euler's equations (conservation of momentum); and (3) boundary conditions at surfaces. Each of these equations is developed below.

In two-dimensional rectangular coordinates, the equation of continuity of the incompressible fluid is (Newman, 1977, p. 59)

$$\frac{\partial u}{\partial x} + \frac{\partial v}{\partial y} = 0. \tag{2-2}$$

Equivalently, in vector notation the divergence of $\mathbf{V}$ is zero,

$$\nabla \cdot \mathbf{V} = 0, \tag{2-3}$$

where ∇ is the vector gradient operator and $(\cdot)$ is the vector dot product.

12

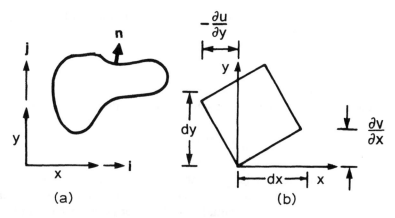

Fig. 2-1 Fluid in a reservoir: (a) body in an ideal fluid reservoir; (b) rotation of a fluid element.

Consider a velocity field that is derived from the differentials of a function $\phi(x, y, t)$ called a *velocity potential*,

$$u = \frac{\partial \phi}{\partial x} \quad \text{and} \quad v = \frac{\partial \phi}{\partial y}. \tag{2-4}$$

In vector terminology, the velocity vector is the gradient of the velocity potential,

$$\mathbf{V} = \mathbf{\nabla}\phi. \tag{2-5}$$

This equation implies that lines of constant ϕ (potential lines) are normal to the velocity vector.

By substituting Eq. 2-4 into the equation of continuity (Eq. 2-2) we see that the velocity potential satisfies a linear equation called the Laplace equation,

$$\frac{\partial^2 \phi}{\partial x^2} + \frac{\partial^2 \phi}{\partial y^2} \equiv \nabla^2 \phi = 0. \tag{2-6}$$

In cylindrical coordinates, the components of fluid velocity are

$$u_r = \frac{\partial \phi}{\partial r} \quad \text{and} \quad u_\theta = \frac{1}{r}\frac{\partial \phi}{\partial \theta} \tag{2-7}$$

and the Laplace equation is

$$\frac{\partial^2 \phi}{\partial r^2} + \frac{1}{r}\frac{\partial \phi}{\partial r} + \frac{1}{r^2}\frac{\partial^2 \phi}{\partial \theta^2} \equiv \nabla^2 \phi = 0, \tag{2-8}$$

where r is the radial coordinate, θ is the angular coordinate, and u_r and u_θ are the corresponding fluid velocities.

The *vorticity* is defined as a measure of the rotational velocity of fluid elements,

$$\omega = \frac{\partial v}{\partial x} - \frac{\partial u}{\partial y}, \qquad (2\text{-}9)$$

as shown in Figure 2-1(b). ω is the sum of the rotational velocities of two adjacent sides of the fluid element. In cylindrical coordinates the vorticity is

$$\omega = \frac{1}{r} \frac{\partial(r u_\theta)}{\partial r} - \frac{1}{r} \frac{\partial u_r}{\partial \theta}. \qquad (2\text{-}10)$$

Some authors define vorticity as the negative of these quantities. Substituting Eq. 2-4 into Eq. 2-9, we see that a potential flow has zero vorticity, $\omega = 0$. That is, a potential flow does not permit rotation of the fluid elements. This result can also be obtained from a vector identity. The vorticity vector is the curl of the velocity vector,

$$\boldsymbol{\omega} = \boldsymbol{\nabla} \times \mathbf{V}. \qquad (2\text{-}11)$$

The vorticity vector in a potential flow is zero since the curl of the gradient of any continuous, differentiable function is identically zero,

$$\boldsymbol{\nabla} \times \mathbf{V} = \boldsymbol{\nabla} \times (\boldsymbol{\nabla} \phi) \equiv \mathbf{0}, \qquad (2\text{-}12)$$

where ($\times$) is the vector cross product. Thus, if a velocity potential exists at a point, then the vorticity must be zero at that point. Conversely, zero vorticity implies the existence of a velocity potential (Newman, 1977, p. 105). A flow with zero vorticity everywhere in the flow field is said to be *irrotational*. In many ideal flows, the vorticity is zero at all but a few points where a differentiable velocity potential ceases to exist. These points are called *singular points* and they are associated with sources of vorticity.

Euler's equations describe the relationship between pressure gradients and motion of an ideal incompressible fluid. The equation was first derived by Leonhard Euler in 1755; it can be expressed in vector form as follows (Landau and Lifshitz, 1959, p. 5):

$$\frac{\partial \mathbf{V}}{\partial t} + (\mathbf{V} \cdot \boldsymbol{\nabla})\mathbf{V} = -\frac{1}{\rho} \boldsymbol{\nabla} p, \qquad (2\text{-}13)$$

where p is static pressure and ρ is fluid density. Using an identity from vector mechanics,

$$\tfrac{1}{2}\boldsymbol{\nabla} V^2 = \mathbf{V} \times (\boldsymbol{\nabla} \times \mathbf{V}) + (\mathbf{V} \cdot \boldsymbol{\nabla})\mathbf{V}, \qquad (2\text{-}14)$$

where $V = (u^2 + v^2)^{1/2}$ represents the magnitude of $\mathbf{V}$. Equation 2-13 can be written

$$\frac{\partial \mathbf{V}}{\partial t} + \tfrac{1}{2}\boldsymbol{\nabla} V^2 - \mathbf{V} \times (\boldsymbol{\nabla} \times \mathbf{V}) = -\frac{1}{\rho} \boldsymbol{\nabla} p. \qquad (2\text{-}15)$$

Substituting the velocity potential, Eq. 2-5, into Eq. 2-15 and using the vector identity from Eq. 2-12, the gradient operator can be brought outside,

$$\mathbf{V}\left(\frac{\partial \phi}{\partial t} + \tfrac{1}{2}V^2 + \frac{p}{\rho}\right) = 0.$$

Thus, the quantity in parentheses is not a function of space; it is constant or a function of time. Setting it equal to a function $F(t)$, we have

$$p = -\rho \frac{\partial \phi}{\partial t} - \tfrac{1}{2}\rho V^2 + F(t). \tag{2-16}$$

This is the generalized Bernoulli equation for static pressure as a function of the velocity potential. The function $F(t)$ is a function of time but it is independent of location in the reservoir; commonly it is included in ϕ. For steady flows, $\partial \phi / \partial t = 0$, $F(t) = $ constant, and the Bernoulli equation reduces to a familiar form, $p + \tfrac{1}{2}\rho V^2 = $ total pressure = constant.

The final equations needed to describe an ideal fluid flow are boundary conditions. The boundary condition for a solid surface is simply that no flow passes through the surface,

$$\mathbf{V} \cdot \mathbf{n} = 0, \qquad \text{on the surface } S, \tag{2-17}$$

where $\mathbf{n}$ is the unit outward vector normal to the surface. If the surface moves with velocity $\mathbf{U}$, then this condition is generalized to

$$\mathbf{V} \cdot \mathbf{n} = \mathbf{U} \cdot \mathbf{n}, \qquad \text{on the surface } S, \tag{2-18}$$

where $\mathbf{U}$ is the local surface velocity vector. Either of these conditions implies that the surface is a *streamline* or *pathline*. A streamline is tangent to the flow. No fluid flows across a streamline. A pathline is the path followed by an element of fluid in an unsteady flow. Mathematically, a stream function ψ is defined in terms of the velocity potential as (Sabersky et al., 1971, p. 180)

$$\frac{\partial \psi}{\partial x} = -\frac{\partial \phi}{\partial y}, \qquad \frac{\partial \psi}{\partial y} = \frac{\partial \phi}{\partial x} \tag{2-19}$$

in rectangular coordinates and

$$\frac{\partial \psi}{\partial \theta} = r\frac{\partial \phi}{\partial r}, \qquad \frac{\partial \psi}{\partial r} = -\frac{1}{r}\frac{\partial \phi}{\partial \theta} \tag{2-20}$$

in cylindrical coordinates. These equations are Cauchy–Reimann conditions. They imply that lines of constant ϕ (potential lines) are orthogonal to lines of constant ψ (streamlines). Note that the boundary conditions of Eqs. 2-17 and 2-18 act only on the normal component of velocity. In a viscous flow, this would also imply zero tangential velocity

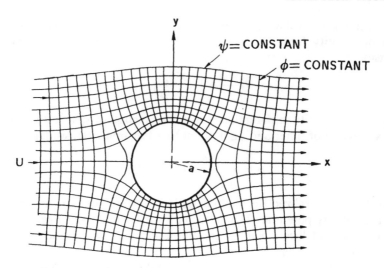

Fig. 2-2 Potential flow over a cylinder.

at the surface (i.e., no slip). In ideal flow there is not necessarily a restriction on the tangential velocity at a surface.

Figure 2-2 shows steady potential flow over a cylinder. The streamlines are orthogonal to the potential lines. Streamlines are concave on the side of greater velocity and the velocity increases in the direction of converging streamlines. The cylinder itself is a streamline. In fact, any of the streamlines can be interpreted as a solid surface, so the same potential flow can be interpreted many ways. For example, the potential flow of Figure 2-2 can be interpreted both as flow over a cylinder and as flow over a semicircular bump in a wall.

The velocity potential for Figure 2-2 is

$$\phi = U\left(r + \frac{a^2}{r}\right)\cos\theta. \tag{2-21}$$

The corresponding stream function is $\psi = U(r - a^2/r)\sin\theta$, where r is the radius from the center of the cylinder, θ is the angle from horizontal, and U is the free stream flow velocity from left to right. The radial and tangential components of the flow velocity are obtained using Eq. 2-7,

$$u_r = \frac{\partial\phi}{\partial r} = U\left(1 - \frac{a^2}{r^2}\right)\cos\theta,$$

$$u_\theta = \frac{1}{r}\frac{\partial\phi}{\partial\theta} = -U\left(1 + \frac{a^2}{r^2}\right)\sin\theta. \tag{2-22}$$

The radial component of velocity is always zero at the radius $r = a$; thus, the cylinder is a streamline. The tangential component of velocity is not zero at the surface of the cylinder, because an ideal flow permits slip relative to a surface.

The pressure on the surface of the cylinder is obtained from the Bernoulli equation (Eq. 2-16),

$$p_{r=a} - p_{r=\infty} = \tfrac{1}{2}\rho U^2 (4 \cos^2 \theta - 3). \tag{2-23}$$

Since this pressure distribution, like the flow pattern, is symmetric about orthogonal planes passing through the center of the cylinder, the ideal flow exerts no net force on the cylinder. The result conflicts with practical experience; the contradiction is called d'Alembert's paradox after J. R. d'Alembert (1717–1783), who formulated potential flow theory. The inability of potential flow to model fluid shearing stress (Eq. 2-12) leads to zero force; this is one of the fundamental limitations of classical potential flow theory. It is possible to overcome this limitation, to a degree, using a numerical technique called *discrete vortex modeling,* which is described in Section 2.4.

A number of potential flows are given in Table 2-1. Many others are given by Milne-Thomson (1968), Newman (1977), Kirchhoff (1985), Kochin et al. (1964), Kennard (1967), Sedov (1965), and Blevins (1984a). Most of the flows in Table 2-1 are the results of singular velocity potentials. That is, there are one or more points in the flow field at which the potential function is infinite and its differentials are not defined. These singularities are labeled *sources, sinks,* or *centers of vorticity*, and they tend to condition the remainder of the flow field.

The potential vortex given in case 4 of Table. 2-1 is in many ways the most interesting of the potential flows. It is the only potential flow in Table 2-1 that does not possess a plane of symmetry. The vortex center is a source of vorticity in an otherwise irrotational fluid field. The strength of the rotational field is measured by its circulation. The circulation of a region is defined as the line integral of the velocity vector around a closed path,

$$\Gamma = \oint \mathbf{V} \cdot d\mathbf{l}, \tag{2-24}$$

where $d\mathbf{l}$ is the vector element of the integration path. We can apply Stokes' theorem in order to evaluate the circulation as an area integral rather than a path integral. Incorporating the vorticity, Eq. 2-11, we have

$$\Gamma = \oint \mathbf{V} \cdot d\mathbf{l} = \int_A (\nabla \times \mathbf{V}) \cdot d\mathbf{A} = \int_A \omega \, dA. \tag{2-25}$$

The element of area $d\mathbf{A}$ has vector direction normal to its surface. Reviewing the flow fields in Table 2-1, we see that, with the exception of the vortex, all of these flows are irrotational and symmetric about at least one axis and hence $\Gamma = 0$. However, integrating Eq. 2-24 around any

Table 2-1 Potential flow fields[a]

Flow field, lines of ψ = constant	Potential function and remarks
1 Uniform flow 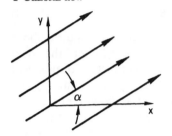	$\phi = ax + by$ $\psi = -bx + ay$ $V = (a^2 + b^2)^{1/2}$ $\alpha = \arctan(b/a)$ (For uniform flow along x axis, $\Phi = U_0 z$.) Can be applied to three dimensions
2 Flow over a wedge 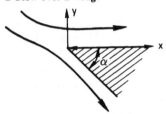	$\Phi = Az^{2/(2-\alpha/\pi)}$ $\phi = Ar^{2/(2-\alpha/\pi)} \cos\left(\dfrac{2\theta}{2-\alpha/\pi}\right)$ $\psi = Ar^{2/(2-\alpha/\pi)} \sin\left(\dfrac{2\theta}{2-\alpha/\pi}\right)$ $V = \dfrac{2A}{2-\alpha/\pi} r^{2/(2-\alpha/\pi)-1}$
3 Source or sink 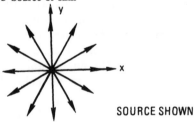 **SOURCE SHOWN**	$\Phi = k \log_e z$ $\phi = k \log_e r$ $\psi = k\theta$ $V = k/r$ Source: $k > 0$ Sink: $k < 0$ (For an axisymmetric source, $\phi = A/r$.)
4 Vortex 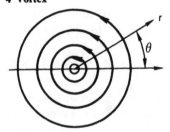	$\Phi = -\dfrac{i\Gamma}{2\pi} \log_e z$ $\phi = \dfrac{\Gamma}{2\pi}\theta$ $\psi = \dfrac{-\Gamma}{2\pi} \log_e r$ Γ = circulation

Table 2-1 *continued*

Flow field, lines of ψ = constant	Potential function and remarks
5 Flow over cylinder with circulation 	$$\Phi = U_0\left(a + \frac{a^2}{z}\right) + \frac{i\Gamma}{2\pi}\log_e\frac{z}{a}$$ Lift on cylinder = $\rho\Gamma U_0$ Γ = circulation
6 Doublet 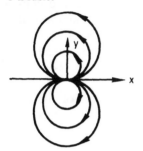	$$\Phi = \frac{m}{2\pi z}$$ $$\phi = \frac{m\cos\theta}{2\pi r}$$ $$\psi = -\frac{m\sin\theta}{2\pi r}$$ (Doublet is a source and sink in close proximity. For an axisymmetric doublet, $\phi = (A\cos\theta)/r^2$)
7 Flow over a cylinder 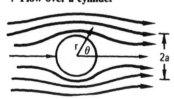	$$\Phi = U_0\left(z + \frac{a^2}{z}\right)$$ $$\phi = U_0\left(r + \frac{a^2}{r}\right)\cos\theta$$ $$\psi = U_0\left(r - \frac{a^2}{r}\right)\sin\theta$$
8 Flow over a sphere 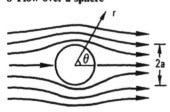	$$\phi = U_0\left(r + \frac{a^3}{2r^2}\right)\cos\theta$$ $$V\big|_{\text{on sphere}} = \tfrac{3}{2}U_0\sin\theta$$ $$p\big|_{\text{on sphere}} - p_0 = \frac{\rho}{2}U_0^2(\tfrac{9}{4}\sin^2\theta - 1)$$

Source: Blevins (1984a); also see Milne-Thomson (1968) and Kirchhoff (1985).

[a]
p	Pressure	ϕ	Potential function
r	Radius from origin	α	Angle, radians
V	Magnitude of flow velocity	θ	Angle from x axis
x, y	Orthogonal coordinates	$\Phi = \phi + i\psi$	Complex potential function
$z = x + iy$	Complex coordinates	ψ	Stream function

streamline of the vortex gives nonzero Γ. The circulation of the vortex is the resultant of vorticity embedded in the singularity at the origin. As a result, a potential vortex has properties that enable it to be used in modeling real fluid flows, as is discussed in Section 2.4.

Exercises

1. Consider a small, rectangular, two-dimensional element of fluid of height dy and length dx, Figure 2-1(b), relaxing the rotation. The lower left-hand corner is located at x_0, y_0. The velocity into the left-hand side is $u(x = x_0)$ and that exiting the right-hand side is $u(x = x_0) + (du/dx)\,dx$. Using this series expansion, and also considering the flow into the horizontal faces, prove the equation of continuity, Eq. 2-2.

2. Prove the vector identities of Eqs. 2-12 and 2-14 by substituting in the expressions for the gradient and curl, Eqs. 2-4, 2-5, and 2-9.

3. The Navier–Stokes equations describing the motion of an incompressible viscous fluid can be written as (White, 1974, p. 455)

$$\frac{\partial \mathbf{V}}{\partial t} + (\mathbf{V} \cdot \nabla)\mathbf{V} = -\frac{1}{\rho}\nabla p + \nu\nabla^2\mathbf{V},$$

where ν is called the kinematic viscosity of the fluid and $\nabla^2 = \partial^2/\partial x^2 + \partial^2/\partial y^2$. What are the similarities and differences between this equation and Euler's equation, Eq. 2-13? Which of the flows in Table 2-1 will satisfy both Euler's equation and the Navier–Stokes equation?

4. Because Laplace's equation is a linear equation, it is possible to superimpose two potential flows and obtain a third potential flow. What is the potential obtained from superimposing steady cylinder flow (case 7 of Table 2-1) and the vortex (case 4 of Table 2-1)? What physical problem does this represent? Sketch the streamlines. What is the static pressure on the circular streamline at $r = a$? Is there a net force on the circular cylinder?

5. Prove that $\psi = U(r - a^2/r)\sin\theta$ is the stream function that corresponds to the potential function of Eq. 2-21.

2.2. ADDED MASS

Our primary reason for studying fluid motion past a body is to compute the force and moment that the fluid exerts on the body. The net fluid force on a body is the sum of the normal pressure and tangential shear stress summed over the surface of the body. Since an ideal fluid is incapable of shear deformation, viscous shear makes no contribution to the force on a body in an ideal fluid. There are a number of practically important fluid phenomena, such as added mass, that are largely independent of viscosity, and these can be analyzed with ideal fluid theory.

The net force and net moment on a body in an ideal fluid are simply the resultants of the pressure on the surface,

$$F_x = -\int_S p\mathbf{n} \, dS, \tag{2-26}$$

$$F_\theta = -\int_S p(\mathbf{r} \times \mathbf{n}) \, dS. \tag{2-27}$$

The unit normal vector outward to the surface is $\mathbf{n}$, $\mathbf{r}$ is a radius vector from a reference point to the element of surface dS, and S is the surface of the body; see Figure 2-1. Substituting the Bernoulli equation (Eq. 2-16) into these equations and rolling the function $F(t)$ into ϕ, we have

$$F_x = \rho \int_S \left(\frac{\partial \phi}{\partial t} + \tfrac{1}{2}V^2 \right) \mathbf{n} \, dS, \tag{2-28}$$

$$F_\theta = \rho \int_S \left(\frac{\partial \phi}{\partial t} + \tfrac{1}{2}V^2 \right) (\mathbf{r} \times \mathbf{n}) \, dS. \tag{2-29}$$

An example will illustrate the properties of these equations.

Consider a *stationary cylinder in an accelerating fluid stream,* shown in Figure 2-3(a). The velocity potential for this flow is the sum of a steady flow (case 1 of Table 2-1) and a doublet (case 6 of Table 2-1). The velocity potential can be written either in cylindrical coordinates (case 7

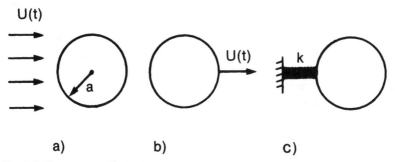

a)   b)   c)

Fig. 2-3 Three cases of cylinder–fluid relative motion: (a) stationary cylinder, moving fluid; (b) moving cylinder; (c) spring-supported cylinder.

of Table 2-1),

$$\phi = U(t)\left(r + \frac{a^2}{r}\right)\cos\theta, \qquad (2\text{-}30)$$

or in rectangular coordinates,

$$\phi = U(t)x + U(t)\frac{a^2 x}{x^2 + y^2}, \qquad (2\text{-}31)$$

where $x = r\cos\theta$ is the horizontal coordinate, $y = r\sin\theta$ is the vertical coordinate, $U(t)$ is the horizontal free stream flow, and a is the cylinder radius. This velocity potential represents both steady and time-dependent free stream flows. Consider U to be time dependent and accelerating. In order to evaluate the fluid force on the cylinder (Eq. 2-28) the two terms in the integrand of Eq. 2-28 that must be evaluated are $\partial\phi/\partial t$ and V^2. Differentiating Eqs. 2-30 and 2-31 with respect to time and utilizing Eq. 2-22, we have

$$\frac{\partial\phi}{\partial t} = \frac{dU}{dt}\left[r + \frac{a^2}{r}\right]\cos\theta = \frac{dU}{dt}x + \frac{dU}{dt}\frac{a^2 x}{x^2 + y^2},$$

$$V^2 = u_r^2 + u_\theta^2 = U^2\left[1 + \left(\frac{a}{r}\right)^4\right] - 2U^2\left(\frac{a}{r}\right)^2\cos 2\theta. \qquad (2\text{-}32)$$

The element of cylinder surface is $dS = a\,d\theta$. The normal vector can be expressed in rectangular coordinates as $\mathbf{n} = \mathbf{i}\cos\theta + \mathbf{j}\sin\theta$. Substituting these into Eq. 2-28 and integrating around the cylinder surface at $r = a$, we have the net force on the cylinder per unit depth,

$$F = \rho \int_{\theta=0}^{\theta=2\pi}\left[\left(\frac{dU}{dt}\right)(a + a)\cos\theta + U^2(1 - \cos 2\theta)\right](\mathbf{i}\cos\theta + \mathbf{j}\sin\theta)a\,d\theta$$

$$= \left(\rho\pi a^2\frac{dU}{dt} + \rho\pi a^2\frac{dU}{dt}\right)\mathbf{i}. \qquad (2\text{-}33)$$

This result has three interesting consequences. First, the force is independent of the mean flow. This is an illustration of d'Alembert's paradox: the net force on a single body without circulation in a steady ideal flow is zero. Second, the fluid force is proportional to fluid acceleration. Third, the first term is a buoyancy force resulting from the mean pressure gradient in the fluid ($dp/dx = \rho\,dU/dt$, Eqs. 2-15, 2-31). This buoyancy force is equal to the displaced fluid mass ($\rho\pi a^2$) times the acceleration of the fluid; it acts in the direction of fluid acceleration. The second component of fluid force on the cylinder is the mass of fluid entrained by the cylinder. This force is called *added mass* or *virtual mass* or *hydrodynamic mass*, and it also acts in the direction of fluid acceleration.

The moment on the cylinder due to acceleration can be computed from Eq. 2-29. If the reference point is chosen to be the center of the cylinder, then the radius vector is $\mathbf{r} = \mathbf{i} \cos \theta + \mathbf{j} \sin \theta$. Since this is identical in direction to the unit normal vector $\mathbf{n}$, the cross product $\mathbf{r} \times \mathbf{n} = 0$. (The magnitude of the cross product in two dimensions is the magnitude of the two vectors times the sine of the subtended angle.) Hence the moment on the cylinder about its center is zero. If the reference point is off center, then a net moment can be induced by steady or accelerated ideal flow (Newman, 1977).

To further explore the nature of added mass, consider a *moving cylinder in a reservoir of otherwise stationary fluid*, shown in Figure 2-3(b). The velocity potential is adapted from the velocity potential for a cylinder in a uniform flow (Eq. 2-31) by deleting the horizontal flow (the term $U(t)x$ in Eq. 2-31) and translating the doublet relative to a moving horizontal frame (Milne-Thomson, 1968, p. 243),

$$\phi = \frac{U(t)a^2[x - X(t)]}{[x - X(t)]^2 + y^2},\qquad (2\text{-}34)$$

where $X(t) = \int U(t)\,dt$ is the position of the cylinder and $U(t)$ is the velocity of the cylinder moving along the x axis in the $+x$ direction. The terms required for evaluation of the force on the cylinder (Eq. 2-8) are found from this potential as the cylinder passes over the origin, $X = 0$,

$$\frac{\partial \phi}{\partial t} = \frac{a^2 x}{x^2 + y^2}\frac{dU}{dt} - \frac{U^2 a^2}{x^2 + y^2} + \frac{2U^2 a^2 x^2}{(x^2 + y^2)^2}, \qquad \text{at } X = 0,$$

$$V^2 = U^2 \frac{a^4}{r^4}, \qquad \text{at } X = 0,$$

$$(2\text{-}35)$$

where $r^2 = x^2 + y^2$. Note that only the first term on the right-hand side of Eq. 2-35 is proportional to acceleration and that the V^2 term does not contribute to the net force, since V^2 is independent of θ. Inserting the above results into Eq. 2-28 and integrating around the cylinder surface, we have the added mass force on the accelerating cylinder per unit depth,

$$\mathbf{F} = -\rho \pi a^2 \frac{dU}{dt}\mathbf{i}. \qquad (2\text{-}36)$$

The added mass force acts to oppose the cylinder's acceleration. The net force is one-half that for the cylinder in an accelerating fluid because there is no net pressure gradient in the fluid.

Finally, consider a *spring-supported cylinder in a reservoir of otherwise stationary fluid*, shown in Figure 2-3(c). The cylinder is free to vibrate in the reservoir of otherwise stationary ideal fluid. The cylinder is acted on only by the added mass force. The equation of motion of the spring-supported cylinder responding to the fluid force (Eq. 2-36) is

$$m \frac{d^2X}{dt^2} + kX = -\rho \pi a^2 \frac{dU}{dt},$$

where $X(t)$ is the position of the cylinder, m is its mass per unit depth, and k is the spring constant per unit depth. The negative term on the right-hand side exists because the added mass force opposes the acceleration of the cylinder. Substituting $U(t) = dX/dt$, this equation can be rearranged to give

$$(m + \rho \pi a^2) \frac{d^2X}{dt^2} + kX = 0. \tag{2-37}$$

The fluid added mass increases the effective structural mass for dynamic analysis. The magnitude of the effect depends on the density of the fluid relative to the mass of the structure. In general, added mass is much more important in relatively dense fluids, such as water, than in gases, such as air.

The added mass for a number of sections and bodies is given in Table 2-2. Many others are given by Blevins (1984b), Chen (1976), Kennard (1967), Newman (1977), Milne-Thomson (1968), and Sedov (1965). Despite the fact that these coefficients are determined from ideal, inviscid, flow analysis, they generally agree with experimental measurements to within 10% for a wide variety of flow and vibration conditions (King et al., 1972; Stelson, 1957; Ackermann and Arbhabhirama, 1964).

In general, added mass can be associated with both translation and rotation of a body. The added mass force on a rigid three-dimensional body accelerating from rest in a reservoir of stationary fluid is the sum of added mass forces associated with all six possible rigid-body accelerations:

$$F_j = -\sum_{i=1}^{6} M_{ij} \frac{dU_i}{dt}. \tag{2-38}$$

M_{ij} is the 6-by-6 added mass matrix. The indices i and j range over three rigid-body translations in the direction of the orthogonal coordinates and three rotations about these coordinates.

Theoretical principles show that the added mass matrix is symmetric, $M_{ij} = M_{ji}$ (Newman, 1977, p. 141). Thus, 21 of the 36 entries can be independent. There are $N(N+1)/2$ independent entries for a body with N degrees of freedom. Symmetries in the body geometry further reduce the number of independent entries. For example, if the reference point for the cylinder is chosen to be its center, then the only nonzero terms in the added mass matrix for two-dimensional motion of the cylinder are $M_{11} = M_{22} = \rho \pi a^2$. However, if the reference point for the cylinder is chosen to lie on the rim of the cylinder at the $+y$ axis, then the

Table 2-2 Added mass for lateral acceleration[a]

Geometry	Added mass
1. Circular cylinder of radius a	$\rho \pi a^2 b$
2. Square section of side $2a$	$1.51 \rho \pi a^2 b$
3. Elliptical section with major radius a	$\rho \pi a^2 b$
4. Flat plate of height $2a$	$\rho \pi a^2 b$ Added mass moment of inertia for rotation about centroid c, $\rho(\pi/8)a^4$.
5. Sphere of radius a	$\frac{2}{3}\rho \pi a^3$
6. Cube of side a	$0.7 \rho a^3$
7. Cylinder in array of fixed cylinders	$\dfrac{\rho \pi D^2 b}{4}\left[\dfrac{(D_e/D)^2 + 1}{(D_e/D)^2 - 1}\right]$ where $D_e/D = (1 + \frac{1}{2}P/D)P/D$

Source: Blevins (1984b), Pettigrew et al. (1988).
[a] ρ is fluid density. The acceleration is left to right. b is the span for two-dimensional sections.

off-diagonal terms $M_{16} = M_{61} = -\rho \pi a^3$ arise because the net force, acting at the centroid, couples to a moment for an off-center reference point.

Exercises

1. A steel pipeline 1 meter in diameter and 0.05 meter thick runs from the sea floor to an offshore platform. Calculate the added mass, per unit length, of the empty pipe in air and in water. The density of steel is $8000 \, kg/m^3$, the density of air is $1.2 \, kg/m^3$, and the density of sea water is $1020 \, kg/m^3$. What is the ratio of added mass to structural mass?

2. Formulas developed in Section 6.3.1 show that the added mass of a circular cylinder in a viscous fluid is approximately $\rho \pi a^2 b[1 + 2(\pi^2 af/v)^{-1/2}]$, where f is the frequency of oscillation in Hertz, a is the radius, b is the span, and v is the kinematic viscosity. Does viscosity increase or decrease added mass? Justify your answer. What contribution does viscosity make to added mass in Exercise 1 if $f = 1 \, Hz$ and $v = 0.0000012 \, m^2/sec$ in water and $0.00002 \, m^2/sec$ in air?

3. By resolving the lateral acceleration of a flat plate (case 4 of Table 2-2) into components normal and tangential to its plane, show that, if the plate is inclined at an angle θ from the x axis, the added mass coefficients are $M_{11} = \rho \pi a^2 \sin^2 \theta$, $M_{22} = \rho \pi a^2 \cos^2 \theta$, $M_{12} = M_{21} = \rho \pi a^2 \sin 2\theta$. What is the added mass moment of inertia (Newman, 1977)?

2.3. FLUID COUPLING

Often one structure is adjacent to another and fluid fills the gap between the two. In this case, the fluid provides not only added mass but also coupling between the structures. When one structure is set into motion, the adjacent structure tends to vibrate. The magnitude of the coupling is a function of the proximity of the structures and the intermediate fluid. Following Fritz (1972), in this section we will use ideal fluid theory to determine the fluid coupling and the resultant modes of free vibration.

Consider two concentric cylinders separated by a gap filled with ideal, incompressible fluid as shown in Figure 2-4. The inner cylinder has radius R_1 and the outer cylinder has radius R_2. Both cylinders maintain their circular sections, but they may move horizontally relative to each other. The horizontal velocity of the inner cylinder is $V_1(t)$ and the horizontal velocity of the outer cylinder is $V_2(t)$. The boundary conditions associated with this motion are that the components of fluid velocity normal to the

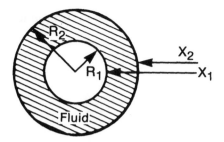

Fig. 2-4 Two concentric cylinders coupled through a fluid-filled gap.

cylinder surfaces conform to the cylinder motion. At the moment the two cylinders are concentric, these boundary conditions are

$$u_r = \frac{\partial \phi}{\partial r} = V_1 \cos \theta, \qquad r = R_1,$$

$$u_r = \frac{\partial \phi}{\partial r} = V_2 \cos \theta, \qquad \text{at } r = R_2. \qquad (2\text{-}39)$$

There is no boundary condition on the tangential component of velocity because an ideal fluid permits slip at a boundary; see Section 2-1.

The velocity potential for the fluid in the annulus must satisfy Laplace's equation (Eq. 2-8). A general solution for Laplace's equation in cylindrical coordinates is

$$\phi(r, \theta, t) = a_i f_i(r) \cos i\theta + b_i f_i(r) \sin i\theta, \qquad (2\text{-}40)$$

where θ is the angle from horizontal, i is an index, and $f(r)$ is a function of the radius r. Substituting this equation into Eq. 2-8, Laplace's equation is reduced to an ordinary differential equation,

$$r^2 f_i'' + r f_i' - i^2 f_i = 0, \qquad (2\text{-}41)$$

where the prime denotes differentiation with respect to r. This equation has solutions of r raised to the ith power, and so a general solution of Laplace's equation is a series of Fourier harmonics,

$$\phi(r, \theta, t) = \sum_{i=-\infty}^{i=+\infty} r^i (a_i \cos i\theta + b_i \sin i\theta). \qquad (2\text{-}42)$$

The index i must be integer $\cdots -2, -1, 0, 1, 2 \cdots$ for the solution to be continuous about the $\theta = 0$ axis. The coefficients a_i are determined by the boundary conditions, Eq. 2-39. Examining Eq. 2-42 in light of the boundary conditions (Eq. 2-39), it is easy to see that this velocity potential can contain only terms proportional to $\cos \theta$,

$$\phi(r, \theta, t) = a_1 r \cos \theta + a_{-1}(1/r) \cos \theta. \qquad (2\text{-}43)$$

This equation is substituted into Eq. 2-39 to determine the coefficients a_1 and a_{-1}. The results are

$$a_1 = \frac{V_2 R_2^2 - V_1 R_1^2}{R_2^2 - R_1^2},$$

$$a_{-1} = \frac{(V_2 - V_1) R_1^2 R_2^2}{R_2^2 - R_1^2}.$$

(2-43a)

Equation 2-7 gives the fluid velocity components $u_r = (-a_{-1}/r^2 + a_1) \cos\theta$ and $u_\theta = -(a_{-1}/r^2 + a_1) \sin\theta$. Thus, the magnitude of the fluid velocity,

$$V^2 = u_r^2 + u_\theta^2 = a_1^2 + \frac{a_{-1}^2}{r^4} - 2\left(\frac{a_{-1}a_1}{r^2}\right) \cos 2\theta,$$

(2-44)

is symmetric about two orthogonal planes.

The fluid forces on the two concentric cylinders are found from Eq. 2-28. The integrand of this equation includes the terms V^2 and $\partial\phi/\partial t$. Since V^2 is doubly symmetric in θ (Eq. 2-44), it cannot contribute to a net lateral force on the cylinders; $\partial\phi/\partial t$ is not doubly symmetric. The velocity potential ϕ is a function of the positions and velocities of the cylinders. $\partial\phi/\partial t$ is found using the chain rule for differentiation,

$$\frac{\partial\phi}{\partial t} = \frac{\partial\phi}{\partial V_1}\frac{dV_1}{dt} + \frac{\partial\phi}{\partial V_2}\frac{dV_2}{dt} + \frac{\partial\phi}{\partial X_1}\frac{dX_1}{dt} + \frac{\partial\phi}{\partial X_2}\frac{dX_2}{dt}.$$

(2-45)

X_1 and X_2 are the lateral positions of the centers of the cylinders from their concentric starting point. The velocity of the cylinders is $V_1 = dX_1/dt$ and $V_2 = dX_2/dt$. The derivatives $\partial\phi/\partial X_1$ and $\partial\phi/\partial X_2$ that appear in the last two terms cannot be evaluated using the velocity potential of Eq. 2-43 because this potential is valid only for $X_1 = X_2 = 0$, but if we consider that the cylinders accelerate from rest, $V_1 = V_2 = 0$, then Eq. 2-45 implies that these terms will not contribute to the force on the cylinders. [See Paidoussis et al. (1984) for a discussion of the influence of terms proportional to cylinder velocity.] For this case, only the first two terms to the right of the equals sign in Eq. 2-45 contribute to net force. The first two terms are evaluated using Eq. 2-43 substituted into Eq. 2-28 and integrated about the surface of each cylinder.

The result is the lateral fluid force on each cylinder for small oscillations about the concentric position,

$$F_1 = -M_1 \ddot{X}_1 + M_{12} \ddot{X}_2,$$

(2-46)

$$F_2 = M_{12} \ddot{X}_1 - M_2 \ddot{X}_2,$$

(2-47)

where the dots ($\cdot$) denote differentiation with respect to time and X_1 and X_2 are the displacements of the cylinders from the origin. The fluid masses and coupling per unit length of cylinder are

$$M_1 = \rho \pi R_1^2 \left(\frac{1 + (R_1/R_2)^2}{1 - (R_1/R_2)^2} \right) = \text{added mass of inner cylinder,}$$

$$M_{12} = \rho \pi R_1^2 \left(\frac{2}{1 - (R_1/R_2)^2} \right) = \text{fluid coupling,} \qquad (2\text{-}48)$$

$$M_2 = \rho \pi R_2^2 \left(\frac{1 + (R_1/R_2)^2}{1 - (R_1/R_2)^2} \right) = \text{added mass of outer cylinder.}$$

If the gap $R_1 - R_2$ becomes greater than the inner cylinder radius R_1, the added mass of the inner cylinder approaches the added mass for a single cylinder, $\rho \pi R_1^2$, and the added mass of the outer cylinder approaches the mass of the fluid it contains. As the gap becomes small, the added mass and coupling become large and the fluid forces tend to cushion one cylinder from the other. This is called a *squeeze film* effect.

Consider that both the inner and outer cylinders are spring supported. We will determine the free vibration of the cylinders. The equations of motion for small displacements about concentricity are found using Eqs. 2-46 and 2-47:

$$m_1 \ddot{X}_1 + k_1 X_1 = -M_1 \ddot{X}_1 + M_{12} \ddot{X}_2,$$
$$m_2 \ddot{X}_2 + k_2 X_2 = M_{12} \ddot{X}_1 - M_2 \ddot{X}_2. \qquad (2\text{-}49)$$

m_1 and m_2 are the cylinder structural masses per unit length and k_1 and k_2 are the corresponding spring constants. These equations can be written in matrix form:

$$\begin{bmatrix} (m_1 + M_1) & -M_{12} \\ -M_{12} & (m_2 + M_2) \end{bmatrix} \begin{Bmatrix} \ddot{X}_1 \\ \ddot{X}_2 \end{Bmatrix} + \begin{bmatrix} k_1 & 0 \\ 0 & k_2 \end{bmatrix} \begin{Bmatrix} X_1 \\ X_2 \end{Bmatrix} = 0. \qquad (2\text{-}50)$$

This linear matrix equation has solutions for the cylinder displacements that are sinusoidal functions of time:

$$X_1 = A_1 \sin \omega t, \qquad X_2 = A_2 \sin \omega t, \qquad (2\text{-}51)$$

where A_1 and A_2 are amplitudes and ω is angular frequency (units of radians per second). (With apologies to the reader, the symbol ω is used for angular velocity, and not vorticity, in this section.) Substituting Eq. 2-51 into Eq. 2-50 gives an eigenvalue problem for the frequency and amplitudes of vibration:

$$\begin{bmatrix} -(m_1 + M_1)\omega^2 + k_1 & M_{12}\omega^2 \\ M_{12}\omega^2 & -(m_2 + M_2)\omega^2 + k_2 \end{bmatrix} \begin{Bmatrix} A_1 \\ A_2 \end{Bmatrix} = 0. \qquad (2\text{-}52)$$

Nontrivial solutions (i.e., solutions other than $A_1 = A_2 = 0$) to this equation exist only if the determinant of the matrix on the left-hand side is zero (Bellman, 1970, p. 372). Setting the determinant to zero gives a polynomial for the two natural frequencies of free vibration:

$$(1 - \alpha)\omega^4 - \omega^2(\omega_1^2 + \omega_2^2) + \omega_1^2\omega_2^2 = 0, \tag{2-53}$$

where $\alpha = M_{12}^2/[(m_1 + M_1)(m_2 + M_2)]$ is a coupling factor. The corresponding mode shapes (i.e., relative amplitudes) are obtained by substituting solutions of this equation back into Eq. 2-52:

$$\frac{A_1}{A_2} = \left[1 - \frac{\omega_2^2}{\omega^2}\right]\left[\frac{m_2 + M_2}{M_{12}}\right]. \tag{2-54}$$

ω_1 and ω_2 are the circular natural frequencies of the inner and outer cylinder, respectively, when the fluid coupling (M_{12}) is absent,

$$\omega_1 = \left(\frac{k_1}{m_1 + M_1}\right)^{1/2}, \qquad \omega_2 = \left(\frac{k_2}{m_2 + M_2}\right)^{1/2}. \tag{2-55}$$

These circular natural frequencies have units of radians per second.

Reviewing Eqs. 2-48 through 2-54, we see that when the fluid density is zero ($\rho = 0$), then the added masses M_1, M_2, and M_{12} are zero, $\alpha = 0$, and the natural frequencies are given by Eq. 2-55. Moreover, the cylinders vibrate independently of one another. However, when the fluid density is nonzero, the coupling factor α is nonzero, the natural frequencies are given by the two roots of Eq. 2-53, and the corresponding mode shapes involve simultaneous, coupled motion of both cylinders. The greater the fluid density, the greater the coupling.

Heat-exchanger tube banks are an example of multiple elastic cylinders (tubes) coupled through a surrounding fluid. Arrays of cylinders can be analyzed using a series expansion, similar to Eq. 2-42, for the velocity potential and then matching boundary motion on the surface of the cylinders. The series is truncated to give a finite matrix for the series

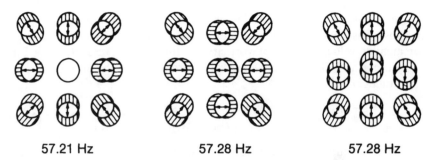

57.21 Hz   57.28 Hz   57.28 Hz

Fig. 2-5 Modes of fluid-coupled arrays of tubes. (Chen, 1976.)

coefficients (see Paidoussis et al., 1984; Chen, 1975). Some results computed by Chen (1976) are given in Figure 2-5 for the free vibration of three modes of an array of closely spaced tubes in a dense fluid. The fluid coupling that gives rise to these modes is most important in dense fluids, such as water, when the gap between tubes is less than a tube radius.

Another application of coupled fluid analysis is the dynamics of pressurized water-cooled nuclear reactors. Here, concentric steel cylindrical shells are coupled through a relatively small water-filled gap. When the shells move without deforming the cross section (the $n = 1$ shell mode), the analysis of this section applies exactly. For higher shell modes, other methods must be used; see reviews by Brown (1982) and Au-Yang (1986). Numerical methods for determining added mass and fluid coupling are discussed by Zienkiewicw (1977), Chilukuri (1987), Pattani and Olson (1988), and Montero de Espinosa and Gallego-Juarez (1984). Added mass of plates and shells is reviewed briefly in Chapter 8, Section 8.2.1. Yeh and Chen (1978) discuss the effect of viscosity on added mass coupling.

Exercises

1. Verify Eqs. 2-46, 2-47, and 2-48.

2. Consider two concentric spheres, coupled by a fluid-filled gap, which move relative to one another as in Figure 2-4 but with θ now considered to be a polar angle. The radius of the inner sphere is R_1 and the radius of the outer sphere is R_2. For an axisymmetric problem, Laplace's equation is given by

$$\frac{\partial^2 \phi}{\partial r^2} + \frac{2}{r}\frac{\partial \phi}{\partial r} + \frac{1}{r^2}\frac{\partial^2 \phi}{\partial \theta^2} + \frac{\cot \theta}{r^2}\frac{\partial \phi}{\partial \theta} = 0.$$

The boundary conditions are identical to those given by Eq. 2-39 if now θ is the polar angle from the horizontal surface. Consider a solution of the form $\phi = f(r) \cos \theta$. Substitute this solution into Laplace's equation to determine a differential equation for $f(r)$ and solve this equation.

3. Continue Exercise 2 by solving the coupled-sphere problem for added mass following the procedure illustrated in this section. The solution is given by Fritz (1972).

2.4. VORTEX MOTION

As shown in Table 2-1, many two-dimensional inviscid flows can be modeled by sources, sinks, or vortices. This leads to the possibility that any two-dimensional, inviscid flow can be modeled by distributions of a large number of singularities. The discrete vortex method (DVM) is a numerical method for modeling two-dimensional, inviscid flows by vortices. The advantage of the DVM is that any incompressible two-dimensional flow can be modeled. The disadvantages are that the technique is limited to two-dimensional flows. Viscous effects, such as boundary layers, can only be approximately represented and substantial computer resources are required.

For flows that are dominated by vortices, it can be useful to consider vorticity (Eq. 2-11) rather than velocity as the primary parameter of the flow. An equation for the transport of vorticity in an incompressible, inviscid fluid can be developed by taking the curl of Euler's equation (Eq. 2-13). For two-dimensional flow, the result is

$$\frac{\partial \omega}{\partial t} + (\mathbf{V} \cdot \mathbf{\nabla})\omega = \frac{D\omega}{Dt} = 0, \tag{2-56}$$

where ω is the magnitude of the vector vorticity, which has direction out of the two-dimensional plane, and $\mathbf{V}$ is the velocity vector. The operator $\partial/\partial t$ is the time derivative relative to a fixed coordinate (Lagrangian frame), and the total derivative, $D/Dt = \partial/\partial t + (\mathbf{V} \cdot \mathbf{\nabla})$, is the time derivative with respect to an observer who moves with the fluid (Eulerian frame).

Equation 2-56 states that the total derivative of vorticity in a two-dimensional, inviscid, incompressible flow is zero. That is, vorticity is neither created nor destroyed; the vorticity in each fluid element will remain constant; the vorticity will be carried along with the surrounding fluid. These are the Helmholtz theorems for two-dimensional vorticity transport. (The situation for three dimensions is more complicated; see Lugt (1983, p. 263).)

Complex numbers are a compact notation for specifying a two-dimensional potential flow. Consider two vortices located on the x axis as shown in Figure 2-6. The *complex velocity potential* for this case is adapted from case 4 of Table 2-2:

$$\Phi(z) = -\frac{i\Gamma_1}{2\pi}\log_e (z - z_1) - \frac{i\Gamma_2}{2\pi}\log_e (z - z_2). \tag{2-57}$$

The vortices are located at z_1 and z_2. Their circulations are Γ_1 and Γ_2. The complex velocity potential is the sum of the potential function and i times the stream function,

$$\Phi(z) = \phi(x, y) + i\psi(x, y), \tag{2-58}$$

where the complex constant $i = (-1)^{1/2}$ and z is the complex coordinate

$$z = x + iy, \tag{2-59}$$

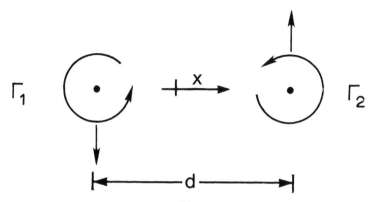

Fig. 2-6 Two vortices of the same circulation.

which transforms the $x-y$ plane into the real–imaginary plane. Thus, the coordinate z carries with it both x and y. Similarly, the complex potential Φ carries with it both the velocity potential ϕ and the stream function ψ. One advantage of the complex potential is the ease of computing the fluid velocity:

$$\frac{d\Phi}{dz} = \frac{\partial\phi}{\partial x} + i\frac{\partial\psi}{\partial x} = \frac{\partial\phi}{\partial(iy)} + i\frac{\partial\psi}{\partial(iy)} = u - iv. \tag{2-60}$$

The horizontal component of velocity is u and the vertical component is v. The derivative of the complex function $\Phi(z)$ with respect to the complex coordinate z exists only if it is independent of differentiation path in the complex plane. That is, the increment dz can be dx or $d(iy)$. This independence of path implies the Cauchy–Reimann conditions, Eq. 2-19. Under these conditions, Φ is called an *analytic function* of z.

The complex velocity is found for the two-vortex problem of Figure 2-7 by applying Eq. 2-60 to Eq. 2-57,

$$u - iv = -\frac{i\Gamma_1}{2\pi(z - z_1)} - \frac{i\Gamma_2}{2\pi(z - z_2)}. \tag{2-61}$$

Now consider local radial coordinates from the center of each vortex,

$$z - z_1 = r_1 e^{i\theta_1}, \qquad z - z_2 = r_2 e^{i\theta_2}, \tag{2-62}$$

where the complex exponential is

$$e^{i\theta} = \cos\theta + i\sin\theta. \tag{2-63}$$

Since a vortex induces no velocity on itself as a consequence of Eq. 2-56, the velocity at each vortex center is due to the neighboring vortex,

$$v(z_1) = \frac{\Gamma_2}{2\pi d}, \qquad v(z_2) = -\frac{\Gamma_1}{2\pi d},$$

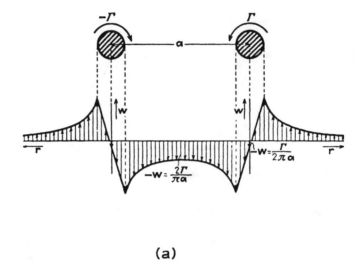

(a)

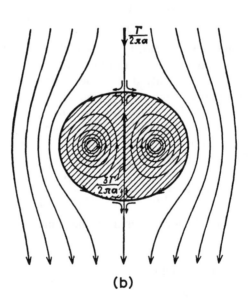

(b)

Fig. 2-7 Two vortices of opposite circulation: (a) induced velocity; (b) streamline (Prandtl and Tietjens, 1934).

where d is the distance between vortex centers. The vortex on the right will experience an upwash and the vortex on the left will experience a downwash. The vortices will be convected and the members of the vortex pair will spin about each other. If the two circulations are equal, $\Gamma_1 = \Gamma_2 = \Gamma$, the vortices will spin about a point midway between their two centers and the frequency of the spin will be $\Gamma/(2\pi^2 d^2)$.

On the other hand, if the two vortices have opposite but equal circulation, then the velocities induced by the vortices on each other are in the same direction. Both vortices are convected in the same direction, perpendicular to a line between their centers (Fig. 2-7). The complex potential for this flow can be adapted from Eq. 2-57 with $\Gamma = \Gamma_1 = -\Gamma_2$,

$$\Phi(z) = \phi + i\psi = \frac{\Gamma}{2\pi}(\theta_2 - \theta_1) + i\frac{\Gamma}{2\pi}\log_e\left(\frac{r_1}{r_2}\right). \tag{2-64}$$

This equation implies that streamlines have constant ratio r_1/r_2. One of these streamlines circles the two vortices in a roughly elliptical shape (Prandtl and Tietjens, 1934). Thus, this potential could be considered in a quasiellipse moving vertically at velocity $\Gamma/(2\pi d)$ as shown in Figure 2-7.

Figure 2-8(a) shows an infinite double staggered row of vortices. Note that the vortices in the first row have opposite signs from those in the bottom row. von Karman (1912) investigated the stability of this pattern by finding a clever closed-form expression for the velocity potential (see Lim and Sirovich, 1988; Lamb, 1945, p. 234; Saffman and Schatzman, 1982; Milne-Thomson, 1968; Kochin et al., 1964 for more recent approaches to von Karman's stability analysis). It is possible to model an arbitrary number of vortices numerically (Sarpkaya, 1989; Rangel and Sirignano, 1989; Kadtke and Campbell, 1987; Aref, 1983).

The complex potential for an assemblage of N vortices located at the points $z_1, z_2, \ldots, z_N$ is a finite series,

$$\Phi(z) = -\frac{1}{2\pi}\sum_{j=1}^{N} \Gamma_j \log_e (z - z_j). \tag{2-65}$$

The associated velocity is found by differentiating this expression,

$$u - iv = -i\sum_{j=1}^{N} \frac{\Gamma_j}{2\pi(z - z_j)}. \tag{2-66}$$

The velocity is defined everywhere except at the vortex centers. The velocity at the center of the k vortex ($z = z_k$) is obtained from the above series with the k term omitted because a vortex does not induce velocity upon its own center as a consequence of Eq. 2-56,

$$u - iv\big|_{z=z_k} = -i\sum_{\substack{j=1 \\ \text{except } j=k}}^{N} \frac{\Gamma_j}{2\pi(z_k - z_j)}. \tag{2-67}$$

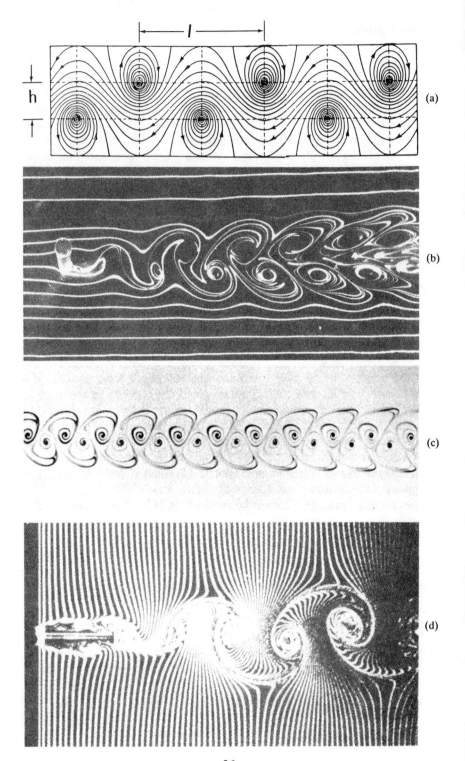

36

This expression states that the velocity at a vortex center is the sum of the velocities induced by its neighbors. It is known as the Biot–Savart law after an analogous expression in electromagnetic theory.

A computational difficulty with Eq. 2-66 is that as a point z approaches a vortex center z_j, the velocity becomes singular and the sum is strongly influenced by numerical round-off error. A practical, but nonrigorous, method for reducing the singularity is to introduce a small finite core about the vortex center. The core used by Spalart and Leonard (1981) and Rottman et al. (1987) gives the following expression for velocity:

$$u(z) - iv(z) = -\frac{i}{2\pi} \sum_{j=1}^{N} \frac{(z - z_j)^* \Gamma_j}{\sigma^2 + [\text{Real}(z - z_j)]^2 + [\text{Im}(z - z_j)]^2}. \quad (2\text{-}68)$$

The vortex core has radius σ. Real denotes 'real part of,' Im denotes 'imaginary part of,' and * denotes complex conjugate. This expression is not singular at a vortex center $z = z_j$. In fact, at a vortex center, the singular term is zero, in accordance with the Biot–Savart law, so there is no need to omit a term from the series. As the core size approaches zero, Eq. 2-68 approaches Eq. 2-66.

A surface in potential flow can be represented by a distribution of surface vortices. The surface is discretized by a finite number of surface points, preferably nearly equally spaced. A vortex is introduced at or near each surface point. The circulation of each of these surface vortices is chosen so that the stream functions at all surface points are equal and so a streamline is created through the points. As we have seen in Section 2.1, a streamline can represent a surface in ideal flow theory.

The circulation of the surface vortices must be chosen to balance the stream function in the external flow and the stream functions induced by the neighboring surface vortices. If there are P surface points and N vortices in the remainder of the flow, then the stream function at the surface point z_p is calculated using Eq. 2-67 as follows:

$$\psi(z_{pi}) - \psi_e(z_{pi}) = -\frac{1}{2\pi} \text{Im} \left\{ \sum_{j=1}^{P} [\Gamma_j \log_e (z_{pi} - z_{pj})] \right\}$$

$$= \sum_{j=1}^{P} a_{ij} \Gamma_j, \quad (2\text{-}69)$$

Fig. 2-8 The vortex street: (a) von Karman's ideal vortex street, $h/l = 0.281$; (b) Bearman's photograph (1987) of the vortex wake of a circular cylinder; (c) Stuber and Gharib's photograph (1988) of the vortex wake of an airfoil; (d) Nakamura and Nakashima's (1986) photograph of the vortex wake of a bridge deck section.

where ψ_e is the stream function due to nonsurface vortices in the free stream. If the stream function at the surface is set to zero, this equation becomes a linear matrix equation for the circulation of the surface vortices,

$$[a_{jk}]\{\Gamma_j\} = -\{\psi_e\},$$

where a_{ij} is a $P \times P$ symmetric matrix, $\{\Gamma_j\}$ is a $P \times 1$ vector, and ψ_e is a $P \times 1$ vector. By inverting the a_{ij} matrix, this equation is solved for the required circulations of the surface vortices. If the surface geometry is fixed, the matrix inversion has to be computed only once. Surface vortex circulations are found by back-substitution for any external flow stream function.

The discrete vortex method is applied as follows: (1) discretize the surface and specify the free stream; (2) compute the circulation of the surface vortices from Eq. 2-69; (3) compute the velocity at each vortex center from Eq. 2-68; (4) convect the vortices, including surface vortices, with this velocity over a small time step Δt (experience shows that Δt should be chosen so that the vortices are convected about the distance between discretized surface points; Prof. H. Aref's, UCSD, results indicate that a variable time step may be required to accurately reproduce the details of vortex interaction); (5) introduce new surface vortices to maintain the surface streamline; (6) calculate forces and moments (van der Vegt, 1988). With the vortices at their new positions, step (2) is repeated and so on as the flow evolves with time. Because new vortices are constantly introduced at the surface, the number of vortices will tend to grow geometrically unless vortices outside the region of interest are merged to hold the total number approximately constant. A review of the method is given by Leonard (1980). Theory is discussed by

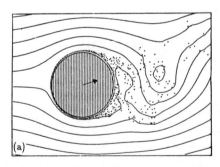
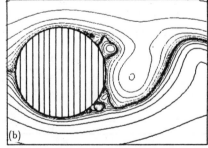

Fig. 2-9 Streamlines computed numerically for flow over stationary circular cylinders using (a) discrete vortex method (author's result) and (b) grid-based numerical solution to Navier–Stokes equation (Tsuboi et al., 1989).

Chorin (1973). Applications are presented by Spalart and Leonard (1981), Stansby and Dixon (1983), Nagano et al. (1983), Blevins (1989), Sarpkaya (1986, 1989), and Smith and Stansby (1989).

Figure 2-9(a) shows vortex shedding from a cylinder calculated by the discrete vortex method. The dots are centers of the discrete vortices. Note that the flow is time dependent and it separates. Because the fluid within the cylinder is held stationary, there is a numerical gradient in the short distance between the cylinder surface and the flow. This is effectively an inviscid boundary layer and it results in separated flow. Figure 2-9(b) is a calculation for the same geometry but this time with a grid-based numerical solution to the Navier–Stokes equation. Note the similarities in the results of the two calculations based on two very different approaches to the same problem.

Exercises

1. Develop a general expression for the motion of two vortices of unequal circulation that are separated by a distance d. Classify the various types of possible behavior in terms of the sign and magnitude of the circulations.

2. Consider a doubly infinite line of vortices located on the real axis at $x = 0$, $+a$, $-a$, $+2a$, $-2a$, $+3a$, $-3a$, ... Write the series expression for the induced velocity field. Show that the series is equivalent to the following expression given by Lamb (1945, p. 224):

$$\Phi(z) = \frac{i\Gamma}{2\pi} \log_e \sin\left(\frac{\pi z}{a}\right).$$

Hint: $\sin u = u(1 - u^2/\pi^2)(1 - u^2/2^2\pi^2) \cdots (1 - u^2/n^2\pi^2) \cdots$.
(Milne-Thomson, 1968, p. 374.)

3. Consider three point vortices of the same circulation set at the vertices of an equilateral triangle. What are their instantaneous velocity vectors? If a fourth vortex of opposite sign but same magnitude is now added at the centroid of the triangle, what are the vortex motions?

4. On a piece of graph paper track the motion of a pair of vortices (a) of the same circulation and (b) of opposite circulation.

5. Write a computer program to track the motion of vortices. Place three vortices along the x axis with coordinates $x_1 = 1$, $y_1 = 0$, $x_2 = -1$, $y_2 = 0$, and $x_3 = y_3 = 0$. Use the following circulations: $\Gamma_1 = \Gamma_2 = 1$, $\Gamma_3 = -1$. With a time step of 0.1 and a core radius of 0.01, follow the subsequent motion for 200 time steps. Explain the

motion of the center vortex. Now restart the problem at the old
locations and circulations but with $x_3 = 0.05$. Note the motion of the
center vortex. Does the small change in initial location make a big
change in final location? This sensitivity to initial conditions is a
route to the dynamics of chaos.

REFERENCES

Ackermann, N. L., and A. Arbhabhirama (1964) "Viscous and Boundary Effects on Virtual
    Mass," Proceedings of the American Society of Civil Engineers, *Journal of the
    Engineering Mechanics Division,* **90,** 123–130.
Aref, H. (1983) "Integrable, Chaotic, and Turbulent Vortex Motion in Two-Dimensional
    Flows," *Annual Review of Fluid Mechanics,* **15,** 345–389.
Au-Yang, M. K. (1986) "Dynamics of Coupled Fluid-Shells," *Journal of Vibration,
    Acoustics and Reliability in Design,* **108,** 339–347.
Bearman, P. W. (1987) personal correspondence and transmittal of photograph of vortex
    wake taken at Imperial College of Science and Technology, London.
Bellman, R. (1970) *Introduction to Matrix Analysis,* 2d ed., McGraw-Hill, New York.
Blevins, R. D. (1984a) *Fluid Dynamics Handbook,* Van Nostrand Reinhold, New York.
────── (1984b) *Formulas for Natural Frequency and Mode Shape,* Robert E. Krieger,
    Malabar, Fla. Reprint of 1979 edition.
────── (1989) "Application of the Discrete Vortex Method to Fluid-Structure Interaction,"
    Journal of Pressure Vessel Technology **113,** 437–445, 1991.
Brown, S. J. (1982) "A Survey of Studies into the Hydrodynamic Response of Fluid
    Coupled Circular Cylinders," *Journal of Pressure Vessel Technology,* **104,** 2–19.
Chen, S-S (1975) "Vibration of Nuclear Fuel Rod Bundles," *Nuclear Engineering and
    Design,* **35,** 399–422.
────── (1976) "Dynamics of Heat Exchanger Tube Banks," ASME Paper 76-WA/FE-28.
    Also see Chen, S-S, and Ho Chung, "Design Guide for Calculating Hydrodynamic
    Mass," Argonne National Laboratory Report ANL-CT-76-45, Argonne, Ill., June 1976.
Chilukuri, R. (1987) "Incompressible Laminar Flow Past a Transversely Vibrating
    Cylinder," *Journal of Fluids Engineering,* **109,** 166–171.
Chorin, A. J. (1973) "A Numerical Study of a Slightly Viscous Flow," *Journal of Fluid
    Mechanics,* **57,** 785–796.
Fritz, R. J. (1972) "The Effect of Liquids on the Dynamic Motions of Immersed Solids,"
    Journal of Engineering for Industry, **94,** 167–173.
Kadtke, J. B., and L. J. Campbell (1987) "Method for Finding Stationary States of Point
    Vortices," *Physical Review A,* **36,** 4360–4370.
Kennard, E. H. (1967) "Irrotational Flow of Frictionless Fluids, Mostly of Invariable
    Density," David Taylor Model Basin Report 2229.
King, R., M. J. Prosser, and D. J. Johns (1972) "On Vortex Excitation of Model Piles in
    Water," *Journal of Sound and Vibration,* **29,** 169–188.
Kirchhoff, R. H. (1985) *Potential Flows, Computer Graphic Solutions,* Marcel Dekker, New
    York.
Kochin, N. E., I. A. Kibel, and N. V. Roze (1964) *Theoretical Hydromechanics,*
    Interscience Publishers, Wiley, New York.

Lamb, H. J. (1945) *Hydrodynamics*, 6th ed., Dover, New York. Reprint of 1932 edition.

Landau, L. D., and E. M. Lifshitz (1959) *Fluid Mechanics*, Pergamon Press, Addison-Wesley, Reading, Mass.

Leonard, A. (1980) "Vortex Methods for Flow Simulation, *Journal of Computational Physics*, **37**, 289–335.

Lim, C. C., and L. Sirovich (1988) "Nonlinear Vortex Trail Dynamics," *Physics of Fluids*, **31**, 991–993.

Lugt, H. J. (1983) *Vortex Flow in Nature and Technology*, Wiley, New York.

Milne-Thomson, L. M. (1968) *Theoretical Hydrodynamics*, 5th ed., Macmillan, New York.

Montero de Espinosa, F., and J. A. Gallego-Juarez (1984) "On the Resonance Frequencies of Water Loaded Circular Plates," *Journal of Sound and Vibration*, **94**, 217–222.

Nagano, S., M. Naito, and H. Takata (1983) "A Numerical Analysis of Two-Dimensional Flow Past a Rectangular Prism by a Discrete Vortex Model," *Computers and Fluids*, **10**, 243–259.

Nakamura, Y., and M. Nakashima (1986) "Vortex Excitation of Prisms with Elongated Rectangular, H, and T Cross-Sections," *Journal of Fluid Mechanics*, **163**, 149–169.

Newman, J. N. (1977) *Marine Hydrodynamics*, The MIT Press, Cambridge, Mass.

Paidoussis, M. P., D. Mavriplis, and S. J. Price (1984) "A Potential-Flow Theory for the Dynamics of Cylinder Arrays in Cross-Flow," *Journal of Fluid Mechanics*, **146**, 227–252.

Pattani, P. G., and M. D. Olson (1988) "Forces on Oscillating Bodies in Viscous Fluids," *International Journal of Numerical Methods in Fluids*, **8**, 519–536.

Pettigrew, M. J., et al. (1988) "Vibration of Tube Bundles in Two-Phase Cross Flow, Part 1," *1988 International Symposium on Noise and Vibration*, Vol. 2, M. P. Paidoussis, ed., ASME, New York.

Prandtl, L., and O. G. Tietjens (1934) *Fundamentals of Hydro- and Aeromechanics*, Dover Publications, New York.

Rangel, R., and W. Sirignano (1989) "The Dynamics of Vortex Pairing and Merging," AIAA Paper 89-0218, AIAA, Washington, D.C.

Rottman, J. W., et al. (1987) "The Motion of a Cylinder of Fluid Released from Rest in Cross Flow," *Journal of Fluid Mechanics*, **177**, 307–337.

Sabersky, R. H., A. J. Acosta, and E. G. Hauptmann (1971) *Fluid Flow: A First Course in Fluid Mechanics*, Macmillan, New York.

Saffman, P. G., and J. C. Schatzman (1982) "An Inviscid Model for the Vortex Street," *Journal of Fluid Mechanics*, **122**, 467–486.

Sarpkaya, T. (1989) "Computational Methods with Vortices," *Journal of Fluids Engineering*, **111**(1), 5–52.

Sarpkaya, T., and C. J. Ihrig (1986) "Impulsively Started Steady Flow about Rectangular Prisms: Experiments and Discrete Vortex Analysis," *Journal of Fluids Engineering*, **108**, 47–54.

Sedov, L. I. (1965) *Two Dimensional Problems in Hydrodynamics and Aerodynamics*, Interscience, New York. Translated from the Russian edition of 1950.

Smith, P. A., and P. K. Stansby (1989) "An Efficient Algorithm for Random Particle Simulation of Vorticity and Heat Transport," *Journal of Computational Physics*, **81**, 349–371.

Spalart, P. R., and A. Leonard (1981) "Computation of Separated Flows by a Vortex Tracing Algorithm," AIAA Paper 81-1246, presented at AIAA 14th Fluid and Plasma Dynamics Conference, Palo Alto, Calif., June 23, 1981; AIAA, New York.

Stansby, P. K., and A. G. Dixon (1983) "Simulation of Flows Around Cylinders by a Lagrangian Vortex Scheme," *Applied Ocean Research*, **5**, 167–175.

Stelson, T. E. (1957) "Virtual Mass and Acceleration in Fluids," ASCE, Paper No. 2870, *Transactions* **81**, 518.

Stuber, K., and M. Gharib (1988) "Experiment on the Forced Wake of an Airfoil Transition from Order to Chaos," AIAA Paper 88-3840-CP, AIAA, Washington, D.C.

Tsuboi, K., et al. (1989) "Numerical Study of Vortex-Induced Vibration of Circular Cylinder in High-Reynolds-Number Flow," AIAA Paper 89-0294, AIAA, Washington, D.C.

van der Vegt, J. J. (1988) "Calculation of Forces and Moments in Vortex Methods," *Journal of Engineering Mathematics*, **22**, 225–238.

von Karman, T. (1912) "Uber den Mechanismuss des Widersstandes den ein bewegter Korper in einen Flussigkeit Erfahrt," *Nachrichten der K. Gesellschaft der Wissenschaften zu Gottingen*, 547–556. See also *Collected Works of Theodore von Karman*, Vol. 1, Butterworths, London, 1956, pp. 331–338.

White, F. M. (1974) *Viscous Fluid Flow*, McGraw-Hill, New York.

Yeh, T. T., and S. S. Chen (1978) "The Effect of Fluid Viscosity on Coupled Tube/Fluid Vibrations," *Journal of Sound and Vibration*, **59**, 453–467.

Zienkiewicw, O. C. (1977) *The Finite Element Methods*, 3rd ed., McGraw-Hill, New York.

Chapter 3

Vortex-Induced Vibration

Structures shed vortices in a subsonic flow. The vortex street wakes tend to be very similar regardless of the geometry of the structure. As the vortices are shed from first one side and then the other, surface pressures are imposed on the structure as shown in Figure 3-1. The oscillating pressures cause elastic structures to vibrate and generate aeroacoustic sounds called *Aeolian tones* (Chapter 9). The vibration induced in elastic structures by vortex shedding is of practical importance because of its potentially destructive effect on bridges, stacks, towers, offshore pipelines, and heat exchangers.

3.1. VORTEX WAKE OF A STATIONARY CIRCULAR CYLINDER

Since ancient times, it has been known that wind causes vortex-induced vibration of the taut wires of an Aeolian harp. According to Rabbinic records, King David hung his kinnor (kithara) over his bed at night where it sounded in the midnight breeze. In the fifteenth century, Leonardo da Vinci sketched a row of vortices in the wake of a piling in a stream (Lugt, 1983). In 1878, Strouhal found that the Aeolian tones generated by a wire in the wind were proportional to the wind speed divided by the wire thickness. He observed that the sound greatly increased when the natural tones of the wire coincided with the Aeolian tones. In 1879, Lord Rayleigh found that a violin string in a chimney draft vibrated primarily across the flow, rather than with the flow. The periodicity of the wake of a cylinder was associated with vortex formation by Benard in 1908 and with the formation of a stable street of staggered vortices by von Karman in 1912 (Chapter 2, Fig. 2-8).

As a fluid particle flows toward the leading edge of a cylinder, the pressure in the fluid particle rises from the free stream pressure to the stagnation pressure. The high fluid pressure near the leading edge impels flow about the cylinder as boundary layers develop about both sides. However, the high pressure is not sufficient to force the flow about the back of the cylinder at high Reynolds numbers. Near the widest section

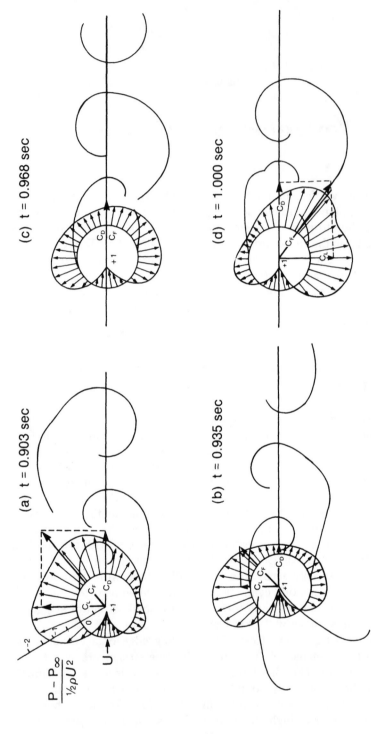

Fig. 3-1 A sequence of simultaneous surface pressure fields and wake forms at Re = 112 000 for approximately one-third of one cycle of vortex shedding (Drescher, 1956).

44

of the cylinder, the boundary layers separate from each side of the cylinder surface and form two shear layers that trail aft in the flow and bound the wake. Since the innermost portion of the shear layers, which is in contact with the cylinder, moves much more slowly than the outermost portion of the shear layers, which is in contact with the free flow, the shear layers roll into the near wake, where they fold on each other and coalesce into discrete swirling vortices (Perry et al., 1982; Williamson and Roshko, 1988). A regular pattern of vortices, called a *vortex street,* trails aft in the wake (Chapter 2, Fig. 2-8). The vortices interact with the cylinder and they are the source of the effects called vortex-induced vibration.

Vortex shedding from a smooth, circular cylinder in a steady subsonic flow is a function of Reynolds number. The Reynolds number is based on free stream velocity U and cylinder diameter D,

$$Re = \frac{UD}{v}, \tag{3-1}$$

where v is the kinematic viscosity of the fluid. The major Reynolds number regimes of vortex shedding from a smooth circular cylinder, summarized by Lienhard (1966), are given in Figure 3-2. At very low Reynolds numbers, below about $Re = 5$, the fluid flow follows the cylinder contours. In the range $5 \leq Re \leq 45$, the flow separates from the back of the cylinder and a symmetric pair of vortices is formed in the near wake. The streamwise length of the vortices increases linearly with Reynolds number, reaching a length of three cylinder diameters at a Reynolds number of 45 (Nishioka and Sato, 1978). As Reynolds number is further increased, the wake become unstable (Huerre and Monkewitz, 1990) and one of the vortices breaks away (Friehe, 1980). A laminar periodic wake of staggered vortices of opposite sign (vortex street) is formed. Roshko (1954) found that between $Re = 150$ and 300, the vortices breaking away from the cylinder become turbulent, although the boundary layer on the cylinder remains laminar.

The Reynolds number range $300 < Re < 1.5 \times 10^5$ is called *subcritical.* In this range, the laminar boundary layers separate at about 80 degrees aft of the nose of the cylinder and the vortex shedding is strong and periodic. In the *transitional* range, $1.5 \times 10^5 < Re < 3.5 \times 10^6$, the cylinder boundary layer becomes turbulent, the separation points move aft to 140 degrees, and the cylinder drag coefficient drops to 0.3 (Farell, 1981; Chapter 6, Fig. 6-9). In the transitional range, laminar separation bubbles and three-dimensional effects disrupt the regular shedding process and broaden the spectrum of shedding frequencies for smooth surface cylinders (Bearman, 1969; Jones et al., 1969; Farell and Blessmann, 1983; Achenbach and Heinecke, 1981).

In the *supercritical* Reynolds number range, $Re > 3.5 \times 10^6$, regular

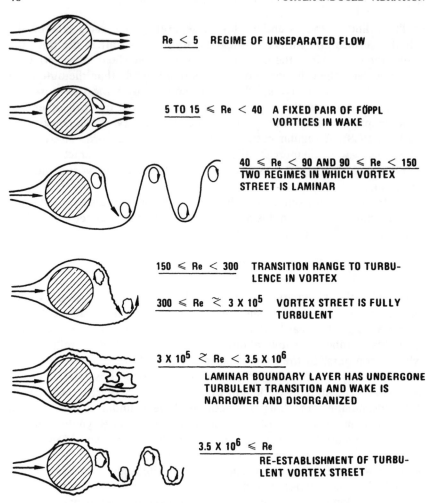

Re < 5   REGIME OF UNSEPARATED FLOW

5 TO 15 ⩽ Re < 40   A FIXED PAIR OF FÖPPL
                       VORTICES IN WAKE

40 ⩽ Re < 90 AND 90 ⩽ Re < 150
TWO REGIMES IN WHICH VORTEX
STREET IS LAMINAR

150 ⩽ Re < 300   TRANSITION RANGE TO TURBU-
                  LENCE IN VORTEX

300 ⩽ Re ≳ 3 X 10⁵   VORTEX STREET IS FULLY
                      TURBULENT

3 X 10⁵ ≳ Re < 3.5 X 10⁶

LAMINAR BOUNDARY LAYER HAS UNDERGONE
TURBULENT TRANSITION AND WAKE IS
NARROWER AND DISORGANIZED

3.5 X 10⁶ ⩽ Re
RE-ESTABLISHMENT OF TURBU-
LENT VORTEX STREET

Fig. 3-2 Regimes of fluid flow across smooth circular cylinders (Lienhard, 1966).

vortex shedding is re-established with a turbulent cylinder boundary layer (Roshko, 1961). Satellite photos of the vortex patterns of wind-driven clouds in the lee of islands make it clear that the vortex shedding persists at Reynolds numbers as high as $Re = 10^{11}$ (Griffin, 1982; Nickerson and Dias, 1981).

von Karman (1912) found that the ratio of the longitudinal to lateral spacing of an ideal, staggered vortex street was $h/l = 0.281$ (Chapter 2, Fig. 2-8). More recent investigations have shown that the lateral spacing necks down to a minimum at a few diameters downstream of the cylinder and then increases (Schaefer and Eskinazi, 1959). Griffin and Ramberg (1975) found a spacing ratio of 0.18 at $Re = 100$ to 500. The ratio of longitudinal spacing (l) to cylinder diameter (D) is nearly constant for most vortex streets (Sarpkaya, 1979). Griffin et al. (1980) found $4.8 < l/D < 5.2$ and that typically the vortices are spaced five cylinder diameters behind one another. Griffin and Ramberg also found that the circulation of the shed vortices was $1.75 < \Gamma/(UD) < 3.2$, where the circulation Γ is defined by Eq. 2-24. These values apply relatively near the cylinder. Downstream, the vorticity decreases (Sarpkaya, 1979) and the staggered vortex pattern expands laterally in accordance with asymptotic laws of turbulent wakes (Cimbala et al., 1988; Rhodi, 1975).

Additional studies of vortex wake dynamics are given by Lim and Sirovich (1988), Saffman and Schatzman (1982), Aref and Siggia (1981), and Griffin (1981)—see also Section 2.4. Reviews of the voluminous literature on vortex shedding from cylinders are given by Griffin (1985), Farell (1981), Sarpkaya (1979), King (1977), Barnett and Cermak (1974), Marris (1964), Morkovin (1964), Bearman (1984), and Mair and Maull (1971). Lugt (1983) and Swift et al. (1980) review vortex motion. Escudier (1987) discusses vortices in ducts and inlets. Discrete vortex modeling is discussed in Chapter 2, Section 2.4.

3.2. STROUHAL NUMBER

Strouhal number (S) is the dimensionless proportionality constant between the predominant frequency of vortex shedding and the free stream velocity divided by the cylinder width,

$$f_s = \frac{SU}{D}, \tag{3-2}$$

where f_s is the vortex shedding frequency in units of hertz (cycles per second), U is the free stream flow velocity approaching the cylinder, and D is the cylinder diameter. U and D must have consistent units. That is, if D is in meters, then U must be in meters per second, or if D is in inches, then U must be in inches per second, and so on.

For cylinders inclined to the flow, Ramberg (1983) and King (1977) found that the shedding frequency varies as $f_s(\theta) = f_s(\theta = 0) \cos \theta$, where θ is the angle of inclination of the cylinder axis from being perpendicular to the flow, for angles up to about 30 degrees. At larger angles, end effects become increasingly important. Experiments show that the oscillations in lift force (force perpendicular to the flow) occur at the shedding frequency, but oscillations in the drag force (force parallel to the flow) occur at *twice* the shedding frequency. This is a consequence of the geometry of the vortex street; see Figures 3-1 and 3-8.

The Strouhal number of a stationary circular cylinder in a subsonic flow is a function of Reynolds number (Eq. 3-1) and, to a lesser extent,

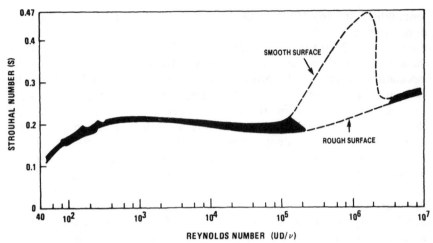

Fig. 3-3 Strouhal number–Reynolds number relationship for circular cylinders (Lienhard, 1966; Achenbach and Heinecke, 1981). $S \approx 0.21$ $(1-21/\text{Re})$ for $40 < \text{Re} < 200$ (Roshko, 1955).

surface roughness and free stream turbulence, as shown in Figure 3.3. The Strouhal number follows the Reynolds number flow regimes of Figure 3-2. In the transitional range, $2 \times 10^5 < \text{Re} < 2 \times 10^6$, Achenbach and Heinecke (1981) found that very smooth surface cylinders had a chaotic, disorganized, high-frequency wake and Strouhal numbers as high as 0.5, while rough surface cylinders (surface roughness $\epsilon/D = 3 \times 10^{-3}$ or greater, where ϵ is a characteristic surface roughness) had organized, periodic wakes with Strouhal numbers of $S \simeq 0.25$. In the transitional Reynolds number regime, vortex-induced vibration of cylinders generally occurs at $S \sim 0.2$ rather than the higher Strouhal numbers suggested by Figure 3-3 (Coder, 1982). Free stream turbulence up to about 10% of the mean flow has relatively little effect on either the oscillating lift coefficient or vortex-induced vibration of circular cylinders, although it can shift the flow regimes downward similarly to an increase in Reynolds number (Torum and Anand, 1985; Gartshore, 1984; Fage and Warsap, 1929; Barnett and Cermack, 1974).

Vortex shedding occurs from multiple cylinders and from a cylinder near a surface. Studies by Torum and Anand (1985), Jacobsen et al. (1984), and Tsahalis (1984) show that vortex shedding from a cylinder above a plane surface parallel to the flow (e.g., a pipe just above the seabed) persists for gaps between the cylinder and the plane as small as 0.5 diameter. Bearman and Zdravkovich (1978) and Buresti and Lanciotti (1979) found that gaps smaller than 0.3 diameter suppress vortex shedding. Similarly, Kiya et al. (1980) found that vortex shedding from a pair of cylinders occurs behind each cylinder individually when the gap between the two is greater than 0.4 diameter (center-to-center distance greater than 1.4 diameters), but for smaller gaps the pair of cylinders behaves as a single body with regard to vortex shedding. Vortex shedding occurs in groups of a few cylinders (Vickery and Watkins, 1962; Zdravkovich, 1985) and in large arrays of cylinders (Weaver and Adb-Rabbo, 1985; Chapter 9, Fig. 9-14). Fitzhugh's (1973) compilation of Strouhal numbers for flow within arrays of cylinders, such as heat-exchanger tube arrays, is given in Figures 3-4 and 3-5. These Strouhal numbers are based on tube diameter and average flow velocity through the minimum area between cylinders. These Strouhal numbers characterize the peak in the turbulence spectrum within the array. In very closely spaced tube arrays with tube center to tube center spacing less than about 1.5 diameters, the distinct frequency associated with shedding degenerates into broad-band turbulence; see Fitzpatrick and Donaldson (1980), Fitzpatrick et al. (1988), Murray et al. (1982), Price and Paidoussis (1989), Weaver et al. (1986) and Sections 7.3 and 9.3.

Noncircular sections shed vortices. Figures 3-6 and 3-7 give Strouhal numbers for noncircular sections and some three-dimensional bodies. Even an airfoil will form a vortex wake with a Strouhal number of

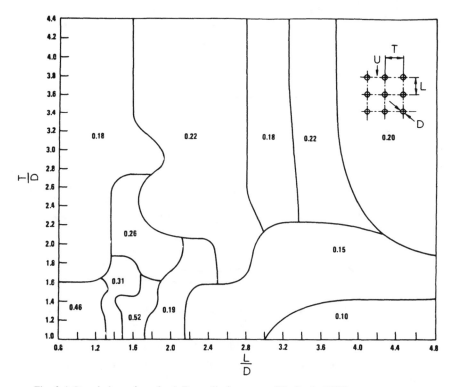

Fig. 3-4 Strouhal numbers for inline cylinder arrays (Fitzhugh, 1973).

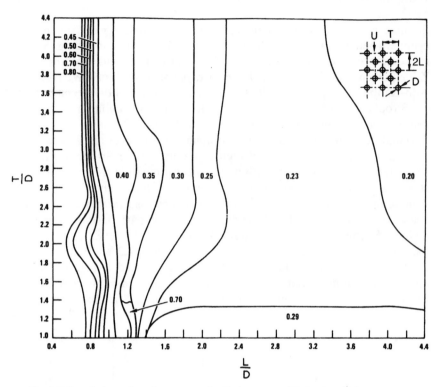

Fig. 3-5 Strouhal numbers for staggered cylinder arrays (Fitzhugh, 1973).

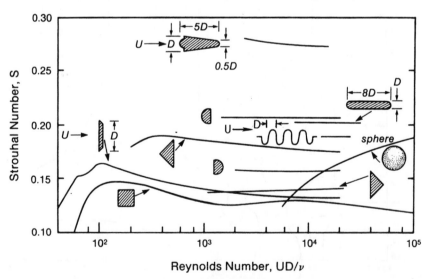

Fig. 3-6 Strouhal numbers for noncircular sections. Flow is left to right (Roshko, 1954; Wardlaw, 1966; Mujumdar and Douglas, 1973; Vickery, 1966; Gerlach, 1972; Toebes and Eagleson, 1961; Okajima, 1982; Achenbach and Heinecke, 1981).

50

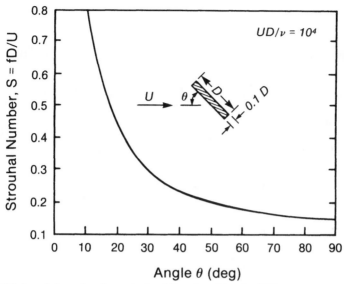

Fig. 3-7 Strouhal number for an inclined flat plate (Novak, 1973).

approximately 0.2 based on the width D of the boundary layers at the trailing edge (Fig. 2-8(c); Stuber, 1988; Parker and Stoneman, 1985). Vortex street wakes tend to be very similar regardless of the geometry of the tripping structure. Compare Figures 2-8(b), 2-8(c), and 9-13. Because the vortex street wake is formed by the interaction of two free shear layers that trail behind a structure, Roshko (1954, 1961), Griffin (1981), and others (Sarpkaya, 1979) have suggested that it is possible to define a "universal" Strouhal number for any bluff section based on the separation distance between shear layers. In any case, if the characteristic dimension D in Eq. 3-2 is defined as the width between the separation points, then the Strouhal number is approximately 0.2 over broad ranges of Reynolds number regardless of the section geometry. This consistency of Strouhal number has allowed vortex shedding to be used as the basis for the design of fluid flow meters (Miller, 1989; Yeh et al., 1983).

The descriptions of vortex shedding presented thus far may give the impression that vortex shedding is a steady, harmonic, two-dimensional process. Unfortunately this is not completely true. Vortex shedding from a stationary cylinder in the higher Reynolds number range does not occur at a single distinct frequency, but rather it wanders over a narrow band of frequencies with a range of amplitudes and it is not constant along the span (Blevins, 1985; Schewe, 1983; Jones et al., 1969). For example, Figure 3-8(b) shows time-history plots of four flush film anemometers mounted on a cylinder and a neighboring microphone. The cylinder is exposed to an air flow at $Re = 4 \times 10^4$. The flush films are spaced along

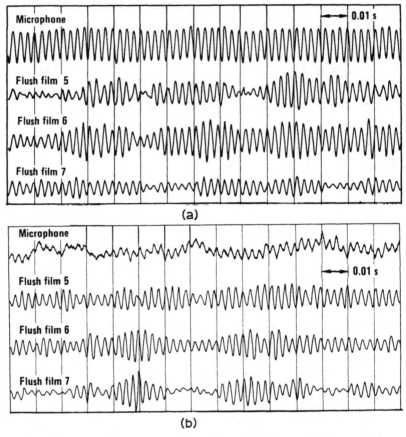

Fig. 3-8 Time-history plots of hot-film sensors on a 0.75 in. (1.9 cm) diameter cylinder exposed to air flow at 108 ft/sec (33 m/sec) (a) with applied sound (300 Pa, 375 Hz) and (b) without Re = 42 000 (Blevins, 1985). Note that the irregular appearance of the vortex shedding cycles.

the cylinder at two-diameter intervals. Note the lack of spanwise correlation and the irregular appearance of the vortex shedding cycles. When intense sound is applied at the nominal shedding frequency in Figure 3-8(a), the shedding does become in-phase and two-dimensional at the sound frequency, but the irregular beat of amplitude persists.

The three-dimensionality of vortex shedding can be characterized by a spanwise correlation length; typical values for stationary cylinders range from 100 or more diameters for laminar vortex streets at Re = 60 to 20 diameters at Re = 100 and 5 diameters for fully turbulent vortex streets at Re = 10^4 (Friehe, 1980; King, 1977). Spanwise coherent cells of vortex shedding, three to four diameters in extent, develop at transitional

Reynolds numbers (Humphreys, 1960). If the free stream flow velocity varies over the cylinder span, these cells also develop and the vortex shedding frequency varies discretely in ladderlike steps along the span with each step, spanning about four diameters (Griffin, 1985; Ramberg, 1983; Rooney and Peltzer, 1981). Also see Fig. 3-32.

Internal flow through bellows-induced vibration and sound at the frequency $f_s = SU/D$, where S is reported to lie between 0.18 and 0.22 if D is defined as the convolution width, that is, the width of a single fold, and U is the average flow velocity through the bellows (Johnson et al., 1979; Gerlach, 1972). Two methods of eliminating these vortex-induced vibrations are (1) to externally damp the convolute vibrations or (2) to provide a continuous liner, anchored at the upstream end, to isolate the high-speed flow from the convolutions (Weaver and Ainsworth, 1988; Gerlach, 1972). There is a toy in which the vortex shedding off corrugated pipe (bellows) induces acoustic tones. Vortex shedding induces vibration in groups of cylinders and arrays of stacks; see Vickery (1981), King and Johns (1976), and Chapter 5. Vortex shedding may be associated with the ovaling oscillation of large hyperbolic cooling towers (Paidoussis et al., 1982; Uematsu and Uchiyama, 1985) and the vibration of bridge decks (Sections 3.7 and 4.4).

The emphasis of this chapter is on predicting vortex-induced vibration of cylinders. However, just as the vortex shedding imposes a force on the cylinder, so the cylinder imposes a force on the fluid. The latter produces aeroacoustic sound with an almost pure tone at the vortex shedding frequency, as discussed in Chapter 9. A free fluid surface will respond to vortex shedding in the fluid below. Rhode (1979) has performed an extensive study of transverse sloshing waves in open channels induced by vortex shedding from bridge piers. Levi (1983) discussed application of vortex shedding concepts to hydraulic and ocean systems. Vortex-induced vibration of cylinders also occurs in oscillatory flows such as those induced by ocean waves over pilings and pipelines. These vibrations are discussed in Chapter 6, Section 6.4.

Exercises

1. What is the predominant vortex shedding frequency in Figure 3-8(a)? Pick one line of this figure. What is the highest frequency (1/time interval) between any two adjacent peaks; what is the lowest? Estimate the difference in phase between the other sensors and the chosen sensor at two different times. Repeat these exercises for Fig. 3-8(b).

2. Calculate the vortex shedding frequency from (a) a telephone pole 10 in. in diameter in a wind with velocity of 20 ft/sec; (b) a marine pipeline 0.1 m in diameter in a current at 3 m/sec; (c) a

telephone wire 1 cm in diameter at 5 m/sec; (d) a wire 1 mm in diameter at 5 m/sec; (e) your middle finger while swimming with your hands open; and (f) a tall square-section building 50 m on a side in a 30 m/sec wind. Please note that the Reynolds number must be calculated first.

3. Continue the sequence of Figure 3-1 by sketching the patterns of vortex shedding at regular time intervals through one cycle. What is the frequency of vortex shedding? At what frequency does the net lift force (force perpendicular to the mean flow) vary? At what frequency does the drag force vary? Is the oscillating component of lift greater than the oscillating component of drag?

3.3. EFFECT OF CYLINDER MOTION ON WAKE

Transverse cylinder vibration (i.e., vibration perpendicular to the free stream), with frequency at or near the vortex shedding frequency, has a large effect on vortex shedding. The cylinder vibration can:

1. Increase the strength of the vortices (Davies, 1976; Griffin and Ramberg, 1975).

2. Increase the spanwise correlation of the wake (Toebes, 1969; Ramberg and Griffin, 1976; Novak and Tanaka, 1977).

3. Cause the vortex shedding frequency to shift to the frequency of cylinder vibration (Bishop and Hassan, 1964). This effect is called lock-in or synchronization, and it can also be produced to a lesser extent if the vibration frequency equals a multiple or submultiple of the shedding frequency.

4. Increase the mean drag on the cylinder (Bishop and Hassan, 1964; Tanida et al., 1973; Sarpkaya, 1978).

5. Alter the phase, sequence, and pattern of vortices in the wake (Zdravkovich, 1982; Ongoren and Rockwell, 1988; Williamson and Roshko, 1988).

Transverse cylinder vibration with frequency at or near the shedding frequency organizes the wake. Vibration increases the correlation of vortex shedding along the cylinder axis as shown in Figure 3-9. The correlation is a measure of the three-dimensionality of the flow in the cylinder wake and a correlation of 1.0 implies two-dimensional flow. Increased transverse vibration amplitude (A_y) also increases the ability of the vibration to lock-in, or synchronize, the shedding frequency. The lock-in band is the range of frequencies over which cylinder vibration controls the shedding frequency. The fundamental lock-in band is shown in Figure 3-10(a). Note that large-amplitude cylinder vibration can shift the vortex shedding frequency by as much as ±40% from the stationary

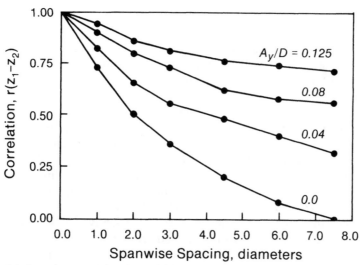

Fig. 3-9 Spanwise correlation for a rigid circular cylinder vibrating at resonance with vortex shedding (Toebes, 1969).

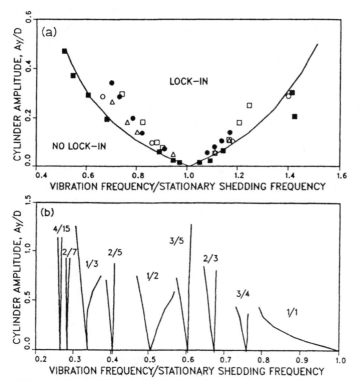

Fig. 3-10 Lock-in bands for synchronization of vortex shedding with transverse cylinder vibration. A_y is amplitude of transverse vibration and D is cylinder diameter. Experimental data: Koopman (1967), □ for Re = 100, ● for Re = 200, and △ for Re = 300; Stansby (1976), ■ for Re = 3600, ○ for Re = 9200; Olinger and Sreenivasan (1988) for subharmonic bands.

cylinder shedding frequency. Subharmonic lock-in bands are shown in Figure 3-10(b). In each band, the natural shedding frequency disappears in favor of the excitation frequency (fundamental band) or a rational multiple of the excitation frequency (harmonic bands).

Cylinder vibration with frequencies near the shedding frequency influences the pattern and phasing of the vortices. Vortices tend to shed from the cylinder at that part of the vibration cycle when the cylinder is near its maximum displacement. There is an abrupt 180 degree phase shift between shedding and cylinder motion as the cylinder vibration frequency passes through the natural shedding frequency (Eq. 3-2) (Stansby, 1976; Ongoren and Rockwell, 1988). Zdravkovich (1982) observed that for vibration frequencies slightly *below* the natural shedding frequency, the vortex is shed from the *side opposite* the side experiencing maximum displacement. For vibration frequencies slightly *above* the shedding frequency, the vortex is shed from the *same side* as the maximum displacement. The phase shift contributes to a hysteresis effect whereby the range of lock-in depends to a degree on whether the shedding frequency is being approached from above or below.

As the amplitude of cylinder vibration is increased beyond approximately one-half diameter, the symmetric pattern of alternate vortices begins to break up (Williamson and Roshko, 1988). It can be seen in Figure 3-11 that at an amplitude of one diameter, three vortices are formed per cycle of vibration instead of the stable pattern of two per cycle at lower amplitudes. This breakup implies that the fluid forces imposed on the cylinder by vortex shedding will be a function of cylinder amplitude and may be self-limiting at large vibration amplitudes.

The average (i.e., steady) drag on a cylinder vibrating at or near the vortex shedding frequency is also a function of vibration amplitude. Drag increases with transverse vibration amplitude. The drag force per unit length on a cylinder of diameter D is

$$F_D = \tfrac{1}{2}\rho U^2 D C_D. \tag{3-3}$$

The drag coefficient C_D increases with A_y at resonance (ρ is the fluid density). This fact, like the lock-in, hysteresis, and resonance at sub and super harmonics, was first discovered by Bishop and Hassan in 1964. Figure 3-12 shows the amplification of the drag due to transverse cylinder motion when the vibration frequency coincides with the shedding frequency.

Three expressions for the increase in the drag coefficient with vibration are

$$\frac{C_D|_{A_y>0}}{C_D|_{A_y=0}} = \begin{cases} 1 + 2.1(A_y/D) & \text{Fig. 3-12 fit} \\ 1 + 1.043(2Y_{\text{rms}}/D)^{0.65} & \text{Vandiver (1983)} \\ 1 + 1.16\{[(1 + 2A_y/D)f_n/f_s] - 1\}^{0.65} & \text{Skop et al. (1977)} \end{cases} \tag{3-4}$$

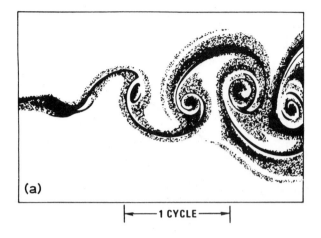

Fig. 3-11 Vortex street behind a cylinder vibrating transverse to the mean flow at resonance with vortex shedding; Re = 190. (a) Stable staggered vortex street; $A_y/D = 0.5$. (b) Unstable pattern with three vortices formed per cycle of vibration; $A_y/D = 1.0$ (Griffin and Ramberg, 1974).

A_y is the amplitude of transverse cylinder motion, i.e., one-half the peak-to-peak extremes measured perpendicular to the flow. C_D at $A_y = 0$ is given by Figure 6-9, Chapter 6. The first expression is a curve fit to the data of Sarpkaya (1978), Tanida et al. (1973), and Torum and Anand (1985) for rigid oscillating cylinders. The second was presented by Vandiver (1983), who found that it accurately predicted the drag on marine cables that vibrated in response to vortex shedding. Here Y_{rms} is the square root of the time average of the square of the amplitude measured at an antinode. For sinusoidal motion, $Y_{rms} = A_y/2^{1/2}$. Skop's expression includes an estimate of the effect off-resonance; it is valid for

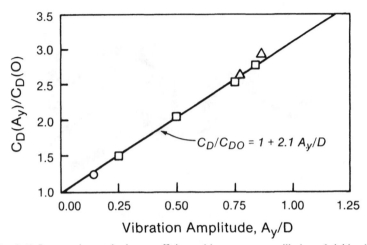

Fig. 3-12 Increase in steady drag coefficient with transverse oscillation of rigid cylinders at frequency equal to the shedding frequency. Experimental data: open square, Sarpkaya (1978) at Re = 8000; open circle, Tanida et al. (1973) at Re = 4000; open triangles, Torum and Anand (1985) at Re = 15 000.

$(1 + 2A_y/D)(f/f_s) > 1$, where f is the vibration frequency and f_s is the stationary cylinder shedding frequency (Eq. 3-2). At resonance, $f_n = f_s$, the three expressions are within 15% of one another.

As can be seen from Figure 3-12, very substantial increases in drag can occur for cylinders vibrating in sympathy with vortex shedding. This can have important consequences for the design of marine pipelines and cables that are exposed to ocean currents. As discussed in Section 3.8, the resonant amplitude of vortex-induced transverse vibration of these structures is often 0.5 to 1 diameter with drag coefficients between 2.5 and 3.0 (Dale and McCandles, 1967; Vandiver, 1983; Torum and Anand, 1985).

Lock-in also occurs, but to a lesser degree than for a circular cylinder, with square, triangular, and D sections and other sections that have sharp corners that fix the points of fluid separation (Feng, 1968; Ongoren and Rockwell, 1988; Bearman and Obasaju, 1982; Nakamura and Yoshimura, 1982; Washizu et al., 1978; Bokaian and Geoola, 1984). Bearman and Obasaju (1982) showed an increase in drag force on a square section with resonant vibration amplitude. Probably all noncircular sections that shed vortices will lock-in, increase spanwise correlation, and increase drag with resonant transverse vibration (see Sections 3.5.3, 3.7, and 4.4).

3.4. ANALYSIS OF VORTEX-INDUCED VIBRATION

As the flow velocity is increased or decreased so that the vortex shedding frequency (f_s) approaches the natural frequency (f_n) of an elastic structure, so that

$$f_n \simeq f_s = \frac{SU}{D}, \qquad \text{or} \qquad \frac{U}{f_n D} \simeq \frac{U}{f_s D} = \frac{1}{S} \simeq 5, \qquad (3\text{-}5)$$

the vortex shedding frequency suddenly locks onto the structure frequency. The resultant vibrations occur at or nearly at the natural frequency of the structure. The locked-in resonant oscillations of the near wake input substantial energy to the structure and large-amplitude vibrations can result. The torsional vibrations of a plate can be seen in Figure 3-13. Toebes and Eagleson (1961) found that these vibrations are sensitive to trailing-edge geometry; see Section 3.5 and Table 10-2. The transverse, translational vibrations of a spring-mounted circular cylinder are shown in Figure 3-14 for two levels of damping. The fundamental transverse vibrations ordinarily occur over the reduced velocity range $4 < U/(f_n D) < 8$, but vibrations also occur at sub and super harmonics of the shedding frequency.

King (1977) found that two distinct regimes of subharmonic resonance

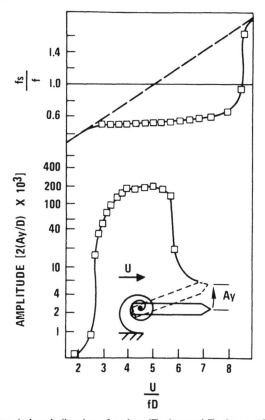

Fig. 3-13 Vortex-induced vibration of a plate (Toebes and Eagleson, 1961).

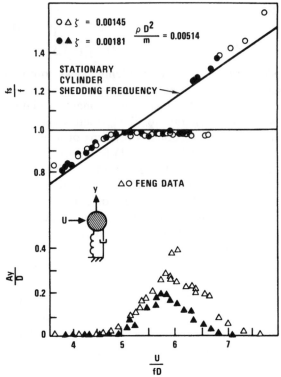

Fig. 3-14 Vortex-induced vibration of a spring-supported, damped circular cylinder. f is the natural frequency of the cylinder (Feng, 1968).

produce inline vibration. In the range $1.5 < U/(f_nD) < 2.5$, two nearly symmetric vortices are shed each cycle; this pattern of vortices is unstable, and downstream the vortices coalesce to form the usual staggered pattern. In the range $2.7 < U/(f_nD) < 3.8$, resonance of the structural frequency occurs with twice the shedding frequency ($2f_s$) and vortices are shed from alternate sides of the cylinder. The inline vibrations in these two regimes have amplitude approximately one-tenth that of the transverse vibrations at $4 < U/(f_n/D) < 8$. (See also Naudascher, 1987; Griffin and Ramberg, 1976.) Durgin et al. (1980) found that transverse vibrations were also induced in the range $12 < U/(f_nD) < 18$ corresponding to resonance with the third subharmonic ($f_s/3$) of vortex shedding.

3.5. MODELS FOR VORTEX-INDUCED VIBRATION

Models for vortex-induced vibration of circular cylindrical structures will be discussed in the following paragraphs. The first is a simple linear harmonic model that does not incorporate feedback effects, but it does serve to develop the appropriate nondimensional parameters and provide a forum for experimental data. The second models vortex shedding as a nonlinear oscillator. Its nonlinear solutions are correspondingly more difficult, but they have the potential for describing a much larger range of phenomena. These models will be compared with experimental data. These data, suitably nondimensionalized, are cast as the final model.

3.5.1. Harmonic Model

Because vortex shedding is a more or less sinusoidal process, it is reasonable to model the vortex shedding transverse force imposed on a circular cylinder as harmonic in time at the shedding frequency:

$$F_L = \tfrac{1}{2}\rho U^2 D C_L \sin{(\omega_s t)}, \tag{3-6}$$

where

   ρ = fluid density (must be in mass units),
   U = free stream velocity,
   D = cylinder diameter,
   C_L = lift coefficient, dimensionless,
   $\omega_s = 2\pi f_s$, circular vortex shedding frequency, radians/sec, where the vortex shedding frequency f_s is given by Eq. 3-2,
   t = time, sec,
   F_L = lift force (force perpendicular to the mean flow) per unit length of cylinder.

This force is applied to a spring-mounted, damped, rigid cylinder, shown in Figure 3-15. The cylinder is restrained to move perpendicular to the flow. The equation of motion of the cylinder is

$$m\ddot{y} + 2m\zeta\omega_y\dot{y} + ky = \tfrac{1}{2}\rho U^2 D C_L \sin{\omega_s t}, \tag{3-7}$$

where

   y = displacement of the cylinder in the vertical plane from its equilibrium position,
   m = mass per unit length of the cylinder, including added mass (see Chapter 2),
   ζ = structural damping factor (see Chapter 8),
   k = spring constant, force/unit displacement,
   $\omega_y = (k/m)^{1/2} = 2\pi f_y$, circular natural frequency of cylinder, and the overdots signify differentiation with respect to time.

Solutions to this linear equation are found by postulating a sinusoidal

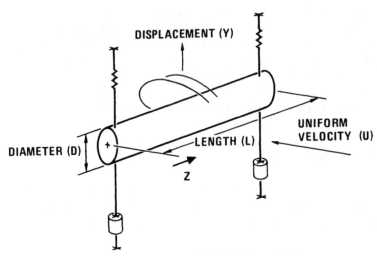

Fig. 3-15 Cylinder and coordinate system. Rigid cylinder mode shown.

steady-state response with amplitude A_y, frequency ω_y, and phase, ϕ,

$$y = A_y \sin(\omega_s t + \phi), \qquad (3\text{-}8)$$

and substituting this equation into Eq. 3-7. The result is (Thompson, 1988)

$$\frac{y}{D} = \frac{\frac{1}{2}\rho U^2 C_L \sin(\omega_s t + \phi)}{k\sqrt{[1-(\omega_s/\omega_y)^2]^2 + (2\zeta\omega_s/\omega_y)^2}}, \qquad (3\text{-}9)$$

where the phase angle is defined by

$$\tan\phi = (2\zeta\omega_s\omega_y)/(\omega_s^2 - \omega_y^2). \qquad (3\text{-}10)$$

The phase angle shifts by 180 degrees as the cylinder passes through $f_s = f_y$.

The response is largest when the shedding frequency approximately equals the cylinder natural frequency, $f_s = f_y$, a condition called resonance. Using Eqs. 3-8 and 3-9, the resonant vibration amplitude is

$$\left.\frac{A_y}{D}\right|_{f_y=f_s} = \frac{\rho U^2 C_L}{4k\zeta} = \frac{C_L}{4\pi S^2 \delta_r}. \qquad (3\text{-}11)$$

The right-hand form of this equation was obtained by incorporating the Strouhal relationship (Eq. 3-2) and the natural frequency, $f_y = (1/2\pi)(k/m)^{1/2}$. The resonant amplitude decreases with increasing reduced damping δ_r, which is defined as the mass ratio times the structural damping factor (Section 1.1.6),

$$\delta_r = \frac{2m(2\pi\zeta)}{\rho D^2}. \qquad (3\text{-}12)$$

m is the mass per unit length including added mass (Table 2-2), ζ is the damping factor (Chapter 8) usually measured in still fluid, ρ is the density of the surrounding fluid, and D is the cylinder external diameter. (The right-hand side of Eq. 3-11 implies that the amplitude at resonance is independent of flow velocity. This is because fixing Strouhal number fixes the relationship between cylinder and fluid frequencies.)

Measurements of the vortex-induced lift coefficient are given in Figure 3-16. $C_L = 1.0$ gives a conservative upper estimate of the locked-in response of nearly all circular cylindrical structures to transverse vortex-induced vibration. However, $C_L = 1.0$ may be overly conservative. Equation 3-11 with $C_L = 1.0$ will estimate amplitudes of several diameters for most cylinders in water, whereas in fact amplitudes exceeding 1.5 diameters are never observed. This suggests that the resonant cylinder motion feeds back into the vortex shedding process to influence the lift coefficient and limit the cylinder response.

Models have been developed for vortex-induced vibration of circular cylinders in the subcritical Reynolds number range that incorporate the amplitude dependence of lift (Basu and Vickery, 1983; Blevins and Burton, 1976). In the Blevins and Burton model, the experimental response data of Vickery and Watkins (1962) and Hartlen et al. (1968) were fitted to a three-term polynomial using Eq. 3-11 to obtain C_L as a function of A_y/D,

$$C_{Le} = a + b\left(\frac{A_y}{D}\right) + c\left(\frac{A_y}{D}\right)^2. \tag{3-13}$$

The coefficients in the curve fit are

$$a = 0.35, \quad b = 0.60, \quad c = -0.93.$$

The standard deviation of the fit is 0.07. Figure 3-17 shows the experimental data and the curve fit. These data clearly show that the lift coefficient is strongly amplitude dependent (Bishop and Hassan, 1964; Blevins, 1972; King, 1977).

As cylinder vibration at resonance with vortex shedding increases in amplitude from zero, the cylinder motion organizes the wake and spanwise correlation increases. The vortex strength increases, and the lift coefficient with it. As the cylinder amplitude increases beyond approximately one-half diameter, the cylinder begins to outrun the shedding vortices and the lift coefficient diminishes. This is reflected by $c < 0$ in the polynomial fit. Large-amplitude oscillation beyond one diameter produces a breakdown in the regular vortex street (Fig. 3-11) and the lift coefficient approaches zero. As a consequence, there is a maximum limiting amplitude of vortex-induced vibration of regular cylinders that is independent of structural damping.

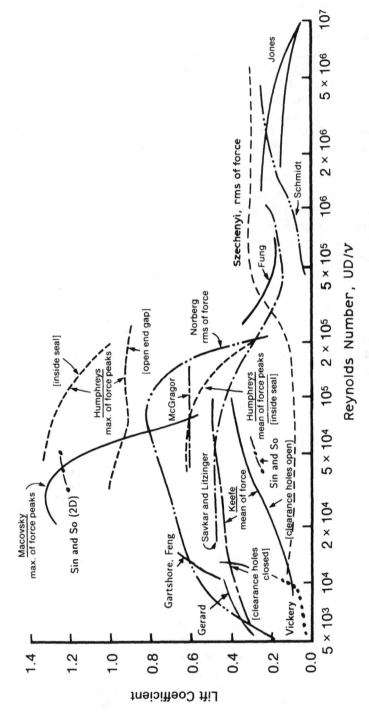

Fig. 3-16 Measurements of vortex-induced lift coefficient on stationary cylinder (Yamaguchi, 1971; Szechenyi, 1975; Jones et al., 1969; Gartshore, 1984; Sin and So, 1987; Savkar and Litzinger, 1982).

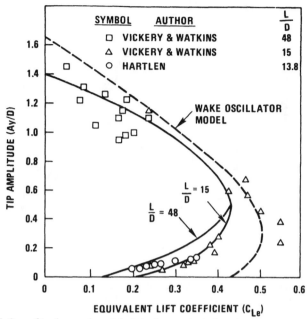

Fig. 3-17 Curvefit of experimental data for a pivoted rod at resonance with vortex shedding (Blevins and Burton, 1976). Data from Vickery and Watkins (1962) and Hartlen et al. (1968).

The amplitude-dependent lift model has been applied to the mode shapes of the rigid, spring-supported cylinder, the pivoted cantilever, and the sine mode. Some limiting cases of the analysis are shown in Table 3.1. l_c is the spanwise correlation length and L is the cylinder length. Numerical results for the amplitude of response as a function of reduced damping for two of these mode shapes are shown in Figures 3-18 and

Table 3-1 Correlation model results for response of three mode shapes

Mode	$\bar{y}(z)$	$C_{LE}{}^a$ $\dfrac{A_y}{D} \ll 1$ $l_c \ll L$	C_{LE} $l_c \gg L$	$\dfrac{A_y}{D}$ $\delta_r \to 0$
Rigid cylinder	1	$a\left(\dfrac{l_c}{L}\right)^{1/2}$	$a + b\dfrac{A_y}{D} + c\left(\dfrac{A_y}{D}\right)^2$	1.0
Pivoted rod	$\dfrac{z}{L}$	$a\left(\dfrac{4l_c}{3L}\right)^{1/2}$	$a + \dfrac{2}{3}b\dfrac{A_y}{D} + \dfrac{c}{2}\left(\dfrac{A_y}{D}\right)^2$	1.4
Sine mode	$\sin\dfrac{\pi z}{L}$	$a\left(\dfrac{\pi^2 l_c}{8L}\right)^{1/2}$	$a + \dfrac{\pi}{4}b\dfrac{A_y}{D} + \dfrac{2}{3}c\left(\dfrac{A_y}{D}\right)^2$	1.2

Source: Blevins and Burton (1976).
[a] $a = 0.35$, $b = 0.60$, $c = -0.93$.

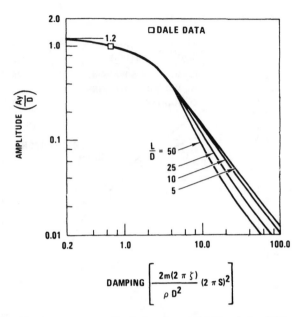

Fig. 3-18 Significant resonant amplitude for sine model $[\bar{y} = \sin(\pi z/L)]$ as a function of damping in comparison with experimental data of Dale et al. (1966) extrapolated with $\zeta = 0.005$.

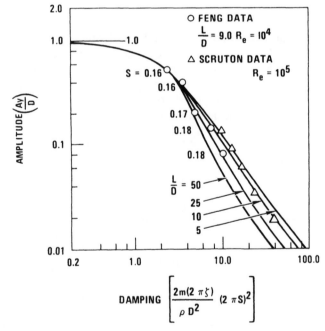

Fig. 3-19 Significant resonant amplitude for a spring-supported rigid cylinder ($\bar{y} = 1$) in comparison with experimental data of Feng (1968) and Scruton (1963).

3-19. The aspect ratio (ratio of cylinder length to diameter) affects resonant response only when the cylinder amplitude is below that required for full spanwise correlation. In the model, the amplitude required for fully correlated flow is taken to be one-half diameter. If a different value were used, it would shift the origin of the fan in Figures 3-18 and 3-19.

Exercises

1. Starting with Eqs. 3-6, 3-7, and 3-8, derive Eqs. 3-9 through 3-12. *Hint*: set $y = A \sin \omega t + B \cos \omega t$, substitute this into Eq. 3-7, set the coefficients of the sine and cosine terms to zero, solve the resultant equations for A and B, then note $|y| = (A^2 + B^2)^{1/2}$.

2. Show the peak response of Eq. 3-9 occurs when $\omega_s = \omega_y(1 - 2\zeta^2)^{1/2}$. Differentiate the amplitude of Eq. 3-9 with respect to ω_s and set the result to zero to determine the frequency at which the maximum amplitude response occurs. How much different is this amplitude than predicted by Eq. 3-11? *Hint*: the peak amplitude occurs at the same frequency as the minimum of the square of the inverse of the amplitude and the latter is far easier to differentiate.

3. Consider the peak responses in Figure 3-14 for the two different values of damping. Compare the prediction given by the right-hand form of Eq. 3-11 with $C_L = 1.0$ and $S = 0.2$ with these resonant amplitudes. What lift coefficient would be required to match the data exactly?

3.5.2. Wake Oscillator Model and Numerical Models

The nature of self-excited vortex shedding suggests that the fluid behavior might be modeled by a simple, nonlinear, self-excited oscillator. This idea was first suggested by Bishop and Hassan (1964), and it has been pursued by Hartlen et al. (1970), Skop and Griffin (1973), and Iwan and Blevins (1974). The Blevins–Iwan model described in the following section utilizes a Van der Pol type equation with a flow variable to describe the effects of vortex shedding. Model parameters are determined by curve-fitting experimental results for stationary and forced cylinders in the Reynolds number range between 10^3 and 10^5. Some of the basic properties of the model will be presented here. More advanced applications of the model are given by Hall and Iwan (1984) and Poore et al. (1986).

The basic assumptions of the model are as follows:

1. Inviscid flow provides a good approximation for the flow field outside the near wake.

2. There exists a well-formed, two-dimensional vortex street with a well-defined shedding frequency.

3. The force exerted on the cylinder by the flow depends on the velocity and acceleration of the flow relative to the cylinder.

The forces on the cylinder are evaluated using the momentum equation for the control volume containing a cylinder as shown in Figure 3-20. The momentum equation is

$$P_y = \frac{dJ_y}{dt} + S_y + F_y, \qquad (3\text{-}14)$$

where F_y is the fluid force on the cylinder, P_y is the pressure force on the control surface in the vertical direction, S_y is the momentum flow through the control surface, and J_y is the vertical momentum within the control volume. The vertical momentum is

$$J_y = \iint_A \rho v \, dx \, dy, \qquad (3\text{-}15)$$

where v is the vertical component of fluid velocity and ρ is the fluid density. A "hidden" fluid variable w is defined in terms of J_y such that

$$J_y = a_0 \rho \dot{w} D^2. \qquad (3\text{-}16)$$

$\dot{w}$ is a measure of the weighted average of the vertical component of flow velocity within the control volume. The dot ($\cdot$) denotes differentiation

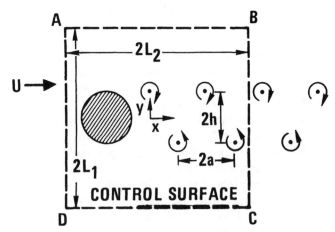

Fig. 3-20 Control volume containing a cylinder shedding vortices.

with respect to time. D is the cylinder diameter and a_0 is a dimensionless proportionality constant determined by experiment.

The far-field flow velocity induced by the vortex street diminishes as $(1/r)$, where r is the distance from the street. Application of the Bernoulli equation, and integration of the fluid pressure along boundaries AB and CD in the limit as L_1 approaches infinity while L_2 remains finite, therefore gives

$$P_y = 0, \tag{3-17}$$

i.e., zero net pressure force on the control volume.

The momentum flow through the control volume is expressible as a line integral along the sides BC and DA of the control volume. Examining the phase of the momentum flow in the control volume, and that across the boundary BC, it can be argued that S_y must lag dw/dt by approximately one-quarter cycle. Hence

$$S_y = K\rho u_t w\left(\frac{t-T}{4}\right)D + \text{correction terms}, \tag{3-18}$$

where T is the period of vortex shedding. The correction terms can be expressed in a power series of w and its time derivatives. For simplicity, only linear and cubic terms in w will be retained in S_y. Assuming that w oscillated harmonically at the vortex shedding frequency, this gives

$$S_y = K\rho u_t \omega_s wD - a_1 \rho UD\dot{w} - a_2 \rho \dot{w}^3 \frac{D}{U}, \tag{3-19}$$

where a_1 and a_2 are dimensionless constants that are assumed to be small compared with K; ω_s is the circular frequency of vortex shedding.

It is assumed that the force between the cylinder and the fluid depends on a weighted average velocity and acceleration of the fluid relative to the cylinder. Hence, there is no fundamental fluid-mechanical distinction between forced motion and elastic cylinder motion in response to vortex shedding. The force on the cylinder by the relative fluid velocity is written in the form of a lift coefficient whose magnitude is proportional to the relative angle between the free stream and the component of incoming flow to the cylinder. For small angles, this angle is $(w-y)/U$. The net force on the cylinder is

$$F_y = a_3 \rho D^2(\ddot{w} - \ddot{y}) + a_4 \rho D(\dot{w} - \dot{y}), \tag{3-20}$$

where a_3 and a_4 are dimensionless constants.

The fluid oscillator is assembled by substituting the component expressions (Eqs. 3-16, 3-17, 3-19, and 3-20) into the momentum equation (Eq. 3-14). This gives a nonlinear, self-excited fluid oscillator equation,

$$\ddot{w} + K'\frac{u_t}{U}\frac{U}{D}\omega_s w = (a_1' - a_4')\frac{U}{D}\dot{w} - a_2'\frac{\dot{w}^3}{UD} + a_3'\ddot{y} + a_4'\frac{U}{D}\dot{y}, \tag{3-21}$$

where $K' = K/(a_0 + a_3)$ and $a'_i = a_i/(a_0 + a_3)$, $i = 1, 2, 3, 4$. If the cylinder is elastically mounted as in Figure 3-15, the cylinder will respond dynamically to the fluid force. Applying the fluid force of Eq. 3-20 to the spring-supported cylinder of Eq. 3-7, the following equation of cylinder motion is produced:

$$\ddot{y} + 2\zeta_T \omega_y \dot{y} + \omega_y^2 y = a_3'' \ddot{w} \frac{U}{D} + a_4'' w \frac{U}{D}, \tag{3-22}$$

where

$a_i'' = \rho D^2 a_i / (m + a_3 \rho D^2)$, $i = 3, 4$,
$\omega_y = (k/m)^{1/2} / (1 + a_3 \rho D^2/m)$,
$\zeta_T = [\zeta (k/m)^{1/2} / \omega_y + \zeta_f] / (1 + a_3 \rho D^2/m)$,
$\zeta_f = a_4 \rho D U / (2 m \omega_y)$.

ω_y is the circular natural frequency of the cylinder in the fluid; k is the spring stiffness per unit length; ζ_T is the total effective damping coefficient, which is composed of a component caused by structural viscous damping (ζ) and a component caused by viscous fluid damping (ζ_f). The fluid damping limits the amplitude of vibration even as the structural damping approaches zero.

Equation 3-21 is a self-excited fluid oscillator equation. The first term on the right-hand side of this equation is a negative damping term that represents the extraction of fluid energy from the free stream and the transformation of that energy into transverse fluid oscillations. The second term from the right on the right-hand side represents a nonlinear fluid governor that limits the amplitude of fluid oscillations. The left-hand side of Eq. 3-21 represents the fluid feedback between the near wake and the boundary layer of the cylinder. The transverse fluid oscillations impose an oscillating fluid force on the cylinder through the terms on the right-side of Eq. 3-22. Just as the fluid imposes a force on the cylinder, so the cylinder imposes an equal but opposite force on the fluid. The force placed on the fluid by the cylinder is represented by the last two terms from the right on the right-hand side of Eq. 3-21.

As the frequency of the fluid oscillations approaches the natural frequency of the cylinder, large-amplitude cylinder motion can be induced; this cylinder motion feeds back into the fluid oscillator. The fluid forces and resultant cylinder amplitude are determined by the interaction of the fluid oscillator and the cylinder motion. Because there is a nonlinear term in the fluid oscillator, substantial motion can also be induced at sub and super harmonics of the fluid oscillation frequency. The model exhibits entrainment—the frequency of vortex shedding from an elastically mounted cylinder is entrained by the natural frequency of structural motion; see Poore et al. (1986).

There is only one nonlinear term in Eqs. 3-21 and 3-22. More nonlinear terms could be employed if a more precise model were desired. Unfortunately, increasing the number of nonlinear terms greatly increases the difficulty of analyzing the model and determining model parameters. The model of Eqs. 3-21 and 3-22 was analyzed using the method of slowly varying parameters; see Blevins (1974), Hall and Iwan (1984), and Poore et al. (1986). The model parameters were fixed by matching experimental measurements of vortex shedding from stationary and forced cylinder motion at resonance with vortex shedding (Iwan and Blevins, 1974). Typical results are $a_1 = 0.44$, $a_2 = 0.2$, $a_4 = 0.38$, $a_3 = 0$.

Comparing Eq. 3-21 with Eq. 3-7, it is easy to see that the natural frequency (the vortex shedding frequency) as given by Eq. 3-21 is

$$\omega_s = \left(K' \frac{u_t}{U} \right) \frac{U}{D}. \tag{3-23}$$

Since u_t/U is approximately constant for a large range of Reynolds numbers, the wake oscillator model states that the natural frequency of the fluid oscillator is proportional to the ratio of free stream velocity to cylinder diameter. This replicates Strouhal's result (Eq. 3-2).

The peak resonant cylinder amplitude of the model can be expressed in terms of a single variable called reduced damping (Eq. 3-12). Iwan (1975) extended the model to predict the resonant amplitude of elastic structures with cylindrical cross sections. Figure 3-21 shows that the results are in good agreement with data over the range $2 \times 10^2 < \text{Re} < 2 \times 10^5$ where experimental data are available.

The equation for the maximum response predicted by the model in Figure 3-21 is given in Table 3-2. For comparison, the expressions of Griffin and Sarpkaya are given; Griffin's expression is a curvefit to data, while Sarpkaya's has an analytical basis. These expressions also appear in

Table 3-2 Expressions for maximum resonant amplitude

Investigator	Predicted displacement amplitude
Wake oscillator, Blevins (1977)	$\dfrac{A_y}{D} = \dfrac{0.07\gamma}{(1.9 + \delta_r)S^2} \left[0.3 + \dfrac{0.72}{(1.9 + \delta_r)S} \right]^{1/2}$
Griffin and Ramberg (1982)	$\dfrac{A_y}{D} = \dfrac{1.29\gamma}{[1 + 0.43(2\pi S^2 \delta_r)]^{3.35}}$
Sarpkaya (1979)	$\dfrac{A_y}{D} = \dfrac{0.32\gamma}{[0.06 + (2\pi S^2 \delta_r)^2]^{1/2}}$
Harmonic model (Eq. 3-11)	$\dfrac{A_y}{D} = \dfrac{C_L}{4\pi S^2 \delta_r}$

[C_L or, equivalently, C_{LE}, given by Table 3-1 or Fig. 3-17; δ_r given by Eq. 3-12; γ given by Table 3-3]

Appendix N of Code Section 3 of the ASME Boiler and Pressure Vessel Code. All three expressions are within about 15% of one another.

γ is a dimensionless mode factor obtained from solution of the wake oscillator model as shown in Section 3.8. A_y in Figure 3-21 and Table 3-2 refers to the maximum amplitude along the span:

$$\gamma = \bar{y}_{\max}(z/l)\left\{\frac{\displaystyle\int_0^L \bar{y}^2(z)\,dz}{\displaystyle\int_0^L \bar{y}^4(z)\,dz}\right\}^{1/2}, \tag{3-24}$$

where $\bar{y}_{\max}(z/l)$ is the maximum value of the mode shape $\bar{y}(z/l)$ over the span extending from $z = 0$ to $z = L$. Values of γ are given in Table 3-3. If the mass of the cylinder is not uniformly distributed, then the equivalent mass per unit length is

$$m = \frac{\displaystyle\int_0^L m(z)\bar{y}^2(z)\,dz}{\displaystyle\int_0^L \bar{y}^2(z)\,dz}, \tag{3-25}$$

where $m(z)$ is the mass per unit length of the cylinder at each spanwise

Table 3-3 Mode shape and geometric factors for wake oscillator model

Structural element	Mode shape[a] $\bar{y}(z/L)$	Natural frequency, ω_y	γ
Rigid cylinder	1	$\sqrt{\dfrac{k}{m}}$	1.000
Uniform pivoted rod	z/l	$\sqrt{\dfrac{3k_\theta}{mL^3}}$	1.291
Taut string or cable	$\sin\dfrac{n\pi z}{L}$	$n\pi\sqrt{\dfrac{T}{mL^2}}$	1.155 for $n = 1, 2, 3, \ldots$
Simply supported uniform beam	$\sin\dfrac{n\pi z}{L}$	$n^2\pi^2\sqrt{\dfrac{EI}{mL^4}}$	1.155 for $n = 1, 2, 3, \ldots$
Cantilevered uniform beam	$(\sin\beta_n L - \sinh\beta_n L)$ $\times (\sin\beta_n z - \sinh\beta_n z)$ $+ (\cos\beta_n l + \cosh\beta_n l)$ $\times (\cos\beta_n z - \cosh\beta_n z)$ $\beta_n^4 = \omega_n^2 m/EI$	$\omega_1 = 3.52\sqrt{\dfrac{EI}{mL^4}}$ $\omega_2 = 22.03\sqrt{\dfrac{EI}{mL^4}}$ $\omega_3 = 61.70\sqrt{\dfrac{EI}{mL^4}}$	$\gamma_1 = 1.305$ $\gamma_2 = 1.499$ $\gamma_3 = 1.537$

Source: Iwan (1975).
[a] m = mass/unit length and includes the appropriate added fluid mass; E = modulus of elasticity; I = area moment of inertia of the section; L = spanwise length.

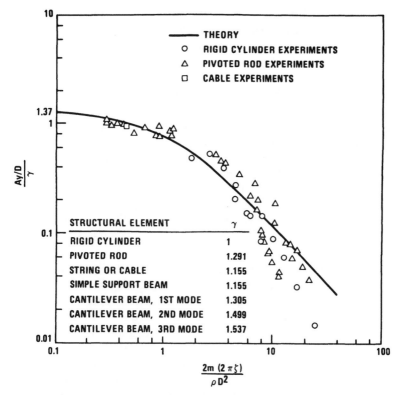

Fig. 3-21 Normalized maximum amplitude of resonant vortex-induced response versus reduced damping. (After Iwan and Blevins, 1974). γ is given in Table 3-3. The theory line is given in Table 3-2.

location. The effect of a spanwise-varying flow velocity is discussed in Section 3.8.

Two regions of Figure 3-21 are of particular interest. First, as structural damping approaches zero, the model predicts that the vortex-induced vibration reaches a maximum limiting amplitude between one and two diameters ($1.37D$ for the rigid spring-supported cylinder). This limit is produced by the reduction in the exciting component of the fluid force with increasing amplitude, possibly caused by the breakdown in the vortex street at large-amplitude vibration (Fig. 3-11). Second, the model predictions are well above the experimental data for amplitudes on the order of one-tenth diameter or less. Spanwise correlation effects reduce the excitation at low amplitudes. Low-amplitude response is probably better predicted by the correlation model of the previous section than the wake oscillator model.

It would be desirable to predict analytically the amplitude of vortex-induced vibration using cylinder surface pressures that were obtained from an exact analysis of the flow field. Ideally, one would solve the time-dependent Navier–Stokes equation in the presence of a vibrating cylinder. Flow separation and vortex formation would emerge naturally from the solution, and the pressure and shear loading on the surface would provide the forcing function for the coupled cylinder motion. With advances in computers, the approach is becoming possible. Two of the numerical solutions methods that have been advanced are (1) divide the flow field into a mesh and then construct finite-element solutions to the Navier–Stokes equation using a turbulence model (Dougherty et al., 1989; Ghia, 1987; Chilukuri, 1987), or (2) use a discrete vortex model based on motion of point vortices (Chapter 2, Section 2.4). Presently, the numerical solutions are two-dimensional. Most Navier–Stokes solutions are limited to Reynolds numbers of a few hundred but higher Reynolds numbers are being attempted (Tsuboi et al., 1989).

The author has made a number of flow-induced vibration calculations using the discrete vortex method (Blevins, 1989); see Figure 2-9 and Section 2.4. He found that the results are in general agreement with data, including the self-limiting nature of vortex-induced vibration, and the vortex wake was well represented. However, quantitatively the results for force applied to the cylinder were not in detailed agreement with laboratory data and the flow-induced vibration predictions were not superior to available techniques. These shortcomings of numerical models may be resolved with increased computing power (van der Vegt, 1988).

Exercises

1. Solve Eq. 3-21 for a stationary cylinder, $y = \dot{y} = \ddot{y} = 0$. Do this by assuming that the fluid oscillator has a harmonic response, $w = A_w \sin \omega_s t$, substituting this solution form into the equation, expanding the cubic term in terms of harmonics, $\sin^3 \theta = \frac{3}{4} \sin \theta - \frac{1}{4} \sin 3\theta$, neglecting the third harmonic term, and equating the coefficients of similar terms to solve for A_w.

2. Review the papers of Iwan and Blevins (1974), Hall and Iwan (1984), and Poore et al. (1986). What are the differences in their solution techniques for the nonlinear equation?

3.5.3. Synthesis of Data and Models

The models discussed in Sections 3.5.1 and 3.5.2 suggest that the appropriate nondimensional parameters that are useful for correlating the vortex-induced response of continuous structures are (1) structural damping factor, ζ, (2) reduced velocity, $U/(f_n D)$, (3) mass ratio, $m/(\rho D^2)$, (4) ratio of stationary cylinder shedding frequency to natural frequency, f_n/f_s, (5) aspect ratio L/D, and (6) Reynolds number, UD/ν. For nonuniform structures, these parameters are evaluated on a mode-by-mode basis, and the equivalent uniform mass (Eq. 3-25) can be used.

With sufficient experimental data, nondimensionalized in terms of the above parameters, predictions can be made directly from data. A procedure for this is illustrated in Figure 3-22. This logic is also used in the computer program of the Preface. Griffin and Ramberg (1982) have found that this approach gives good predictions for marine systems.

The procedure of Figure 3-22 is most accurate for long, slender ($L/D > 10$), single cylinders responding in steady uniform flow in the subcritical Reynolds number range. It tends to be conservative for other Reynolds number ranges and other geometries. In the transitional Reynolds number range (Figs. 3-2, 3-3), vortex shedding is more disorganized, especially for smooth surface cylinders ($\epsilon/D < 0.003$, where ϵ is a characteristic surface roughness; see Achenbach and Heinecke, 1981). Wootton (1968) found that the amplitude of vortex-induced vibration for smooth, moderately damped cylinders, $6 < \delta_r < 25$, was reduced by a factor of 4 in the transitional regime $0.2 \times 10^6 < \mathrm{Re} < 1.3 \times 10^6$ from that in either the subcritical or supercritical regime. It is the author's experience in steady flow and that of Sumer and Fredsoe (1988) in oscillating flow that there is no reduction in the transitional regime for rough-surface cylinders.

Many cylinders of practical importance are inclined (i.e., yawed) with respect to the mean flow. King (1977) and Ramberg (1983) found that the inclination does not appreciably reduce inline or transverse vortex-induced vibration for angles of inclination at least as large as 30 degrees. Inclination does reduce the component of flow velocity normal to the cylinder axis and the vortex shedding frequency (see Section 3.2); hence resonances are postponed to higher velocities than would occur with a cylinder perpendicular to the flow.

While the procedure of Figure 3-22 is based on circular cylinders, it is the author's experience that it provides useful estimates of the vortex-induced response of noncircular sections such as rectangles, squares, bridge decks, and so on, when no other data are available. For noncircular sections, the dimension D is chosen to be the vertical height of the section that characterizes the distance between the separating free shear layers that will form the vortex street. Similarly, Eq. 3-6 with Figure 3-16 can provide rough estimates of the vortex-induced forces on noncircular sections.

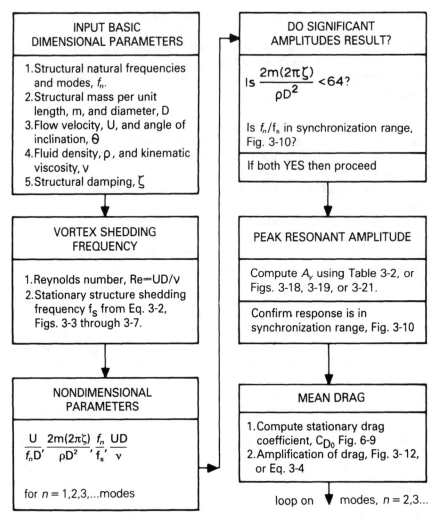

Fig. 3-22 A flow chart for determining amplitude and drag in vortex-induced vibration.

Exercises

1. Consider a free-standing smoke stack 4.5 ft (1.5m) in dia-meter that has been welded from 0.25-in. (0.63-cm) sheet steel. The stack has a height of 125 ft (40 m) and its base is anchored to bedrock. Because of the absence of joints that expend energy, the damping is estimated to be only $\zeta = 0.001$ (Fig. 8-15). Show that that natural frequency of the stack in the fundamental cantilever mode is 0.9 Hz. At 20°C, air has a density of 0.075 lb/ft^3

$(1.2 \, \text{kg/m}^3)$ and a kinematic viscosity of $0.0001 \, \text{ft}^2/\text{sec}$ $(0.000015 \, \text{m}^2/\text{sec})$. Steel has a density of $0.3 \, \text{lb/in.}^3$ $(7.97 \, \text{gm/cc})$ and a modulus of $30 \times 10^6 \, \text{psi}$ $(190 \times 10^9 \, \text{Pa})$. What is the reduced damping (Eq. 3-12)? At what wind velocity will the first mode resonate with vortex shedding? Will these amplitudes be destructive? Design three alternatives for reducing the vibration (Section 3.6).

2. Predict the vortex-induced response of the rectangle given in Figure 1-3 by assuming that it is equivalent to a circular cylinder with diameter equal to the depth D of the rectangle. Use the procedure of Figure 3-22 and compare your results with the data of Figure 1-3.

3.6. REDUCTION OF VORTEX-INDUCED VIBRATION

The amplitude of resonant vortex-induced vibration, and the associated magnification of steady drag, can be substantially reduced by modifying either the structure or the flow as follows:

1. *Increase reduced damping.* If the reduced damping (Eq. 3-12) can be increased, then the amplitude of vibration will be reduced as predicted by the formulas of Table 3-2 or Figure 3-18, 3-19, or 3-21. In particular, if the reduced damping exceeds about 64,

$$\frac{2m(2\pi\zeta)}{\rho D^2} > 64, \tag{3-26}$$

then peak amplitudes at resonance are ordinarily less than 1% of diameter and are ordinarily negligible in comparison with the deflection induced by drag. Reduced damping can be increased by either increasing structural damping or increasing structural mass. Increased damping can be achieved by permitting scraping or banging between structural elements, by using materials with high internal damping such as viscoelastic materials, rubber, and wood, or by using external dampers. The Stockbridge damper (Chapter 4, Fig. 4-18(a)) has been used to reduce vortex-induced vibration of powerlines (Hagedorn, 1982). Walshe and Wootton (1970) and Scanlan and Wardlaw (1973) have suggested a variety of damping devices—see Chapter 8.

2. *Avoid resonance.* If the reduced velocity is kept below 1,

$$\frac{U}{f_n D} < 1, \tag{3-27}$$

where f_n is the natural frequency of the structure in the mode of interest, the inline and transverse resonances are avoided. This is ordinarily achieved by stiffening the structure. Stiffening is often most practical for smaller structures.

3. *Streamline cross section.* If separation from the structure can be minimized, then vortex shedding will be minimized and drag will be reduced. Streamlining the downstream side of a structure ordinarily requires a taper of 6 longitudinal for each unit lateral, or an included angle of the taper no bigger than 8 to 10 degrees, to be effective. The NACA 0018 airfoil has been used for a streamline fairing. Hanko (1967) discusses the reduction of vortex-induced vibration of a pier by tapering. Gardner (1982) discussed streamlined oil pipe. Toebes and Eagleson (1961) found that the vortex-induced vibrations of a plate were suppressed by streamlining the trailing edge. See Chapter 10, Table 10-2, for

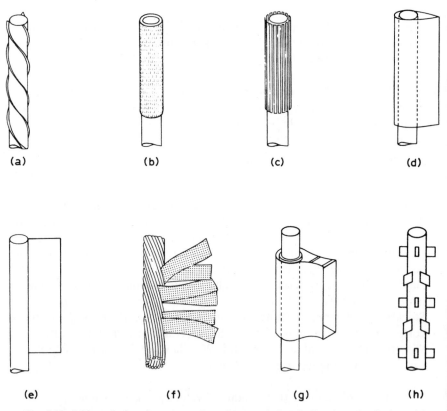

Fig. 3-23 Add-on devices for suppression of vortex-induced vibration of cylinders: (a) helical strake; (b) shroud; (c) axial slats; (d) streamlined fairing; (e) splitter; (f) ribboned cable; (g) pivoted guiding vane; (h) spoiler plates.

the effect of various trailing edges. Streamlining is most effective when the direction of the flow is fixed relative to the structure and the structure has sufficient stiffness to avoid flutter (Chapter 4).

4. *Add a vortex suppression device.* Zdravkovich (1981), Wong and Kokkalis (1982), Hafen and Meggitt (1971), Every et al. (1982), and Rogers (1983) review add-on devices for suppression of vortex-induced vibration of cylindrical structures in wind and marine applications. Seven devices of proven effectiveness are shown in Figure 3-23. These devices were developed through ad hoc experiments. They act by disrupting or preventing the formation of an organized, two-dimensional vortex street.

Design Guidelines for Vortex Suppression Devices

Design guidelines for optimal devices are as follows. The design dimensions and drag coefficients are based on the plain cylinder outside diameter D (Eq. 3-3).

Helical strake, Figure 3-23(a)

- Height of strakes: 0.05 to 0.12 D; 0.1 D is widely used
- Number of strakes: 3 in a parallel helix pattern
- Pitch of strakes: 3.6 to 5 D

Comment: Wilson and Tinsley (1989) recommend 0.1 D high strakes spaced at 120 degree intervals with a 60 degree helix angle from the cylinder axis. Sharp-edged helical strakes were invented by C. Scruton and D. E. Walshe (Patent 3,076,533, Feb. 5, 1963). They are widely used. Typical drag coefficient at Re = 10^5, $C_D = 1.35$. [Halkyard and Grote (1987) report that two wraps of 0.1 D diameter rope with an 18 D pitch gave good suppression for a marine pipeline, but Airey et al. (1988) found that a single wrap of 0.15 D rubber with 4.6 D pitch was ineffective.]

Perforated shroud, Figure 3-23(b)

- Shroud outside diameter: 1.25 D
- Percent open area: 30 to 40
- Hole geometry: square, 0.07 D on a side, or round, 0.125 D dia.
- Holes in a circumferential ring: 32

Comment: Perforated shroud consists of a relatively thin metal cylinder that is held off the cylinders by struts. The outside diameter of the shroud is approximately 1.25 cylinder diameters (D). Typical drag coefficient at Re = 10^5, $C_D = 0.91$.

Axial slats, Figure 3-23(c)

- Slat shroud outside diameter: 1.29 D
- Slat width: 0.09 D
- Percent open area: 40
- Number of slats around circumference: 25 to 30

Comment: Wong and Kokkalis (1982) report that if the two forwardmost and the two aftmost slats are removed, performance is improved and drag is reduced 9%. Slats, like shrouds, are held off the cylinder by struts or annular frames. Typical drag coefficient at $Re = 10^5$, $C_D = 1.05$.

Streamlined fairing, Figure 3-23(d)

- Overall length (nose to tail): 3 to 6 D
- Profile: airfoil or tapered wedge

Comment: Streamlined fairings avoid separation. If the flow direction is variable, then they must pivot on the cylinder. Failure to pivot can lead to large side-forces and instability. Gardner (1982) utilized pivoting fairings to protect oil pipelines in the Amazon delta. Typical drag coefficient, based on diameter, at $Re = 10^5$, $C_D \sim 0.1$ to 0.3.

Splitter, Figure 3-23(e)

- Length: 4 to 5 D to be effective

Comment: See Sallet (1970, 1980), Apelt and West (1975), and Unal and Rockwell (1987). Typical drag coefficient at $Re = 10^5$, $C_D = 1.0$.

Ribbon or hair cable, Figure 3-23(f)

- Ribbon width: 1 to 2 D
- Ribbon length: 6 to 10 D (rear of cylinder to tip of ribbon)
- Ribbon thickness: 0.05 D
- Ribbon spacing: on 1 to 3 D centers
- Ribbon material: polyurethane film

Comment: By separating the outer strands of a marine cable, urethane or other compliant plastic ribbons can be doubled through the wire strands. Ribboned cable will pass over a shieve and can be wound on an ordinary cable drum, which is impossible for a cable with a streamlined fairing. See Hafen and Meggitt (1971), who also give the results of Blevins (1971). One version is manufactured commercially by Zippertubing Co., Los Angeles, California. Typical drag coefficient at $Re = 10^5$, $C_D = 1.5$.

Guiding vane, Figure 3-23(g)

- Plate length beyond rear of cylinder: $1\,D$
- Lateral separation between trailing edges: $0.9\,D$
- Method of attachment: sliding bearing allowing pivot. Rogers (1983) reports that the guiding vanes provide complete suppression of vortex-induced vibration in ocean environments. They are a shortened form of a pivoting streamlined fairing. Typical drag coefficient $C_D = 0.33$ at $\mathrm{Re} = 10^5$.

Spoiler plates, Figure 3-23(h)

- Spoiler plate size: square, $D/3$ on a side
- Number of spoiler plates: 4 in a circumferential ring
- Axial distance between plates: $2/3\,D$

Comment: Stansby et al. (1986) found spoiler plates reduce resonant vortex-induced response by 70%. Drag coefficient is unknown.

Stepped cylinder

Brooks (1987) has modified a drilling riser (Fig. 3-26(b)) by attaching 50 ft (16 m) long buoyancy modules (Fig. 3-27(a)) at 100 ft (33 m) intervals, leaving 50 ft (16 m) intermediate intervals of bare pipe. This creates a stepped cylinder with single and double diameters. He reports no vortex-induced vibration in the 2135 ft (700 m) deep water drilling in a high current area of the North Sea. Walker and King (1988) found sufficient changes in diameter in a stepped cylinder reduce resonant response and suppress spanwise correlation of vortex shedding.

The amplitude of resonant vibration with a vortex-suppression device installed depends on the reduced damping, or, in other words, it depends on the amplitude of the plain cylinder before the device is installed. Optimal helical strakes, perforated shrouds, ribbon cable, spoilers, and axial slats can reduce resonant vortex-induced response by 70–90% of the plain-cylinder response. The increase in effective diameter and surface area with these devices entails an increase in drag over a stationary cylinder of approximately 15–50% but often less drag than a resonantly vibrating cylinder (Eq. 3-4). Streamlined fairings or guided vanes can be expected to reduce vortex-induced response by 80% or more and give a 50% or better reduction in drag but at a cost of increased complexity.

3.7. THE PROBLEM WITH BRIDGE DECKS

Wind has been a common cause of failure of bridges. Full-scale and model tests have shown that vibration of bridges in wind occurs with mode shapes and frequencies that differ little from the natural modes and frequencies of the bridge. The wind speeds that excite bridge deck vibrations are approximately proportional to the natural frequencies; hence bridges with low natural frequencies, such as slender suspension or cable-stayed bridges, are the most vulnerable (Scruton, 1981). Almost without exception, the modes involved are either vertical plunge (i.e., vertical bending) or torsion. In vertical bending, the roadway deck moves up and down; on a suspension bridge, the two cables displace equally and in step. In torsional vibration, the deck twists and the cables displace out of phase.

Any section, including the bridge deck, will shed vortices and respond to the vortex-induced forces (Bearman, 1984; Nakamura and Yoshimura, 1982; Komatsu and Kobayashi, 1980). In addition, noncircular sections are subject to aerodynamic instability produced by quasi-steady aerodynamic forces (Chapter 4, see Section 4.4). At low reduced velocities, $U/(fD) < 10$ where D is the vertical depth of the section, the unsteady vortex shedding forces and quasi-steady instability forces act at the same or similar frequencies and so it is often impossible to experimentally differentiate between vortex-induced and the quasi-steady aerodynamic instability (Ericsson and Reding, 1988; Bearman and Luo, 1988; also see Fig. 4-13). Both vortex shedding and aerodynamic instability induce bridge vibration (Wardlaw, 1988), and in many cases it has not been possible to identify which is the dominant cause of vibration (Wardlaw and Blevins, 1980).

The Tacoma Narrows bridge failed in 1940 in a 42 mph (68 km/hr) wind. First it vibrated in a 0.62 Hz vertical mode corresponding to a reduced velocity of approximately $U/(fD) \approx 12$; then it switched to a 0.23 Hz torsional mode, where it eventually failed (Steinman and Watson, 1957). U is the wind velocity, D is the depth of the deck, and f is the natural frequency in hertz. The failure has been variously ascribed to stall-flutter by Parkinson (1971), vortex-induced vibration by Theodore von Karman (Farquharson et al., 1949), and aerodynamic instability by Steinman and Watson (1957). Figure 4-15, Chapter 4, shows the results of model tests on this bridge deck. Note the vertical and torsional modes of response and the lock-in of the shedding frequency with the natural frequencies. Other bridges have had similar failure modes (Scruton, 1981; Steinman and Watson, 1957; Farquharson et al., 1949).

Figure 3-24 shows a typical amplitude response of a bridge deck as a function of reduced velocity. The range A of this figure is

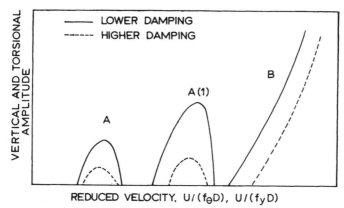

Fig. 3-24 Typical response of a bridge deck in torsion and vertical plunge to wind (Scruton, 1981).

limited-amplitude response occurring over a small wind speed range that is characteristic of vortex-induced vibration. Typically, vertical bending vibrations occur at reduced velocities in the range $1.5 < U/(fD) < 8$, where f is the frequency of the mode in hertz and D is the vertical depth of the bridge deck (Section 4.4; Komatsu and Kobayashi, 1980; Scruton, 1981; Fig. 4-15). This may be followed by a second mode (A1). Increasing damping diminishes these oscillations. Range B is indicative of a galloping or stall-flutter instability because the vibrations increase continuously with velocity. Often the instability occurs after the A and A1 ranges, but this is not always so. Increasing damping often shifts the B range of onset of instability to higher velocities, but again this is not always so. For example, in Chapter 1, Figure 1-3 shows that damping has little effect on the onset of instability of a rectangular section, possibly because of the regulating effect of the third harmonic of vortex shedding (Novak, 1971; Durgin et al., 1980).

Scruton (1981) summarizes methodology for minimizing vibration of bridge decks. The single most reliable method is to increase the stiffness and frequency of vertical plunge and torsional modes so that reduced velocities are maintained below 1. This can be accomplished by using a torsionally stiff, diagonally braced truss or, alternatively, a fairly deep, closed box section. The original Tacoma Narrows bridge deck was a shallow unbraced H section fabricated from solid plates with the vertical height being only 8 ft (2.4 m) for a 2800 ft (850 m) span. Its successor is a substantial open truss, 33 feet (10 m) deep and 50% heavier than the original (Steinman and Watson, 1957). At low wind speeds, the Golden Gate Bridge in San Francisco responded primarily at 0.13 Hz in a vertical mode that is symmetric about midspan, but with increasing wind speed it switched to an asymmetric vertical-torsion coupled mode at 0.1 Hz

(Vincent, 1962; Tanaka and Davenport, 1983). During a 70 mph
(110 km/hr) wind storm in 1951, it suffered peak-to-peak vibration
amplitudes of 12 ft (3.5 m) and torsional peak-to-peak amplitudes of 22
degrees (Vincent, 1958). Subsequent torsional stiffening suppressed the
coupled torsional vibration.

The second method of minimizing bridge deck vibrations is to use
open web type construction to minimize wind pressure differential-
induced forces. Grillage road decks, open hand rails, and open trusses
are examples of this (Advisory Board on the Investigation of Suspension
Bridges, 1955). The third approach to minimizing the vibration is to
streamline the section with fairings to minimize unsteady forces of
separated flow (see Section 3.6).

Figure 3-25 shows the benefit of streamlining and open construction on
the Longs Creek, Canada cable stayed bridge. Shortly after fabrication,
the bridge was observed to vibrate vertically as much as 8 in. (12 cm) at
its natural frequency of 0.6 Hz during winds speeds over a limited range
of 25 to 30 mph (40 to 48 km/hr) (Wardlaw and Ponder, 1970). No bridge
motion was observed for winds exceeding 35 mph (56 km/hr). There was
no significant torsional motion. As a result of 1/30 scale model tests on a
spring-supported representative section, 10 ft (3 m) fairings, shown in

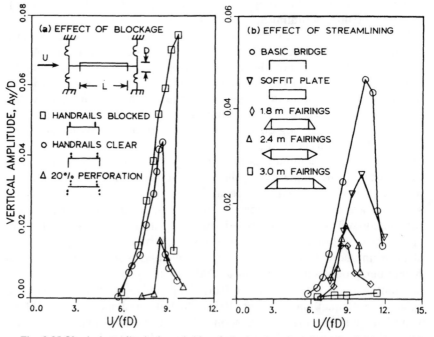

Fig. 3-25 Vertical amplitude for a bridge deck showing the effect of rail blockage (a)
and fairings (b) (Wardlaw and Ponder, 1970; Scanlan and Wardlaw, 1973).

Figure 3-25(b), were installed. These have eliminated the bridge motion. Wind-tunnel tests on dynamically scaled two-dimensional, spring-supported sections such as shown in Figure 3-25 have often been used for bridge decks (also see Figs. 4-12, 4-15, Chapter 1, and Section 4.4). A more expensive alternative is to test a complete three-dimensional dynamic model of the entire bridge in a simulated boundary layer. See Chapter 7, Section 7.4.3.

Exercises

1. Review the data on the Longs Creek bridge descriptions given in this section and Figure 3-25. Review Sections 4.2 and 4.4. Is vortex-shedding or instability the dominant cause of the bridge motion? Provide specific arguments by estimating the shedding frequency and comparing Figure 3-25 with Figure 3-14.

2. Repeat Exercise 1 for the spring-supported rectangular section of Figure 1-3. Predict the onset of galloping instability of this section using Eq. 4-16 and $\partial C_y / \partial \alpha \approx 2$.

3.8. EXAMPLE: MARINE CABLES AND PIPELINES

A towed hydrophone is shown in Figure 3-26(a) and an offshore oil production riser (pipe from a reservoir) is shown in Figure 3-26(b). In both cases a relatively flexible, tensioned cylinder is exposed to a current. Typical cross sections are shown in Figure 3-27. The mass ratio of these sections can be computed as

$$\frac{m}{\rho D^2} = \frac{(\pi/4)(\rho_s \alpha D^2 + \rho D^2)}{\rho D^2} = \frac{\pi}{4}\left(\frac{\alpha \rho_s}{\rho + 1}\right) = \text{order of 1 to 10},$$

where α is the fraction of the section occupied by steel, typically 5% to 80%, ρ_s is the density of steel (8000 kg/m^3), and ρ is the density of sea water (1025 kg/m^3). The mass per unit length, m, includes added mass equal to the displaced mass of the cylinder (Chapter 2). Thus, the mass ratio for these structures is typically between 1 and 10. If the structural damping factor is estimated at 1% (better estimates are given in Chapter 8), then the reduced damping is relatively small:

$$\delta_r = \frac{2m(2\pi\zeta)}{\rho D^2} = \text{order 1 or less}.$$

Figures 3-18, 3-19, and 3-21 imply that transverse vibration amplitudes as

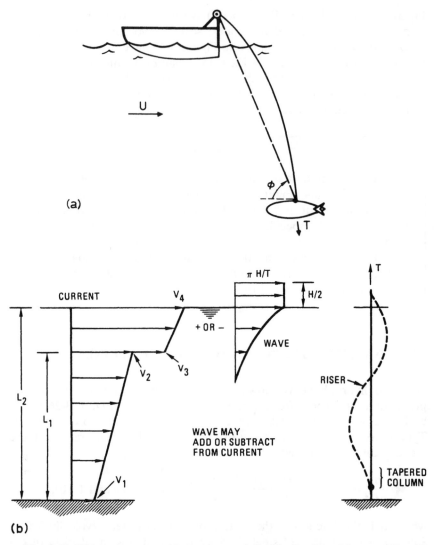

Fig. 3-26 Offshore systems subject to vortex-induced vibration: (a) tow cable; (b) offshore riser.

high as 1 to 1.5 diameters can be experienced at resonance with vortex shedding. These amplitudes are largely independent of reduced damping at this low value of reduced damping. Large-amplitude vibration persists over a range of reduced velocity $4.5 < U/(f_n D) < 7.5$, where f_n is the natural frequency of the nth mode (Griffin and Ramberg, 1982).

The mean tension dominates the bending stiffness of cables and deep-water risers. The natural frequencies can often be well approximated by the natural frequencies of a straight tensioned string.

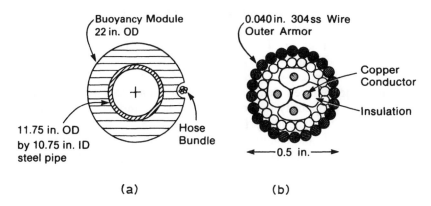

Fig. 3-27 Cross sections of the systems in Fig. 3-26: (a) deep-water sales riser; (b) double armoured cable.

These frequencies are

$$f_i = \frac{i}{2L} \left(\frac{T}{m}\right)^{1/2} \quad \text{Hz}, \qquad i = 1, 2, 3, \ldots \qquad (3\text{-}28)$$

and the corresponding mode shapes are sinusoidal:

$$\bar{y}_i(z) = \sin\left(\frac{i\pi z}{L}\right),$$

where T is the mean tension, L is the span between tie-downs, and m is the mass per unit length, including added mass.

For example, for a typical North Sea deep-water riser shown in Figures 3-26(b) and 3-27(a), the buoyancy module has an outside diameter $D = 22$ in., and the central pipe is 11.75 in. in outside diameter with a 10.75 in. inside diameter. The net weight is 200 lb/ft. The base is located 40 ft off the seafloor, the mean water level is 1200 ft off the seabed, and the gimballed upper tie-down is on a platform at 1300 ft off the seafloor. The net tension at the top is 89000 lb. The added mass of the riser is 169 lb/ft, so the net mass is 364 lb/ft or 11.3 slug/ft in mass units. Using the above equation, the natural frequencies are estimated to be

$$f_i = 0.0341 \, i \quad \text{Hz} \qquad i = 1, 2, 3, 4 \ldots$$

based on a length $L = 1260$ ft. That is, in the fundamental mode, the riser period $(T = 1/f)$ is 29.3 sec per cycle. In the higher modes, the frequency spacing between modes remains constant but the *relative* spacing between modes decreases.

The riser is exposed to both waves and current. A typical current profile, shown in Figure 3-26(b), has $V_1 = 0.9$ ft/sec, $V_2 = 1.2$ ft/sec, $V_3 = V_4 = 2.3$ ft/sec, and a wave height of $H = 13$ ft and a wave period of

7.8 sec. The wave is cyclic and so it can add to or substract from the current velocity (vortex-induced vibration due to waves is further discussed in Chapter 6). The Reynolds number (Eq. 3-1) based on a velocity of 2 ft/sec and $v = 1.5 \times 10^{-5}$ ft²/sec is 240 000, at the upper end of the subcritical range.

Because the reduced damping is low, a broad lock-in band can be expected. The propensity for resonance with the modes can be assessed by computing the shedding frequency (Eq. 3-2), drawing a ±40% entrainment band, and comparing the result with the natural frequencies as shown in Figure 3-28. As can be seen from this figure, mode 1 is free

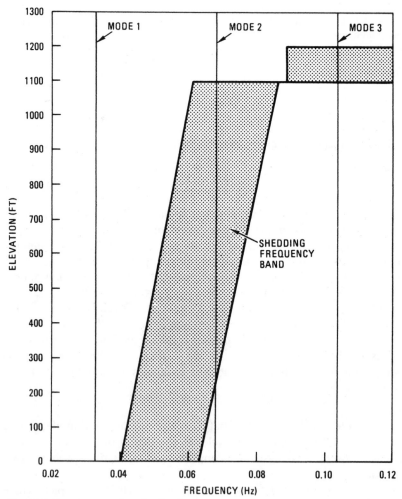

Fig. 3-28 Entrainment of modes for the deep-water riser.

of resonance, mode 2 is potentially resonant along the majority of its length, and mode 3 may be resonant with wave action.

A very conservative estimate of the vortex-induced response would be to apply Table 3-2 to modes 2 and 3. However, because portions of the riser span are not resonant, the results will be very conservative. A less-conservative approach is to utilize one of the vortex models (Section 3.5) to estimate the effect of nonuniform shedding. We will take this approach using the wake oscillator model of Section 3.5.2.

In order to apply the models to nonuniform flow, the span of the riser is divided into segments that are within the resonance band and the remainder, which is outside the resonance band. Define a parameter $s(z)$ that defines the spanwise extent of the resonance band:

$$s(z) = \begin{cases} 1, & \text{if } \alpha f_s < f_i < \beta f_s \text{ (lock-in)} \\ 0, & \text{(no lock-in)} \end{cases},$$

where α and β specify the lock-in band and f_s is the stationary cylinder vortex shedding frequency (Eq. 3-2). Typically $\alpha = 0.6$, $\beta = 1.4$ (Fig. 3-10) for lightly damped, large-amplitude cylinders. The parameter $s(z)$ will step between 0 and 1 along the span, depending on the current profile and the mode.

The structure will respond resonantly to vortex shedding along the part of its span in the lock-in band. Outside of that part, structural oscillations will be damped by the fluid. The equation of motion of the riser is

$$\frac{\partial^2}{\partial x^2}\left[EI\frac{\partial^2 Y}{\partial x^2}\right] - T\frac{\partial^2 Y}{\partial x^2} + c\frac{\partial Y}{\partial t} + m\frac{\partial^2 Y}{\partial t^2} = F_y$$

$$= a_4\rho DU\left(\frac{\partial W}{\partial t} - \frac{\partial Y}{\partial t}\right)s - \tfrac{1}{2}\rho UDC_D\frac{\partial Y}{\partial t}(1-s). \quad (3\text{-}29)$$

This partial differential equation includes stiffness (EI), tension (T), and structural damping (c) effects on the left-hand side and fluid oscillator excitation and fluid damping effects on the right-hand side. It is a generalization of Eq. 3-22. The fluid excitation acts only over the resonant portion and the fluid damping acts over the nonresonant portion of the span. (The linearized fluid damping term is developed in Chapters 4 and 8.)

The fluid oscillator equation is generalized from Eq. 3-21:

$$\frac{\partial^2 W}{\partial t^2} + \omega_s^2 W = (a_1' - a_4')\frac{U}{D}\frac{\partial W}{\partial t} - \frac{a_2'}{UD}\left[\frac{\partial W}{\partial t}\right]^3 + a_4'\frac{U}{D}\frac{\partial Y}{\partial t}. \quad (3\text{-}30)$$

The transverse displacement $Y(z, t)$ and the fluid oscillator parameter

$W(z, t)$ vary along the span. For response in a single resonant mode,

$$Y(z, t) = A_y \bar{y}(z) y(t),$$

$$W(z, t) = W_0 s \bar{y}(z) w(t).$$

z is the coordinate along the span and $\bar{y}(z)$ is the mode shape. The spanwise mode shape of the fluid oscillator is assumed to coincide with that of the structure, but the fluid oscillator only exists within the spanwise extent of the resonant band. That is, resonant excitation is applied only within the resonant band.

Solutions are developed in three steps. First, the structural natural frequency and mode shape are determined using either closed form or numerical techniques. Second, the partial differential equations are reduced to ordinary differential equations using the Glerkin energy technique with the calculated mode shapes and natural frequencies by substituting the two previous solution forms into the differential equations, multiplying by the mode shapes, and integrating over the span to determine the generalized force and response; see Appendix A. This casts the spanwise dependence into integrals over the span. For example, the structural oscillator becomes

$$\ddot{y} + 2\zeta_T \omega_n \dot{y} + \omega_n^2 y = a_4'' \left(\frac{U_s}{D} \right) \dot{w}.$$

Finally, this equation can be solved simultaneously with the fluid oscillator equation by conventional nonlinear techniques.

The total damping coefficient is the sum of structural damping, fluid damping associated with the wake oscillator model, and fluid damping in the nonresonant portions:

$$\zeta_T = \zeta_s + \frac{a_4 \rho D U_s}{2\omega_n m} + \frac{\rho D C_D U_{1-s}}{4m\omega_n}.$$

ζ_s is the structural damping factor, m is the equivalent mass per unit length (Eq. 3-25 for nonuniform distributions), and the velocities U_s and U_{s-1} are defined in terms of integrals over the mode shape,

$$U_s = \int_0^L U(z) \bar{y}^2(z) s \, dz \bigg/ \int_0^L \bar{y}^2(z) \, dz, \qquad (3\text{-}31)$$

$$U_{1-s} = \int_0^L U(z) \bar{y}^2(z)(1-s) \, dz \bigg/ \int_0^L \bar{y}^2(z) \, dz, \qquad (3\text{-}32)$$

over the cylinder span. In general, these integrals are evaluated numerically for known mode shapes and current profiles.

The solutions for cylinder displacement now depend on spanwise properties through the integrals. The solution for the maximum cylinder displacement along the span is

$$\frac{Y_{max}}{D} = \frac{0.07\,\gamma'}{\delta_{rT}}\left[\frac{U_s}{f_n D}\right]^2\left[0.3 + 0.72\left(\frac{U_s}{f_n D}\right)\frac{1}{\delta_{rT}}\right]^{1/2}, \qquad (3\text{-}33)$$

where the total reduced damping is the sum of structural damping and two fluid damping components,

$$\delta_{rT} = \frac{2m(2\pi\zeta_s)}{\rho D^2} + 0.38\frac{U_s}{f_n D} + \frac{C_D U_{1-s}}{2f_n D}, \qquad (3\text{-}34)$$

and the mode shape parameter γ' is

$$\gamma' = \bar{y}_{max}\left[\left[\int_0^L \bar{y}^2(z)\,dz\right]\left[\int_0^L \frac{s}{U(z)}\bar{y}^4(z)\,dz\int_0^L U(z)\bar{y}^2(z)s\,dz\right]^{-1/2}, \qquad (3\text{-}35)$$

where $\bar{y}_{max}$ refers to the maximum modal displacement along the span so that Y_{max} gives the displacement at that point.

For uniform flow at resonance, $s(z) = 1$ over L, $U_s = U$, $U_{1-s} = 0$, and $U_s/(f_n D) = 1/S \approx 5$; then $\gamma' = \gamma$ (Eq. 3-24), $\delta_{rT} = \delta_r + 1.9$, and the above equation reduces to the corresponding equation in Table 3-2 for uniform flow. For nonuniform flows where $s = 0$ over at least part of the span, $\delta_{rT} > \delta_r + 1.9$, $U_s/(f_D D) < 5$, and the response is reduced from that of uniform flow. The power of the present formulation is that it allows consideration of these nonuniform flows. Attached masses can be included and, with some additional effort, can be extended to nonuniform diameters (see Iwan, 1981; Humphries, 1988; Walker and King, 1988).

Marine cables, such as tow cables, also have low natural frequencies and low reduced damping, and they are exposed to appreciable flow velocities (Fig. 3-26(a)). As a result, resonance results in large-amplitude vibration with a broad lock-in band. As flow velocity is increased, the cable passes from one resonance to another, as shown in Figure 3-29.

As a consequence of its large amplitude, the drag of the cable increases (Section 3.3). Drag coefficients as high as 3.0 have been measured (Vandiver, 1983; Dale and McCandles, 1967). Because of the mean drag, the mean cable position distorts into a catenary; see review by Casarella and Parsons (1970). The combination of high load, resulting in high induced tension, and high fluctuating tension can lead to tow cable failure after a few hours. Ribbon and streamlined fairings (Figs. 3-23(d) and (f)) have been successfully attached to fairing to reduce both the mean drag and the vibration. Experiments by King (1977) and Ramberg (1983) show that the amplitude of vortex-induced vibration is not significantly decreased at angles up to about 30 degrees from perpendicular. Blevins (1971) found some decrease in amplitude at an angle of 45 degrees.

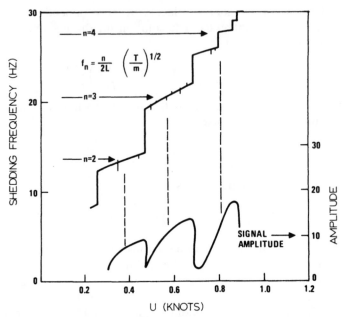

Fig. 3-29 Frequency characteristics of a 3 ft long, 0.1 in. diameter, flexible cable (Dale et al., 1966).

The cables oscillate both in the lift and drag directions and the vibration induces a fluctuating tension in the cable. Figure 3-30 shows the spectrum of acceleration transverse to and parallel to the mean flow, and of fluctuating tension for a 0.35 in. diameter cable, 17 ft long, under 1200 lb tension and angles at 45 degrees to the 6 knot (10.1 ft/sec) flow. The dominant transverse vibration of the cable is at 48 Hz, which corresponds very well to the predicted shedding frequency of an inclined cylinder,

$$f_s = SU \frac{(\cos \theta)}{D} = 49 \, \text{Hz}, \qquad \text{for } S = 0.2,$$

and the fourth ($i = 4$) vibration mode of the cable. The dominant response inline with the flow is at twice this frequency and fluctuating tension is induced at one, two, and three times the shedding frequency. The combined response in both directions results in the cable tracing out a figure-of-eight pattern at resonance as shown in Figure 3-31 with strong coupling between transverse and inline vibration (Vandiver and Jong, 1987).

Finally, it is interesting to note that the three-dimensionality of vortex shedding persists even with locked-in vibration of a wire. Figure 3-32 is a photograph obtained using smoke visualization in the wake of a vibrating

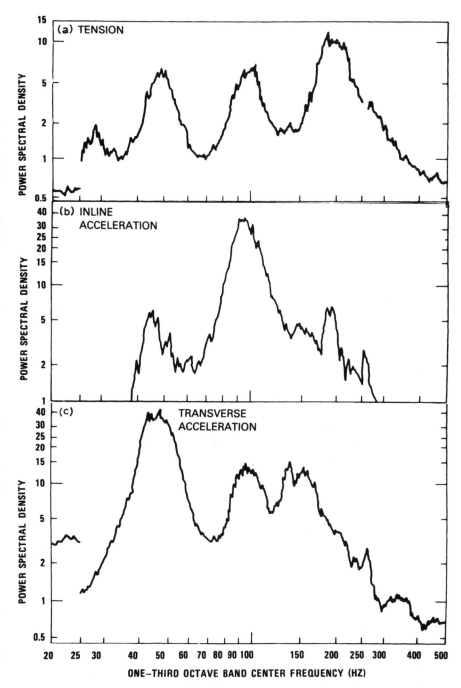

Fig. 3-30 Spectra of acceleration and tension of 0.35 in. diameter cable in a water channel in a 6 knot current; Re = 30 000. (a) Tension; (b) inline acceleration; (c) transverse acceleration (Blevins, 1971).

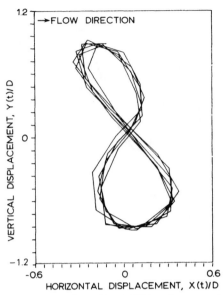

Fig. 3-31 Displacement patterns for a $D = 1.631$ in. diameter pipe, 75 ft long, at an antinode at resonant vortex-induced vibration in a 2 ft/sec current; $Re = 18\,000$ (Vandiver, 1983).

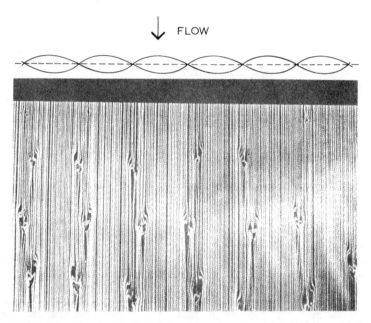

Fig. 3-32 Smoke wire visualization of the wake of a wire with locked-in vortex-induced vibration in a higher mode. The wire runs horizontally left to right. The flow is vertically downward. The spacing between individual vortices is equal to the distance between dashes noticeable in some of the streaklines. The line of blobs originates at the nodes of the vibrating wire (Photograph courtesy of M. Hammache and M. Gharib, 1989).

wire at resonance with vortex shedding. Three-dimensional blobs form in streamwise-staggered rows with the center of the rows located directly aft of the nodes of the vibrating wire (Van Atta et al., 1988). These blobs are about 25 times larger than the spacing between individual vortices. They are probably produced by the 180 degree phase shift across the nodes of the vibrating wire.

REFERENCES

Achenbach, E., and E. Heinecke (1981) "On Vortex Shedding from Smooth and Rough Cylinders in the Range of Reynolds Numbers 6×10^3 to 5×10^6," *Journal of Fluid Mechanics*, **109**, 239–251.

Advisory Board on the Investigation of Suspension Bridges (1955) "Aerodynamic Stability of Suspension Bridges," *American Society of Civil Engineers Transactions*, **120**, 721–781.

Airey, R. C., G. C. Hartnup, and D. Stewart (1988) "A Study of Two Vortex Suppression Devices for Fitting to Marine Risers," in *Proceedings of the Seventh International Conference on Offshore Mechanics and Artic Engineering*, Vol. II, J. S. Chung (ed.), American Society of Mechanical Engineers, New York, pp. 245–251.

Apelt, C. J., and G. S. West (1975) "The Effects of Wake Splitter Plate on Bluff-Body Flow in the Range $10^4 < R < 5 \times 10^4$, Part 2," *Journal of Fluid Mechanics*, **71**, 145.

Aref; H., and E. D. Siggia (1981) "Evolution and Breakdown of a Vortex Street in Two Dimensions," *Journal of Fluid Mechanics*, **109**, 435–463.

Barnett, K. M., and J. E. Cermak (1974) "Turbulence Induced Changes in Vortex Shedding from a Circular Cylinder," Technical Report CER73-74KMB-JEC27, Engineering Research Center, Colorado State University.

Basu, R. I., and B. J. Vickery (1983) "Across Wind Response of Structures of Circular Cross Section," *Journal of Wind Engineering and Industrial Aeronautics*, **12**, 49–97.

Bearman, P. W. (1969) "On Vortex Shedding from a Circular Cylinder in the Critical Reynolds Number Regime," *Journal of Fluid Mechanics*, **37**, 577–586.

—— (1984) "Vortex Shedding from Oscillating Bluff Bodies," *Annual Review of Fluid Mechanics*, **16**, 195–222.

Bearman, P. W., and S. C. Luo (1988) "Investigation of the Aerodynamics of a Square Section Cylinder by Forced Oscillation," *Journal of Fluids and Structures*, **2**, 161–176.

Bearman, P. W., and E. D. Obasaju (1982) "An Experimental Study of Pressure Fluctuations on Fixed and Oscillating Square-Section Cylinders," *Journal of Fluid Mechanics*, **119**, 297–321.

Bearman, P. W., and M. M. Zdravkovich (1978) "Flow Around a Circular Cylinder Near a Plane Boundary," *Journal of Fluid Mechanics*, **89**, 33–47.

Benard, H. (1908) "Formation de Centres de Giration a l'Arriere d'un Obstacle en Mouvement," *Comptes Rendus*, **146**.

Bishop, R. E. D., and A. Y. Hassan (1964) "The Lift and Drag Forces on a Circular Cylinder in a Flowing Fluid," *Proceedings of the Royal Society (London) Series A*, **277**, 51–75.

Blevins, R. D. (1971) "Vortex-Induced Vibration of Ribbon and Bare Cable," unpublished David Taylor Model Basin Report.

—— (1972) "Vortex Induced Vibration of Circular Cylindrical Structures," ASME Paper No. 72-WA/FE-39. American Society of Mechanical Engineers, New York.

—— (1974) "Flow-Induced Vibration," Thesis presented to California Institute of Technology, Pasadena, Calif.

—— (1984) *Applied Fluid Dynamics Handbook*, Van Nostrand Reinhold, New York.

—— (1985) "The Effect of Sound on Vortex Shedding from Cylinders," *Journal of Fluid Mechanics*, **161**, 217–237.

—— (1989) "Application of the Discrete Vortex Technique to the Fluid–Structure Interaction," in *Flow-Induced Vibration—1989*, M. K. Au-Yang (ed.), Vol. 154, American Society of Mechanical Engineers, New York, pp. 131–140.

Blevins, R. D., and T. E. Burton (1976) "Fluid Forces Induced by Vortex Shedding," *Journal of Fluids Engineering*, **95**, 19–24.

Bokaian, A. R., and F. Geoola (1984) "Hydroelastic Instabilities of Square Cylinders," *Journal of Sound and Vibration*, **92**, 117–141.

Brooks, I. H. (1987) "A Pragmatic Approach to Vortex-Induced Vibrations of a Drilling Riser," Paper OTC 5522, 19th Annual Offshore Technology Conference, Houston, Texas.

Buresti, G., and A. Lanciotti (1979) "Vortex Shedding from Smooth and Roughened Cylinders in Cross-Flow Near a Plane Surface," *Aeronautical Quarterly*, **30**, 305–321.

Casarella, M. J., and M. Parsons (1970) "A Survey of Investigations on the Configuration and Motion of Cable Systems Under Hydrodynamic Loading," *Marine Technology Society Journal*, **4**(4), 27–44.

Chen, Y. N. (1972) "Fluctuating Lift Forces of Karman Vortex Streets on Single Circular Cylinders and in Tube Bundles, Parts 1 and 2," *Journal of Engineering for Industry*, **93**, 603–628.

Chilukuri, R. (1987) "Incompressible Laminar Flow Past a Transversely Vibrating Cylinder," *Journal of Fluids Engineering*, **109**, 166–170.

Cimbala, J. M., et al. (1988) "Large Structure in the Far Wakes of Two-Dimensional Bluff Bodies," *Journal of Fluid Mechanics*, **190**, 265–298.

Coder, D. W. (1982) "The Strouhal Number of Vortex Shedding from Marine Risers in Currents at Supercritical Reynolds Numbers," Paper 4318, 14th Annual Offshore Technology Conference, Houston, Texas.

Dale, J. R., and J. M. McCandles (1967) "Water Drag Effects of Flow Induced Cable Vibrations," U.S. Naval Air Development Center, Johnsville, Pa. Report No. NADC-AE-6731.

Dale, J., H. Nenzel, and J. McCandles (1966) "Dynamic Characteristics of Underwater Cables-Flow Induced Transverse Vibration," U.S. Naval Air Development Center, Johnsville, Pa. Report NADC-AV-6620.

Davies, M. E. (1976) "A Comparison of the Wake Structure of a Stationary and Oscillating Bluff Body, Using a Conditional Averaging Technique," *Journal of Fluid Mechanics*, **75**, 209–231.

Dougherty, N., et al. (1989) "Time-Accurate Navier–Stokes Computations of Unsteady Flows: The Karman Vortex Street," AIAA Paper 89-0144, AIAA, Washington, D.C.

Drescher, H. (1956) "Messung der auf querongestromte Zylinder ausgeubten seitlich veranderten Drucke," *Zeitschrift für Flugwissensschaft*, **4**, 17–21.

Durgin, W. W., P. A. March, and P. J. Lefebvre (1980) "Lower Mode Response of Circular Cylinders in Cross Flow," *Journal of Fluids Engineering*, **102**, 183–189.

Ericsson, L. E., and J. P. Reding (1988) "Fluid Mechanics of Dynamic Stall, Part I, Unsteady Flow Concepts," *Journal of Fluids and Structures*, **2**, 1–33.

Escudier, M. (1987) "Confined Vortices in Fluid Machinery," *Annual Review of Fluid Mechanics*, **19**, 27–52.

Every, M. J., R. King, and D. S. Weaver (1982) "Vortex-Excited Vibrations of Cylinders and Cables and Their Suppression," *Ocean Engineering*, **9**, 135–157.

Fage, A., and J. H. Warsap (1929) "The Effects of Turbulence and Surface Roughness on Drag of a Circular Cylinder," British Aerodynamics Research Council, Rep. Memo. 1283.

Farell, C. (1981) "Flow Around Circular Cylinders: Fluctuating Loads," *ASCE Journal of Engineering Mechanics*, **107**, 565–588. Also see *Closure*, **109**, 1153–1156 (1983).

Farell, C., and J. Blessmann (1983) "On Critical Flow around Smooth Circular Cylinders," *Journal of Fluid Mechanics*, **136**, 375–391.

Farquharson, F. B., et al. (1949) *Aerodynamic Stability of Suspension Bridges with Special Reference to the Tacoma Narrows Bridge*, The University of Washington Press, Seattle.

Feng, C. C. (1968) "The Measurement of Vortex-Induced Effects in Flow Past Stationary and Oscillating Circular and D-Section Cylinders," M.A.Sc. Thesis, University of British Columbia.

Fitzhugh, J. S. (1973) "Flow Induced Vibration in Heat Exchangers," Oxford University Report RS57 (AERE P7238). Also *Proceedings of UKAEA/NPL International Symposium on Vibration Problems in Industry, Keswick, April 1973*, Paper 427.

Fitzpatrick, J. A., et al. (1988) "Strouhal Numbers for Flows in Deep Tube Arrays," *Journal of Fluids and Structures*, **2**, 145–160.

Fitzpatrick, J. A., and I. S. Donaldson (1980) "Row Depth Effects on Turbulence Spectra and Acoustic Vibrations in Tube Banks," *Journal of Sound and Vibration*, **73**, 225–237.

Friehe, C. A. (1980) "Vortex Shedding from Cylinders at Low Reynolds Numbers," *Journal of Fluid Mechanics*, **100**, 237–241.

Gardner, T. N. (1982) "Deepwater Drilling in High Current Environment," 14th Annual Offshore Conference, Houston, Tex., paper 4316.

Gartshore, I. S. (1984) "Some Effects of Upstream Turbulence on the Unsteady Lift Forces Imposed on Prismatic Two Dimensional Bodies," *Journal of Fluids Engineering*, **106**, 418–424.

Gerlach, C. R. (1972) "Vortex Excitation of Metal Bellows," *Journal of Engineering for Industry*, **94**, 87–94.

Ghia, K. N. (ed.) (1987) *Forum on Unsteady Flow Separation*, FED-Vol. 52, American Society of Mechanical Engineers, ASME, New York.

Griffin, O. M. (1981) "Universal Similarity in the Wakes of Stationary and Vibrating Bluff Structures," *Journal of Fluids Engineering*, **103**, 52–58.

——— (1982) "Vortex Streets and Patterns," *Mechanical Engineering*, **104**(3), 56–61.

——— (1985) "Vortex Shedding from Cables and Structures in a Shear Flow: A Review," *Journal of Fluids Engineering*, **107**, 298–306.

Griffin, O. M., and S. E. Ramberg (1974) "The Vortex Street Wakes of Vibrating Cylinders," *Journal of Fluid Mechanics*, **66**, 553–576.

——— (1975) "On Vortex Strength and Drag in Bluff-Body Wakes," *Journal of Fluid Mechanics*, **69**, 721–728.

——— (1976) "Vortex Shedding from a Cylinder Vibrating inline with an Incidence Uniform Flow," *Journal of Fluid Mechanics*, **75**, 257–271.

——— (1982) "Some Recent Studies of Vortex Shedding with Application to Marine Tubulars and Risers," *Journal of Energy Resources Technology*, **104**, 2–13.

Griffin, O. M., S. E. Ramberg, and M. E. Davies (1980) "Calculation of the Fluid Dynamic Properties of Coherent Vortex Wake Patterns," in *Vortex Flows*, W. L. Swift, ed., ASME, New York.

Hafen, B. E., and D. J. Meggitt (1971) "Cable Strumming Suppression," Civil Engineering Laboratory, Naval Construction Battalion Center, Port Hueneme, Calif., Report TN-1499, DN787011.

Halkyard, J. E., and P. B. Grote (1987) "Vortex-Induced Response of a Pipe at Supercritical Reynolds Numbers," Paper 5520, 19th Annual Offshore Technology Conference, Houston, Texas.

Hagedorn, P. (1982) "On the Computation of Damped Wind-Excited Vibrations of Overhead Transmission Lines," *Journal of Sound and Vibration*, **83**, 253–271.

Hall, S. A., and W. D. Iwan (1984) "Oscillations of a Self-Excited Nonlinear System," *Journal of Applied Mechanics*, **51**, 892–898.

Hanko, Z. G. (1967) "Vortex Induced Vibration at Low-Head Weirs," Proceedings of the American Society of Civil Engineers, *Journal of the Hydraulics Division,* **93**, 255–270.

Hartlen, R. T., W. D. Baines, and I. G. Currie (1968) "Vortex Excited Oscillations of a Circular Cylinder," University of Toronto Report UTME-TP 6809. Also see Hartlen, R. T., and I. G. Currie (1970) "Lift Oscillation Model for Vortex-Induced Vibration," Proceedings ASCE, *Journal of the Engineering Mechanics Division,* **96**, 577–591.

Huerre, P., and P. A. Monkewitz (1990) "Local and Global Instabilities in Spatially Developing Flow," *Annual Review of Fluid Mechanics,* **22**, 473–538.

Humphreys, J. S. (1960) "On a Circular Cylinder in a Steady Wind at Transition Reynolds Number," *Journal of Fluid Mechanics,* **9**, 603–612.

Humphries, J. A. (1988) "Comparison Between Theoretical Predictions for Vortex Shedding in Shear Flow and Experiments," in *Proceedings of the Seventh International Conference on Offshore Mechanics and Artic Engineering,* Vol. II, J. S. Chung (ed.), American Society of Mechanical Engineers, New York, pp. 203–209.

Iwan, W. D. (1975) "The Vortex Induced Oscillation of Elastic Structural Elements," *Journal of Engineering for Industry,* **97**, 1378–1382.

——— (1981) "The Vortex-Induced Oscillation of Non-Uniform Structural Systems," *Journal of Sound and Vibration,* **79**, 291–301.

Iwan, W. D., and R. D. Blevins (1974) "A Model for Vortex-Induced Oscillation of Structures," *Journal of Applied Mechanics,* **41**, 581–586.

Jacobsen, V., et al. (1984) "Cross-Flow Vibration of a Pipe Close to a Rigid Boundary," *Journal of Energy Resources Technology,* **106**, 451–457.

Johnson, J. E., et al. (1979) "Bellows Flow-Induced Vibration," Southwest Research Institute Final Report Contract No. NAS8-31994, SwRI Project No. 02-4548. Prepared for NASA.

Jones, G. W., J. J. Cincotta, and R. W. Walker (1969) "Aerodynamic Forces on a Stationary and Oscillating Circular Cylinder at High Reynolds Numbers," National Aeronautics and Space Administration Report NASA TR R-300.

King, R. (1977) "A Review of Vortex Shedding Research and its Application," *Ocean Engineering,* **4**, 141–171.

——— (1977) "Vortex Excited Oscillations of Yawed Circular Cylinders," *Journal of Fluids Engineering,* **99**, 495–502.

King, R., and D. J. Johns (1976) "Wake Interaction Experiments with Two Flexible Circular Cylinders in Flowing Water," *Journal of Sound and Vibration,* **45**, 259–283.

Kiya, M., et al. (1980) "Vortex Shedding from Two Circular Cylinders in Staggered Arrangement," *Journal of Fluids Engineering,* **102**, 166–182.

Komatsu, S., and H. Kobayashi (1980) "Vortex-Induced Oscillation of Bluff Cylinders," *Journal of Wind Engineering and Industrial Aerodynamics,* **6**, 335–362.

Koopman, G. H. (1967) "The Vortex Wakes of Vibrating Cylinders at Low Reynolds Numbers," *Journal of Fluid Mechanics,* **28**, 501–512.

Levi, E. (1983) "A Universal Strouhal Law," *ASCE Journal of Engineering Mechanics,* **109**, 718–727. Also see *Discussion,* **110**, 839–845 (1984).

Lienhard, J. H. (1966) "Synopsis of Lift, Drag and Vortex Frequency Data for Rigid Circular Cylinders," Washington State University, College of Engineering, Research Division Bulletin 300.

Lim, C. C., and L. Sirovich (1988) "Nonlinear Vortex Trail Dynamics," *Physics of Fluids,* **31**, 991–998.

Lugt, H. J. (1983) *Vortex Flow in Nature and Technology,* Wiley, New York.

Mair, W. A., and D. J. Maull (1971) "Bluff Bodies and Vortex Shedding—A Report on Euromech 17," *Journal of Fluid Mechanics,* **45**, 209–224.

Marris, A. W. (1964) "A Review on Vortex Streets, Periodic Wakes, and Induced Vibration Phenomena," *Journal of Basic Engineering,* **86,** 185–194.

Miller, R. W. (1989) *Flow Measurement Engineering Handbook,* 2d ed., McGraw-Hill, New York.

Morkovin, M. V. (1964) "Flow Around a Circular Cylinder," Symposium on Fully Separated Flows, in *Proceedings of the American Society of Mechanical Engineers, Engineering Division Conference, Held in Philadelphia, May 18–20, 1964,* A. G. Hansen, ed., ASME, New York, pp. 102–118.

Mujumdar, A. S., and W. J. Douglas (1973) "Vortex Shedding from Slender Cylinders of Various Cross Sections," *Journal of Fluids Engineering,* **95,** 474–476.

Murray, B. G., W. B. Bryce, and G. Rae (1982) "Strouhal Numbers in Tube Arrays," Third Keswick Conference on Vibration in Nuclear Plant, Keswick, United Kingdom.

Nakamura, Y., and T. Yoshimura (1982) "Flutter and Vortex-Induced Excitation of Rectangular Prisms in Pure Torsion in Smooth and Turbulent Flows," *Journal of Sound and Vibration,* **84,** 305–317.

Naudascher, E. (1987) "Flow-Induced Streamwise Vibrations of Structures," *Journal of Fluids and Structures,* **1,** 265–298.

Nickerson, E. C., and M. A. Dias (1981) "On the Existence of Atmospheric Vortices Downwind of Hawaii during the HAMEC Project," *Journal of Applied Meteorology,* **20,** 868–873.

Nishioka, M., and H. Sato (1978) "Mechanism of Determination of the Shedding Frequency of Vortices Behind a Cylinder at Low Reynolds Number," *Journal of Fluid Mechanics,* **89,** 49–60.

Novak, J. (1973) "Strouhal Number and Flat Plate Oscillation in an Air Stream," *Acta Technica Csav,* **4,** 372–386.

Novak, M. (1971) "Galloping and Vortex Induced Oscillation of Structures," *Proceedings of the Conference on Wind Effects on Buildings and Structures,* Tokyo, Japan.

Novak, M., and H. Tanaka (1977) "Pressure Correlations on a Vibration Cylinder," in *Proc. 4th Int. Conf. on Wind Effects on Buildings and Structures, Heathrow, 1975,* K. J. Eaton, ed., Cambridge University Press, London, pp. 227–232.

Okajima, A. (1982) "Strouhal Numbers of Rectangular Cylinders," *Journal of Fluid Mechanics,* **123,** 379–398.

Olinger, D. J., and K. R. Sreenivasan (1988) "Nonlinear Dynamics of the Wake of an Oscillating Cylinder," *Physical Review Letters,* **60,** 797–800.

Ongoren, A., and D. Rockwell (1988) "Flow Structure from an Oscillating Cylinder," *Journal of Fluid Mechanics,* **191,** 197–245.

Paidoussis, M. P., S. J. Price, and H. C. Suen (1982) "Ovaling Oscillation of Cantilevered and Clamped-Clamped Cylindrical Shells in Cross Flow: An Experimental Study," *Journal of Sound and Vibration,* **83,** 533–553.

Parker, R., and S. A. T. Stoneman (1985) "Experimental Investigation of the Generation and Consequences of Acoustic Waves in an Axial Flow Compressor; Large Axial Spacing between Blade Rows," *Journal of Sound and Vibration,* **99,** 169–182.

Parkinson, G. V. (1971) "Wind-Induced Instability of Structures," *Philosophical Transactions of the Royal Society of London,* **A269,** 395–409.

Peltzer, R. D., and D. M. Rooney (1985) "Vortex Shedding in a Linear Flow from a Vibrating Marine Cable with Attached Masses," *Journal of Fluids Engineering,* **107,** 61–66.

Perry, A. E., et al. (1982) "The Vortex Shedding Process Behind Two-Dimensional Bluff Bodies," *Journal of Fluid Mechanics,* **116,** 77–90.

Poore, A. B., E. J. Doedel, and J. E. Cermak (1986) "Dynamics of the Iwan-Blevins Wake Oscillator Model," *International Journal of Non-Linear Mechanics,* **21,** 291–302.

Price, S. J., and M. P. Paidoussis (1989) "The Flow-Induced Response of a Single Flexible

Cylinder in an In-Line Array of Rigid Cylinders," *Journal of Fluids and Structures,* **3,** 61–82.

Ramberg, S. E. (1983) "The Effects of Yaw and Finite Length upon the Vortex Wakes of Stationary and Vibrating Cylinders," *Journal of Fluid Mechanics,* **128,** 81–107.

Ramberg, S. E., and O. M. Griffin (1976) "Velocity Correlation and Vortex Spacing in the Wake of a Vibrating Cable," *Journal of Fluids Engineering,* **98,** 10–18.

Rayleigh, J. W. S. (1879) "Acoustical Observations II," *Philosophical Magazine,* **7,** 149–162. Contained in *Scientific Papers,* Dover, New York, 1984.

Rhode, F. G. (1979) "Self-Excited Oscillatory Surface Waves around Cylinders," Vol. 29, Mitteilungen, Institut fur Wasserbau und Wasserwirtschaft, Reinisch-Westfalische Technische Hochschule Aachen, ISSN 0343-1045, Aachen, West Germany.

Rhodi, W. (1975) "A Review of Experimental Data of Uniform Density Free Turbulence Boundary Layers," in *Studies in Convection,* Vol. 1, B. E. Launder, ed., Academic Press, London.

Rogers, A. C. (1983) "An Assessment of Vortex Suppression Devices for Production Risers and Towed Deep Ocean Pipe Strings," Paper 4594, 15th Annual Offshore Technology Conference, Houston, Texas.

Rooney, D. M., and R. D. Peltzer (1981) "Pressure and Vortex Shedding Patterns Around a Low Aspect Ratio Cylinder in a Sheared Flow at Transitional Reynolds Numbers," *Journal of Fluids Engineering,* **103,** 88–96.

Roshko, A. (1954) "On the Drag and Vortex Shedding Frequency of Two-Dimensional Bluff Bodies," National Advisory Committee for Aeronautics Report NACA TM 3159, July. Also see "On the Wake and Drag of Bluff Bodies," *Journal of Aeronautical Science,* **22,** 124–135 (1955).

—— (1961) "Experiments on the Flow Past a Circular Cylinder at Very High Reynolds Number," *Journal of Fluid Mechanics,* **10,** 345–356.

Saffman, P. G., and J. C. Schatzman (1982) "An Inviscid Model for the Vortex-Street Wake," *Journal of Fluid Mechanics,* **122,** 467–486.

Sallet, D. W. (1970) "A Method of Stabilizing Cylinders in Fluid Flow," *Journal of Hydronautics,* **4,** 40–45.

—— (1980) "Suppression of Flow-Induced Motions of a Submerged Moored Cylinder," in *Practical Experiences with Flow-Induced Vibrations,* E. Naudascher and D. Rockwell, eds, Springer-Verlag, New York.

Sarpkaya, T. (1978) "Fluid Forces on Oscillating Cylinders," *Journal of the Waterway Port, Coastal and Ocean Division,* **104,** 275–290.

—— (1979) "Vortex-Induced Oscillations," *Journal of Applied Mechanics,* **46,** 241–258.

Savkar, S. D., and T. A. Litzinger (1982) "Buffeting Forces Induced by Cross Flow through Staggered Arrays of Cylinders," General Electric, Corporate Research and Development, Report No. 82RD238.

Scanlan, R. H., and R. L. Wardlaw (1973) "Reduction of Flow Induced Structural Vibrations," in *Isolation of Mechanical Vibration, Impact, and Noise,* AMD Vol. 1, J. C. Snowden and E. E. Ungar, eds., ASME, New York.

Schaefer, J. W., and S. Eskinazi (1959) "An Analysis of the Vortex Street Generated in a Viscous Fluid," *Journal of Fluid Mechanics,* **6,** 241–260.

Schewe, G. (1983) "On the Force Fluctuations Acting on a Circular Cylinder in Crossflow from Subcritical up to Transcritical Reynolds Numbers," *Journal of Fluid Mechanics,* **133,** 265–285.

Scruton, C. (1963) "On the Wind Excited Oscillation of Stacks, Towers and Masts," in *Proceedings of the Conference on Wind Effects on Buildings and Structures, Held in Teddington, England, June 1963,* National Physical Laboratory.

—— (1981) "An Introduction to Wind Effects on Structures," *Engineering Design Guide 40,* Published for the Design Council, British Standards Institution and the Council of Engineering Institutions, Oxford University Press, Oxford.

Sin, V. K., and R. M. C. So (1987) "Local Force Measurements on Finite Span Cylinders in Cross Flow," *Journal of Fluids Engineering*, **109**, 136–143.

Skop, R. A. and O. M. Griffin (1973) "An Heuristic Model for Determining Flow-Induced Vibrations of Offshore Structures," Paper OTC 1843, Offshore Technology Conference, April 1973, Houston, Tex.

Skop, R. A., O. M. Griffin, and S. E. Ramberg (1977) "Strumming Predictions for the SEACON II Experimental Mooring," Paper OTC 2884, Offshore Technology Conference, May 1977, Houston, Tex.

Stansby, P. K. (1976) "The Locking-on of Vortex Shedding due to the Cross Stream Vibration of Circular Cylinders in Uniform and Shear Flows," *Journal of Fluid Mechanics*, **74**, 641–665.

Stansby, P. K., J. N. Pinchbeck, and T. Henderson (1986) "Spoilers for Suppression of Vortex-Induced Oscillations," *Applied Ocean Research*, **8**, 169–173.

Steinman, D. B., and S. R. Watson (1957) *Bridges and Their Builders*, Dover, New York.

Strouhal, V. (1878) "Uber eine Besondere Art der Tonerregung," *Annalen der Physik und Chemie (Leipzig)*, **V**, 217–251.

Williams–Stuber, K., and M. Gharib, (1990) "Transition from Order to Chaos in the Wake of an Airfoil," *Journal of Fluid Mechanics*, Vol. 213, 29-57.

Sumer, B. M., and J. Fredsoe (1988) "Vibration of Cylinders at High Reynolds Numbers," in *Proceedings of the Seventh International Conference on Offshore Mechanics and Artic Engineering*, J. S. Chung (ed.), American Society of Mechanical Engineers, New York, pp. 211–222.

Swift, W. L., P. S. Barna, and C. Dalton (1980) *Vortex Flows*, American Society of Mechanical Engineers, Symposium presented at Winter Annual Meeting, Chicago Ill., Nov. 16–21, ASME, New York.

Szechenyi, E. (1975) "Supercritical Reynolds Simulation of Two-Dimensional Flow Over Cylinders," *Journal of Fluid Mechanics*, **70**, 529–542.

Tanaka, H., and A. G. Davenport (1983) "Wind-Induced Response of Golden Gate Bridge," *ASCE Journal of the Engineering Mechanics Division*, **109**, 296–312.

Tanida, Y., A. Okajima, and Y. Watanabe (1973) "Stability of a Circular Cylinder Oscillating in Uniform Flow or Wake," *Journal of Fluid Mechanics*, **61**, 769–784.

Thomson, W. T. (1988) *Theory of Vibration with Applications*, 3d ed., Prentice-Hall, Englewood Cliffs, N.J.

Toebes, G. H. (1969) "The Unsteady Flow and Wake Near an Oscillating Cylinder," *ASME Journal of Basic Engineering*, **91**, 493–502.

Toebes, G. H., and P. S. Eagleson (1961) "Hydroelastic Vibrations of Flat Plates Related to Trailing Edge Geometry," *ASME Journal of Basic Engineering*, **83**, 671–678.

Torum, A., and N. M. Anand (1985) "Free Span Vibrations of Submarine Pipelines in Steady Flows-Effect of Free-Stream Turbulence on Mean Drag Coefficients," *Journal of Energy Resources Technology*, **107**, 415–420.

Tsahalis, D. T. (1984) "Vortex-Induced Vibrations of a Flexible Cylinder Near a Plane Boundary Exposed to Steady and Wave-Induced Currents," *Journal of Energy Resources Technology*, **106**, 206–213.

Tsuboi, K. et al. (1989) "Numerical Study of Vortex-Induced Vibration of Circular Cylinder in High-Reynolds-Number Flow," AIAA Paper 89-0294, AIAA, Washington, D.C.

Uematsu, Y., and K. Uchiyama (1985) "An Experimental Investigation of Wind Induced Ovalling Oscillations of Thin, Cylindrical Shells," *Journal of Wind Engineering and Industrial Aerodynamics*, **18**, 229–243.

Unal, M. F., and D. Rockwell (1987) "On Vortex Formation from a Cylinder, Control by Splitter Plate Interference," *Journal of Fluid Mechanics*, **190**, 513–529.

Vandiver, J. K. (1983) "Drag Coefficients of Long-Flexible Cylinders," Paper OTC 4490, Offshore Technology Conference, Houston, Tex, May 2–5.

Vandiver, J. K., and J. Y. Jong (1987) "The Relationship Between In-Line and Cross-Flow Induced-Vibration of Cylinders," *Journal of Fluids and Structures*, **1**, 381–399.

Van Atta, C., M. Gharib, and M. Hammache (1988) "Three-Dimensional Structure of Ordered and Chaotic Vortex Streets Behind Circular Cylinders at Low Reynolds Numbers," *Fluids Dynamics Research*, **3**, 127–132.

van der Vegt, J. J. (1988) "A Variationally Optimized Vortex Tracing Algorithm for Three-Dimensional Flow Around Solid Bodies," Ph.D. Thesis, Techniche Universiteit Delft, Amsterdam.

Vickery, B. J. (1966) "Fluctuating Lift and Drag on a Long Cylinder of Square Cross Section," *Journal of Fluid Mechanics*, **25**, 481.

―――― (1981) "Across-Wind Buffeting in a Group of Four In-Line Model Chimneys," *Journal of Wind Engineering and Industrial Aerodynamics*, **8**, 177–193.

Vickery, B. J., and R. D. Watkins (1962) "Flow-Induced Vibration of Cylindrical Structures," in *Proceedings of the First Australian Conference*, held at the University of Western Australia.

Vincent, G. S. (1958) "Golden Gate Vibration Studies," *ASCE Journal of the Structural Division*, **84**, Paper 1817.

―――― (1962) "Golden Gate Bridge Vibration Studies," *Transactions of the American Society of Civil Engineers*, **127**, 667–707.

von Karman, T. (1912) "Uber den Mechanismuss des Widersstandes den ein bewegter Korper in einen Flussigkeit Erfahrt," *Nachrichten der K. Gesellschaft der Wissenschaften zu Gottingen*, 547–556. See also *Collected Works of Theodore von Karman*, Vol. 1, Butterworths, London, 1956, pp. 331–338.

Walker, D., and R. King (1988) "Vortex Excited Vibrations of Tapered and Stepped Cylinders," in *Proceedings of the Seventh International Conference on Offshore Mechanics and Artic Engineering*, Vol. II, J. S. Chung (ed.), American Society of Mechanical Engineers, New York, pp. 229–234.

Walshe, D. E. (1962) "Some Measurements of the Excitation Due to Vortex Shedding of a Smooth Cylinder of Circular Cross Section," National Physical Laboratory Aero Report 1062.

Walshe, D. E., and L. R. Wootton (1970) "Preventing Wind Induced Oscillations of Structures of Circular Section," *Proceedings of the Institution of Civil Engineers (London)*, **47**, 1–24.

Wardlaw, R. L. (1966) "On Relating Two-Dimensional Bluff Body Potential Flow to the Periodic Vortex Wake," *DME/NAE Quarterly Bulletin*, **2**.

―――― (1988) "The Wind Resistant Design of Cable Staged Bridges," in *Cable Staged Bridges*, C. C. Vlstrupl (ed.), American Society of Civil Engineers, New York, pp. 46–61.

Wardlaw, R. L., and R. D. Blevins (1980) "Discussion on Approaches to the Suppression of Wind-Induced Vibration of Structures," in *Practical Experiences with Flow-Induced Vibration*, E. Naudascher, and D. Rockwell, eds., Springer-Verlag, New York, pp. 671–672.

Wardlaw, R. L., and C. A. Ponder (1970) "Wind Tunnel Investigation of the Aerodynamic Stability of Bridges," National Aeronautical Establishment Report LTR-LA-47, Ottawa, Canada.

Washizu, K., et al. (1978) "Aeroelastic Instability of Rectangular Cylinders in Heaving Mode," *Journal of Sound and Vibration*, **59**, 195–210.

Weaver, D. S., and Abd-Rabbo (1985) "A Flow Visualization Study of a Square Array of Tubes in Water Cross Flow," *Journal of Fluids Engineering*, **107**, 354–363.

Weaver, D. S., and P. Ainsworth (1988) "Flow-Induced Vibrations in Bellows," in *1988 International Symposium on Flow-Induced Vibration and Noise*, M. P. Paidousis (ed.), Vol. 4, American Society of Mechanical Engineers, New York, pp. 205–214.

Weaver, D. S., J. A. Fitzpatrick, and M. ElKashlan (1986) "Strouhal Numbers in Heat

Exchanger Tube Arrays in Cross Flow," in *Flow-Induced Vibration—1986,* Symposium Proceedings presented at 1986 Pressure Vessel and Piping Conference, Chicago, Ill., July 20–24, 1986, ASME Publication PVP—Vol. 104, New York.

Williamson, C. H. K., and A. Roshko (1988) "Vortex Formation in the Wake of an Oscillating Cylinder," *Journal of Fluids and Structures,* **2,** 355–381.

Wilson, J. E., and J. C. Tinsley (1989) "Vortex Load Reduction; Experiments in Optimal Helical Strake Geometry for Rigid Cylinders," *Journal of Energy Resources Technology,* **3,** 72–76.

Wong, H. Y., and A. Kokkalis (1982) "A Comparative Study of Three Aerodynamic Devices for Suppressing Vortex-Induced Oscillation," *Journal of Wind Engineering and Industrial Aerodynamics,* **10,** 21–29.

Wootton, L. R. (1968) "The Oscillations of Model Circular Stacks due to Vortex Shedding at Reynolds Numbers from 10^5 to 3×10^6," National Physical Laboratory, Aerodynamics Division NPL Aero Report 1267.

Yamaguchi, T., et al. (1971) "On the Vibration of the Cylinder by Karman Vortex," *Mitsubishi Heavy Industries Technical Journal,* **8**(1), 1–9. Also see Chen (1972).

Yeh, T. T., B. Robertson, and W. M. Mattar (1983) "LDV Measurements Near a Vortex Shedding Strut Mounted in a Pipe," *Journal of Fluids Engineering,* **105,** 185–194.

Zdravkovich, M. M. (1981) "Review and Classification of Various Aerodynamic and Hydrodynamic Means for Suppressing Vortex Shedding," *Journal of Wind Engineering and Industrial Aerodynamics,* **7,** 145–189.

—— (1982) "Modification of Vortex Shedding in the Synchronization Range," *Journal of Fluids Engineering,* **104,** 513–517.

—— (1985) "Flow Induced Oscillations of Two Interfering Circular Cylinders," *Journal of Sound and Vibration,* **101,** 511–521.

Chapter 4

Galloping and Flutter

A structure with noncircular cross section experiences a fluid force that changes with orientation to the flow. As the structure vibrates, its orientation changes and the fluid force oscillates. If the oscillating fluid force tends to increase vibration, the structure is aerodynamically unstable and very large-amplitude vibration can result. In this chapter, stability analysis is developed for galloping of civil engineering structures and the flutter of aircraft wings.

4.1. INTRODUCTION TO GALLOPING

All noncircular cross sections are susceptible to galloping and flutter. Ice-coated power lines gallop in winter winds (Cheers, 1950; Edwards and Madeyski, 1956; Richardson et al., 1965; Hunt and Richards, 1969), as do bridge decks (Parkinson, 1971; Simiu and Scanlan, 1986). Marine structures can gallop in ocean currents (Bokaian and Geoola, 1984). Wings flutter (Bisplinghoff et al., 1955). Stall-flutter of turbine blades produces large amplitudes of oscillation (Sisto, 1953; Pigott and Abel, 1974; Yashima and Tanaka, 1978).

Galloping vibrations and aircraft flutter arise from fluid forces induced by structural vibration in a fluid flow. Differences between galloping and flutter lie largely in the historical usage of the terms. *Flutter* is aerospace terminology for coupled torsion–plunge instability of airfoil structures. *Galloping* is the term favored by civil engineers for one-degree-of-freedom instability of bluff structures in winds and currents.

Galloping of a pivoted D-section cylinder with a flat face normal to the wind was described by Lanchester (1907). In 1930, den Hartog (1956) associated the instability with the aerodynamic coefficients of the cross section and described a toy with a spring-supported D-section facing into a fan. Sisto (1953) described the instability of stalled turbine blades. Parkinson and Brooks (1961) generalized the graphical approach of Scruton (1960) to determine the limiting amplitude of galloping. Novak (1969) and Novak and Tanaka (1974) explored the effect of turbulence and nondimensionalized Parkinson and Brook's result. Blevins and Iwan

(1974) explored two-degree-of-freedom effects. Obasaju (1983), Parkinson and Wawzonek (1981), and Nakamura and Matsukawa (1987) have explored coupled vortex shedding–galloping and the limitations of quasisteady theory.

Most galloping and flutter analysis utilizes quasisteady fluid dynamics; i.e., the fluid force on the structure is assumed to be determined solely by the instantaneous relative velocity, so the fluid forces can be measured in wind-tunnel tests on stationary models held at various angles. The quasisteady assumption is valid only if the frequency of periodic components of fluid force, associated with vortex shedding or time-lag effects, is well above the vibration frequency of the structure, $f_s \gg f_n$. This requirement is often met at higher reduced velocities such that

$$\frac{U}{f_n D} > 20, \tag{4-1}$$

where U is the free stream velocity, f_n is the natural frequency of vibration, and D is the width of the cross section normal to the free stream (Bearman et al., 1984; Fig. 4-13). Unfortunately, many structures of practical importance have galloping-like instabilities in the range of reduced velocities $1 < U/(f_n D) < 20$, where the quasisteady assumption is questionable and vortex-induced vibrations may occur as described in Sections 3.7 and 4.4.

4.2. GALLOPING INSTABILITY

4.2.1. Plunge Stability

Figure 4-1 shows a spring-supported building model exposed to a steady flow of velocity U and density ρ. The spring has stiffness k_y per unit length. The steady fluid dynamic forces on the section are the lift force and the drag force per unit length:

$$F_L = \tfrac{1}{2}\rho U^2 D C_L, \tag{4-2}$$

$$F_D = \tfrac{1}{2}\rho U^2 D C_D. \tag{4-3}$$

The lift force acts perpendicularly to the mean flow and the drag force is the component of force parallel to the mean flow. The width D is a dimension used as a reference to nondimensionalize the lift and drag aerodynamic coefficients C_L and C_D. Ordinarily, the fluid force is measured at various angles of attack in a wind tunnel and then resolved into lift and drag coefficients using Eqs. 4-2 and 4-3.

The stability of the model of Figure 4-1 is studied by developing a quasisteady model for the aerodynamic force and then examining its

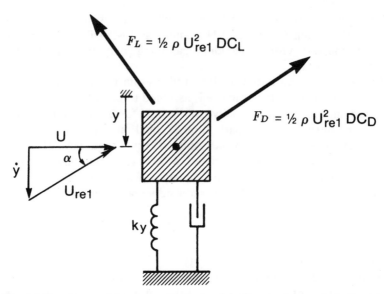

Fig. 4-1 One-degree-of-freedom galloping model. The steady flow is horizontal with magnitude U.

response during small perturbations from equilibrium. When the model translates downward, the angle of the flow relative to the mode is

$$\alpha = \arctan (\dot{y}/U). \qquad (4\text{-}4)$$

α is called the angle of attack. Here $\alpha = 0$ has been referenced to the equilibrium position $y = 0$. The aeronautical sign convention is used for α and y. Rotating the model clockwise in a flow from left to right increases the angle of attack. The vertical displacement y is positive downwards and lift is upward, as is customary in aircraft flutter analysis. The reader should be warned that not all authors use this convention.

The velocity of the fluid relative to the moving model is the sum of the free stream velocity and the induced velocity,

$$U_{rel}^2 = \dot{y}^2 + U^2, \qquad (4\text{-}5)$$

where $\dot{y} = dy/dt$ is the vertical velocity. F_y is the vector resultant of lift and drag in the vertical plane, positive downward,

$$F_y = -F_L \cos \alpha - F_D \sin \alpha = \tfrac{1}{2}\rho U^2 D C_y, \qquad (4\text{-}6)$$

where the vertical force coefficient is

$$C_y = -\frac{U_{rel}^2}{U^2}(C_L \cos \alpha + C_D \sin \alpha). \qquad (4\text{-}7)$$

C_y, like C_L and C_D, is a function of shape, angle of attack, and Reynolds number.

For small angles of attack, α, U_{rel}, and C_y can be expanded in power series:

$$\alpha = \frac{\dot{y}}{U} + O(\alpha^2) \qquad U_{rel} = U + O(\alpha^2), \tag{4-8}$$

$$C_y(\alpha) = C_y|_{\alpha=0} + \frac{\partial C_y}{\partial \alpha}\Big|_{\alpha=0} \alpha + O(\alpha^2),$$

$$= -C_L|_{\alpha=0} - \left[\frac{\partial C_L}{\partial \alpha} + C_D\right]_{\alpha=0} \alpha + O(\alpha^2), \tag{4-9}$$

where $o(\alpha^2)$ means terms proportional to α^2 and higher powers of α have been neglected. Equation 4-7 implies that the slope of the vertical force curve at zero angle of attack is

$$\frac{\partial C_y}{\partial \alpha}\Big|_{\alpha=0} = -\left(\frac{\partial C_L}{\partial \alpha} + C_D\right)_{\alpha=0}. \tag{4-10}$$

At zero angle of attack ($\alpha = 0$) the vertical force coefficient is the negative of the lift coefficient $C_y = -C_L$, since the vertical force is defined positive downward and lift is positive upward.

The equation of motion for the spring-supported, damped model responding to the aerodynamic force (Eqs. 4-6 through 4-10) is

$$m\ddot{y} + 2m\zeta_y\omega_y\dot{y} + k_y y = F_y = \tfrac{1}{2}\rho U^2 D C_y,$$

$$= -\tfrac{1}{2}\rho U^2 D C_L|_{\alpha=0} + \tfrac{1}{2}\rho U^2 D \frac{\partial C_y}{\partial \alpha}\Big|_{\alpha=0}\left(\frac{\dot{y}}{U}\right) + O(\alpha^2). \tag{4-11}$$

The mass per unit length, including added mass (Section 2.2), is m and the damping factor due to dissipation within the structure is ζ_y. (Chapter 8, Section 8.4). If only the terms of order α and smaller are retained, the result is a linear equation that governs the stability of small vertical displacements about $y = 0$,

$$m\ddot{y} + 2m\omega_y\left(\zeta_y - \frac{\rho U D}{4m\omega_y}\frac{\partial C_y}{\partial \alpha}\Big|_{\alpha=0}\right)\dot{y} + k_y y = -\tfrac{1}{2}\rho U^2 D C_L|_{\alpha=0}. \tag{4-12}$$

The natural frequency in radians per second is $\omega_y = 2\pi f_y = (k_y/m)^{1/2}$, where f_y is the natural frequency of plunge in cycles per second (hertz). The term in parentheses ( ) is the net damping factor of vertical motion,

$$\zeta_T = \zeta_y - \frac{U}{4\omega_y D}\frac{\rho D^2}{m}\frac{\partial C_y}{\partial \alpha}\Big|_{\alpha=0}, \tag{4-13}$$

which is the sum of structural and aerodynamic components.

The solution to Eq. 4-12 is the sum of steady and oscillatory displacements,

$$y = \frac{\frac{1}{2}\rho U^2 DC_L|_{\alpha=0}}{k_y} + A_y e^{-\zeta_T \omega_y t} \sin\left[\omega_y (1 - \zeta_T^2)^{1/2} t + \phi\right]. \quad (4\text{-}14)$$

The oscillatory component can either increase with time (unstable vibration) or decrease with time (stable vibration, Fig. 1-2) depending on the sign of the net damping coefficient ζ_T. Vibrations will decay with time for all angles of attack for which $\zeta_T > 0$, which is the case if the slope of the vertical force coefficient is negative. Thus, the model will be *stable* if (Eqs. 4-13, 4-10)

$$\frac{\partial C_y}{\partial \alpha} < 0, \quad \text{or equivalently} \quad \frac{\partial C_L}{\partial \alpha} + C_D > 0, \quad \text{for stability}, \quad (4\text{-}15)$$

and potentially unstable otherwise (Den Hartog, 1956). A section can become unstable in plunge only if the lift coefficient decreases at a rate that exceeds the drag coefficient as the angle of attack increases (Fig. 4-2).

Any noncircular section will have ranges of α where $\partial C_y/\partial \alpha > 0$ and the section is potentially unstable. Instability will occur if ζ_T (Eq. 4-13) passes through zero and becomes negative. By setting ζ_T to zero, the critical velocity for onset of plunge galloping instability becomes

$$\frac{U_{\text{crit}}}{f_y D} = \frac{4m(2\pi\zeta_y)}{\rho D^2} \Big/ \frac{\partial C_y}{\partial \alpha}. \quad (4\text{-}16)$$

f_y is the plunge natural frequency in hertz and ζ_y is the damping factor (Chapter 8).

Table 4-1 gives the slope of the vertical force coefficients for various sections. A wide variety of sections including square (Parkinson and

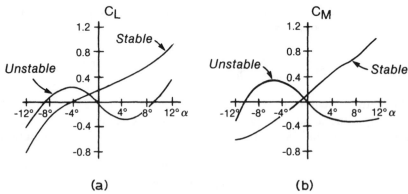

(a)   (b)

Fig. 4-2 Sections that are stable and potentially unstable to galloping: (a) lift; (b) moment. (ASCE, 1961.)

Table 4-1 Slope of vertical force coefficient for various sections in a steady flow

Section	$\partial C_y / \partial \alpha$[a] Smooth flow	Turbulent flow[b]	Reynolds number
Square $D \times D$	3.0	3.5	10^5
Rectangle $\frac{2}{3}D$ wide, D tall	0.	−0.7	10^5
Rectangle $D/2$ wide, D tall	−0.5	0.2	10^5
Rectangle $D/4$ wide, D tall	−0.15	0.	10^5
Rectangle $\frac{2}{3}D$ tall, D wide	1.3	1.2	66 000
Rectangle $D/2$ tall, D wide	2.8	−2.0	33 000
Rectangle $D/4$ tall, D wide	−10.	—	2 000–20 000
Flat rectangle D wide	−6.3	−6.3	$>10^3$
Airfoil D chord	−6.3	−6.3	$>10^3$
D-section D	−0.1	0.	66 000
Half-circle section	−0.5	2.9	51 000
Angle/chevron section D	0.66	—	75 000

Source: Richardson et al. (1965), Parkinson and Brooks (1961), Slater (1969), Nakamura and Mizota (1977), Nakamura and Tomonari (1977). See Figs. 4-4, 4-9, 4-11, and 4-22 for additional data.

[a] α is in radians; flow is left to right. $\partial C_y / \partial \alpha = -\partial C_L / \partial \alpha - C_D$. C_y based on the dimension D. $\partial C_y / \partial \alpha < 0$ for stability.

[b] Approximately 10% turbulence.

Smith, 1964), rectangular (Novak and Tanaka, 1974), right angle (Slater, 1969), D-section (Novak and Tanaka, 1974), channel (Mahrenholtz and Bardowicks, 1980), octagonal (Mahrenholtz and Bardowicks, 1980), polygonal (Mahrenholtz and Bardowicks, 1980), stalled airfoil (Sisto, 1953), H-section (Tai et al., 1976), and irregular (Cheers, 1950; Novak et al., 1978) are prone to instability and are associated with practical problems of galloping.

However, for the most part, the value of $\partial C_y / \partial \alpha$ in the unstable ranges is not large. This fact, and the relatively high stiffness (proportional to $f_y m$) of most common structures such as buildings and girders, leads to a high velocity for the onset of galloping (Eq. 4-16). This explains why only lightweight, lightly damped, flexible structures such as signs, chimneys, towers, suspension bridge decks, and ice-coated power lines gallop in high winds.

Exercises

1. Prove that Eq. 4-14 is the solution to Eq. 4-12.
2. The slope of the lift coefficient of an airfoil is $\partial C_L / \partial \alpha = 2\pi$ and the drag coefficient is $C_D \approx 0.01$ for angles of attack below about 8 degrees. Is the airfoil stable to galloping? Under what conditions might it gallop? *Hint*: Read ahead to Section 4.4.
3. Explain the difference between vortex-induced vibration and galloping.

4.2.2. Torsion Stability

Figure 4-3 shows a spring-supported, damped section that is constrained to rotate about a pivot. J_θ is the polar mass moment of inertia about the

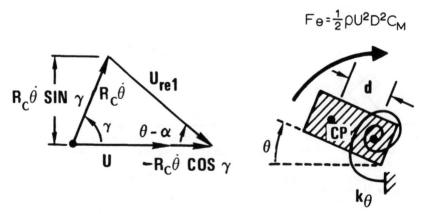

$$F_\theta = \tfrac{1}{2}\rho U^2 D^2 C_M$$

Fig. 4-3 Galloping torsion model (not showing damper parallel to spring).

pivot, including the added mass moment of inertia of the fluid, and k_θ is the torsional spring constant per unit length of the section.

In plunge galloping, the changes in angle of attack induced by vibration are a function of the vertical velocity (Eq. 4-3). In torsional galloping, the angle of attack changes with the angular position θ and also with the angular velocity $d\theta/dt$. The angular velocity $d\theta/dt$ induces angles of attack that vary over the section: a positive $d\theta/dt$ induces a downdraft forward of the pivot (negative angle of attack) and an updraft aft of the pivot (positive angle of attack). Therefore, an approximation is used to simulate the effect of angular velocity on the flow field: a reference point on the cross section at a radius R_c and an angle γ with respect to the pivot (Fig. 4-3) are chosen for evaluation of the angle of attack induced by $d\theta/dt$. The angle of attack and relative velocity at the reference point, from Figure 4-3, are

$$\alpha = \theta - \arctan \left[\frac{R_c \dot{\theta} \sin \gamma}{(U - R_c \dot{\theta} \cos \gamma)} \right]$$

$$\simeq \theta - \frac{R\dot{\theta}}{U}, \qquad \text{for } \alpha \ll 1, \tag{4-17}$$

$$U_{\text{rel}}^2 = (R_c \theta \sin \gamma)^2 + (U - R_c \dot{\theta} \cos \gamma)^2$$

$$\simeq U^2, \qquad \text{for } \alpha \ll 1, \tag{4-18}$$

where $R = R_c \sin \gamma$ is the characteristic radius. If $R > 0$, the reference point for velocity induced by the angular velocity is forward of the pivot; if $R < 0$, it is aft of the pivot. For torsion of rectangles about the centroid, Nakamura and Mizota (1975) chose R to be half the length of the rectangles corresponding to the instantaneous angle of attack at the leading edge. For torsion of right-angle sections facing into the wind about the apex of the angle, Slater (1969) chose R to be one-half the fore and aft length of the section. For flutter of airfoils (Section 4.5), R is chosen to give the angle of attack at a point three-quarters of the way back from the leading edge.

The torque per unit length on the cross section about the pivot is

$$F_\theta = \tfrac{1}{2}\rho U^2 D^2 C_m, \tag{4-19}$$

where the torque coefficient C_m is related to the steady torque coefficient C_M measured in wind-tunnel tests by

$$C_m = C_M \frac{U_{\text{rel}}^2}{U^2}. \tag{4-20}$$

U_{rel} is the instantaneous value of the relative fluid velocity at the reference point on the cross section.

The elastically mounted section of Figure 4-3 responds dynamically to the applied fluid torque. The equation of motion of the torsional response of the section is

$$J_\theta \ddot{\theta} + 2 J_\theta \zeta_\theta \omega_\theta \dot{\theta} + k_\theta \theta = \tfrac{1}{2} \rho U^2 D^2 C_m$$

$$= \tfrac{1}{2} \rho U^2 D^2 \left(C_M|_{\alpha=0} + \frac{\partial C_M}{\partial \alpha} \bigg|_{\alpha=0}^{\alpha} + \cdots \right). \quad (4\text{-}21)$$

This equation can be linearized for small angles of attack, $\alpha \ll 1$, as shown in the second equality of Eqs. 4-17 and 4-18. By utilizing the first two terms in the series with some rearranging, the result becomes

$$J_\theta \ddot{\theta} + \left(2 J_\theta \zeta_\theta \omega_\theta + \tfrac{1}{2} \rho U R D^2 \frac{\partial C_M}{\partial \alpha} \right) \dot{\theta} + \left(k_\theta - \tfrac{1}{2} \rho U^2 D^2 \frac{\partial C_M}{\partial \alpha} \right) \theta = 0. \quad (4\text{-}22)$$

This equation has two modes of instability. First, it can display a steady instability called *divergence* when the sum of the structural and the aerodynamic torsional stiffness terms falls to zero (see Section 4.5.2). Second, it can torsionally gallop when the coefficient of the $d\theta/dt$ term passes through zero. This instability can occur only if $R \, \partial C_M/\partial \alpha < 0$.

Table 4-2 Slope of moment coefficient for various sections rotating about geometric center[a]

Section	$\partial C_M/\partial \alpha$ [b]	Reynolds number
▢ ↕D	−0.18	10^4–10^5
2D, + 1D	−0.64	5×10^3–5×10^5
4D, + 1D	−18.	2×10^3–2×10^4
5D, + 1D	−26.	2×10^3–2×10^4
D, ○ + , D/4 ⟷ a	$\dfrac{2\pi a}{D}$ [c]	$>10^3$

[a] Nakamura and Mizota (1975). Also see Fig. 4-9.
[b] α is in radians; flow is left to right.
[c] For angles up to about 8 degrees.

Values of $\partial C_M / \partial \alpha$ for various sections are given in Table 4-2. The minimum velocity for the onset of torsional galloping instability is

$$\frac{U}{f_\theta D} = -\frac{4J_\theta (2\pi \zeta_\theta)}{\rho D^3 R} \bigg/ \frac{\partial C_M}{\partial \alpha}\bigg|_{\alpha=0}, \qquad (4\text{-}23)$$

where $f_\theta = (k_\theta / J_\theta)^{1/2} / (2\pi)$ is the torsional natural frequency in hertz and ζ_θ is the torsional structural damping factor. The influence of the aerodynamic force on the natural frequency has been neglected here; i.e., it is assumed that the velocity that produces instability is a small fraction of the velocity that produces divergence; see Section 4.5.2.

The usefulness of torsional stall-flutter or torsional galloping theory is to illustrate the potential for instability rather than provide accurate estimates of onset of instability. Tests on rectangular and bridge deck sections by Nakamura and Mizota (1975), Studnickova (1984), Washizu et al. (1978), and Nakamura and Yoshimura (1982) demonstrate that the torsional galloping stability criterion of Figure 4-2 and Eq. 4-23 is at best limited to compact section and higher reduced velocities and is often unreliable otherwise. There seem to be two shortcomings of the theory. First, approximating the effect of torsional velocity on angle of attack with a single characteristic point is not viable for bluff or complex sections (it is viable for airfoil sections at small angles of attack (Section 4.5.2; Dowell et al., 1978; Fung, 1969)). Second, torsional flow-induced vibrations of bluff sections are particularly sensitive to the vortices thrown off by the rotating section and quasisteady theory cannot predict the associated unsteady vortex forces (Nakamura and Matsukawa, 1987); see Section 4.4.

4.3. GALLOPING RESPONSE

4.3.1. One-Degree-of-Freedom Response

If the flow velocity exceeds the critical velocity for the onset of galloping, the energy input to the structure by the flow exceeds the energy dissipated by structural damping. The result is unstable galloping vibration. The amplitude of unstable galloping is limited by nonlinearities in the fluid force or by nonlinearities in the structure. Iwan (1973) discusses the effect of nonlinear structures on galloping. In this section, the effect of the nonlinear aerodynamic forces on a linear structure is calculated using the methods of Parkinson and Brooks (1961) and Novak (1969) to determine the steady plunge galloping response.

It is possible to express the vertical force coefficient directly as a polynomial in $\dot{y}/U$, since angle of attack α is a function of this parameter:

$$C_y(\alpha) = a_0 + a_1\left(\frac{\dot{y}}{U}\right) + a_2\left(\frac{\dot{y}}{U}\right)^2 + a_3\left(\frac{\dot{y}}{U}\right)^3 + \cdots, \qquad (4\text{-}24)$$

where the first two coefficients are determined by the vertical lift force and its slope at zero angle of attack,

$$a_0 = -C_L, \qquad a_1 = \frac{\partial C_y}{\partial \alpha} = -\frac{\partial C_L}{\partial \alpha} - C_D, \qquad \text{at } \alpha = 0. \qquad (4\text{-}25)$$

The remainder of the coefficients are determined by curve-fitting experimental data for $C_y(\alpha)$ over an appropriate range of α.

If the section is symmetric about a line in the direction of the wind through the center of the section, only odd harmonics a_1, a_3, etc., in the series are nonzero. Even harmonics can be retained if they are made odd. For example, the term $\dot{y}\,|\dot{y}|/U^2$ is an odd term containing the second power of velocity. Curve-fitted coefficients are given in Table 4-3 for symmetric sections, where some even harmonics have been retained using this device. The vertical force coefficients of various sections are shown in Figure 4-4 as functions of angle of attack.

The equation of motion of the model responding to the nonlinear vertical force is

$$m\ddot{y} + 2m\zeta_y\omega_y\dot{y} + k_y y = \tfrac{1}{2}\rho U^2 D C_y = \tfrac{1}{2}\rho U^2 D\left[a_0 + a_1\left(\frac{\dot{y}}{U}\right) + a_2\left(\frac{\dot{y}}{U}\right)^2 + \cdots\right].$$

$$(4\text{-}26)$$

This nonlinear, self-excited oscillator equation can be integrated numerically, or it can be solved approximately using asymptotic techniques or energy arguments (Richardson, 1988). Here the asymptotic method of slowly varying parameters will be employed (see Minorsky, 1962; Struble, 1962; Nayfeh and Moot, 1980) to construct approximate transient and steady-state solutions and the stability of those solutions will be examined. The accuracy of this asymptotic procedure is greatest when the aerodynamic and damping forces are small compared with the inertial and spring forces so that the model is a slightly perturbed oscillator.

Solutions to Eq. 4-26 are assumed to have time-varying amplitude and phase:

$$y = A_y(t)\cos\Phi, \qquad \text{where } \Phi = \omega_y t + \phi(t). \qquad (4\text{-}27)$$

The amplitude $A_y(t)$ and the phase $\phi(t)$ are assumed to vary only slightly over one cycle of vibration, so the velocity is well approximated by

$$\dot{y} = -\omega_y A_y \sin\Phi, \qquad (4\text{-}28)$$

which implies that combinations of derivatives of the amplitude and

Table 4-3 Aerodynamic coefficients in polynomial curve fit for vertical force coefficient[a]

Coefficients	3/2 rectangle Smooth flow	3/2 rectangle Turbulent flow	2/3 rectangle Smooth flow	2/3 rectangle Turbulent flow	D-section Smooth flow	D-section Turbulent flow	Square Smooth flow
a_1	0.	0.74285	1.9142	1.833	−0.097431	0.	2.69
a_2	−3.2736 + 1[b]	−0.24874	3.4789 + 1	5.2396	4.2554	−0.74824	0.
a_3	7.4467 + 2	1.7482 + 1	−1.7097 + 2	−1.4518 + 2	−2.8835 + 1	5.4705	−1.684
a_4	−5.5834 + 3	−3.6060 + 2	−2.2074 + 1	3.1206 + 2	6.1072 + 1	−6.3595	0.
a_5	1.4559 + 4	2.7099 + 3	0.	0.	−4.8006 + 1	2.6844	6.27 + 3
a_6	8.1990 + 3	−6.4052 + 3	0.	0.	1.2462 + 1	−0.3903	0.
a_7	−5.7367 + 4	−1.1454 + 4	0.	0.	0.	0.	−5.99 + 3
a_8	−1.2038 + 5	6.5022 + 4	0.	0.	0.	0.	0.
a_9	3.37363 + 5	−6.6937 + 4	0.	0.	0.	0.	0.
a_{10}	2.0118 + 5	0.	0.	0.	0.	0.	0.
a_{11}	−6.7549 + 5	0.	0.	0.	0.	0.	0.

Sources: Novak and Tanaka (1974), Novak (1969).

[a] C_y based on D.

[b] Re = 5 × 10^4. Flow left to right. The notation +2, etcetera, denotes powers of 10: for example, 3.445 + 2 = 344.5.

115

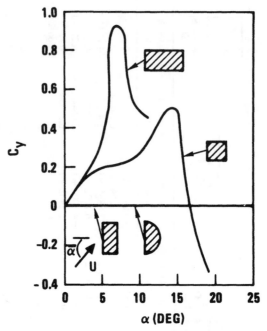

Fig. 4-4 Aerodynamic force coefficients. Re = 33 000 to 66 000 (Parkinson and Brooks, 1961).

phase are zero,

$$\dot{A}_y \cos \Phi - A_y \dot{\phi} \sin \Phi = 0. \tag{4-29}$$

If Eqs. 4-27 and 4-28 are substituted into Eq. 4-26, the result is

$$-m\omega_y \dot{A}_y \sin \Phi - m A_y \omega_y \dot{\phi} \cos \Phi + (k_y - m\omega_y^2) A_y \cos \Phi$$
$$= -2m\zeta_y \omega_y \dot{y} + \tfrac{1}{2}\rho U^2 D C_y. \tag{4-30}$$

Since the natural frequency is $\omega_y = (k_y/m)^{1/2}$, the last term on the left-hand side of this equation vanishes. If the equation is multiplied through by $\sin \Phi$ and Eq. 4-29 is multiplied through by $-m\omega_y \cos \Phi$ and the two equations are added, the result is

$$\omega_y m \dot{A}_y = (2m\zeta_y \omega_y \dot{y} - \tfrac{1}{2}\rho U^2 D C_y) \sin \Phi. \tag{4-31}$$

Since A_y and ϕ vary slowly over one cycle, they can be considered constant as Eq. 4-31 is averaged over one cycle by integrating from $\Phi = 0$ to 2π:

$$\frac{dA_y}{dt} = -\frac{1}{2\pi m \omega_y} \int_0^{2\pi} (2m\zeta_y \omega_y^2 A_y \sin \Phi + \tfrac{1}{2}\rho U^2 D C_y) \sin \Phi \, d\Phi. \tag{4-32}$$

Similarly, if Eq. 4-30 is multiplied through by $\cos \Phi$ and Eq. 4-29 is multiplied through by $m\omega_y \sin \Phi$ and the two equations are added and averaged over one cycle, then

$$\frac{d\phi}{dt} = -\frac{1}{2\pi\omega_y A_y m} \int_0^{2\pi} (2m\zeta_y\omega_y^2 A_y \sin \Phi + \tfrac{1}{2}\rho U^2 D C_y) \cos \Phi \, d\Phi. \quad (4\text{-}33)$$

Approximate transient solution to Eq. 4-26 is obtained by simultaneously integrating Eqs. 4-32 and 4-33 with time using Eqs. 4-28 and 4-24. The result is the time history of amplitude (A_y) and phase (ϕ) from an initial condition.

Steady-state solutions can be obtained by continuing the integration until $dA_y/dt = 0$. There is a more direct approach: dA_y/dt is plotted against A_y as shown in Figure 4-5. Zero crossing corresponds to steady-state solutions. Steady-state solutions can be stable or unstable. If $d\dot{A}_y/dA_y$ is greater than zero at a zero crossing, the solution is unstable because small perturbations will tend to cause the amplitude to grow. If $d\dot{A}_y/dA_y < 0$ at a zero crossing, perturbations in A_y will diminish in time and the solution is stable.

For example, if a cubic polynomial is used to approximate the vertical force coefficient for the square section (Fig. 4-1), then

$$C_y(\alpha) = \frac{a_1\dot{y}}{U} + a_3\left(\frac{\dot{y}}{U}\right)^3, \quad (4\text{-}34)$$

where the coefficients are approximately $a_1 = 2.7$ and $a_3 = -31$. (Only

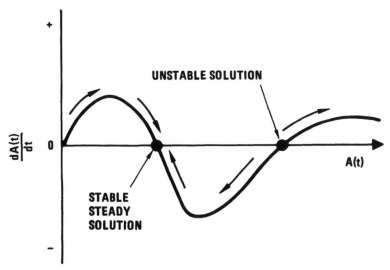

Fig. 4-5 Stability of steady galloping. If $dA/dt > 0$, aerodynamic excitation energy exceeds damping dissipation; if $dA/dt < 0$, damping dissipation exceeds aerodynamic excitation energy.

odd powers are included, since the section is symmetric and the vertical force coefficient is asymmetric about $\alpha = 0$.) Steady-state solutions are found by substituting Eq. 4-34 and Eq. 4-28 into Eq. 4-32 with $dA_y/dt = 0$:

$$0 = \int_0^{2\pi} \left\{ \left[2m\zeta_y \omega_y^2 A_y - \tfrac{1}{2}\rho U^2 D \frac{a_1 A_y \omega_y}{U} + \frac{\tfrac{3}{4}a_3 A_y^3 \omega_y^3}{U^3} \right] \sin \Phi \right.$$

$$\left. + \tfrac{1}{8}a_3 \rho U^2 D A_y^3 \omega_y^3 \frac{\sin 3\Phi}{U^3} \right\} \sin \Phi \, d\Phi. \qquad (4\text{-}35)$$

The resultant cubic polynomial,

$$A_y \left\{ 2m\zeta_y \omega_y - \tfrac{1}{2}\rho D \left[\frac{a_1 U + \tfrac{3}{4}a_3 A_y^2 \omega_y^2}{U} + \frac{\tfrac{1}{8}a_3 c\rho D A_y^2 \omega_y^2}{U} \right] \right\} = 0, \quad (4\text{-}36)$$

has three solutions. The equilibrium solution $A_y = 0$ is stable only if the free stream velocity U is less than the critical velocity (Eq. 4-16). If U exceeds the critical velocity, then $A_y = 0$ is unstable, but the solution corresponding to the zero of the quantity in braces { } is stable. By using a clever nondimensionalization discovered by Novak (1969), this solution is expressed in a simple form,

$$A = \left[4(1 - Ua_1)\frac{U}{3a_3} \right]^{1/2}, \qquad (4\text{-}37)$$

where the nondimensionalized amplitude A and velocity U,

$$A = \frac{A_y}{D} \frac{\rho D^2}{4m\zeta_y}, \qquad U = \frac{U}{f_y D} \frac{\rho D^2}{4m(2\pi\zeta_y)}, \qquad (4\text{-}38)$$

allow for a complete representation of the plunge galloping response of any one-dimensional system in terms of only two parameters. Unfortunately, this nondimensionalization does not apply to torsional galloping, multidimensional systems, or interactions with vortex shedding (Olivari, 1983; Bearman et al., 1987).

Figure 4-6 shows experimental data for the plunge galloping response of a square section (Fig. 4-1) in comparison with the third-order curve fit (Eq. 4-34) and seventh-order curve fit. Parkinson and Smith (1964) have shown that the portion of the curve between $U = 0.4$ and $U = 0.7$ contains two stable solutions and one unstable solution. The hysteresis is produced by the change in slope of the $C_y(\alpha)$ curve at $\alpha = 7$ degrees (Fig. 4-4). The third-order fit cannot reproduce this change in slope but the seventh-order fit does. Inflections in the vertical force coefficient curve produce multiple solutions in the galloping response.

The form of the vertical force curves and the associated galloping response for square and rectangular sections are shown in Figure 4-7. For

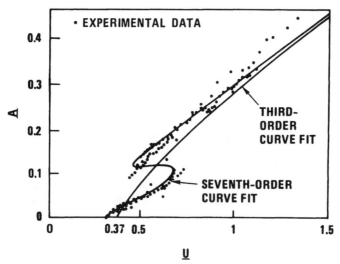

Fig. 4-6 Experimental data and response of a square section (Fig. 4-1). Re = 4000 to 20 000 (After Parkinson and Smith, 1964).

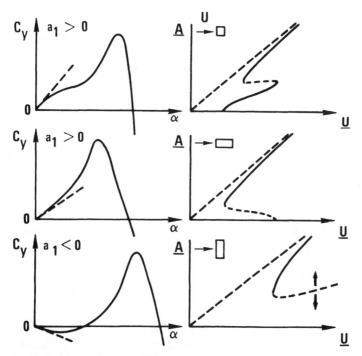

Fig. 4-7 Typical lateral force coefficients for rectangular sections and the corresponding types of galloping response: a_1 given by Eq. 4-25 (Novak, 1971).

119

tall rectangles (small length-to-height ratio), no transverse force develops, $\partial C_y / \partial \alpha \leq 0$, and they are stable (also see Table 4-1). However, tall rectangles will gallop like a *hard oscillator*; i.e., an appreciable impulse is required to induce galloping. The square section is unstable at rest and gallops spontaneously. Low, long rectangles (large ratio of length parallel to flow-to-height) also gallop as hard oscillators up to length-to-height ratios of about 3; beyond this, the rectangles are stable (Parkinson, 1971; Washizu et al., 1978). Their stability is associated with reattachment of the shear layer that separates near the leading edge.

Galloping analysis can be applied to continuous structures through modal analysis. This is accomplished by developing the partial differential equation describing the structure, selecting the mode most susceptible to instabilities (ordinarily the fundamental mode), and using the Glerkin technique to reduce the partial differential equation to an ordinary differential equation for each mode. This equation will have the same form as Eq. 4-26 if displacement is restricted to plunge. This procedure is illustrated in Section 4.7, in Appendix A, by Novak and Tanaka (1974), and by Skarachy (1975). Ordinarily, the response of continuous structures is very similar to the one-degree-of-freedom model provided that the natural frequencies of the structure are well separated and there is no inertial coupling. If two or more modes of the structure are nearly equal or if the torsion and plunge modes are inertially coupled because the center of gravity does not coincide with the shear center, then coupled galloping can arise, as discussed in Sections 4.3.2 and 4.5.2.

Exercises

1. Consider a transmission line that has been coated with ice so that a D-shaped cross section is created with the flat face into the wind. The width of the flat face (diameter D) is 2.5 in (6.4 cm). The fundamental natural frequency is 0.5 Hz and the weight is 1.1 lb/ft (1.64 kg/m). The structural damping of transmission lines is low, $\zeta_y = 0.005$. The air density is 0.075 lb/ft³ (1.2 kg/m³). What is the mass parameter $m/(\rho D^2)$? What is the stability at $\alpha = 0$? Describe the galloping response amplitude as a function of wind velocity for wind velocities between 0 and 100 ft/sec (30 m/sec) by scaling Figure 4-12. What amplitude is required to trigger the oscillations? How does turbulence influence the result (see Table 4-3)? Answers are given by Novak and Tanaka (1974).

2. Verify Eq. 4-37. By integrating Eq. 4-35, determine the numerical value of the coefficient c in Eq. 4-36.

4.3.2. Two-Degree-of-Freedom Response

Most structures are free to translate and rotate. Rotation and displacement are aerodynamically coupled through the angle of attack. Torsion and displacement can also be inertially coupled if the elastic axis of the cross section does not coincide with the center of mass. The *elastic axis,* also called the center of twist or locus of shear centers, is the point on the cross section of a slender elastic structure where an applied force produces no torsion and a torsion produces no displacement.

A two-degree-of-freedom structure that both translates vertically and rotates is shown in Figure 4-8. The structure consists of a cross section of arbitrary shape with structural stiffness in plunge (k_y) and torsion (k_θ) and viscous structural damping in both plunge and torsion (ζ_y and ζ_θ). The equations of motion for the two-degree-of-freedom model shown in Figure 4-8 are shown in Appendix B to be

$$m\ddot{y} + 2m\zeta_y\omega_y\dot{y} + S_x\ddot{\theta} + k_y y = F_y = \tfrac{1}{2}\rho U^2 D C_y, \qquad (4\text{-}39)$$

$$J_\theta\ddot{\theta} + 2J_\theta\zeta_\theta\dot{\theta} + S_x\ddot{y} + k_\theta\theta = F_M = \tfrac{1}{2}\rho U^2 D^2 C_M, \qquad (4\text{-}40)$$

where the mass per unit length, mass moment of inertia, and lateral position of the center of gravity are

$$m = \int_A \mu\, d\xi\, d\eta, \qquad J_\theta = \int_A (\xi^2 + \eta^2)\mu\, d\xi\, d\eta, \qquad S_x = \int_A \xi\mu\, d\xi\, d\eta. \tag{4-41}$$

μ is the density and ξ and η are coordinates fixed to the cross section. m

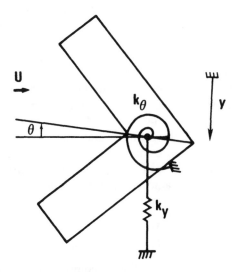

Fig. 4-8 Elastically supported two-degree-of-freedom, uncoupled angle section.

is the mass per unit length, S_y/m is the lateral position of the center of gravity measured from the elastic axis, J_θ is the polar mass moment of inertia of the section, and A denotes the cross-sectional area. Added fluid mass should be included in both m and J_θ (Section 2.2). J_y and J_θ are the damping due to internal energy dissipation of the structure (Chapter 8).

The angle of attack includes the effects of both torsion and plunge of the cross section. For small angles, the angle of attack is

$$\alpha = \theta - \frac{R\dot{\theta}}{U} + \frac{\dot{y}}{U}. \tag{4-42}$$

α is a function of both the torsional and plunge motion. The two-degree-of-freedom motion is coupled aerodynamically through the angle of attack. It is also coupled inertially through the term S_x, which induces torsion as a result of plunge and vice versa. The system can be inertially decoupled through the use of principal coordinates as discussed in Appendix B provided the two natural frequencies of the system are well separated. The system natural frequencies are given by Eq. B-17. For this case the onset of instability is given by

$$U = \text{minimum} \begin{cases} \dfrac{-4m\zeta_y\omega_y - 4c_2^2 J_\theta \zeta_\theta \omega_\theta}{\rho D(Rc_2 + 1)(c_2 D\, \partial C_M/\partial\alpha - \partial C_y/\partial\alpha)} \\[2ex] \dfrac{-4c_1^2 m\zeta_y\omega_y - 4J_\theta \zeta_\theta \omega_\theta}{\rho D(R + c_1)(-c_1\, \partial C_y/\partial\alpha + D\, \partial C_M/\partial\alpha)} \end{cases}. \tag{4-43}$$

As the inertial coupling (S_x) decreases, the coupling terms c_1 and c_2 (Eq. B-19) approach zero and the coupled instability (Eq. 4-43) reduces to the individual plunge and torsional instability criteria (Eqs. 4-16 and 4-23).

Blevins and Iwan (1974) have investigated the case where the inertial coupling is zero ($S_x = 0$) but the torsion and plunge natural frequencies are close. They used cubic fits to the nonlinear coefficients, $C_y = a_1\alpha + a_3\alpha^3$ and $C_M = b_1\alpha + b_3\alpha^3$, as shown in Figure 4-9. Their results show that the solutions fall into two classes, depending on the ratio of the natural frequencies in torsion and plunge.

The first class of solutions is valid when ω_y and ω_θ are not approximately equal or in the ratio of small integers, i.e., when $i\omega_y$ is not equal to $j\omega_\theta$, where i and j are small integers, $1, 2, 3, \ldots$. The stability of these uncoupled solutions can be specified by two nondimensional parameters that have the same form as the nondimensional parameters used by Novak (1969) (Eq. 4-38):

$$U_y^* = \frac{\rho D^2}{4m} \frac{U}{\omega_y D} \frac{a_1}{\zeta_y}, \qquad U_\theta^* = \frac{\rho D^4}{4J_\theta} \frac{U}{\omega_\theta D} \frac{R}{D} \frac{b_1}{\zeta_\theta}, \tag{4-44}$$

where U_y^* is the ratio of the linear component of the destabilizing aerodynamic force to the damping force for a section in pure plunge, and

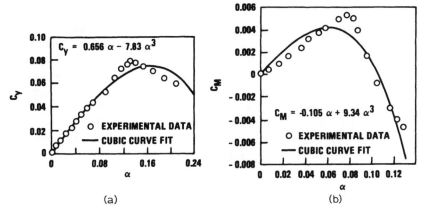

Fig. 4-9 Vertical force coefficients (a) and moment coefficients (b) for right-angle section. Experimental results from Slater (1969).

U_θ is the ratio of the linear component of the destabilizing aerodynamic force to the damping force for a section in pure torsion. These solutions, shown in Table 4-4, are characterized by either a steady plunge amplitude A_y and no torsion, or by a steady amplitude in torsion A_θ and no plunge, or by trivial solution $y = \theta = 0$.

Table 4-4 Galloping response for the case $\omega_y/\omega_\theta = \frac{1}{3}$, 1, or $3 + O(\zeta)^a$

Nature of solution	Amplitude of steady-state response	Stability criteria[a]
Equilibrium	$A_y = 0$	$1 - \dfrac{1}{U_y^*} < 0$
	$A_\theta = 0$	$1 - \dfrac{1}{U_\theta^*} < 0$
Plunge[b]	$\dfrac{A_y}{D} = \left[-\dfrac{4U_y}{3a_3}\left(a_1 U_y - \dfrac{2\zeta_y}{n_y} \right) \right]^{1/2}$	$1 - \dfrac{1}{U_y^*} > 0$
	$A_\theta = 0$	$2\dfrac{b_3 a_1}{a_3 b_1}\left(1 - \dfrac{1}{U_\theta^*} \right) > 1 - \dfrac{1}{U_\theta^*}$
Torsion	$A_y = 0$	$1 - \dfrac{1}{U_\theta^*} > 0$
	$A_\theta = \left[-\dfrac{4U_\theta}{3b_3}\left(\dfrac{b_1 rU_\theta - 2\zeta_\theta/n_\theta}{r^3 + rU_\theta^2} \right) \right]^{1/2}$	$2\dfrac{b_1 a_3}{b_3 a_1}\left(1 - \dfrac{1}{U_\theta^*} \right) > 1 - \dfrac{1}{U_y^*}$
Plunge and torsion	$A_y = 0(1)$ $A_\theta = 0(1)$	Always unstable for $a_3 < 0$, $rb_3 < 0$

Source: Blevins and Iwan (1974).

[a] $r = R/D$; $n_\theta = \rho D^4/(2I_\theta)$; $n_y = \rho D^2/(2m)$; $U_y = U/(\omega_y D)$; $U_\theta = U/(\omega_\theta D)$. Note that the stability criteria $(1 - 1/U_y^*) < 0$ and $(1 - 1/U_\theta^*) < 0$ are identical to Eqs. 4-16 and 4-23, respectively.

[b] This plunge amplitude is identical to Eq. 4-37.

The second class of solutions arises when ω_y and ω_θ are approximately equal or when they are approximately multiples of small integers. Both torsion and plunge are simultaneously excited in this class of solutions. The interaction occurs through the linear and nonlinear aerodynamic force terms. Although the strongest interaction between torsion and plunge occurs when $\omega_y \simeq \omega_\theta$, as shown in Figure 4-10, a small degree of interaction also occurs when $\omega_y/\omega_\theta \simeq \frac{1}{3}, 3$. Some aspects of these predictions have been confirmed by experiment. Modi and Slater (1983) made experiments on the right-angle section of Figure 4-9 with no inertial coupling. The frequency ratio was $\omega_y/\omega_\theta = 2.92$. This frequency ratio is

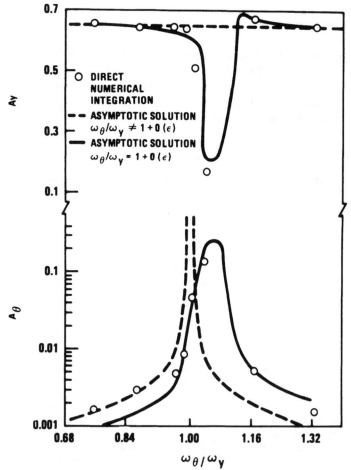

Fig. 4-10 Normalized response of the right-angle section, Figure 4-8, in the vicinity of $\omega_\theta/\omega_y \simeq 1$, $U_y = 5$ (Blevins and Iwan, 1974).

sufficiently different from 3 (that is, 2.92 differs from 3.0 by more than the bandwidth of the lightly damped oscillator; see Chapter 8, Eq. 8-39), so that there is essentially no torsion-plunge interaction (Blevins and Iwan, 1974). Modi and Slater (1983) confirmed this by finding that the angle section responded as either uncoupled torsion or uncoupled plunge. They also found strong coupling of torsion with vortex shedding, and this is discussed in the following section.

This coupled torsion–plunge instability is essentially identical to the flutter instability that arises in aircraft structures; see Section 4.5.

Exercise

1. Show that Eq. 4-43 reduces to the uncoupled stability criterion if $S_x = 0$.

4.4. VORTEX SHEDDING, TURBULENCE, AND GALLOPING

The major limitation of the galloping theory outlined thus far is that the aerodynamic coefficients are assumed to vary only with angle of attack, but experience shows that the coefficients are affected by turbulence and vortex shedding.

Laneville and Parkinson (1971), Nakamura and Yoshimura (1982), and Novak and Tanaka (1974) have found that turbulence in the mean flow can either reduce or increase the tendency toward galloping instability of bluff sections. As an example of the effect of turbulence, consider the D-section, with flat face into the airstream. The D shape is similar to the contour of ice-coated power lines. Simulated power lines fitted with D-shaped fairings gallop under natural conditions of wind (Novak and Tanaka, 1974). Force coefficients in smooth and turbulent flow are given in Figures 4-11 and 4-22. In smooth flow, the D-section is stable in the range $0 < \alpha < 30$ degrees. In turbulent flow, the section has initial neutral stability but becomes unstable at $\alpha = 11$ degrees and oscillations can be triggered by an initial perturbation at much lower velocity as shown in Figure 4-12.

The quasisteady assumption used in galloping theory (Eq. 4-1) requires that the vortex shedding frequency (Chapter 3, Eq. 3-2) be well above the natural frequency so that the fluid responds quickly to any structural motion. (Novak and Tanaka, 1974; Parkinson, 1971). From forced vibration measurements on square and rectangular sections, Otauki et al. (1974), Bearman et al. (1987), and Washizu et al. (1978) concluded that the reduced velocity $[U/(f_n D)$, where D is the height of the section and f_n is the natural frequency of plunge] must exceed 20 and

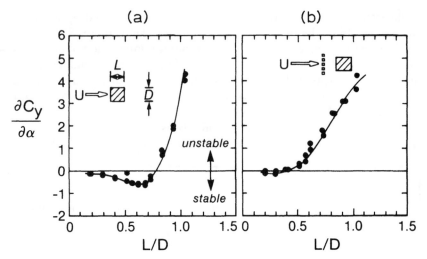

Fig. 4-11 Experimental measurements of slope of vertical force coefficient in smooth and turbulent flow at Re = 10^5, based on D, as a function of the height-to-length ratio of a rectangular section; C_y based on D (Nakamura and Tomonari, 1977).

that the amplitude of vibration cannot exceed 0.1 to 0.2 D for application of the quasisteady theory; see Figure 4-13.

A bluff section sheds vortices at or near the natural vortex shedding frequency (Chapter 3) and forms a vortex street wake. A transversely vibrating section also throws off vortices at the vibration frequency, creating a second organized vortex wake system that exerts vortex forces

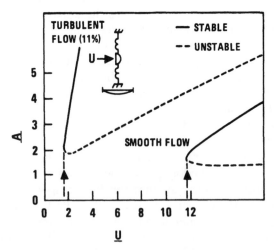

Fig. 4-12 Galloping response curve for a D-section in smooth and turbulent flow (Novak and Tanaka, 1974).

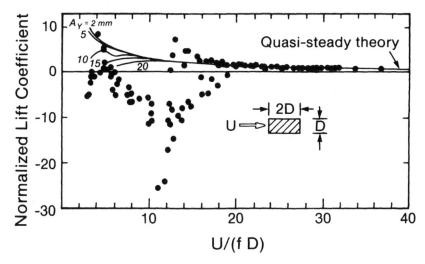

Fig. 4-13 Experimental data for aerodynamic force coefficients on a vertically oscillating rectangular section (Washizu et al., 1978).

that are synchronized with but out of phase with the motion (Bearman et al., 1987; Bearman and Luo, 1988; Williamson and Roshko, 1988). The two vortex systems coexist. They can merge into a single system at resonance with the natural vortex shedding. Both vortex shedding and galloping can excite square sections (Bearman et al., 1987; Parkinson and Wawzonek, 1981; Obasaju, 1983), rectangular sections (Novak, 1971; Olivari, 1983; Bokaian and Geoola, 1984), angle sections (Modi and Slater, 1983), bridge decks (Wardlaw and Ponder, 1970; Konishi et al., 1980), and stalled air foils (Zaman et al., 1989). These structures respond to vortex shedding near the vortex resonance, $U/(f_yD) \approx 5$, and classical galloping at higher reduced velocities, $U/(f_yD) > 20$. The onset of galloping can be brought into or below the vortex shedding resonance by reducing damping. However, the galloping theory predictions become relatively poor at lower reduced velocities because the quasisteady forces induced by changes in angle of attack are obscured by unsteady vortex shedding forces acting at the same or similar frequencies.

Torsional oscillation can be induced by both galloping and vortex shedding. For example, the angle section of Figure 4-14 clearly shows both torsional vortex-induced vibration and an amplitude-limited instability that may be labeled galloping. Suspension bridge decks are particularly susceptible to wind-induced torsional vibration (Steinman and Watson, 1957; ASCE, 1955, 1961; Chapter 3, Section 3.7). Results of model tests on the Tacoma Narrows Bridge, plotted in Figure 4-15, clearly show that the vortex shedding equals the natural frequency at the onset of large-amplitude vibrations, but the build-up of torsional amplitude

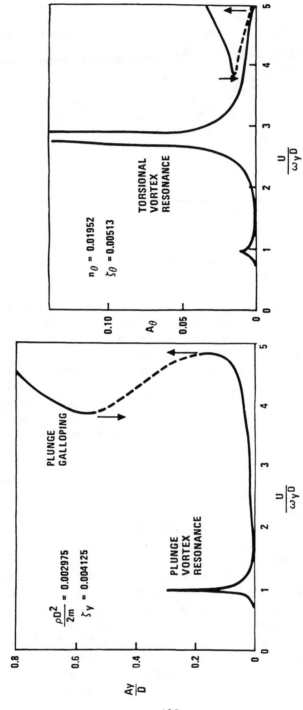

Fig. 4-14 Combined displacement and torsional galloping and vortex-induced vibration of the angle section (Slater, 1969; Modi and Slater, 1983).

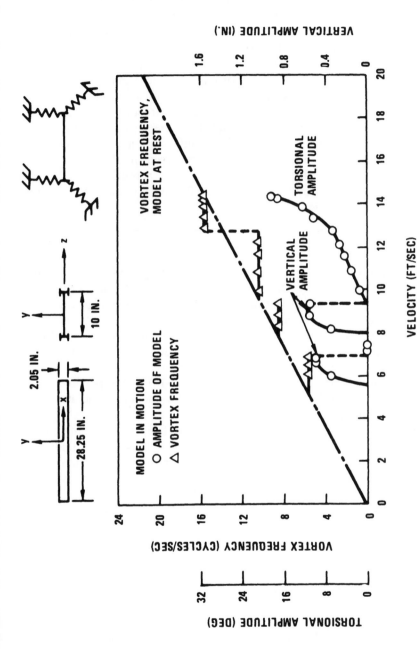

Fig. 4-15 Test results for spring-supported model of the Tacoma Narrows Bridge (ASCE, 1961).

129

suggests an instability as well. Probably vortex shedding provides an initial amplitude required for hard oscillator galloping and then the torsion of sharp-edged sections synchronizes the shedding as well as increases the magnitude of the shed vortices; see Scanlan (1971), Nakamura and Nakashima (1986), and Shiraishi and Matsumoto (1983).

The most reliable approach to accounting for unsteady and coupled vortex effects on noncircular sections is to scale from test results on models or comparable structures. The American Society of Civil Engineers Task Force on Cable Suspension Structures (ASCE, 1977) recommends that scaled wind-tunnel tests be made to ensure the adequacy of the structure when sufficient information is not available to ensure stability. Scaled model testing is discussed in Chapters 1 and 7, Section 7.4.3. Numerous practical examples of aeroelastic wind-tunnel model tests are given by Naudascher and Rockwell (1980) and Reinhold (1982). See also Section 3.7.

Exercises

1. The response of a spring-mounted building model is shown in Chapter 1, Figure 1-3. Describe three phenomena that influence the response. Using the models developed for vortex shedding response in Chapter 3, predict the vortex shedding response near $(U/f_yD) = 5$. Using galloping theory, predict galloping response for one level of damping using $\partial C_y/\partial a = 2.3$. The onset of galloping near $U/(f_yD) = 12$ appears to be independent of damping, which is inconsistent with galloping theory (Eq. 4-16). One possible explanation is that the third harmonic of vortex shedding regulates the onset of galloping. Do you agree?

2. Free stream turbulence, up to about 10%, has little influence on vortex shedding or vortex-induced vibration of circular cylinders, but turbulence has a large influence on aerodynamic forces and flow-induced vibration of rectangular sections. Can you explain this? (See Section 3.2, or Gartshore, 1984.)

4.5. FLUTTER

4.5.1. Forces on an Airfoil

Flutter is the term applied to a class of aeroelastic phenomena of aircraft structures. It includes torsion–plunge coupled instability ("flutter"), engine rotor–aerodynamic vibration ("whirl flutter"), control reversal, buffeting, dynamic load redistribution, single-degree-of-freedom oscillation ("stall flutter"), and unstable torsion ("divergence"). The first aircraft to flutter was the monoplane built by Professor Samuel Langley which experienced torsional divergence of the wing as it was launched from a house boat on the Potomac in 1903, ten days before the Wright brothers' first powered flight. The stiff, cross-braced structure of the Wrights' biplane had no such problems (Bisplinghoff et al., 1955; Gordon, 1978). Flutter remains a primary criterion in licensing new aircraft (FAA, 1985).

In contrast to the situation with bluff sections, accurate theories are available for the aerodynamic forces on airfoils in subsonic and supersonic flow. These theories are reviewed in books on flutter by Bisplinghoff and Ashley (1962), Fung (1969), Garrick (1969), Dowell (1974), Forsching (1974), Dowell et al. (1978), and Hajela (1986). In this section, the fundamentals of divergence and flutter are developed using linear aerodynamic theory and contrasted with the galloping instability of the previous sections.

Consider the spring-supported wing section shown in Figure 4-16. This two-degree-of-freedom spring-supported model represents a wing with a single bending and a single torsional mode. The equations of motion for this section are shown in Appendix B to be

$$m\ddot{y} + 2m\zeta_y\omega_y\dot{y} + S_x\ddot{\theta} + k_y y = F_y', \qquad (4\text{-}45)$$

$$J_\theta\ddot{\theta} + 2J_\theta\zeta_\theta\dot{\theta} + S_x\ddot{y} + k_\theta\theta = F_\theta', \qquad (4\text{-}46)$$

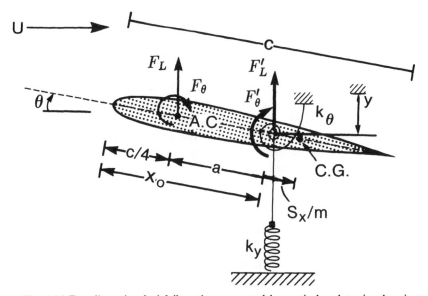

Fig. 4-16 Two-dimensional airfoil section supported by vertical and torsional springs. A.C. is the aerodynamic center and C.G. denotes the center of mass.

where y is the vertical displacement at the elastic axis, positive downward; F_y' is the vertical aerodynamic force per unit span at the elastic axis, positive downward; θ is the angle of twist, positive clockwise in a flow from left to right; F_θ' is the aerodynamic moment per unit span at the elastic axis, positive clockwise; k_y and k_θ are spring constants per unit span; and ζ_y and ζ_θ are the damping factors. The *elastic axis,* sometimes called the shear center, center of flexure, or center of twist, is the point about which a vertical static force produces displacement but no torsion (Timoshenko and Goodier, 1951). Because the center of mass (c.g.) does not generally coincide with the elastic axis, these equations are coupled inertially through the term S_x (Eq. 4-41, Appendix B), which is m times the lateral position of the center of mass aft of the elastic axis.

The mass per unit span (m) and the mass moment of inertia about the elastic axis per unit span (J_θ) include added mass (Chapter 2, Section 2.2). These are approximately the same for an airfoil as for a flat plate of the same chord (Table 2-2): $m_a = \rho\pi c^2/4$, $J_a = \rho\pi c^4/128$, where ρ is the air density and c is the airfoil *chord*, the distance between leading edge and trailing edge, and J_a has been computed about a point halfway between the leading and trailing edge. Typically, the added mass of air is 20% of the mass of an aircraft wing (Section 4.7).

The steady lift, drag, and moment per unit span on an airfoil are expressed in terms of the lift coefficient C_L, the drag coefficient C_D, and the moment coefficient C_M,

$$F_L = \tfrac{1}{2}\rho U^2 c C_L, \qquad F_D = \tfrac{1}{2}\rho U^2 c C_D, \qquad F_\theta = \tfrac{1}{2}\rho U^2 c^2 C_M. \qquad (4\text{-}47)$$

U is the freestream velocity and c is the chord (length) of the airfoil. The lift, drag, and moment coefficients for airfoils are a function of the angle of attack α between the airfoil axis and the flow and of Reynolds number. For three-dimensional wings, they are functions of the aspect ratio of the wind as well (Abbott and von Doenhoff, 1959).

Figure 4-17 shows typical lift, drag, and moment coefficients for an airfoil. Note that the lift coefficient is a linear function of angle of attack up to about the maximum lift and that the drag and moment coefficients are very small in this range. For angles up to about ±8 degrees, for most airfoil these coefficients are

$$C_L = 2\pi \sin \alpha,$$

$$C_D = \text{order of } 0.01, \qquad\qquad (4\text{-}48)$$

$$C_M = \text{order of } 0.01.$$

The drag and moment coefficients are dependent on the shape of the airfoil, but the lift coefficient and its slope are nearly independent of the airfoil shape for angles of attack below about 8 degrees. At large angles, the flow separates from the back of the airfoil and drag increases sharply

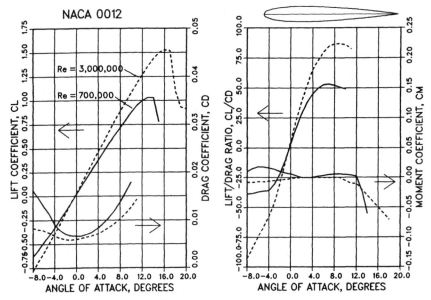

Fig. 4-17 Aerodynamic coefficients for the NACA 0012 (12% thick) symmetric airfoil. The drag and moment coefficients are given by the lower sets of curves in each figure. The moment coefficient is referenced to the quarter-chord point (Miley, 1982).

as lift falls. This condition is known as *stall* in aeronautical terminology. In bluff body aerodynamics, it is called *separated flow*.

The net vertical force on the airfoil is the vector sum of lift and drag in the vertical plane (Fig. 4-1, Eqs. 4-5, 4-6). However, at small angles of attack the lift is much larger than drag (Eq. 4-48) so the vertical force is almost entirely due to lift. (This is not true for bluff sections.)

$$F'_y = F_y \approx -F_L.$$ (4-49)

The negative sign arises because lift is positive upward and the vertical motion is positive downward. By convention, the aerodynamic moment on an airfoil is generally referenced to the *aerodynamic center,* the point where aerodynamic moment is minimum; it is generally one-quarter of the distance from the leading to the trailing edge. The moment at the elastic axis is the sum of the moment at the aerodynamic center and the moment induced by the offset (a) between the aerodynamic center and the elastic axis,

$$F'_\theta = F_\theta + aF_L \approx aF_L.$$ (4-50)

Because the moment coefficient at the aerodynamic center is small (Eq. 4-48), the resultant moment is dominated by lift. If the aerodynamic center (quarter-chord point, Fig. 4-16) is forward of the elastic axis, *a* is positive; otherwise it is negative.

Exercise

1. Compare and contrast the assumptions concerning aerodynamic forces in flutter analysis (Eqs. 4-47, 4-59, and 4-60) and in galloping analysis (Eqs. 4-2, 4-3, and 4-6). Does galloping analysis apply to airfoils? Does airfoil analysis apply to bluff sections?

4.5.2. Flutter Stability

Static stability—divergence. Consider the response of the spring-supported airfoil of Figure 4-16 to *steady* aerodynamic forces. The steady equations of motion are adapted from Eqs. 4-45 and 4-46 by neglecting all the time derivative ($\dot{}$) terms and incorporating Eqs. 4-47 through 4-50:

$$k_y y = -\tfrac{1}{2}\rho U^2 c(2\pi \sin \alpha), \qquad k_\theta \theta = \tfrac{1}{2}\rho U^2 ca(2\pi \sin \alpha). \qquad (4\text{-}51)$$

For *steady* analysis, the angle of attack equals the angle of torsion, $\alpha = \theta$ if we reference both angles to the angle of zero lift. Using this convention and incorporating the small-angle approximation $\sin \alpha = \alpha$ gives the following equations for stability of small perturbations in torsion and plunge:

$$y = -\frac{\pi\rho U^2 c\theta}{k_y}, \qquad (k_\theta - \rho U^2 ca\pi)\theta = 0. \qquad (4\text{-}52)$$

The equilibrium solution $y = \theta = 0$ is stable if the quantity in parentheses $(k_\theta - \rho U^2 ca\pi)$ is positive, so that the torsional aerodynamic moment is held in check by the torsional spring. If the flow velocity is increased to the point where the quantity in parentheses in Eq. 4-52 falls to zero, large torsional deformations will result from any slight perturbation. This condition is called *divergence*. It occurs at the following velocity:

$$U_{\text{diverge}} = \left[\frac{k_\theta}{\pi\rho ca}\right]^{1/2}. \qquad (4\text{-}53)$$

k_θ is the torsional spring constant per unit span, i.e., the torsional spring constant of the wing divided by the span of the wing.

Divergence can be prevented by incorporating sufficient stiffness in the design that the divergence airspeed (Eq. 4-53) is not exceeded in flight. The cross-bracing of early biplanes prevented divergence of their wings (Gordon, 1978). Another approach to divergence stability is to move the elastic axis forward of the aerodynamic center so that the distance a (Fig. 4-16) is negative and the aerodynamic moment assists, rather than opposes, the spring torsional moment. This is the principle behind putting tails on the rear of kites and aircraft.

If the wing or aircraft is free, that is, $k_\theta = 0$, then the appropriate moment arm (a) for divergence stability is the distance between the aerodynamic center and the center of gravity. Stability can be achieved by moving the center of gravity forward of the aerodynamic center by adding mass to the nose or by moving the aerodynamic center aft of the center of gravity by adding a tail. Primitive man accomplished this by using a stone head on his spears and adding feathers to the tail of his arrows. It is called *arrow stability*.

One-degree-of-freedom dynamic stability—stall flutter. Here we will examine the one-degree-of-freedom dynamic stability of airfoils in vertical motion (plunge) alone and then in torsion alone. Using *quasisteady* aerodynamic theory for the vertical force coefficient (Eqs. 4-6 through 4-9), the equation of vertical motion (Eq. 4-45) without torsion ($\theta = 0$) for small angles of attack describes dynamic stability in plunge,

$$m\ddot{y} + 2m\zeta_y\omega_y\dot{y} + k_y y = -\tfrac{1}{2}\rho U c\left(\frac{\partial C_L}{\partial \alpha} + C_D\right)_{\alpha=0}\dot{y}. \tag{4-54}$$

The component of the aerodynamic force induced by plunge acts as a stabilizing aerodynamic damper if the quantity in parentheses on the right side of this equation is positive,

$$\frac{\partial C_L}{\partial \alpha} + C_D > 0 \qquad \text{for stability.} \tag{4-55}$$

At small angles of attack, $\partial C_L/\partial\alpha = 2\pi$ and $C_D > 0$ (Eq. 4-48) and airfoils are stable in plunge. However, if the angle of attack exceeds about ± 8 degrees, the airfoil *stalls* (Fig. 4-17); lift decreases with increasing angle of attack, $\partial C_L/\partial\alpha < 0$; Eq. 4-55 is violated; and an instability called *stall flutter* can occur. Stall flutter requires an initial amplitude to trigger the instability or an equilibrium position such that the airfoil is initially stalled; see Sisto (1953) and Fung (1969). Stall flutter is identical to plunge galloping in the hard oscillator form (Section 4.3.1; middle row of Fig. 4-7).

Consider torsion without plunge at small angles of attack. As noted in Section 4.2.2, unsteady aerodynamic forces play a large role in torsional stability. Fung (1969, pp. 194–195) presents a suitable quasisteady model for torsional forces on an airfoil at small angles of attack. With Fung's *quasisteady* aerodynamic forces, the equation of motion of torsion, Eq. 4-46 with $y = 0$, becomes

$$J_\theta\ddot{\theta} + 2J_\theta\zeta_\theta\dot{\theta} + k_\theta\theta = \tfrac{1}{2}\rho U^2 c^2\left(\frac{x_o}{c} - \frac{1}{4}\right)\frac{\partial C_L}{\partial \alpha}\theta$$

$$- \tfrac{1}{2}\rho U c^3\left[\frac{\pi}{8} + \left(\frac{x_o}{c} - \frac{1}{4}\right)\left(\frac{x_o}{c} - \frac{3}{4}\right)\right]\dot{\theta}. \tag{4-56}$$

x_0 is the distance of the elastic axis aft of the leading edge (Fig. 4-16). The quantity $(x_0 - c/4)$ is the distance between the elastic axis and the aerodynamic center. The quantity $(3/4)c - x_o$ is the distance between the elastic axis and the three-quarter chord point. Because the quantity in square brackets is always positive, the aerodynamic moment proportional to angular velocity provides aerodynamic damping. Thus, an airfoil is *dynamically* stable in torsion but *statically* unstable in torsion if the velocity exceeds the divergence velocity (Eq. 4-53).

Torsion–plunge coupled instabilities—flutter. An airfoil can become dynamically unstable in plunge if large displacements induce stall and statically unstable in torsion if the velocity exceeds the velocity for onset of divergence. An airfoil can also become unstable in a coupled torsion–plunge mode called *classical flutter*.

The equations of motion of coupled plunge and torsion motion (Eqs. 4-45 and 4-46) are utilized for flutter analysis with linear models for the aerodynamic vertical force and moment. Using *steady* aerodynamic analysis, the angle of torsion equals the angle of attack, $\alpha = \theta$, and the vertical aerodynamic force (Eq. 4-49) is produced by lift

$$F_y = -F_L = -\tfrac{1}{2}\rho U^2 c\left[C_L|_{\alpha=0} + \left.\frac{\partial C_L}{\partial \alpha}\right|_{\alpha=0} \theta + O(\alpha^2)\right]. \qquad (4\text{-}57)$$

The linear equations of coupled torsion and plunge motion are obtained by applying this force to Eqs. 4-45 and 4-46 using Eq. 4-50 and neglecting damping ($\zeta_y = \zeta_\theta = 0$) and the mean deformations,

$$m\ddot{y} + S_x\ddot{\theta} + k_y y = -\tfrac{1}{2}\rho U^2 c\left(\frac{\partial C_L}{\partial \alpha}\right)\theta, \qquad (4\text{-}58)$$

$$J_\theta\ddot{\theta} + S_x\ddot{y} + k_\theta\theta = \tfrac{1}{2}\rho U^2 ca\left(\frac{\partial C_L}{\partial \alpha}\right)\theta. \qquad (4\text{-}59)$$

The onset of instability is found by assuming small perturbations in y and θ, about the equilibrium position $y = \theta = 0$,

$$y = A_y e^{\lambda t}, \qquad (4\text{-}60)$$

$$\theta = A_\theta e^{\lambda t}, \qquad (4\text{-}61)$$

where λ, A_y, and A_θ are constants and t is time. This solution form is substituted into Eqs. 4-58 and 4-59 to give the following matrix equation:

$$\begin{bmatrix} m\lambda^2 + k_y & S_x\lambda^2 + \tfrac{1}{2}\rho U^2 c(\partial C_L/\partial \alpha) \\ S_x\lambda^2 & J_\theta\lambda^2 + k_\theta - \tfrac{1}{2}\rho U^2 ca(\partial C_L/\partial \alpha) \end{bmatrix}\begin{pmatrix} A_y \\ A_\theta \end{pmatrix} = 0. \qquad (4\text{-}62)$$

Solutions other than the trivial solution $A_y = A_\theta = 0$ exist only if the determinant of the coefficient matrix on the left is zero (Bellman, 1970).

Setting the determinant to zero gives a stability polynomial for λ,

$$C_0\lambda^4 + C_2\lambda^2 + C_4 = 0, \tag{4-63}$$

where the coefficients of the polynomial are

$$C_0 = mJ_\theta - S_x^2, \qquad C_4 = k_y\left[k_\theta - \tfrac{1}{2}\rho U^2 ca\left(\frac{\partial C_L}{\partial \alpha}\right)\right],$$

$$C_2 = m\left[k_\theta - \tfrac{1}{2}\rho U^2 ca\left(\frac{\partial C_L}{\partial \alpha}\right)\right] + k_y J_\theta - \tfrac{1}{2}\rho U^2 c\left(\frac{\partial C_L}{\partial \alpha}\right)S_x. \tag{4-64}$$

An exact solution exists for λ,

$$\lambda = \pm\left(\frac{[-C_2 \pm (C_2^2 - 4C_0 C_4)^{1/2}]}{2C_0}\right)^{1/2}. \tag{4-65}$$

There are four roots, some of which will be complex. Their sign depends on the sign and magnitude of the coefficients $C_0, \ldots, C_4$. C_0 is always positive since $J_\theta \geq S_x^2/m$ (Appendix B). C_4 is positive for velocities below the onset of divergence (Eq. 4-53). Only positive values of C_2 are of consequence in flutter analysis (Dowell et al., 1978, p. 79). If λ has negative real part, small perturbations will diminish in time and the equilibrium solution $y = \theta = 0$ will be stable. If λ has positive real part, perturbations will grow in time. We can numerically track solutions as functions of velocity, or any other parameter, and determine the onset of flutter instability from the neutral stability criterion Real $(\lambda) = 0$.

Pines (1958; also see Dowell et al., 1978, p. 78) solved Eq. 4-65 for the velocity at $\lambda = 0$ that produces onset of flutter,

$$\tfrac{1}{2}\rho U^2|_{\text{flutter}} = \frac{[-E \pm (E^2 - 4DF)^{1/2}]}{2D}, \tag{4-66}$$

where

$$D = \left((ma + S_x)c\frac{\partial C_L}{\partial \alpha}\right)^2, \qquad F = (mk_\theta + k_y J_\theta)^2 - 4(mJ_\theta - S_x^2)k_y k_\theta,$$

$$E = [-2(ma + S_x)(mk_\theta + k_y J_\theta) + 4(mJ_\theta - S_x^2)ak_y]c\frac{\partial C_L}{\partial \alpha}. \tag{4-67}$$

Equation 4-66 has two values. In order for flutter to occur, at least one of these must be positive and real. If both are positive and real, flutter occurs at the smallest real value. If neither is positive or real, no flutter occurs. As shown in Figure 4-18, flutter is a combination of the torsional and vertical modes with phase and amplitude that extracts energy from the flow when, as we have seen, either mode acting alone would be stable. At the onset of flutter, the natural frequencies of torsional and plunge modes unite to form a single frequency-coupled mode that does not exist without flow.

Fig. 4-18 Plunge-torsion flutter mode. Lift inputs energy into plunge (bending) and torsion lags by 90 degrees (Dat, 1971).

The flutter velocity depends on the geometrical relationships between the aerodynamic center, elastic center, and center of mass. One way of preventing flutter is to move the center of gravity to the elastic axis so that the plunge and torsional modes are not inertially coupled. This is called *mass balancing*. There is also no flutter when the center of mass is forward of the elastic axis, $S_x \le 0$ (Pines, 1958). This and other relationships for divergence and flutter are given in Table 4-5.

The flutter analysis used in Eqs. 4-57 through 4-67 is based on *steady* aerodynamic theory, i.e., the angle of torsion equals the angle of attack $\alpha = \theta$. The effect of airfoil velocity on the instantaneous angle of attack can be included by using *quasisteady* analysis with the approximation $\alpha = \theta + \dot{y}/U + R\dot{\theta}/U$ (Eqs. 4-42, 4-56; Fung, 1969, pp. 194–195). More advanced theories include the effect of the time lag between airfoil motion and imposition of force (Theodorsen, 1935; also see Garrick, 1969, and Fung, 1969, pp. 212–216). The time lag is associated with growth of circulation at small angles of attack and with the formation of vortices at large angles of attack (Naumowicz et al., 1989; Zaman et al., 1989; Ericsson and Reding, 1988). See Section 4.4.

Advances in flutter theory are discussed by Venkatesan and Friedmann (1986) and Dowell et al. (1978). Experimental aspects of flutter are discussed in NASA SP-415, "Flutter Testing Techniques," 1976, and by

Table 4-5 Flutter and divergence sensitivities of airfoils at small angles of attack to geometry and mass

	Center of gravity AFT of elastic axis	Center of gravity FORWARD of elastic axis
Aerodynamic center FORWARD of elastic axis	Flutter Divergence[a]	No flutter Divergence
Aerodynamic center AFT of elastic axis	Flutter No divergence	No flutter No divergence

[a] This geometry is shown in Fig. 4-16.

Ghiringhelli et al. (1987). Suppression of flutter through active control is discussed by Ostroff and Pines (1982) and Murthy (1979). Dowell and Ilgamov (1988) consider nonlinear effects in flutter. Reed (1966) reviews whirl flutter.

Exercises

1. Starting with Eqs. 4-45 and 4-46, derive Eqs. 4-62 through 4-65.
2. Clearly explain the differences between flutter, galloping, and vortex-induced vibration.
3. An airfoil with a chord of 6 ft (2 m) has its elastic axis located midway between the leading edge and the trailing edge. The torsional spring constant is $k_\theta = 1100$ ft-lb/rad/ft (5000 N-m/radian m) per unit length. The density of air is 0.075 lb/ft³ (1.2 kg/m³). What is the velocity for onset of divergence?
4. Using the programs mentioned in the Preface, verify Table 4-5 using the parameters of Table 4-6 as a starting point.

4.6. PREVENTION OF GALLOPING AND FLUTTER

Galloping and flutter are prevented by ensuring that the structure does not cross the stability boundaries for galloping or flutter given by Eqs. 4-16, 4-34, 4-43, 4-53, and 4-66. This can be accomplished by the following means.

1. *Stable aerodynamic contours.* If the slope of the aerodynamic force coefficients is stable (Fig. 4-2), then a bluff structure is stable, at least at small amplitudes. The unstable section of an ice-coated cable can be changed into a stable circular section by melting the ice with resistance heating. Table 4-1 and Figure 4-11 show that rectangles that have narrow face into the wind are more stable than those that have broad face into the wind, although both may experience torsional instability (Figs. 3-23 and 4-15). Table 4-5 shows that moving the aerodynamic center aft of the elastic center enhances the stability of airfoils at small angles of attack. This can be accomplished by adding a tail. Achieving aerodynamic stability of buildings and bridges is tricky at best because of the coupled vibration modes and multiple wind directions. Scanlan (1971) notes that it has apparently been impossible to create an economically practical bridge deck that is completely free of galloping or vortex-induced vibration. Many suspension bridges, including the Golden Gate in San Francisco, are stable by relatively narrow margins; see Section 3.7 and Tanaka and Davenport (1983).

2. *Stiffness or mass.* The velocity for onset of galloping (Eq. 4-16, 4-23, or 4-43) and flutter increases with the square root of both stiffness and mass, other effects being equal,

$$U_{\text{critical}} \sim m^{1/2}k^{1/2},\tag{4-68}$$

so adding either stiffness or mass will increase stability. Increasing the thickness of structural members increases both quantities and this is a very effective means of increasing stability. The Tacoma Narrows Bridge, which replaced the one that failed under wind action in 1940, is 50% heavier than the original, utilizes a truss four times deeper than the original, and has shown no tendency toward instability (Steinman and Watson, 1957). Increasing stiffness can move a structure above vortex resonance (Section 3.6). Addition of mass alone can be effective. Rowbottom (1979, 1981) shows that addition of mass to overhead power lines can suppress galloping. Addition of mass to the forward end of an airfoil can suppress flutter (Table 4-5). The critical velocity for the onset of flutter is minimum when the uncoupled, i.e. $S_x = 0$, natural frequencies of torsion and plunge are equal. Flutter velocity can be increased by increasing the separation of the torsion and plunge frequencies.

3. *Damping.* The velocity for the onset of galloping instability is proportional to structural damping. Cables, bridges, and towers tend to have low structural damping that can be raised by the introduction of material or structural dampers; see Chapter 8, Sections 8.4 and 8.5. Figure 4-19 shows two dampers. The Stockbridge damper of Figure

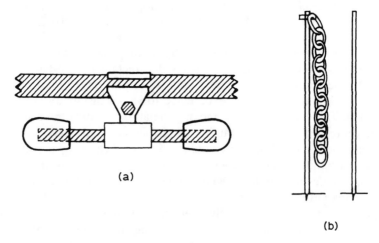

(a)

(b)

Fig. 4-19 Two damping devices: (a) a Stockbridge damper for powerlines; (b) a chain damper for masts and towers. Also see Chapter 8, Fig. 8-26.

4-19(a) consists of a piece of steel cable with weights at its ends, which is clamped to the power line and roughly tuned to the frequency of powerline vibration. It dissipates energy in its loose strands and tends to detune the system; see Dhotarad et al. (1978) and EPRI (1979). The chain impact damper of Figure 4-19(b) dissipated energy by inelastic collision with the surrounding structure. These simple devices are effective if their mass is approximately 5% of the mass of the damped structure. Flutter is relatively uninfluenced by damping because flutter instability arises from phase shifts between coupled modes rather than energy dissipation within a single mode.

4.7. EXAMPLE: FLUTTER ANALYSIS OF 1927 RYAN NYP

Figure 4-20 shows the wing rib section of Charles Lindbergh's 1927 Ryan NYP, *Spirit of St. Louis*. The wing ribs are spaced at 12 in (0.3 m) intervals along the 46 ft (14 m) wing span. Each wing extends 21 ft (6.4 m) outward from the fuselage. The chord is a constant 7 ft (2.1 m). The ribs are constructed of spruce as are the two full-length spars that provide the torsional and bending stiffness of the wing. The wing is covered with doped fabric. A flutter analysis is performed on this aircraft by developing the structural mass and stiffness properties of the wing and then applying the analysis of the previous section.

The density of spruce is $0.014 \, \text{lb/in}^3$ (0.4 gm/cc) and each rib section (Fig. 4-20) has a mass of 0.75 lb (0.34 kg). The forward spar has a mass of 3.36 lb/ft (5 kg/m) and the aft spar has a mass of 2.18 lb/ft (3.25 kg/m). The fabric is estimated to be 0.5 lb/ft (0.75 kg/m) and thus the total structural mass of the wing is 6.79 lb/ft (10.1 kg/m). Using an air density of $0.075 \, \text{lb/ft}^3$ ($1.2 \, \text{kg/m}^3$), the added mass of the wing is estimated as $m_a = \rho \pi c^2 / 4 = 2.89 \, \text{lb/ft}$ (4.31 kg/m), which is 30% of the total mass,

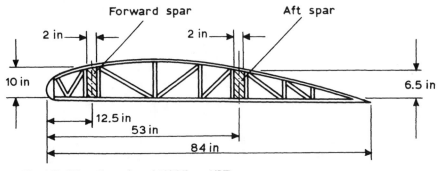

Fig. 4-20 Wing rib section of 1927 Ryan NYP.

Table 4-6 Flutter parameters of 1927 Ryan NYP

Parameter	Symbol	US units	SI units
Airfoil	—	Clark Y	Clark Y
Chord	c	84 in	2.13 m
Span of each wing	L	21 ft	6.4 m
Aspect ratio	AR	6.67	6.67
Slope of lift coefficient	$\partial C_L/\partial \alpha$	4.81	4.81
Characteristic span[a]	L_c	12.6 ft	3.84 m
Maximum speed	U_{max}	129 mph	57.8 m/sec
Air density	ρ	0.075 lb/ft^3	1.2 kg/m^3
Structural mass per unit span	m_s	0.57 lb/in	10.1 kg/m
Added mass per unit span	m_a	0.24 lb/in	4.3 kg/m
Total mass per unit span	m	0.807 lb/in	14.4 kg/m
Distance of aerodynamic center aft of L.E.	$c/4$	21 in	0.533 m
Distance of elastic center aft of L.E.	x_0	22.0 in	0.558 m
Distance aerodynamic center forward of Elas. center	a	1.0 in	0.0254 m
Distance C.G. aft of elastic center	S_x/m	11.7 in	0.297 m
Polar mass moment of inertia per unit span	J_θ	392 lb-in^2/in	4.52 kg-m^2/m
Bending stiffness at L_c	K_y	258 lb/in	4.52 × 10^6 N/cm
Torsional stiffness at L_c	K_θ	128000 in-lb/rad	1.46 × 10^4 N-m/rad
Bending stiffness per unit span	k_y	1.02 lb/in-in	7060 N/m-m
Torsional stiffness per unit span	k_θ	507 in-lb/rad-in	2280 N-m/rad-m
Uncoupled bending natural frequency, f_y	$(1/2\pi)(k_y/m)^{1/2}$	3.52 Hz	3.52 Hz
Uncoupled torsional natural frequency, f_θ	$(1/2\pi)(k_\theta/J_\theta)^{1/2}$	3.56 Hz	3.56 Hz
Divergence velocity, Eq. 4-53	U_{diver}	394 ft/sec	120 m/sec
Flutter velocity, Eq. 4-66	$U_{flutter}$	62.5 ft/sec	19 m/sec

[a] Span used in calculation of bending and torsional stiffness.

142

including added mass, of 9.68 lb/ft (14.4 kg/m) as shown in Table 4-6. The center of mass is calculated to be 33.7 in (0.856 m) aft of the leading edge.

The spars are contained by the ribs and they deform in unison as the wing bends and twists. The AIAA (1987) gives an expression for the location of the elastic axis due to bending of two rectangular sections that are held together by a web,

$$e = \frac{bI_2}{I_1 + I_2} = 8.82 \text{ in } (0.224 \text{ m}). \tag{4-69}$$

$I_1 = 166.7 \text{ in}^4$ $(6.94 \times 10^{-5} \text{ m}^4)$ is the area moment of inertia of the forward spar; $I_2 = 45.7 \text{ in}^4$ $(1.9 \times 10^{-5} \text{ m}^4)$ is the area moment of inertia of the aft spar; $b = 41 \text{ in}$ (104 cm) is the center-to-center distance between the spars; and e is the distance of the center of flexure aft of the centroid of the forward spar. The fabric contributes slightly to the torsional stiffness and it is estimated to move the net flexural center aft another inch to 22 in (55.8 cm) aft of the leading edge. The aerodynamic center is at one-quarter chord, just forward of the elastic axis.

The polar mass moment of inertia of the wing section is computed about the elastic axis and it includes structural and aerodynamic components. The structural components are dominated by the two spars and fabric to give 242 in²-lb/in (2.79 kg-m²/m). The added mass moment of inertia $J_a = \rho \pi c^4 / 128 = 53$ in²-lb/in (0.612 kg-m²/m) about a point equidistant between the leading and trailing edges and 150 in²-lb/in (1.72 kg-m²/m) about the elastic axis give a total polar mass moment of inertia per unit span of $J_\theta = 392$ in²-lb/in (4.52 kg-m²-m²/m).

The torsional and bending stiffness of the wind are calculated using beam theory. Since the wing is a continuous cantilever, the appropriate stiffness should be obtained from modal analysis of the continuous system as indicated in Appendix A in order to reduce the continuous system to the two-mode, two-dimensional model of Figure 4-16. Here we will use an approximation by calculating the wing stiffness at a characteristic span of $L_c = 0.6L = 12.6$ ft (3.84 m). The bending and torsional stiffness of the cantilever wing span at the characteristic span are

$$K_y = \frac{3E(I_1 + I_2)}{L_c^3} = 258 \text{ lb/in } (4.52 \times 10^4 \text{ N/m}),$$

$$K_\theta = \frac{CG}{L_c} = 128\,000 \text{ in-lb/rad } (1.45 \times 10^4 \text{ N-m/rad}),$$

where $E = 1.4 \times 10^6$ psi $(9.6 \times 10^9 \text{ Pa})$ is the modulus of elasticity of spruce, $G = 538\,000$ psi $(3.7 \times 10^9 \text{ Pa})$ is the shear modulus, and $C = 36.1 \text{ in}^4$ $(1.5 \times 10^{-5} \text{ m}^4)$ is the torsional constant. I_1 and I_2 are the moments of inertia of the spars. Dividing by the 21 ft (252 in, 6.4 m) span of the wing gives the distributed stiffness

$$k_y = \frac{K_y}{L} = 1.02 \text{ lb/in-in (7060 N/m-m)},$$

$$k_\theta = \frac{K_\theta}{L} = \text{in-lb/rad-in (2270 N-m/rad-m)}.$$

(4-70)

The distributed stiffness is the equivalent stiffness that the spars provide to each spanwise section.

The slope of the lift coefficient is a function of the aspect ratio of the wing. Von Mises (1959) gives the following approximate expression for the slope:

$$\frac{\partial C_L}{\partial \alpha} = \frac{2\pi}{1 + (2/AR)} = 4.81,$$

(4-71)

where $AR = 6.57$ is the aspect of the wing.

Reviewing Table 4-6 in comparison with Table 4-5, we see that since the aerodynamic center is forward of the elastic axis, the wing is susceptible to divergence, and because the center of gravity is aft of the elastic center, the wing is susceptible to flutter. The divergence velocity is calculated from Eq. 4-53. The flutter velocity is calculated from Eq. 4-56. The results in Table 4-6 show that the minimum velocity for onset of instability is 42.5 mph (19 m/sec, 62.5 ft/sec), which is within the maximum design speed. The designers of the *Spirit of St. Louis* recognized this possibility, and they stiffened the wing with two diagonal struts from the bottom of the fuselage that intercept the wing spars about 15 ft (5 m) from the fuselage. These spars provide a fourfold increase in bending and torsion stiffness, raising the flutter velocity above the aircraft design speed.

4.8. EXAMPLE: GALLOPING OF A CABLE

Often in the United States, Europe, the USSR, and Canada, power transmission line cables are coated with ice during winter months. The ice coating can form an unstable cross section that gallops in winter winds (see Cheers, 1950; Edwards and Madeyski, 1956; Richardson et al., 1965; Hunt and Richards, 1969; Novak et al., 1978). The following analysis is mostly adapted from Richardson et al. (1965). Reviews of transmission line galloping are given by the Electric Power Research Institute (1979), Dubey (1978), and Dhotarad et al. (1978).

An electrical power cable sags into a catenary between two supporting towers. The sag and the catenary strongly influence the natural frequencies and mode shapes of vibration (Blevins, 1984; Irvine and

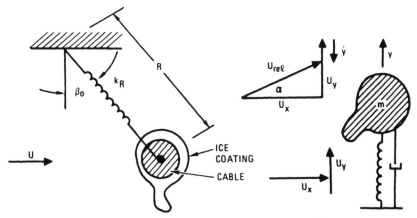

Fig. 4-21 Spring-supported, two-dimensional cable model and its representation as single-degree-of-freedom system with vertical and horizontal components of flow; k_R represents the elasticity of the cable.

Caughey, 1974). Figure 4-21 shows a section of cable catenary. Note that with the cable catenary, the structural axes formed with respect to the plane of the catenary are not aligned with the direction of the wind. The wind on the cable pushes the catenary to an angle β_0. If the lift component is negligible with respect to the drag, it can be easily shown that the blow-back angle β_0 is as follows:

$$\beta_0 = \arctan\left[\frac{\rho D U^2 C_D}{2mg}\right], \qquad (4\text{-}72)$$

where g is the acceleration due to gravity, C_D is the drag coefficient of the cable at the angle β_0, m is the mass per unit length of the cable with its ice coating, and D is the cable diameter, which is used as the characteristic dimension in forming the aerodynamic coefficients (Eq. 4-1).

The cable catenary can vibrate in several modes: (1) as a simple pendulum about $\beta = \beta_0$ at constant radius R; (2) on the radii of constant β_0 with R oscillating; (3) in torsion; or (4) in a combined mode. The torsion mode is discussed in Section 4.3.2. The pendulum and radial mode galloping instability will be further explored in this section by generalizing the approach of Section 4.2.

It is simpler and more general to analyze the spring-supported model shown on the right of Figure 4-21 than to directly approach the cable problem. The mean wind does not align with the structural axes. The axes x and y are relative to the model structural axes. The angle of flow relative to the model in Figure 4-21 is

$$\alpha = \arctan\left[\frac{(U_y - \dot{y})}{U_x}\right],$$

where U_x is the component of mean flow in the x direction and U_y is the component of flow in the y direction. The magnitude of the flow velocity relative to the model is

$$U_{\text{rel}}^2 = (U_y - \dot{y})^2 + U_x^2,$$

and the vertical force on the model is

$$F_y = \tfrac{1}{2}\rho U^2 D C_y,$$

where $U^2 = U_x^2 + U_y^2$ and the vertical force coefficient is given by Eq. 4-6. Instability can be produced by the change in the vertical force with motion of the model. In order to evaluate the effect of small motions on the vertical force, C_y is expanded in a series,

$$C_y = C_y(y = 0) + \frac{\partial C_y}{\partial \dot{y}}\dot{y} + O(\dot{y}^2).$$

The first term produces static deformation that does not directly enter the stability analysis. The second term is easily evaluated using the chain rule for derivatives:

$$\frac{\partial C_y(U_{\text{rel}}, \alpha)}{\partial \dot{y}} = \frac{\partial C_y}{\partial U_{\text{rel}}}\frac{\partial U_{\text{rel}}}{\partial \dot{y}} + \frac{\partial C_y}{\partial \alpha}\frac{\partial \alpha}{\partial \dot{y}}.$$

The result is

$$\frac{\partial C_y}{\partial \dot{y}}(\dot{y} = 0) = \frac{2\sin\alpha_0}{U}(C_L\cos\alpha_0 + C_D\sin\alpha_0)$$

$$+ \frac{\cos\alpha_0}{U}\left[\frac{\partial C_L}{\partial \alpha}\cos\alpha_0 - C_L\sin\alpha_0 + C_D\cos\alpha_0 + \frac{\partial C_D}{\partial \alpha}\sin\alpha_0\right],$$

where C_D, C_L, and their derivatives are evaluated at $\alpha = \alpha_0$, and α_0 is the angle of attack for no motion:

$$\alpha_0 = \arctan\left(\frac{U_y}{U_x}\right).$$

The linearized equation of motion of the model shown in Figure 4-21 for small vertical deformations about the equilibrium position is

$$m\ddot{y} + \left[2m\zeta\omega_y - \tfrac{1}{2}\rho U^2 D\frac{\partial C_y(\dot{y} = 0)}{\partial \dot{y}}\right]\dot{y} + ky = 0.$$

This equation has the stable solution $y = 0$ only if the coefficient of the $\dot{y}$ term is positive. The onset of instability occurs as this term passes through zero. Using this equation and previous equations, the onset of instability is calculated to occur at

$$\frac{U}{f_y D} = -\frac{4m(2\pi\zeta)}{\rho D^2}$$

$$\times \frac{1}{\sin^2 \alpha_0[2C_D + (C_L + \partial C_D/\partial\alpha)\cot\alpha_0 + (C_D + \partial C_L/\partial\alpha)\cot^2\alpha_0]},$$

$$(4\text{-}73)$$

where f_y is the natural frequency of the system. This equation applies exactly to galloping of uniform continuous systems in a single mode if the velocity is constant along the cable span. If the mass or flow velocity varies along the cable span, then the equivalent flow velocity and mass are given by Eqs. A-23 and A-16 of Appendix A.

Three limiting cases of the previous equation can be analyzed:

1. $U_y = 0$. If the flow is perpendicular to the structural axes, then $\alpha_0 = 0$ and Eq. 4-73 reduces to Eq. 4-16. The section can be *unstable* only if

$$\frac{\partial C_L}{\partial\alpha} + C_D < 0.$$

This is the den Hartog stability criterion.

2. *Radial cable mode.* The cable can vibrate in a radial mode along the line defined by $\beta_0 = $ constant. For this mode,

$$U_y = -U\sin\beta_0, \quad \text{and} \quad U_x = U\cos\beta_0.$$

These equations imply that

$$\sin\alpha_0 = -\sin\beta_0, \quad \cot\alpha_0 = -\cot\beta_0.$$

By substituting this equation into Eq. 4-73, the right-hand side can be positive and the system may become unstable only if

$$2C_D - \left(C_L + \frac{\partial C_D}{\partial\alpha}\right)\cot\beta_0 + \left(C_D + \frac{\partial C_L}{\partial\alpha}\right)\cot^2\beta_0 < 0.$$

3. *Pendulum cable mode.* The cable catenary can swing with a constant radius about the equilibrium position $\beta = \beta_0$. For this condition,

$$U_y = U\cos\beta_0, \quad \text{and} \quad U_x = U\sin\beta_0.$$

These equations imply that

$$\sin\alpha_0 = \cos\beta_0, \quad \cot\alpha_0 = \tan\beta_0.$$

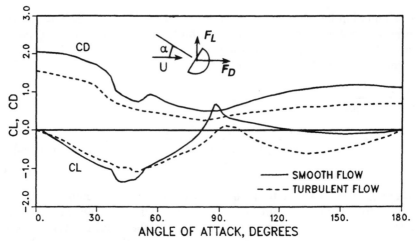

Fig. 4-22 Lift and drag coefficients for D-section in smooth and 11% turbulent flow; Re = 9×10^4 (Novak and Tanaka, 1974).

If these equations are substituted into Eq. 4-73, then for a possible pendulum mode instability,

$$2C_D + \left(C_L + \frac{\partial C_D}{\partial \alpha} \right) \tan \beta_0 + \left(C_D + \frac{\partial C_L}{\partial \alpha} \right) \tan^2 \beta_0 < 0.$$

Of course, if the natural frequencies of any of the three possible vibration modes are close (i.e., if the natural frequencies lie within the bandwidth $2\zeta f$ of each other), then instability in a combined mode is possible. The combined mode analysis is similar to the two-degree-of-freedom analysis in Section 4.3.2. The combined mode instability criterion will be a function of the difference in the natural frequency between the modes as well as the aerodynamic derivatives.

These stability criteria are of more theoretical than practical interest for ice-coated power lines because it is difficult to determine the aerodynamic derivatives of an ice coating in advance; and as the wind varies, so does the catenary angle β_0 and the angle of attack α_0. Practical methods for reducing the vibration are discussed in Section 4.6 and by the Electric Power Research Institute (1979).

Exercise

1. For what equilibrium angles is the D-section cylinder (Fig. 4-22) potentially unstable?

REFERENCES

Abbott, I. H., and A. E. von Doenhoff (1959) *Theory of Wing Sections,* Dover, New York.

AIAA Aerospace Engineer Design Guide (1987) AIAA, New York.

American Society of Civil Engineers (1955) "1952 Report of the Advisory Board on the Investigation of Suspension Bridges," *Transactions of the American Society of Civil Engineers,* **120,** Paper No. 2761, 721–781.

—— (1961) "Wind Forces on Structures, Final Report," *Transactions of the American Society of Civil Engineers,* **126,** 1124–1198.

—— (1977) "Tentative Recommendations for Cable-Stayed Bridge Structures," *ASCE Journal of the Structural Division,* **103,** 929–939.

Bearman, P. W., and S. C. Luo (1988) "Investigation of the Aerodynamics of a Square-Section Cylinder by Forced Oscillation," *Journal of Fluids and Structures,* **2,** 161–176.

Bearman, P. W., et al. (1984) "Experiments on Flow-Induced Vibrations of a Square-Section Cylinder," in *Symposium on Flow-Induced Vibrations,* Vol. 1, M. P. Paidoussis, ed., ASME, New York. Also see *Journal of Fluids and Structures,* **1,** 19–34 (1987).

Bellman, R. (1970) *An Introduction to Matrix Analysis,* McGraw-Hill, New York.

Bisplinghoff, R. L., and H. Ashley (1962) *Principles of Aeroelasticity,* Wiley, New York. Reprinted by Dover, New York, 1975.

Bisplinghoff, R. L., H. Ashley, and R. L. Halfman (1955) *Aeroelasticity,* Addison-Wesley, Reading, Mass.

Blevins, R. D. (1984) *Formulas for Natural Frequency and Mode Shape,* Robert E. Krieger Publishing Company, Malabar, Fla. Reprint of 1979 edition.

Blevins, R. D., and W. D. Iwan (1974) "The Galloping Response of a Two-Degree-of-Freedom System," *Journal of Applied Mechanics,* **41,** 1113–1118.

Bokaian, A., and F. Geoola (1984) "Effects of Vortex-Resonance on Nearby Galloping Instability," *ASCE Journal of Engineering Mechanics,* **111,** 591.

Cheers, F. (1950) "A Note on Galloping Conductors," National Research Council of Canada Report MT-14.

Dat, R. (1971) "Expose D'ensemble sur les Vibrations Aeroelastique," *La Houille Blanche,* No. 5, 391–399.

Den Hartog, J. P. (1956) *Mechanical Vibrations,* 4th ed., McGraw-Hill, New York. Reprinted by Dover, New York, 1984.

Dhotarad, M. S., N. Ganesan, and B. V. A. Rao (1978) "Transmission Line Vibrations," *Journal of Sound and Vibration,* **60,** 217–237.

Dowell, E. H. (1974) *Aeroelasticity of Plates and Shells,* Sijthoff & Noordhoff International Publishers, The Netherlands.

Dowell, E. H. et al. (1978) *A Modern Course in Aeroelasticity,* Sijthoff & Noordhoff International Publishers, The Netherlands.

Dowell, E. H., and M. Ilgamov (1988) *Studies in Nonlinear Aeroelasticity,* Springer-Verlag, New York.

Dubey, R. N. (1978) Vibration of Overhead Transmission Lines," *Shock and Vibration Digest,* **10,** 3–6.

Edwards, A. T., and A. Madeyski (1956) "Progress Report on the Investigation of Galloping," *Transactions of the American Institute of Electrical Engineers,* **75,** 666–683.

Electric Power Research Institute (1979) *Transmission Line Reference Book, Wind-Induced Conductor Motion,* Electric Power Research Institute, Palo Alto, Calif.

Ericsson, L. E., and J. P. Reding (1988) "Fluid Mechanics of Dynamic Stall, Part I, Unsteady Flow Concepts," *Journal of Fluids and Structures,* **2,** 1–33.

Federal Aviation Administration (1985) "Means of Compliance with Section 23.629, Flutter," Advisory Circular No: 23.629-1A.

Forsching, H. W. (1974) *Fundamentals of Aeroelasticity* (in German), Springer-Verlag, Berlin.

Fung, Y. C. (1969) *An Introduction to the Theory of Aeroelasticity,* Dover, New York.

Garrick, I. E. (1969) *Aerodynamic Flutter,* Vol. V, in AIAA Selected Reprint Series, New York.

Gartshore, I. S. (1984) "Some Effects of Upstream Turbulence on the Unsteady Lift Forces Imposed on Prismatic Two Dimensional Bodies," *Journal of Fluids Engineering,* **106,** 418–424.

Ghiringhelli, G. L. (1987) "A Comparison of Methods Used for the Identification of Flutter from Experimental Data," *Journal of Sound and Vibration,* **119,** 39–51.

Gordon, J. E. (1978) *Structures or Why Things Don't Fall Down,* Plenum, New York, p. 268.

Hajela, P. (ed.) (1986) *Recent Trends in Aeroelasticity, Structures and Structural Dynamics,* Papers from the R. L. Bisplinghoff Memorial Symposium, University of Florida Press, Gainesville, Fla.

Hunt, J. C. R., and D. J. W. Richards (1969) "Overhead Line Oscillations and the Effect of Aerodynamic Dampers," *Proceedings of the Institute of Electrical Engineers (London),* **116,** 1869–1874.

Irvine, H. M., and T. K. Caughey (1974) "The Linear Theory of Free Vibrations of a Suspended Cable," *Proceedings of the Royal Society (London) Series A,* **341,** 299–315.

Iwan, W. D. (1973) "Galloping of Hysteristic Structures," *ASCE Journal of the Engineering Division,* **99,** 1129–1146.

Konishi, I., et al. (1980) "Vortex Shedding Oscillations of Bridge Deck Sections," in *Practical Experiences with Flow-Induced Vibrations,* E. Naudascher and D. Rockwell, eds., Springer-Verlag, New York, pp. 619–632.

Lanchester, F. W. (1907) *Aerodynamics,* A. Constable & Co., London.

Laneville, A., and G. V. Parkinson (1971) "Effects of Turbulence on Galloping Bluff Cylinders," presented at the Third International Conference on Winds Effects on Buildings and Structures, Tokyo, Japan, Sept. 6–11, 1971.

Mahrenholtz, O., and H. Bardowicks (1980) "Wind-Induced Oscillations of Some Steel Structures," in *Practical Experiences with Flow Induced Vibrations,* E. Naudascher and D. Rockwell, eds., Springer Verlag, New York, pp. 643–670. Also see "Aeroelastic Problems at Masts and Chimneys," *Journal of Industrial Aerodynamics,* **4,** 261–272 (1979).

Meirovitch, L. (1967) *Analytical Methods in Vibrations,* Macmillan, New York.

Miley, S. J. (1982) "A Catalog of Low Reynolds Number Airfoil Data for Wind Turbine Applications," Department of Aerospace Engineering, Texas A&M University, College Station, Tex.

Minorsky, N. (1962) *Nonlinear Oscillations,* Van Nostrand, Princeton, N.J. Reprinted by Robert E. Kreiger Publishing Company, Malabar, Fla., 1983.

Modi, V. J., and J. E. Slater (1983) "Unsteady Aerodynamics and Vortex Induced Aeroelastic Instability of a Structural Angle Section," *Journal of Wind Engineering and Industrial Aerodynamics,* **11,** 321–334.

Murthy, P. N. (1979) "Some Recent Trends in Aircraft Flutter Research," *Shock and Vibration Digest,* **11**(5), 7–11.

Nakamura, Y., and T. Matsukawa (1987) "Vortex Excitation of Rectangular Cylinders with a Long Side Normal to the Flow," *Journal of Fluid Mechanics,* **180,** 171–191.

Nakamura, Y., and T. Mizota (1975) "Torsional Flutter of Rectangular Prisms," *ASCE Journal of the Engineering Mechanics Division,* **101,** 125–142.

Nakamura, Y., and Y. Tomonari (1977) "Galloping of Rectangular Prisms in a Turbulent Flow," *Journal of Sound and Vibration,* **52,** 233–241.

Nakamura, Y., and T. Yoshimura (1982) "Flutter and Vortex Excitation of Rectangular Prisms in Pure Torsion in Smooth and Turbulent Flows," *Journal of Sound and Vibration,* **84,** 305–317.

Nakamura, Y., and M. Nakashima (1986) "Vortex Excitation of Rectangular, H and T Cross Sections," *Journal of Fluid Mechanics* **163**, 149–169.

Naudascher, E., and D. Rockwell (eds.) (1980) *Practical Experiences with Flow Induced Vibrations*, Springer-Verlag, New York.

Naumowicz, T., et al. (1989) "Aerodynamic Investigation of Delta Wings with Pitch Amplitude," AIAA Paper 88-4332, AIAA, Washington, D.C.

Nayfeh, A. H., and D. T. Moot (1980) *Nonlinear Oscillations*, Wiley-Interscience, New York.

Novak, M. (1969) "Aeroelastic Galloping of Prismatic Bodies," *ASCE Journal of the Engineering Mechanics Division*, **96**, 115–142.

———— (1971) "Galloping and Vortex Induced Oscillations of Structures," in *Proceedings of the Third International Conference on Wind Effects on Buildings and Structures*, Tokyo, Japan, Sept. 6–11, 1971, Paper IV-16, p. 11.

Novak, M., and H. Tanaka (1974) "Effect of Turbulence on Galloping Instability," *ASCE Journal of the Engineering Mechanics Division*, **100**, 27–47.

Novak, M., A. G. Davenport, and H. Tanaka (1978) "Vibration of Towers due to Galloping of Iced Cables," *ASCE Journal of the Engineering Mechanics Division*, **104**, 457–473.

Obasaju, E. D. (1983) "Forced Vibration Study of the Aeroelastic Instability of a Square Section Cylinder near Vortex Resonance," *Journal of Wind Engineering and Industrial Aerodynamics*, **12**, 313–327.

Olivari, D. (1983) "An Investigation of Vortex Shedding and Galloping Induced Vibration on Prismatic Bodies," *Journal of Wind Engineering and Industrial Aerodynamics*, **11**, 307–319.

Ostroff, A. J., and S. Pines (1982) "Application of Modal Control to Wing-Flutter Suppression," NASA Technical Paper 1983, Langley Research Center.

Otauki, Y., et al. (1974) "A Note on the Aeroelastic Instability of a Prismatic Bar with Square Section," *Journal of Sound and Vibration*, **34**, 233–248.

Parkinson, G. V. (1971) "Wind-Induced Instability of Structures," *Philosophical Transactions of the Royal Society of London*, **A269**, 395–409.

Parkinson, G. V., and N. P. H. Brooks (1961) "On the Aeroelastic Instability of Bluff Cylinders," *Journal of Applied Mechanics*, **28**, 252–258.

Parkinson, G. V., and J. D. Smith (1964) "The Square Prism as an Aeroelastic Non-Linear Oscillator," *Quarterly Journal of Mechanics and Applied Mathematics*, **17**, 225–239.

Parkinson, G. V., and M. A. Wawzonek (1981) "Some Considerations of Combined Effects of Galloping and Vortex Resonance," *Journal of Wind Engineering and Industrial Aerodynamics*, **8**, 135–143.

Pigott, R., and J. M. Abel (1974) "Vibrations and Stability of Turbine Blades at Stall," *Journal of Engineering for Power*, **96**, 201–208 .

Pines, S. (1958) "An Elementary Explanation of the Flutter Mechanism," *Proceedings National Specialists Meeting on Dynamics and Aeroelasticity*, Institute on the Aeronautical Sciences, Ft. Worth, Tex.

Reed, W. H. (1966) "Propeller-Rotor Whirl-Flutter: A State-of-the-Art Review," *Journal of Sound and Vibration*, **4**, 526–544.

Reinhold, T. A., (ed.) (1982) *Wind Tunnel Modeling for Civil Engineering Applications*. Cambridge University Press, Cambridge.

Richardson, A. S. (1988) "Predicting Galloping Amplitudes," *ASCE Journal of Engineering Mechanics*," **14**, 716–723 and 1945–1952.

Richardson, A. S., J. R. Martucelli, and W. S. Price (1965) "Research Study on Galloping of Electrical Power Transmission Lines," in *Proceedings of the First International Conference on Wind Effects on Buildings and Structures*, Vol. II, Held in Teddington, England, pp. 612–686.

Rowbottom, M. D. (1979) "The Effect of an Added Mass on the Galloping of an Overhead Line," *Journal of Sound and Vibration*, **63**, 310–313.

Rowbottom, M. D. (1981) "The Optimization of Mechanical Dampers to Control Self-Excited Galloping Oscillations," *Journal of Sound and Vibration*, **75,** 559–576.

Scanlan, R. H. (1971) "The Suspension Bridge: Its Aeroelastic Problems," ASME Paper No. 71-Vibr-38.

—— (1980) "On the State of Stability Considerations for Suspension-Span Bridges Under Wind," in *Practical Experiences with Flow-Induced Vibrations*, E. Naudascher, ed., Springer-Verlag, New York, pp. 595–618.

Scruton, C. (1960) "Use of Wind Tunnels in Industrial Aerodynamic Research," AGARD Report 309.

Shiraishi, N., and M. Matsumoto (1983) "On Classification of Vortex-Induced Oscillation and its Application for Bridge Structures," *Journal of Wind Engineering and Industrial Aerodynamics*, **14,** 419–430.

Simiu, E., and R. H. Scanlan (1986) *Wind Effects on Structures*, 2nd ed., Wiley, New York.

Sisto, F. (1953) "Stall-Flutter in Cascades," *Journal of Aeronautical Sciences*, **20,** 598–604.

Skarachy, R. (1975) "Yaw Effects on Galloping Instability," *ASCE Journal of the Engineering Mechanics Division*, **101,** 739–754.

Slater, J. E. (1969) "Aeroelastic Instability of a Structural Angle Section," Ph.D. Thesis, University of British Columbia.

Steinman, D. B., and S. R. Watson (1957) *Bridges and Their Builders*, Dover, New York.

Struble, R. A. (1962) *Nonlinear Differential Equations*, McGraw-Hill, New York.

Studnickova, M. (1984) "Vibrations and Aerodynamic Stability of a Prestressed Pipeline Cable Bridge," *Journal of Wind Engineering and Industrial Aerodynamics*, **17,** 51–70.

Tai, J., C. T. Grove, and J. A. Robertson (1976) "Aeroelastic Response of Square and H-Sections in Turbulent Flows," ASME Paper 76-WA/FE-19, New York.

Tanaka, H., and A. G. Davenport (1983) "Wind-Induced Response of Golden Gate Bridge," *ASCE Journal of Engineering Mechanics Division*, **109,** 296–312.

Theodorsen, T. (1935) "General Theory of Aerodynamic Instability and the Mechanism of Flutter," NACA Report 496.

Timoshenko, S., and J. N. Goodier (1951) *Theory of Elasticity*, McGraw-Hill, New York, p. 334.

Venkatesan, C., and P. P. Friedmann (1986) "A New Approach to Finite-State Modeling of Unsteady Aerodynamics," *AIAA Journal*, **24,** 1889–1897.

Von Mises, R. (1959) *Theory of Flight*, Dover, New York, p. 243.

Wardlaw, R. L., and R. D. Blevins (1980) "Discussion on Approaches to the Suppression of Wind-Induced Vibration of Structures," in *Practical Experiences with Flow-Induced Vibration*, E. Naudascher, and D. Rockwell, eds., Springer-Verlag, New York, pp. 671–672.

Wardlaw, R. L., and C. A. Ponder (1970) "Wind Tunnel Investigation of the Aerodynamic Stability of Bridges," National Aeronautical Establishment Report LTR-LA-47, Ottawa, Canada.

Washizu, K., et al. (1978) "Aeroelastic Instability of Rectangular Cylinders in a Heaving Mode," *Journal of Sound and Vibration*, **59,** 195–210.

Williamson, C. H. K., and A. Roshko (1988) "Vortex Formation in the Wake of an Oscillating Cylinder," *Journal of Fluids and Structures*, **2,** 355–381.

Yashima, S., and H. Tanaka (1978) "Torsional Flutter in Stalled Cascade," *Journal of Engineering for Power*, **100,** 317–325.

Zaman, K. B., D. J. McKinzie, and C. L. Rumsey (1989) "A Natural Low-Frequency Oscillation of the Flow Over an Airfoil Near Stalling Conditions," *Journal of Fluid Mechanics*, **202,** 403–442.

Chapter 5

Instability of Tube and Cylinder Arrays

When a cylinder in an array of cylinders is displaced, the flow field shifts, changing the fluid forces on the cylinders. These fluid forces can induce instability if the energy input by the fluid force exceeds the energy expended in damping. The cylinders, or tubes, generally vibrate in oval orbits. In tube-and-shell heat exchangers, these vibrations are called *fluid elastic instability*. In arrays of power transmission lines, the vibrations are called *wake galloping*.

Fluid elastic instability is a dominant cause of tube failure in heat exchangers (see Halle et al., 1987; Stevens-Guille, 1974). Similarly, wake galloping instability of powerlines is a major concern in transmission-line design (Electric Power Research Institute, 1979). One can anticipate that arrays of pipelines rising from offshore oil wells will experience instability as the search for petroleum moves into deep-water, high-current tracks (Overvik et al., 1983).

5.1. DESCRIPTION OF FLUID ELASTIC INSTABILITY

In 1966, Roberts found that the fluid jets formed behind a single row of cylinders in cross flow (Fig. 5-1) coalesce in pairs. The pairing switched as a cylinder was displaced fore and aft, inducing an abrupt change in drag on alternate cylinders. Roberts found that a row of elastically mounted tubes responded to these forces and vibrated at large amplitude beyond a critical velocity. Roberts attributed the instability to the time lag between the tube displacement and the imposition of the destabilizing fluid force. Connors (1970) observed the instability of a row of tubes suspended on piano wires in a wind tunnel. He considered the instability to be the result of displacement-induced fluid forces generated by interaction between adjacent tubes as they vibrated in synchronous oval orbits.

One consequence of Connors' displacement mechanism is that a single flexible tube surrounded by rigid tubes is predicted to be stable (Blevins, 1974). Price et al. (1987) found this to be the case in a rotated square array with pitch-to-diameter ratio of 2.12, but Weaver and Yeung (1983) found that a single flexible tube could become unstable in a square array

153

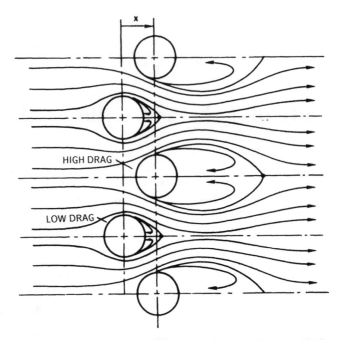

Fig. 5-1 Coupling of jets in the wake of a row of cylinders (Roberts, 1966).

of tubes with pitch-to-diameter ratio of 1.5. This suggests that both
Connors' displacement and Roberts' time-lag mechanisms may be
operative and that simple models are not sufficient to explain all aspects
of tube array instability. See reviews by Paidoussis and Price (1988), S. S.
Chen (1987), Paidoussis (1987), and Weaver and Fitzpatrick (1987).

Figure 5-2(a) shows the response of an array of metallic tubes to water
flow. The initial hump is attributable to vortex shedding (Chapter 3).
Figure 5-2(b) shows the response of an array of plastic tubes to air flow.
In both cases, the instability results in very large-vibration amplitudes
once a critical velocity is exceeded. In water, the onset of instability tends
to coincide with the vortex resonance with the tube natural frequency. In
air, the onset of instability generally occurs well above the vortex
resonance.

Once a critical cross flow velocity is exceeded, the vibration amplitude
increases very rapidly with the flow velocity U, usually as U^n where $n = 4$
or more, compared with an exponent $n \approx 1.5$ below the critical velocity
(see Chapter 7, Section 7.3.2, Eq. 7-55). The amplitude is not steady but
rather beats in time in a pseudorandom fashion (Chen and Jendrzejczyk,
1981). The tubes vibrate at or very near the tube natural frequency in the
natural modes of vibration (Chapter 2, Fig. 2-5; Chapter 7, Fig. 7-10).

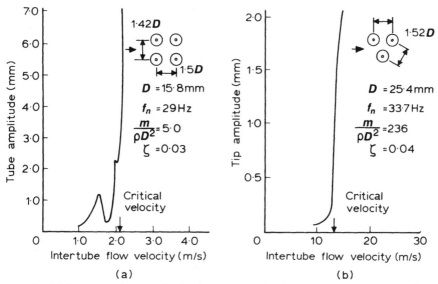

Fig. 5-2 Response of arrays of tubes to cross flow. (a) Brass tubes in water flow (Chen et al., 1978). (b) Plastic tubes in air flow (Soper, 1983).

The maximum amplitudes are often limited by clashing against adjacent tubes. Practically speaking, unstable tube motion causes failure.

As shown in Figure 5-3, the tubes generally vibrate in oval orbits somewhat synchronized with neighboring tubes. Orbits vary in shape from nearly straight lines to circles (Connors, 1970; Chen and Jendrzejczyk, 1981). As the tubes whirl in their orbits, they extract energy

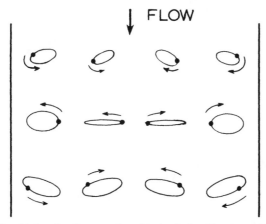

Fig. 5-3 Orbits of tube vibration past the critical velocity (Tanaka and Takahara, 1981).

from the flow and this, in part, induces instability (Section 5.2). Restricting the motion or introducing differences in natural frequency between one or more tubes sometimes increases the critical velocity for onset of instability (Weaver and Lever, 1977; Weaver and Korogannakis, 1983; Chen and Jendrzeczyk, 1981). Increases in critical velocity up to 60% have been obtained by detuning adjacent tubes, but in other cases no increase at all occurs, and a single flexible tube surrounded by rigid tubes may or may not become unstable depending apparently on the details of the array (Tanaka and Takahara, 1981; Weaver and Yeung, 1983; Price and Paidoussis, 1986; Price et al., 1987; Blevins et al., 1981). Often the onset of instability is more gradual in a tube array with tube-to-tube frequency differences than in an array of identical tubes, as shown in Figure 5-4.

Commercial heat-exchanger tube arrays are based on simple geometric patterns as shown in Figure 5-5. For tube arrays based on squares or equilateral triangles, the tube center-to-center spacing is referred to as pitch P. For more general arrangements, both a transverse center-to-center spacing T and the longitudinal center-to-center spacing L are specified. A reference fluid velocity U for flow through an array of tubes or cylinders is the average velocity through the minimum gap between tubes. For 30-degree triangle and square arrays (Fig. 5-5), the gap velocity is larger by the ratio $P/(P-D)$ than the free stream velocity approaching the tube array, where D is the tube outside diameter. Many authors use this reference gap velocity for all tube arrays

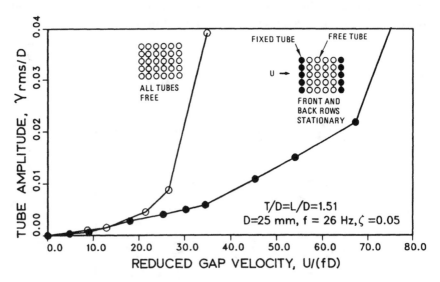

Fig. 5-4 Tube vibration amplitude as a function of gap flow velocity for (a) all tubes free and (b) front and rear rows held rigid; $m/\rho D^2 = 370$ (Blevins et al., 1981).

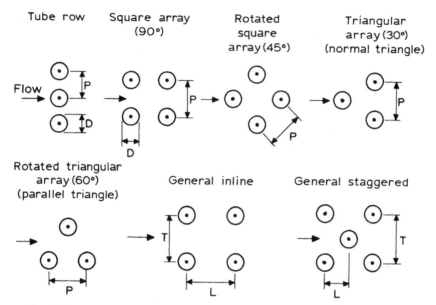

Fig. 5-5 Tube array pattern definitions.

based on a constant pitch P. For general staggered arrays, the minimum gap can be either the transverse gap or the diagonal gap. The diagonal gap is smaller if $L/D < [(2T/D) + 1]^{1/2}$.

The dimensionless analysis of Chapter 1 implies that the onset of instability is governed by the following dimensionless groups: reduced velocity $U/(fD)$, mass ratio $m/(\rho D^2)$, damping factor ζ, tube-to-tube spacing P/D and pattern, Reynolds number UD/ν, and upstream turbulence. The critical velocity for onset of instability can be expressed nondimensionally as a power function of these dimensionless groups,

$$\frac{U_{\text{crit}}}{f_n D} = C\left(\frac{m}{\rho D^2}\right)^a (2\pi\zeta)^b \left(\frac{P}{D}\right)^c \left(\frac{UD}{\nu}\right)^d \left(\frac{u'}{U}\right)^e \ldots, \qquad (5\text{-}1)$$

where a, b, c, d, e, and C are approximately constant over limited ranges of the parameters. f_n is the natural frequency of the cylinder or tube. Tanaka and Takahara (1981), Paidoussis (1982), Weaver and El-Kashlan (1981), and Weaver and Fitzpatrick (1987) suggest that the exponents a and b are between 0 and 0.5 and they may be dependent on other parameters.

The available data do not clarify the dependence of the critical velocity on Reynolds number or turbulence level. For example, the data of Franklin and Soper (1977) show that turbulence *decreases* critical velocity, while the data of Southworth and Zdravkovich (1975) show that turbulence *increases* critical velocity. Many individual studies (see, for example, Soper, 1983, and Yeung and Weaver, 1983) indicate that instability is dependent on the array geometry and, to a lesser degree, tube-to-tube spacing. However, when large amounts of data are assembled, at best a weak dependence on array geometry or spacing emerges. Thus, it seems reasonable to utilize a simpler expression than Eq. 5-1 to fit available data and consider the scatter about the fitted curve as an indication of the uncertainty in the accuracy of the prediction.

Data for critical velocity were fitted to the following equation form, which is suggested by theoretical analysis (Section 5.2):

$$\frac{U_{\text{crit}}}{f_n D} = C \left[\frac{m(2\pi\zeta)}{\rho D^2} \right]^a. \tag{5-2}$$

D = outside diameter,

f_n = natural frequency of tube or cylinder in hertz (Blevins, 1979); ordinarily the lowest-frequency mode is most susceptible,

m = mass per unit length of tube, including internal fluid and external added mass (Table 2-2),

U_{crit} = velocity averaged across the minimum gap between tubes that corresponds to onset of instability (Fig. 5-2),

ζ = tube damping factor; if the tube has some intermediate supports, then ζ often lies between 0.01 and 0.03 (Chapter 8, Section 8.4.4, and Fig. 8-19); when vibration amplitude is large, support plate interaction greatly increases, and damping can rise to 0.05 and more; if there are no intermediate supports, then ζ can be as low as 0.001 (Section 8.3.2),

ρ = fluid density; in two-phase flow an equivalent homogeneous density (Eq. 5-4) is used.

The coefficient C and exponent a are obtained by fitting experimental data. One hundred and seventy measurements of critical velocity for arrays with various patterns and pitch-to-diameter ratios between 1.1 and 2.0 in both single-phase air and water are shown in Figure 5-6(a). The kink in the data at $m(2\pi\zeta)/(\rho D^2) \approx 0.7$ suggests that two different mechanisms may be operative, so each range was fitted separately. A least-squares curve fit of these data gives (Blevins, 1984a)

Mass damping	C_{mean}	$C_{90\%}$	a
$m(2\pi\zeta)/(\rho D^2) < 0.7$	3.9	2.7	0.21
$m(2\pi\zeta)/(\rho D^2) > 0.7$	4.0	2.4	0.5

C_{mean} is the mean fitted line. $C_{90\%}$ is a statistical lower bound. For $m(2\pi\zeta)/(\rho D^2) > 0.7$, there are sufficient data to fit array patterns (Fig. 5-5) by type. Using an exponent of $a = 0.5$, as suggested by displacement

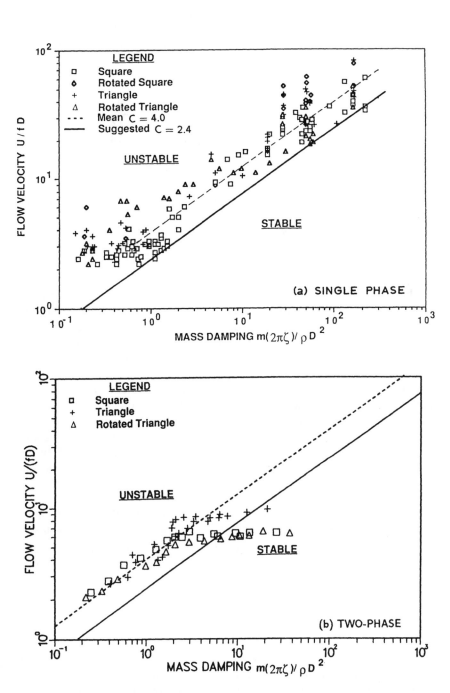

Fig. 5-6 Onset of instability as a function of mass damping parameter: (a) in single-phase fluids (Blevins, 1984a; redrawn with correction by T. M. Mulcahy in 1988); (b) in two-phase mixtures (Pettigrew et al., 1988).

159

mechanism theory (following section) and the previous fit, the coefficients C are as follows:

C	Triangle	Rotated triangle	Rotated square	Square	All arrays	Tube row
C_{mean}	4.5	4.0	5.8	3.4	4.0	9.5
$C_{90\%}$	2.8	2.3	3.5	2.4	2.4	6.4

The mean fitted line for all arrays is shown by the dashed line in Figure 5-6. There is scatter about the fitted line. The scatter is large enough that statistical arguments suggest that the results for various arrays may not significantly differ from one another.

Gaussian statistics imply that 90% of the data points will fall above a line 1.28 standard deviations below the mean. This line is denoted by $C_{90\%}$. For example, for all arrays the standard deviation of the fit is 32% of C_{mean}, thus $C_{90\%} = (1 - 1.28 \times 0.32)C_{mean} = 2.4$. $C = 2.4$ is suggested by the American Society of Mechanical Engineers Boiler and Pressure Vessel Code, Section 3, Appendix N for use in design. It is shown by the solid line in Figure 5-6. While data for tube rows (a line of tubes perpendicular to the flow, Fig. 5-5) were not included in the fit for all arrays, the data suggest $C_{90\%} = 6.4$ is appropriate for tube row design with $m(2\pi\zeta)/(\rho D^2) > 5$.

Cylinder or tube arrays in water generally have $m(2\pi\zeta)/(\rho D^2) < 2$ and critical reduced velocities $1 < U_{crit}/(fD) < 5$ (Fig. 5-6; Weaver and Fitzpatrick, 1987). In this reduced velocity range, substantial vortex-induced vibrations are predicted. As a result, vortex shedding and its associated time-lag effects are intertwined with instability in water flow, although the presence of adjacent tubes and turbulence within the bundle tends to diminish vortex-induced response (Chapter 3, Fig. 3-21; also Chapter 7, Fig. 7-12, Chapter 9, Fig. 9-14).

Fluid elastic instability occurs in two-phase flow (Heilker and Vincent, 1981; Pettigrew et al., 1984). Fifty-five data points of critical velocity measured by Pettigrew et al. (1988) for air–water mixtures and normal triangle, rotated triangle, and square arrays with pitch-to-diameter ratios of 1.32 and 1.47 are shown in Figure 5-6(b). The void fractions ranged from 5% to 99% air. The void fraction is

$$\epsilon_g = \frac{\dot{V}_g}{\dot{V}_g + \dot{V}_l}, \tag{5-3}$$

where $\dot{V}_g$ is the volumetric flow of gas and $\dot{V}_l$ is the volumetric flow of liquid. The homogeneous density used in the calculation of the mass-damping parameter is the average density of the two-phase mixture on a volumetric basis,

$$\rho = \rho_l(1 - \epsilon_g) + \rho_g\epsilon_g, \tag{5-4}$$

where ρ_l is the density of the liquid phase and ρ_g is the density of the gas phase. The corresponding homogeneous velocity is $U = (\dot{V}_l + \dot{V}_g)/A$, where A is the reference cross-sectional area through which the flow passes.

Exercises

1. Plot the critical velocities of Figure 5-2 on Figure 5-6.

2. Inspect Figure 5-3 closely. What is the phase relationship between adjacent tubes in a row, between every other tube in a row, and between a tube and the tube behind it?

.3. A steam condenser consists of an array of titanium heat-exchanger tubes with 0.75 in (19 mm) outside diameter and 0.035 in (0.89 mm) wall thickness, spaced on an equilateral triangular grid with center-to-center spacing of 1.0 in (25.4 mm). The tubes are supported at 23.75 in (603 mm) intervals by passing through over-sized holes in support plates. During a blow-down condition, steam at 400 ft/sec (122 m/sec) free field velocity and a temperature of 440°F (227°C) approaches the tube array at a density of 0.006 lb/ft³ (0.088 kg/m³). The tubes can either be empty or filled with water (density = 60 lb/ft³, 882 kg/m³). Assuming that the supports act as pinned joints, and using titanium properties of $E = 14 \times 10^6$ psi (94×10^9 Pa) and density = 0.15 lb/in³ (4.54 gm/cm³), calculate the tube natural frequencies and show that the fundamental natural frequency of the empty tube is 135 Hz. Determine the steam velocity between tubes and calculate $U/(f_1 D)$ for both empty and water-filled tubes. Assuming a damping of $\zeta = 0.01$, calculate the onset of instability. Will there be instability? What will happen to the tubes?

5.2. THEORY OF FLUID ELASTIC INSTABILITY

Analysis of fluid elastic instability requires an accurate theoretical description of the unsteady fluid forces imposed on a tube in an array of vibrating tubes (Fig. 5-3). This very difficult problem has not yet been fully resolved. In fact, none of the current models are capable of predicting the instability of an untested heat exchanger tube array with better accuracy than the band of data in Figure 5-6 (see Mulcahy et al., 1986). However, theory does provide insight into the nature of the instability and estimates for instability when data are not available.

The displacement mechanism model for fluid elastic instability hypothesizes that the fluid force is linear with cylinder displacement; two force coefficients must be determined experimentally. S. S. Chen (1983; see Eq. 5-31) includes linear displacement, velocity, and acceleration terms in his model. These coefficients will vary with tube pattern and measuring the large number of coefficients requires extensive testing. Moreover, the kink in the curve fit and data of Figure 5-6 imply that more than one fluid mechanical mechanism may be active in producing the onset of instability (Paidoussis and Price, 1988). In approximate terms, the instability for $m(2\pi\zeta)/\rho D^2 > 0.7$ can be associated with a displacement mechanism and that at smaller values of the mass damping parameter with a time-lag, negative damping, or, perhaps, vortex shedding mechanism as reviewed in the following sections.

5.2.1. Displacement Mechanism

A two-dimensional model for a tube row is shown in Figure 5-7. (The terms tube and cylinder are interchangeable in this chapter since it is the flow over the exterior of the cylinder or tube that is of concern.) x_i and y_i represent the longitudinal and transverse displacement of the i tube from its equilibrium position in the row; k_x and k_y are the stiffnesses of the spring supports parallel and normal to the free stream flow; ζ_x and ζ_y are the coefficients of viscous damping of each cylinder parallel and normal to the free stream.

If a tube is slightly displaced in a tube row, the flow pattern changes and the steady fluid force on the tube changes. The flow pattern through a tube array is a function of the positions of the tubes relative to each other, so it is reasonable to assume that the change in fluid force on one displaced tube is a function of its displacements relative to the displacements of the other tubes. It is also reasonable to assume that a

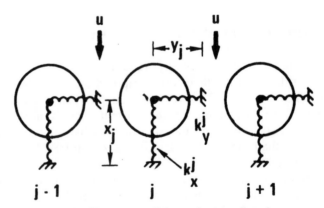

Fig. 5-7 Tube row model (dampers parallel to springs not shown).

tube in a regular array of tubes will interact primarily with the nearest neighboring tubes. Thus, the change in steady fluid force per unit length on the j tube in the x and y directions ($F^j_{x,y}$) can be written as a function of the j tube displacements (x_j, y_j) relative to the displacements of the neighboring $j + 1$ and $j - 1$ tubes:

$$F^j_x = \rho U^2 g_x(x_{j+1} - x_j, x_j - x_{j-1}, y_{j+1} - y_j, y_j - y_{j-1})/4,$$

$$F^j_y = \rho U^2 g_y(x_{j+1} - x_j, x_j - x_{j-1}, y_{j+1} - y_j, y_j - y_{j-1})/4, \qquad (5\text{-}5)$$

where ρ is fluid density and U is the flow velocity referenced to the minimum nominal gap between tubes. The functions g_x and g_y have units of 1/length and they express the variation in fluid force on a tube due to motion relative to neighboring tubes. Theoretical expressions for these functions, based on potential flow theory, have been developed by Balsa (1977), Lever and Weaver (1982), and Paidoussis et al. (1984). These functions have been measured experimentally by Connors (1970), Tanaka and Takahara (1981), Hara (1989), and Price and Valerio (1989) for particular tube arrays.

Symmetry of the tube row requires that the fluid force have certain symmetries for small tube displacements about the equilibrium position $x = y = 0$. As a consequence,

$$\frac{\partial g_{x,y}}{\partial(x_j - x_{j-1})_{x=y=0}} = \mp \frac{\partial g_{x,y}}{\partial(x_{j+1} - x_j)_{x=y=0}} = K_{x,y}, \qquad (5\text{-}6)$$

$$\frac{\partial g_{x,y}}{\partial(y_j - y_{j-1})_{x=y=0}} = \pm \frac{\partial g_{x,y}}{\partial(y_{j+1} - y_j)_{x=y=0}} = C_{x,y}. \qquad (5\text{-}7)$$

The notation $g_{x,y}$, and so on, means each expression represents two equations, a first one with the x components and then a second with the y components. The plus and minus signs are used in Eq. 5-6 for the y and x components, respectively, and in Eq. 5-7 for the x and y components, respectively.

In order to gain physical understanding of Eqs. 5-6 and 5-7, the three tube rows shown in Figure 5-8 are considered. In diagram (a), an end tube is displaced slightly normal to the free stream. If the fluid force on the middle tube corresponds to the vector (a) for diagram (a), then rotating the row 180 degrees about an axis parallel to the free stream produces the pattern (b) and the corresponding force vector (b). Thus, for the displacement from (a) to (b), the y component of the fluid force on the middle tube changes sign, but the x component is unchanged. A similar transformation occurs from (c) to (d). These symmetries imply Eqs. 5-6 and 5-7.

If the fluid force varies sufficiently smoothly with cylinder displacement, then only linear terms in the fluid force function (Eq. 5-5)

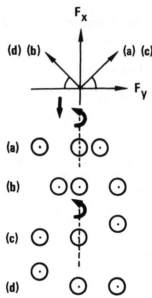

Fig. 5-8 Symmetries in the fluid force on a tube row.

need to be retained in the stability analysis. The linearized component of fluid force on the j tube (Eq. 5-5) in a power series using Eqs. 5-6 and 5-7:

$$F_x^j = \tfrac{1}{4}\rho U^2[K_x(x_j - x_{j-1}) - K_x(x_{j+1} - x_j) + C_x(y_j - y_{j-1}) + C_x(y_{j+1} - y_j)],$$
(5-8)

$$F_y^j = \tfrac{1}{4}\rho U^2[K_y(x_j - x_{j-1}) + K_y(x_{j+1} - x_j) + C_y(y_j - y_{j-1}) - C_y(y_{j+1} - y_j)].$$
(5-9)

The linear equations describing elastic motion of the j tube parallel and perpendicular to the free stream in response to these forces are

$$m\ddot{x}_j + 2m\zeta_x^j\omega_x^j\dot{x}_j + k_x^j x_j = \tfrac{1}{4}\rho U^2[K_x(-x_{j+1} - x_{j-1} + 2x_j)$$
$$+ C_x(y_{j+1} - y_{j-1})], \qquad (5\text{-}10)$$

$$m\ddot{y}_j + 2m\zeta_y^j\omega_y^j\dot{y}_j + k_y^j y_j = \tfrac{1}{4}\rho U^2[C_y(-y_{j+1} - y_{j-1} + 2y_j)$$
$$+ K_y(x_{j+1} - x_{j-1})]; \qquad (5\text{-}11)$$

ω_x^j and ω_y^j are the circular natural frequencies of the tube; m is the mass per unit length including added mass (Chapter 2); ζ_x and ζ_y are damping factors, which, in general, will include both structural and fluid dynamic components (Chapter 8).

It is shown theoretically in Appendix A that this system of differential equation is identical to that generated for a row of tubes bending elastically between supports if a single spanwise mode is excited in both coordinate directions. If the flow velocity varies along the span, the effective flow velocity is given by

$$U^2 = \frac{\displaystyle\int_0^L U^2(z)\bar{y}^2(z)\,dz}{\displaystyle\int_0^L \bar{y}^2(z)\,dz},$$ (5-12)

where $U(z)$ is the flow velocity at each spanwise point z, and $\bar{y}(z)$ is the spanwise mode shape extending from $z = 0$ to $z = L$. If the tube has multiple supports, the span $z = 0$ to $z = L$ refers to the entire length of tube, not just the distance between supports. Franklin and Soper (1977), Connors (1978), and Weaver and Goyder (1989) report reasonable agreement of data with this expression. If the mass varies along the span, the effective mass is given by Eq. A-16 of Appendix A. If the flow is inclined to the tube axis, the normal component of flow velocity should be used.

Pettigrew et al. (1978) report that in one case the eighth mode of vibration in a multispan heat-exchanger tube bank became unstable before the fundamental tube mode. The instability of the higher mode was the result of nonuniform flow distribution, which produces a higher effective velocity (Eq. 5-12) for a higher mode than for the fundamental mode.

Stability analysis of the tube motion (Eqs. 5-10 and 5-11) is considerably simplified using an assumed mode of tube motion. As shown in Figure 5-3, the tubes tend to vibrate in semisynchronized, oval orbits. This mode of tube-to-tube vibration in Figure 5-3 is

$$x_{j+1} = -x_{j-1}, \qquad y_{j+1} = -y_{j-1}.$$ (5-13)

That is, every other tube is 180 degrees out of phase across a row of tubes. This mode has frequently been observed (Connors, 1970; Zdravkovich and Namork, 1979; Tanaka and Takahara, 1981). A second simplifying assumption is that the coefficients $K_x = C_y = 0$. As can be seen from Eqs. 5-10 and 5-11, with the mode of Eq. 5-13, terms with the coefficients K_x and C_y do not input energy into the tubes. They serve only to shift natural frequency, and experiments have shown that the shifts of natural frequency are very small.

With these two assumptions, the equations describing the motion of the j tube parallel to the flow and of the $j + 1$ tube perpendicular to the flow are

$$m\ddot{x}_j + 2m\zeta_x^j \omega_x^j \dot{x}_j + k_x^j x_j = \tfrac{1}{2}\rho U^2 C_x y_{j+1},$$ (5-14)

$$m\ddot{y}_{j+1} + 2m\zeta_y^{j+1} \omega_y^{j+1} \dot{y}_{j+1} + k_y^{j+1} y_{j+1} = -\tfrac{1}{2}\rho U^2 K_y x_j.$$ (5-15)

These are two equations in two unknowns. Solutions are sought such that the displacements either grow or decay exponentially in time:

$$x_j = \bar{x}_j e^{\lambda t}, \qquad y_{j+1} = \bar{y}_{j+1} e^{\lambda t}, \qquad (5\text{-}16)$$

where $\bar{x}_j$, $\bar{y}_{j+1}$, and λ are constants. Substituting Eq. 5-16 into Eqs. 5-14 and 5-15 results in equations that can be put into matrix form:

$$\begin{bmatrix} m\lambda^2 + 2\zeta_x^j m\omega_x^j \lambda + k_x^j & -\rho U^2 C_x/2 \\ \rho U^2 K_y/2 & m\lambda^2 + 2\zeta_y^{j+1} m\omega_y^{j+1} \lambda + k_y^{j+1} \end{bmatrix} \begin{Bmatrix} \bar{x}_j \\ \bar{y}_{j+1} \end{Bmatrix} = 0. \quad (5\text{-}17)$$

For solutions other than the zero solution $\bar{x} = \bar{y} = 0$, the determinant of the matrix must be zero (Bellman, 1970). Setting the determinant to zero gives

$$(m\lambda^2 + 2\zeta_x^j m\omega_x^j \lambda + k_x^j)(m\lambda^2 + 2\zeta_y^{j+1} m\omega_y^{j+1} \lambda + k_y^{j+1}) + \tfrac{1}{4}\rho^2 U^4 C_x K_y = 0.$$
$$(5\text{-}18)$$

Expanding this expression gives a fourth-order polynomial in λ,

$$\lambda^4 + a_1 \lambda^3 + a_2 \lambda^2 + a_3 \lambda + a_4 = 0, \qquad (5\text{-}19)$$

where the coefficients $a_1, \ldots, a_4$ are functions of ρ, U, ζ, ω, C_x, and K_y. Whiston and Thomas (1982) have shown that a similar polynomial results when this analysis is applied to arrays of tubes as well as tube rows.

For stability of the zero solution, λ must have only negative real parts so that small-amplitude perturbations die out in time. The Hurwitz criterion (Bellman, 1970) gives conditions for this:

$$a_1 > 0, \qquad a_4 > 0, \qquad a_1 a_2 > a_3, \qquad a_3(a_1 a_2 - a_3) > a_1^2 a_4. \quad (5\text{-}20)$$

For positive damping and $C_x K_y > 0$, these criteria can be expressed as a critical minimum velocity for the onset of instability (Blevins, 1974):

$$\frac{U_{\text{crit}}}{2\pi (f_x^j f_y^{j+1})^{1/2} D} = \frac{(2m/\rho D^2)^{1/2}}{(C_x K_y)^{1/4}} \left(\frac{\zeta_x^j \zeta_y^{j+1} f_x^j f_y^{j+1}}{(\zeta_x^j f_x^j + \zeta_y^{j+1} f_y^{j+1})^2} \right)^{1/4}$$

$$\times \left[\left(\frac{f_y^{j+1}}{f_x^j} - \frac{f_x^j}{f_y^{j+1}} \right)^2 + 4 \left(\frac{\zeta_y^{j+1}}{f_y^{j+1}} + \frac{\zeta_x^j}{f_x^j} \right) (\zeta_x^j f_x^j + \zeta_y^{j+1} f_y^{j+1}) \right]^{1/4}. \quad (5\text{-}21)$$

This complex expression is explored by considering simplified cases.

If the damping in the x and y directions is equal, $\zeta_x = \zeta_y = \zeta$, and relatively small, and the natural frequencies are close but not equal $[f_x/f_y = 1 + O(\zeta)$, where $O(\zeta)$ means order of $\zeta]$, then the critical reduced velocity falls to

$$\frac{U_{\text{crit}}}{(f_x f_y)^{1/2} D} = \frac{2^{3/2} \pi}{(C_x K_y)^{1/4}} \left(\frac{m}{\rho D^2} \right)^{1/2} \left[\left(1 - \frac{f_x}{f_y} \right)^2 + 4\zeta^2 \right]^{1/4}. \quad (5\text{-}22)$$

This equation predicts that the critical velocity for onset of instability increases with damping and with detuning between tubes. Both effects have been seen in the experimental data discussed in the previous section.

If all the tubes are identical in frequency and damping so that $f_x = f_y = f$ and $\zeta_x = \zeta_y = \zeta$, then Eq. 5-21 reduces to a simple expression that was first derived by Connors (1970):

$$\frac{U_{\text{crit}}}{fD} = C\left(\frac{m(2\pi\zeta)}{\rho D^2}\right)^{1/2}, \tag{5-23}$$

where $C = 2(2\pi)^{1/2}/(C_x K_y)^{1/4}$. For a tube row with a center-to-center spacing of $1.41\,D$, Connors measured $(C_x K_y)^{1/4} = 0.508$. This equation agrees well with tube array data taken for values of the mass damping parameter $m(2\pi\zeta)/\rho D^2 > 1.0$, which usually corresponds to tubes exposed to gas flows (section 5.1).

Equations 5-21 and 5-22 predict that differences in natural frequency between tubes will increase the critical velocity. The displacement theory also leads to the prediction that if a single flexible tube is surrounded by rigid tubes (i.e., if either f_x or f_y is infinite), then the flexible tube is always stable. However, experiments (Section 5.1) show that frequency differences between tubes do not always increase the critical velocity, and a single tube surrounded by rigid tubes can become unstable. Moreover, in dense fluids, notably water, the exponent of $m(2\pi\zeta)/(\rho D^2)$ is less than the predicted $\frac{1}{2}$; a value of 0.2 is a better fit to the data of Section 5.1. These shortcomings of the displacement theory suggest that additional fluid phenomena must influence tube array instability.

Exercises

1. Using Eqs. 5-6 and 5-7, derive the equations of motion for two adjacent identical, flexible tubes in a tube row in which all other tubes are rigid. There are four equations, two for each tube.

2. Solve the equations in Exercise 1 for the critical velocity. Assume that the damping and natural frequencies of both tubes in both directions are identical and that $K_x = C_y = 0$. Note that it is not necessary to assume a mode shape. The answer is given by Blevins (1974).

3. Equations 5-21 through 5-23 predict that the critical velocity is approximately proportional to the square root of the mass damping parameter if the tubes have the same natural frequencies, but Figure 5-6 shows that this relationship fails if the mass damping is below about 1. A possible explanation is that at low mass damping the added mass coupling becomes significant, and fluid-coupled vibration modes arise (Section 2.3). Is it possible that these coupled modes introduce frequency differences that explain the decrease in slope for low-mass damping?

5.2.2. Velocity Mechanism

The displacement mechanism (Eqs. 5-6 and 5-7) assumes that the fluid responds instantaneously to a change in tube position. In fact, there will be a small but finite time lag. Some estimates for the time lag between tube motion and fluid force are

$$\tau = \begin{cases} 10\,D/U & \text{Roberts (1966)} \\ P/U & \text{Lever and Weaver (1982)} \\ D/U & \text{Price and Paidoussis (1986)} \end{cases} \quad (5\text{-}24)$$

Roberts' results come from experimental observation of jet switch behind a row of tubes (Fig. 5-1) with centers spaced 1.5 diameters apart; U is the free stream velocity upstream of the tube row. In both the Lever and Weaver and Price and Paidoussis expressions, U is the velocity between tubes and time lag refers not to jet switching but to the time delay of displacement-induced forces in arrays of tubes. These time delays are of the same order as the period of vortex shedding (Chapter 3, Eq. 3-2; period = 1/frequency), which suggests that coherent vortical fluid structures thrown off by the vibrating cylinders strongly influence the fluid forces; see Sections 4.4 and 3.7.

Another fluid phenomenon not included in the displacement mechanism is fluid damping. As a tube vibrates in the flow field, the velocity of the fluid relative to the tube is increased and the result is, in part, a component of fluid damping. The fluid damping for a single vibrating cylinder is shown in Chapter 8, Section 8.2.2, using quasisteady theory, to result in the following components of fluid damping:

$$2\pi\zeta_x = \frac{C_D}{2}\frac{\rho D^2}{m}\frac{U}{fD}, \qquad 2\pi\zeta_y = \frac{C_D}{4}\frac{\rho D^2}{m}\frac{U}{fD}. \quad (5\text{-}25)$$

C_D is the tube drag coefficient based on the velocity U and the tube diameter D. The drag on a tube is $F_D = \frac{1}{2}\rho U^2 D C_D$. For an array of tubes, drag can be related to the pressure drop by $F_D = T\,\Delta p$, where T is the transverse tube-to-tube spacing perpendicular to the flow and Δp is the pressure drop across a single row of tubes. Blevins (1979) and Roberts (1966) have shown that inclusion of these fluid damping terms in the displacement instability formulation causes the exponent a in Eq. 5-2, to tend to values less than 0.5 at the lower values of mass damping, as suggested by Figure 5-6.

Figure 5-9 shows the transverse drag-induced damping in comparison with experimental measurement. The value $C_D = 0.8$ has been used in the figure based on expected pressure drop through the array. Theory (Eq. 5-25) is an excellent predictor of the damping up to about 2 m/sec flow velocity. At this point the damping decreases sharply and approaches a negative value as the tube array approaches instability.

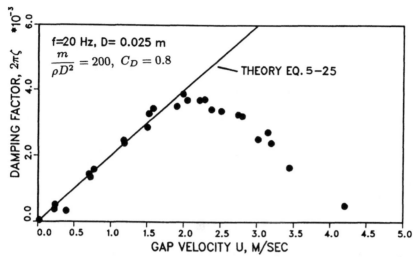

Fig. 5-9 Experimental data of Weaver and El-Kashlan (1981) for transverse fluid dynamic damping of a tube in a rotated triangle array (Fig. 5-5) with $P/D = 1.375$ in comparison with theory.

Goyder and Teh (1984) found a similar result. This suggests that there is a negative damping effect, possibly as a result of the time lag (Eq. 5-24).

Consider a single flexible tube surrounded by rigid tubes. Equation 5-11 with $y_{j+1} = y_{j-1} = x_{j+1} = x_{j-1} = 0$ is the equation of motion in the transverse direction for the j tube with all other tubes stationary:

$$m\ddot{y} + 2m\zeta_y\omega_y\dot{y} + k_y y = \tfrac{1}{2}\rho U^2 C_y y. \qquad (5\text{-}26)$$

This equation is stable for relatively small fluid forces (right-hand side), because the fluid force acts in phase with the displacement and this fluid force is small compared with the structural stiffness $k_y y$. However, if the fluid force does not act instantaneously but rather lags the displacement by a small time delay τ, then the fluid force will also have a component in phase with velocity (i.e., in phase with damping), and this component can be destabilizing.

If the motion is harmonic, then the displacement is 90 degrees out of phase with the velocity:

$$y = A_y \sin \omega t, \qquad \dot{y} = A_y\omega \cos \omega t. \qquad (5\text{-}27)$$

The fluid force in Eq. 5-26 is

$$F_y = \tfrac{1}{2}\rho U^2 C_y y = \tfrac{1}{2}\rho U C_y A_y \sin \omega t, \qquad (5\text{-}28)$$

but if the fluid force lags the displacement, then a component of fluid force acts in phase with velocity,

$$F_y = \tfrac{1}{2}\rho U^2 C_y A_y \sin\left[\omega(t-\tau)\right]$$

$$= \tfrac{1}{2}\rho U^2 D C_y A_y (\sin \omega t \cos \omega \tau - \cos \omega t \sin \omega \tau),$$

$$= \tfrac{1}{2}\rho U^2 C_y [y \cos \omega \tau - (1/\omega)\dot{y} \sin \omega \tau]. \tag{5-29}$$

The time lag is approximately $\tau = D/U$ (Eq. 5-24). With this substitution, the previous equation becomes

$$F_y = \tfrac{1}{2}\rho U^2 C_y [y \cos \omega \tau - \dot{y}(1/\omega)\sin \omega D/U]. \tag{5-30}$$

Inserting this equation in Eq. 5-26, neglecting the in-phase component, and setting $\omega = \omega_j$ gives

$$m\ddot{y}_j + [2m\zeta_y^j \omega_y^j + \tfrac{1}{2}\rho U^2 C_y (1/\omega_j) \sin \omega_j D/U]\dot{y}_j + k_y^j y_j = 0. \tag{5-31}$$

This equation possesses an instability provided that the quantity in square brackets passes through zero so that the net damping passes through zero and becomes negative. With the bracket term set to zero, an instability is predicted at a reduced velocity of

$$\frac{U^2}{\omega_j^2 D^2}\sin\frac{\omega_j D}{U} = -\frac{4m\zeta_y}{C_y\rho D^2}. \tag{5-32}$$

Because of the periodicity of the sine function, this equation predicts multiple ranges of instability provided C_y is negative. Price and Paidoussis (1986) found that C_y is negative in many cases. Figure 5-10 shows their theory for a single flexible cylinder in comparison with data for tube arrays with all cylinders flexible.

A general linear theory for fluid force on a vibrating tube in an array of elastic tubes includes terms that depend on (1) tube acceleration (added mass and added mass coupling, Chapter 2); (2) tube velocity (fluid damping and time lag effects); and (3) tube displacement (displacement mechanism). Following S. S. Chen (1983, 1987), the general linear expression for fluid force on the i tube that interacts with the j surrounding tubes is

$$F_i = -\rho D^2 \sum_j (\alpha_{ij}\ddot{x}_j + \beta_{ij}\ddot{y}_j) - \rho D U \sum_j (\alpha'_{ij}\dot{x}_j + \beta'_{ij}\dot{y}_j)$$

$$+ \rho U^2 \sum_j (\alpha''_{ij}x_j + \beta''_{ij}y_j). \tag{5-33}$$

If the i tube interacts primarily with the two nearest tubes, there are 18 coefficients a, β, α', β', α'', β'' in this expression. Some of these can be determined analytically using potential flow and the fluids models discussed in Section 5.2; most must be measured experimentally. In general, the coefficients are strongly dependent on tube spacing and

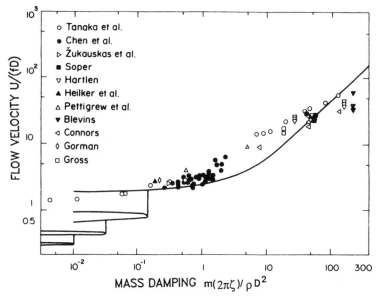

Fig. 5-10 Theory for instability of tube array in comparison with data. The Price and Paidoussis (1986) theory is based on a single flexible tube surrounded by fixed tubes.

pattern (see Price and Paidoussis, 1986). When all these coefficients are measured by a series of reasonably elaborate experiments, the work of Tanaka and Takahara (1981) suggests that good agreement with experiment for the onset of instability is obtained. A possible pitfall is that this theory cannot model vortex shedding and discrete vortex effects on tube vibration (Sections 2.4, 3.7, 4.4, and 9.3).

5.3. PRACTICAL CONSIDERATIONS FOR HEAT EXCHANGERS

Figure 5-11 shows two common types of tube-and-shell exchangers. Figure 5-11(a) shows a heat-recovery boiler. Water in the lower vessel ("mud drum") is convected vertically through tubes that are exposed to a flow of high-temperature gas. The water boils and the stream is collected in the upper vessel ("steam drum"). Figure 5-11(b) shows a chemical process heat exchanger. Here the shell is cylindrical with straight tubes running between tube sheets. The shell side flow is directed across the tubes by a series of baffles that also serve to support the tubes.

In both designs, the flow over the tubes has a component perpendicular to the tube axis and this component of flow can induce fluid elastic instability. It is the author's experience that fluid elastic instability is a primary cause of flow-induced tube failure although

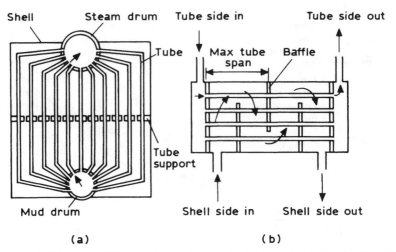

Fig. 5-11 Tube and shell heat-exchanger designs: (a) heat-recovery boiler (gas flow into plane); (b) chemical process heat exchanger.

turbulent buffeting can also contribute to long-term wear (Section 7.3.2). There are three modes of tube failure (Fig. 5-12): (1) if the stress levels in the tube root are above the fatigue limit, a circumferential crack will grow about the tube; (2) large-amplitude vibration results in clashing of adjacent tubes at midspan, and this will wear flats in neighboring tubes; (3) if there is clearance between a tube and its support, large-amplitude tube motion can wear a groove in the tube at a support.

Practically speaking, heat exchangers are designed to maximize heat transfer and minimize cost. Flow-induced vibration is a secondary design consideration and vibration analysis is at best limited to design of tube supports to prevent instability. The natural frequency of tubes is inversely proportional to the unsupported span (Blevins, 1979). Decreasing the

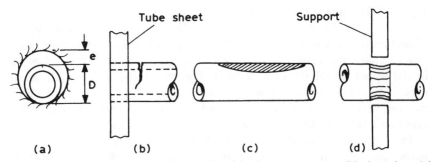

Fig. 5-12 Tube support and failure modes: (a) tube support gap e; (b) circumferential crack at tube sheet; (c) flat worn at midspan by clashing; (d) groove worn at support (In part from Pettigrew et al., 1978).

unsupported tube span increases tube natural frequencies, and the critical velocity for onset of instability can be raised above the shell-side flow velocity. Tubes with rolled-in or welded-in ends without any intermediate supports have very low damping, typically $\zeta = 0.001$, while those with some intermediate supports have damping at least a factor of 10 higher, $\zeta \approx 0.01$ (Chapter 8, Sections 8.3.2, and 8.4.4). Thus, the benefits of adding at least one intermediate support are very great in terms of both frequency and damping.

Usually tubes pass through oversized holes in the intermediate supports as shown in Figure 5-12. Wear can occur between the tube and the support. Tests by Ko and Basista (1984), Ko (1987), Blevins (1985), and Cha et al. (1987) show that the rate of wear *increases* with (1) increasing size of the diametrical gap e (Fig. 5-12(a)), (2) increasing amplitude of tube vibration and amplitude of exciting force, (3) increasing time as wear increases the gap, and (4) decreasing width of the support baffle. Heat-exchanger tube wear is the result of either sliding–fretting as the tube slides around the circumference of the hole or impact wear as the tube strikes and rebounds off the support (Ko, 1987). In general, if impacting occurs, the associated wear rate is much greater than the sliding wear rate.

At least some tubes will "float" in the center of the support hole. These tubes will vibrate as if support were absent, that is, in a lower mode, until the vibration amplitude grows to the point where the tube impacts the support (Goyder, 1985; Halle et al., 1981; Fricker, 1988). The impact velocity is predictable from sinusoidal motion of the tube. The impact results in a surface contact stress. If that stress exceeds the fatigue capability of the material, there will be impact wear (Engel, 1978; Blevins, 1984b; Ko and Basista, 1984). In contrast, sliding wear depends on local relative motion (Hofmann et al., 1986; Ko, 1987).

Both impacting and sliding wear can be minimized by minimizing tube motion relative to the support. This can be accomplished by either loading the tube against a support or minimizing the diametrical gap e. It is desirable to maximize tube–support contact by having the tube support thickness comparable to the tube diameter or larger. Experience and theory (Blevins, 1979, 1984b; Gordon and Lebret, 1988) suggest that for tubes between 0.4 in and 3 in (10 mm and 70 mm) in diameter, the diametrical gaps should be approximately 0.020 in (0.5 mm) or smaller. Unfortunately, a small gap with broad support can encourage crevice corrosion in water systems by restricting flushing. A zero gap can be obtained by welding in the tubes. This eliminates wear but it also greatly reduces damping, so the net effect is not beneficial. Near-zero gaps can be obtained by preloading tubes against the support using a spacer such as the "corkscrew" device patented by Eisinger (1980). Flat and formed bars and steel concrete-reinforcing rods have successfully been inserted between tubes to suppress unstable tube motion (Weaver and Schneider, 1983; Au-Yang, 1987; Horn et al., 1988; Small and Young, 1980; Boyer and Pase, 1980).

5.4. VIBRATION OF PAIRS OF CYLINDERS

5.4.1. Description of Two-Cylinder Flows

Consider two closely spaced, elastic circular cylinders in a flow perpendicular to their axes. Practical applications of this situation include two tandem struts in biplanes, twin chimneys, jetties, pilings, bundles of electrical transmission lines, and groups of free-standing cooling towers. The two cylinders can be arranged one behind the other (i.e., in tandem), side by side, or in a staggered arrangement relative to the free stream flow. Experiments and reviews by Chen (1986), Zdravkovich (1984, 1985), King and Johns (1976), and the Electric Power Research Institute (1979) suggest that there are three mechanisms for flow-induced vibration of pairs of cylinders in cross flow: (1) vortex-induced vibration when cylinder center-to-center spacing exceeds about 1.4 diameters; (2) a fluid elastic instability of the general type observed in tube arrays (Sections 5.1 and 5.2) for cylinder spacing between 1.1 and 8 diameters, and (3) wake galloping when the cylinders are spaced greater than 8 diameters and the downstream cylinder falls in the wake of the upstream cylinder.

Figure 5-13 shows the vortex wakes for various cylinder arrangements. The frequency of vortex shedding is

$$f_s = \frac{SU}{D} \quad \text{Hz,} \qquad (5\text{-}34)$$

where S is the dimensionless Strouhal number given in Figure 5-14. The callouts in this figure are the Strouhal number for shedding from a cylinder at a distance from a fixed cylinder. A single vortex street is formed behind the cylinder pair when the center-to-center spacing between the cylinders is less than about 1.4 diameters (Zdravkovich, 1985; Kiya et al., 1980). When the cylinders are in tandem (one behind the other), there is little or no vortex shedding behind the upstream cylinder for center-to-center spacing up to about 3.9 diameters because of stagnant fluid trapped between the cylinders. In side-by-side arrangements with spacing less than about 4 diameters, bistable, asymmetric jets form behind the cylinders, as shown in Figure 5-1 (Kim and Durbin, 1988). At spacing greater than 4 diameters, a vortex street forms behind each cylinder, more or less independently of the other cylinder; see Chapter 3. Lam and Cheung (1988) have shown that similar phenomena occur behind groups of cylinders.

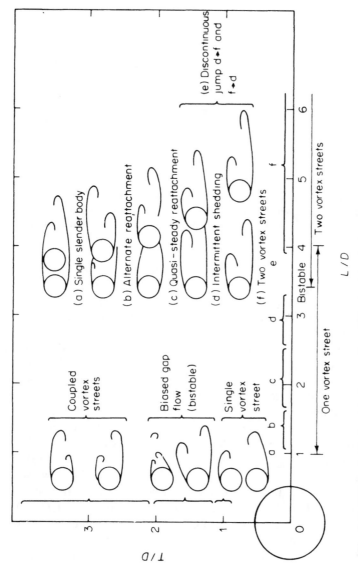

Fig. 5-13 Wakes of cylinder pairs in cross flow (Zdravkovich, 1985).

175

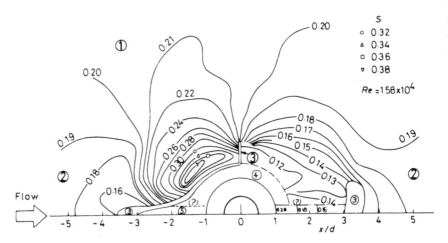

Fig. 5-14 Strouhal numbers for a cylinder at a certain distance from a fixed cylinder. Region 1, Strouhal number is higher than for a single cylinder; Region 2, Strouhal number is less than for a single cylinder; Region 2, Strouhal number is less than for a single cylinder; Region 3, bistable shedding; Region 4, single vortex wake; Region 5, no vortex shedding (Kiya et al., 1980).

The vortex-induced vibrations occur when the shedding frequency (Eq. 5-32) approximately equals the cylinder natural frequency, $f_s \sim f_n$, a situation that occurs near a reduced velocity of $U/(f_n D) = 1/S$. In general, since each cylinder in the pair has its own shedding frequency, the response of the pair will have two peaks, with each cylinder at its own resonance. Often cylinder pairs are coupled by a rigid spacer and then only a single peak emerges from the cylinder pair, as shown in Figure 5-15. Johns and King and Zdravkovich found that the vortex-induced response of closely spaced cylinders in tandem could exceed the vortex-induced response of a single isolated cylinder of identical properties by as much as a factor of 2 to 3 (single-cylinder vortex-induced response is discussed in Chapter 3). The downstream cylinder generally experiences the greatest oscillating force (Arie et al., 1983). When the cylinders are restricted to vibration in line with the flow, subharmonic response occurs in the drag direction with twice the frequency of vortex shedding at reduced velocities of the order $U/(f_n D) \sim 1/(2S) \sim 2$. Typically, the cross flow response is twice the inline response and the largest amplitudes occur at reduced velocities between 4 and 8 (Zdravkovich, 1985). Vickery and Watkins (1962) found similar behavior with four cylinders in tandem.

Instability can result in large-amplitude vibration. Tests by Zdravkovich (1984) and Jendrzejczyk et al. (1979) for closely spaced tube pairs suggest that the instability is at least qualitatively similar to the fluid elastic instability of cylinder arrays discussed in Sections 5.1 and 5.2. At least for side-by-side arrangements (Fig. 5-15), there is an abrupt

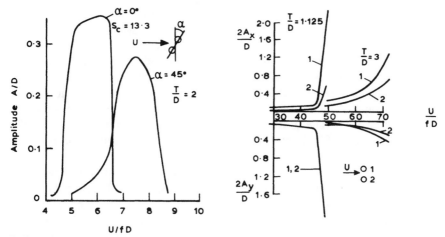

Fig. 5-15 Response of coupled cylinder pairs. (a) Vortex-induced response for $m(2\pi\zeta)/\rho D^2 = 13.3$ for $\alpha = 0$ degree and 10.3 for $\alpha = 45$ degrees; (b) instability for $m(2\pi\zeta)/\rho D^2 = 80$ (Zdravkovich, 1984).

transition into instability. A more gradual transition occurs for tandem arrangements. Their data and the data of Connors (1970) and Halle and Lawrence (1977) for tube rows suggest that an exponent of $a = \frac{1}{2}$ and a coefficient C between 5 and 10 is appropriate in Eq. 5-2 for predicting onset of instability in closely spaced cylinder pairs and cylinder rows for center-to-center spacing up to about 4 diameters in air.

For spacing greater than about 4 diameters, there is little interaction between two cylinders unless the downstream cylinder falls in the wake of the upstream cylinder, as discussed in the following section.

5.4.2. Wake Galloping

Side-by-side cylinders interact fluid-mechanically only if the center-to-center spacing is less than about 5 diameters. In tandem arrangements, interaction occurs only if the downstream cylinder is in the wake of the upstream cylinder (Zdravkovich, 1977, 1984). Some time-averaged properties of the turbulent wake of a cylinder are as follows (Blevins, 1984c):

Half-width to one-half centerline deficit $\quad b = 0.23[C_D D(x + x_0)]^{1/2}$,

Centerline deficit velocity $\quad u_d(y = 0) = 1.2U\left(\dfrac{C_D D}{x + x_0}\right)^{1/2}$,

Deficit velocity profile

$$\frac{u_d}{U} = e^{-0.69y^2/b^2}. \tag{5-35}$$

The half-width b is the distance from the centerline for the deficit velocity u_d (i.e., the reduction in velocity from the free stream velocity U) to decay to one-half the centerline value (see Fig. 5-16); $C_D \simeq 1$ is the drag coefficient based on the cylinder diameter D and free stream velocity U (Chapter 6, Fig. 6-9); x is the distance downstream from the cylinder center; y is the transverse distance; and x_0 is the upstream location of the virtual origin of the wake, $x_0 \simeq 6\,D$. The boundary of the wake is located at about $y \sim 5b$. These are time-averaged mean properties. In fact, the cylinder wake is filled with evolving vortices as discussed in the previous chapter.

The drag on a cylinder in the wake of an upstream cylinder is reduced, and lift forces pull the cylinders together. As shown in Figure 5-17, these forces diminish with increasing distance between the cylinders as the wake melds back into the free stream flow. The lowest drag on the downstream cylinder occurs when the downstream cylinder is in line with the upstream cylinder. The greatest lift occurs when the center of the downstream cylinder is about $T \sim 1.3b$, just outside the wake half-width.

The wake-induced forces can encourage oscillation of the downstream cylinder in an oval orbit, as shown in Figure 5-18. The cylinder moves downstream from the upstream-most point of its orbit under the influence of the large drag force on the outside of the wake. The cylinder then moves inward, encouraged by the lift force within the wake. The cylinder next moves upstream against relatively low drag within the wake, moving toward the edge of the wake, then the cycle repeats. The oscillation orbit inputs energy into the cylinder and so vibration is encouraged. From this

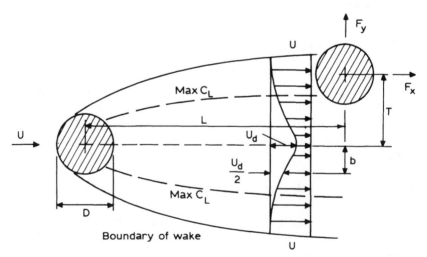

Fig. 5-16 Time-averaged properties of the turbulent wake of a cylinder with a downstream cylinder in the wake.

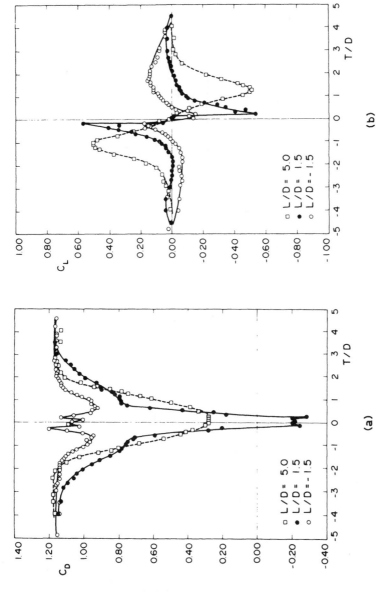

Fig. 5-17 Lift and drag coefficients (Eq. 5-36) for a cylinder in the wake of another cylinder at Re = 5.3×10^4 (Price and Paidoussis, 1984).

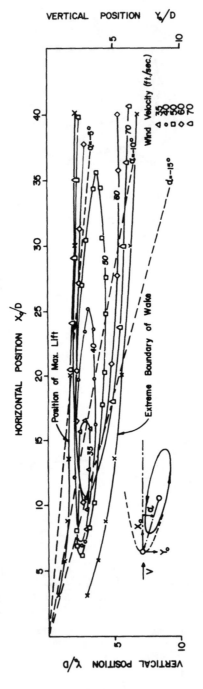

Fig. 5-18 Wake galloping stability boundaries and orbits (Cooper, 1973).

description, one can deduce that the orbits must be counterclockwise in the lower half of the wake in a flow from left to right and clockwise in the upper half of the wake. Moreover, vibration can occur only in the region between lines of maximum lift ($|y| \sim 1.3b$, where y now denotes the center of the downstream cylinder relative to the upstream cylinder) and the extremities of the wake ($|y| \sim 5b$), and not too far downstream of the upstream conductor or the wake-induced forces will be insufficient to encourage the vibration.

The Electric Power Research Institute (1979) reports that these wake galloping oscillations are a relatively common occurrence in bundles of powerline conductors when the downstream conductor falls in the wake of the upstream conductor and the spacing between conductors is between about 10 and 20 diameters. Several modes of conductor motion are shown in Figure 5-19. Wake galloping has also been introduced in the form of traveling waves with wavelengths of 150 to 300 ft (50 to 100 m). Wake galloping of conductors usually occurs in strong winds, 15 to 40 mph (7 to 18 m/sec), over relatively flat terrain. Subspan vibration amplitudes as high as 0.5 m, sufficient to cause adjacent conductors to

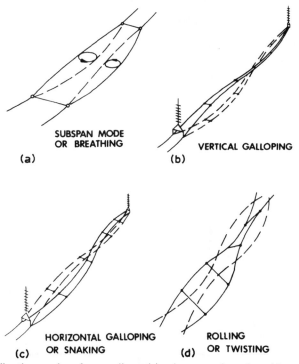

Fig. 5-19 Vibration modes of power lines: (a) subspan mode or breathing; (b) vertical galloping; (c) horizontal galloping or snaking; (d) rolling or twisting (EPRI, 1979).

clash, have been observed, although the damage induced by wake galloping is usually not severe.

The lift and drag forces on an elastically mounted cylinder in the wake of an upstream fixed cylinder (Figs. 5-16 and 5-17) arise from the mean flow plus velocity and pressure gradients in the wake. The vertical force, F_y, and the horizontal force, F_x, are written in terms of force coefficients, C_y and C_x, respectively,

$$F_x = \tfrac{1}{2}\rho U^2 D C_x, \qquad F_y = \tfrac{1}{2}\rho U^2 D C_y. \tag{5-36}$$

These coefficients can be expressed in terms of the lift and drag coefficients that express forces perpendicular and parallel to the mean force (Fig. 5-20),

$$C_x = (C_D \cos \alpha - C_L \sin \alpha) \frac{U_{rel}^2}{U^2}$$

$$C_y = (C_L \cos \alpha + C_D \sin \alpha) \frac{U_{rel}^2}{U^2}. \tag{5-37}$$

Where the lift and drag coefficients are defined by

$$F_L = \tfrac{1}{2}\rho U^2 D C_D, \qquad F_D = \tfrac{1}{2}\rho U^2 D C_D. \tag{5-38}$$

With a stationary cylinder, $\alpha = 0$, $C_x = C_D$, and $C_y = C_L$ because the lift force and vertical force coincide as do the drag and horizontal forces on the cylinder. But when the cylinder vibrates, the flow relative to the

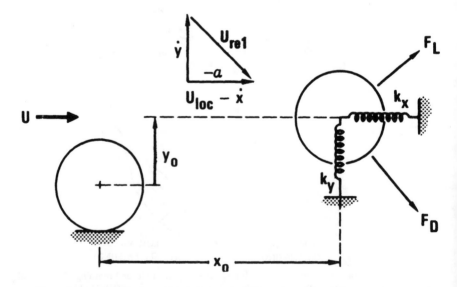

Fig. 5-20 Elastically mounted cylinder in the wake of another cylinder.

cylinder oscillates, $\alpha \neq 0$, and the lift and vertical forces no longer coincide.

The procedure of resolving the force on the vibrating cylinder into horizontal and vertical force coefficients is identical to that previously employed in the galloping analysis of Section 4.2.1, except that here both horizontal and vertical motion are included and y is defined as positive upward. The angle of attack of the flow relative to the cylinder is

$$\alpha = \arctan \frac{\dot{y}}{(U_{\text{loc}} - \dot{x})} \tag{5-39}$$

to account for the change in angle of attack induced by conductor motion. The velocity of the flow relative to the conductor is

$$U_{\text{rel}}^2 = \dot{y}^2 + (U_{\text{loc}} - \dot{x})^2. \tag{5-40}$$

U_{loc} is the local fluid velocity in the wake. The lift and drag coefficients are functions of the position of the downstream cylinder relative to the upstream cylinder and the Reynolds number.

For relatively small motions of the downstream cylinder, Eqs. 5-36 through 5-39 may be expanded in terms of small displacements of the cylinder, x and y, from its nominal position x_0 and y_0. Thus,

$$C_y = C_L + \alpha C_D + O(\alpha^2), \qquad C_x = C_D - \alpha C_L + O(\alpha^2), \tag{5-41}$$

$$C_D(x, y) = C_D(x_0, y_0) + \frac{\partial C_D}{\partial x} x + \frac{\partial C_D}{\partial y} y + O(x^2, y^2),$$

$$C_L(x, y) = C_L(x_0, y_0) + \frac{\partial C_L}{\partial x} x + \frac{\partial C_L}{\partial y} y + O(x^2, y^2), \tag{5-42}$$

$$\alpha = \frac{-\dot{y}}{U_{\text{loc}}} + O\left(\frac{\dot{y}\dot{x}}{U}\right), \qquad U_{\text{rel}}^2 = U_{\text{loc}}^2 - 2U_{\text{loc}}\dot{x} + O(\dot{y}^2). \tag{5-43}$$

C_L and C_D and their derivatives on the right-hand side of these equations are evaluated at the equilibrium position. For stability analysis, only linear terms are retained. The equations of motion are found by substituting Eqs. 5-41 through 5-43 into the fluid force, Eqs. 5-36 through 5-38, and applying these forces to a spring-supported cylinder:

$$m\ddot{x} + 2m\zeta_x\omega_x\dot{x} + k_x x$$
$$= \tfrac{1}{2}\rho U^2 D\left(C_D + \frac{\partial C_D}{\partial x} x + \frac{\partial C_D}{\partial y} y + C_L \frac{\dot{y}}{U_{\text{loc}}} - 2C_D \frac{\dot{x}}{U_{\text{loc}}}\right), \tag{5-44}$$

$$m\ddot{y} + 2m\zeta_y\omega_y\dot{y} + k_y y$$
$$= \tfrac{1}{2}\rho U^2 D\left(C_L + \frac{\partial C_L}{\partial x} x + \frac{\partial C_L}{\partial y} y - C_D \frac{\dot{y}}{U_{\text{loc}}} - 2C_L \frac{\dot{x}}{U_{\text{loc}}}\right). \tag{5-45}$$

The mean drag of the wind blows the cable catenary at an angle, called the blow-back angle, from vertical. As a result, Simpson (1971a,b) and Tsui (1977) include spring stiffness terms on the right-hand sides of these equations to account for the stiffness-induced coupling; also see Section 4.8. Tsui and Tsui (1980) and Simpson and Flower (1977) include motion of the upstream cylinder.

Equations 5-44 and 5-45 are very similar to the equations of motion for fluid elastic instability in tube arrays (Eqs. 5-10 and 5-11). In both cases, a displacement mechanism induces the fluid forces and couples the horizontal and vertical motions, and instability will result when the flow velocity exceeds a critical velocity as the cylinder whirls in an oval orbit. There are two approaches to solution of Eqs. 5-43 and 5-44: (1) integrate the systems of equations directly, using numerical techniques, and continue until a steady solution is achieved (Price and Piperni, 1986); (2) perform a stability analysis as follows.

Following Tsui (1977), for stability analysis, solutions to Eqs. 5-44 and 5-45 are sought in exponential form,

$$x = \bar{x}e^{\lambda t}, \qquad y = \bar{y}e^{\lambda t}, \tag{5-46}$$

where $\bar{x}$, $\bar{y}$, and λ are independent of time. Substituting this solution into Eq. 5-44 and 5-45 results in an equation that can be put in matrix form as was done for fluid elastic instability of tube arrays, Section 5.2.1, Eqs. 5-14 through 5-20. The matrix equation has nontrivial solutions only if the determinant of the coefficient matrix is zero. This results in a polynomial in λ,

$$\lambda^4 + a_1\lambda^3 + a_2\lambda^2 + a_3\lambda + a_4 = 0, \tag{5-47}$$

where $a_1, \ldots, a_4$ are functions of U, m, ρ, and C_L, C_D and their derivatives; see Simpson (1971), Price (1975), and Tsui (1977).

An instability is possible for certain cylinder positions and velocities such that λ has positive real roots. Simpson (1971b) suggests that instability is possible only if

$$\frac{\partial C_L}{\partial x}\frac{\partial C_D}{\partial y} < 0. \tag{5-48}$$

There can be no instability along the centerline of the wake because here both derivatives are zero; nor can instability exist outside of the wake for the same reason. Within the wake, $\partial C_D/\partial x$ is positive in the upper half and negative in the lower half (Fig. 5-17), so instability is possible only in those regions where $\partial C_L/\partial x$ has the opposite sign. This occurs on either side of the wake between the line of maximum lift coefficient and the extremities of the wake, a region bounded by $1.3b < |y| < 5b$, where b is given by Eq. 5-35. Figure 5-18 shows the boundaries of potential instability and data reported for oscillation. Although this region of potential instability broadens downstream, the aerodynamic lift force decreases with distance downstream, and practical observations of instability are limited to cylinder separations of about 25 diameters. Coupling in the cable suspension between horizontal and vertical motion, introduced by mean drag deflection, tends to suppress vibration in the upper half of the wake.

The instability is sensitive to the ratio of natural frequencies in the two coordinate directions. Price and Piperni (1986), Tsui (1977), and Simpson (1971) have shown that small changes in frequency can trigger or inhibit the instability. Price and Piperni (1986) have shown that the instability is relatively insensitive to damping. Bokaian (1989) uses quasisteady analysis to predict the amplitude of vibration.

Exercises

1. Consider a pair of powerline conductors 0.8 in (2.0 cm) in diameter in a wind of 32 ft/sec (10 m/sec). The kinematic viscosity of air is 1.6×10^{-4} in^2/sec (1.5×10^{-5} m^2/sec) and $C_D = 1$. What is the Reynolds number? Plot the boundaries of the wake of the upstream conductor. What is the centerline velocity in the wake as a function of downstream distance? Plot the regions of potential instability.

2. As noted in Section 5.2, fluid elastic instability of arrays of multiple elastic cylinders is relatively insensitive to frequency differences between cylinders and relatively sensitive to damping, while the wake galloping of cylinder pairs is just the opposite. Can you explain this?

5.5. EXAMPLE: TUBE INSTABILITY IN A HEAT EXCHANGER

Consider the segmental baffle heat exchanger design shown in Figure 5-21. Water enters the shell side through a nozzle, then flows in a serpentine pattern over seven baffles and exits on the same side. Some of the dimensions of this test heat exchanger are:

- Tube length = 140.75 in (3.58 m)
- Shell inside diameter = 23.25 in (591 mm)
- Baffle thickness = 0.375 in (9.5 mm)
- Baffle cut = 5.81 in (147 mm)
- Baffle hole diameter = 0.768 in (19.5 mm)
- Baffle spacing = 17.59 in (447 mm)
- Tube outside diameter = 0.75 in (19.1 mm)

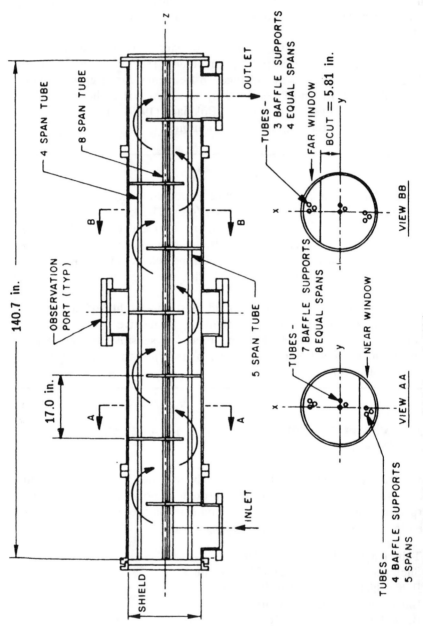

Fig. 5-21 Tube and shell heat exchanger (Mulcahy et al., 1986).

186

- Tube inside diameter $= 0.652$ in (16.56 mm)
- Tube material density $= 0.307$ lb/in^3 (8.5 g/cm^3)
- Tube modulus of elasticity $= 16 \times 10^6$ lb/in^2 (110×10^9 Pa)
- Tube pattern $= 30$ degree triangular (Fig. 5-5)
- Pitch : diameter ratio $= 1.25$
- Shell side fluid $=$ water
- Fluid density $= 62.4$ lb/ft^3 (1 gm/cm^3)
- Fluid kinematic viscosity $= 1.2 \times 10^{-5}$ ft^2/sec (10^{-5} m^2/sec)

There are 499 brass tubes in this heat exchanger, which has been tested by Mulcahy et al. (1986).

The potential for flow-induced instability is found by utilizing Figure 5-6. The correlation presented in this figure is utilized in the following steps: (1) Calculate the mass damping parameter $m(2\pi\zeta)/(\rho D^2)$; (2) from the figure, determine the value of reduced velocity $U/(fD)$ for onset of instability; (3) calculate the tube natural frequency f; and (4) relate the critical unstable velocity U_{crit} to the shell-side flow through the heat exchanger. These steps will be followed in order.

The mass per unit length of the tubes, m, includes both the mass of internal fluid and the added mass due to the external fluid. Consider the tubes to be empty. The mass per unit length of the bare tubing is easily calculated to be 0.033 lb/in (6.0 g/cm). The added mass of water is equal to the mass of water displaced by the tube times an added mass coefficient (Chapter 2, Section 2.2). For a pitch-to-diameter ratio of 1.25, Chapter 2, Table 2-3, suggests an added mass coefficient of 1.6, while Mulcahy et al. (1986) suggests values between 1.5 and 1.7. Using an average value of 1.6 gives an added mass of 0.025 lb/in (4.56 g/cm). Thus, the total mass per unit length is $m = 0.0606$ lb/in (10.8 g/cm).

The damping of tubes is discussed in Chapter 8. Measurements on these tubes showed a range of damping from 1% to 5% of critical with damping decreasing with increasing frequency. A typical value of $\zeta = 0.02$ will be used in this example. With this information, the mass damping parameter is calculated to be

$$\frac{m(2\pi\zeta)}{\rho D^2} = 0.374.$$

Entering Figure 5-6 with this value suggests that a lower bound on the critical reduced velocity is

$$\frac{U_{crit}}{fD} = 2,$$

but the data suggest that values as high as 7 have been observed in this mass-damping range, while Eq. 5.2, using average curve-fit parameters of $a = 2.1$ and $C = 3.9$, gives an average value of 3.2. These values are shown in Table 5-1.

Table 5-1 Data for heat-exchanger example

Stability criteria	$\dfrac{U_{crit}}{fD}$	U_{crit} (m/sec) $f = 38$ Hz	Q_{crit} (m³/sec)	
			$A = 0.057$ m²	Mulcahy (1986) observed instability
Minimum	2.	1.49	0.078 ⎫	
Average	3.2	2.14	0.12 ⎬	0.20
Maximum	7.0	5.21	0.27 ⎭	

The natural frequency of the tubes bending between the baffles is given by the following formula (Blevins, 1979):

$$f = \frac{\lambda^2}{2\pi L^2} \left(\frac{EI}{m}\right)^{1/2} \ \text{Hz.}$$

L is the distance between the tube supports; E is the tube modulus of elasticity; I is the tube moment of inertia for bending, 0.00666 in⁴ (0.277 cm⁴); and λ is a dimensionless parameter that is a function of vibration mode and the baffle spacing. In this heat exchanger, there are tubes with eight spans that pass through seven baffles ($L = 17.59$ in, 44.7 cm), tubes "in the far window" with four spans ($L = 35.18$ in, 89.4 cm) that pass through only four baffles, and tubes "in the rear window" with five unequal spans ($L_{max} = 35.18$ in, 89.4 cm) that pass through five baffles. The tubes tend to pivot freely at the oversized holes in the baffles, but their extremities are clamped at the tube sheets. Taking into account the clamping of the extreme ends gives $\lambda = 3.393$ for the four-span tubes, $\lambda = 3.210$ for the eight-span tubes, and $\lambda = 3.5$ for the five-span tubes. Using these values and utilizing a consistent set of units (i.e., a set of units in which Newton's law is obeyed) gives the following fundamental natural frequencies for the tubes (Mulcahy et al., 1986):

	Four-span tubes	Five-span tubes	Eight-span tubes
Theory	38.7 Hz	41.1 Hz	138.7 Hz
Experiment	38–40 Hz	—	—

In general, since all spans experience similar cross-flow velocities, the lowest-frequency tubes, the four-span tubes, are expected to be most prone to fluid elastic instability.

The cross-flow velocity for instability must be related to overall shell-side flow through the heat exchanger. This would require a detailed evaluation of the flow field and tube mode shape to evaluate the integral of Eq. 5-12. A simple estimate can be obtained using conservation of mass

$$Q = UA,$$

where Q is the volume flow rate, U is the fluid velocity, and A is the flow area. An estimate of the average cross-flow area is the distance between baffles, less the baffle thickness, times the length of the straight edge of a baffle, times the fraction of free flow area between tubes to total area. In units of inches, this calculation gives

$$A = (17.59 - 0.375) \times 19.8 \times (0.25/1.25) = 68.18 \text{ in}^2 = 0.044 \text{ m}^2.$$

Thus, a flow rate of $35 \text{ ft}^3/\text{sec}$ ($1 \text{ m}^3/\text{sec}$) through the heat exchanger would yield an effective velocity of 74.5 ft/sec (22.7 m/sec) using this effective area. There is some additional area associated with flow leakage around and through the baffles. This typically increases the effective area by 30%, $A_{\text{eff}} = 88.6 \text{ in}^2$ (0.057 m^2). [Mulcahy et al. (1986) perform a more detailed calculation of the effective flow area and cross-flow velocity using a numerical flow-modeling method. Alas, there is a substantial error in this part of their calculation.]

Mulcahy's observed onset of instability of $7.44 \text{ ft}^3/\text{sec}$ ($0.21 \text{ m}^3/\text{sec}$) agrees well with the present calculation if an average or maximum value of the critical reduced velocity is used with the calculated effective flow area. The instability appeared in the four-span tubes, which have the longest unsupported span of any tubes in the bundle. The critical velocity for these tubes can be raised a factor of four by providing additional midspan support. See Section 5.3.

REFERENCES

Arie, M. et al. (1983) "Pressure Fluctuations on the Surface of Two Cylinders in Tandem Arrangement," *Journal of Fluids Engineering*, **105**, 161–167.

Au-Yang, M. K. (1987) "Development of Stabilizers for Steam Generator Tube Repair," *Nuclear Engineering Design*, **103**, 189–197.

Balsa, T. F. (1977) "Potential Flow Interactions in an Array of Cylinders in Cross-Flow," *Journal of Sound and Vibration*, **50**, 285–303.

Bearman, P. W., and A. J. Wadcock (1973) "The Interaction Between a Pair of Circular Cylinders Normal to a Stream," *Journal of Fluid Mechanics*, **61**, 499–511.

Bellman, R. (1970) *Introduction to Matrix Analysis*, McGraw-Hill, New York, p. 253.

Blevins, R. D. (1974) "Fluid Elastic Whirling of a Tube Row," *Journal of Pressure Vessel Technology*, **96**, 263–267.

────── (1979a) "Fluid Damping and the Whirling Instability," in *Flow Induced Vibrations*, ASME, New York.

────── (1979b) "Formulas for Natural Frequency and Mode Shape," Van Nostrand Reinhold, New York. Reprinted by R. Kreiger, Malabar, Fla., 1984.

────── (1979c) "Fretting Wear of Heat Exchanger Tubes," *Journal of Engineering for Power*, **101**, 625–633.

────── (1984a) "Discussion of 'Guidelines for the Instability Flow Velocity of Tube Arrays in Cross Flow'," *Journal of Sound and Vibration*, **97**, 641–644.

────── (1984b) "A Rational Algorithm for Predicting Vibration-Induced Damage to Tube and Shell Heat Exchangers," in *Symposium on Flow-Induced Vibrations*, Vol. 3, ASME, New York, pp. 87–104.

────── (1984c) *Applied Fluid Dynamics Handbook*, Van Nostrand Reinhold, New York.

———— (1985) "Vibration-Induced Wear of Heat Exchanger Tubes," *Journal of Engineering Materials and Technology*, **107**, 61–67.

Blevins, R. D., R. J. Gilbert, and B. Villard (1981) "Experiments on Vibration of Heat Exchanger Tube Arrays in Cross Flow," Paper B6/9, 6th International Conference on Structural Mechanics in Reactor Technology, Paris, France.

Bokaian, A. (1989) "Galloping of a Circular Cylinder in the Wake of Another," *Journal of Sound and Vibration*, **128**, 71–85.

Boyer, R. C., and G. K. Pase (1980) "The Energy-Saving NESTS Concept," *Heat Transfer Engineering*, **2**(1), 19–27.

Cha, J. H., M. W. Wambsganss, and J. A. Jendrzejczyk (1987) "Experimental Study on Impact/Fretting Wear in Heat Exchanger Tubes," *Journal of Pressure Vessel Technology*, **109**, 265–274.

Chen, S. S. (1983) "Instability Mechanisms and Stability Criteria of a Group of Cylinders Subjected to Cross Flow, Parts I and II," *Journal of Vibration, Acoustics, Stress and Reliability in Design*, **105**, 51–58, 253–260.

———— (1984) "Guidelines for the Instability Flow Velocity of Tube Arrays in Crossflow," *Journal of Sound and Vibration*, **93**, 439–455.

———— (1986) "A Review of Flow-Induced Vibration of Two Cylinders in Cross Flow," *Journal of Pressure Vessel Technology*, **108**, 382–393.

———— (1987) "A General Theory for Dynamic Instability of Tube Arrays in Cross Flow," *Journal of Fluids and Structures*, **1**, 35–53.

Chen, S. S., and J. A. Jendrzejczyk (1981) "Experiments on Fluid Elastic Instability in Tube Banks Subjected to Liquid Cross Flow," *Journal of Sound and Vibration*, **78**, 355–381.

Chen, S. S., J. A. Jendrzejczyk, and W. H. Lin (1978) "Experiments on Fluid Elastic Instability in Tube Banks Subject to Liquid Cross Flow," Argonne National Laboratory Report ANL-CT-78-44, Argonne, Ill.

Connors, H. J. (1970) "Fluidelastic Vibration of Tube Arrays Excited by Cross Flow," Paper presented at the Symposium on Flow Induced Vibration in Heat Exchangers, ASME Winter Annual Meeting.

———— (1978) "Fluidelastic Vibration of Heat Exchanger Tube Arrays," *Journal of Mechanical Design*, **100**, 347–353.

Cooper, K. R. (1973) "Wind Tunnel and Theoretical Investigations into the Aerodynamic Stability of Smooth and Stranded Twin Bundled Power Conductors," NRC (Canada) Laboratory Technical Report LA-117.

Eisinger, F. L. (1980) "Prevention and Cure of Flow-Induced Vibration Problems in Tubular Heat Exchangers," *Journal of Pressure Vessel Technology*, **102**, 138–145.

Electric Power Research Institute (1979) *Transmission Line Reference Book, Wind-Induced Conductor Motion*, Electrical Power Research Institute.

Engel, P. A. (1978) *Impact Wear of Materials*, Elsevier, New York.

Franklin, R. E., and B. M. H. Soper (1977) "An Investigation of Fluidelastic Instabilities in Tube Banks Subjected to Fluid Cross-Flow," Paper F6/7 in *Proceedings of Conference on Structural Mechanics in Reactor Technology*, San Francisco, Calif.

Fricker, A. J. (1988) "Numerical Analysis of the Fluidelastic Vibration of a Steam Generator Tube with Loose Supports," in *1988 International Symposium on Flow-Induced Vibration and Noise*, M. P. Paidoussis, ed., Vol. 5, ASME, New York, pp. 105–120.

Godon, J. L., and J. Lebret (1988) "Influence of the Tube-Support Plate Clearance on Flow-Induced Vibration in Large Condensers," in *1988 International Symposium on Flow-Induced Vibration and Noise*, M. P. Paidoussis, ed., Vol. 5, ASME, New York, pp. 177–186.

Goyder, H. G. D. (1985) "Vibration of Loosely Supported Steam Generator Tubes," in ASME *Symposium on Thermal Hydraulics and Effects of Nuclear Steam Generators and Heat Exchangers*, S. M. Cho et al., eds., HTD Vol. 51, ASME, New York, pp. 35–42.

Goyder, H. G. D., and C. E. Teh (1984) "Measurement of the Destabilising Forces on a Vibration Tube in Fluid Cross Flow," in *Symposium on Flow-Induced Vibrations*, Vol. 2, M. P. Paidoussis et al., eds., ASME, New York, pp. 151–164.

Halle, H., and W. P. Lawrence (1977) "Crossflow-Induced Vibration of a Row of Circular Cylinders in Water," ASME Paper 77-JPGC-NE-4.

Halle, H., J. M. Chenoweth, and M. W. Wambsganss (1987) "DOE/ANL/HTRI Heat Exchanger Tube Vibration Data Bank," Argonne National Laboratory Technical Memorandum ANL-CT-80-3, Addendum 7, Argonne, Ill.

—— (1981) "Flow-Induced Vibration Tests of Typical Industrial Heat Exchanger Configurations," ASME Paper 81-DET-37.

Hara, F. (1989) "Unsteady Fluid Dynamic Forces Acting on a Single Row of Cylinders Vibrating in a Cross Flow," *Journal of Fluids and Structures*, **3**, 97–113.

Hartlen, R. T. (1974) "Wind Tunnel Determination of Fluid-Elastic Vibration Thresholds for Typical Heat-Exchanger Patterns," Ontario Hydro Report 74-309-K, Toronto, Ontario.

Heilker, W. J., and R. A. Vincent (1981) "Vibration in Nuclear Heat Exchangers Due to Liquid and Two-Phase Flow," *Journal of Engineering for Power*, **103**, 358–366.

Hofmann, P. J., T. Schettler, and D. A. Steininger (1986) "Pressurized Water Reactor Generator Tube Fretting and Fatigue Wear Characteristics," ASME Paper 86-PVP-2.

Horn, M. J., et al. (1988) "Staking Solutions to Tube Vibration Problems," in *1988 International Symposium on Flow-Induced Vibration and Noise*, M. P. Paidoussis, ed., Vol. 5, ASME, New York, pp. 187–200.

Jendrzejczyk, J. A., S. S. Chen, and M. W. Wambsganss (1979) "Dynamic Response of a Pair of Circular Tubes Subjected to Liquid Cross Flow," *Journal of Sound and Vibration*, **67**, 263–273.

Kim, H. J., and P. A. Durbin (1988) "Investigation of the Flow Between a Pair of Circular Cylinders in the Flopping Regime," *Journal of Fluid Mechanics*, **196**, 431–448.

King, R., and D. J. Johns (1976) "Wake Interaction Experiments with Two Flexible Circular Cylinders in Flowing Water," *Journal of Sound and Vibration*, **45**, 259–283.

Kiya, M., et al. (1980) "Vortex Shedding from Two Circular Cylinders in Staggered Arrangement," *Journal of Fluids Engineering*, **102**, 166–173.

Ko, P. L. (1987) "Metallic Wear—A Review, with Special References to Vibration-Induced Wear in Power Plant Components," *Tribology International*, **20**, 66–78.

Ko, P. L., and H. Basista (1984) "Correlation of Support Impact Force and Fretting-Wear for a Heat Exchanger Tube," *Journal of Pressure Vessel Technology*, **106**, 69–77.

Lam, K., and W. C. Cheung (1988) "Phenomena of Vortex Shedding and Flow Interference of Three Cylinders in Different Equilateral Arrangements," *Journal of Fluid Mechanics*, **196**, 1–26.

Lever, J. H., and D. S. Weaver (1982) "A Theoretical Model for Fluid Elastic Instability in Heat Exchanger Tube Bundles," *Journal of Pressure Vessel Technology*, **104**, 147–158.

Modi, V. J., and J. E. Slater (1983) "Unsteady Aerodynamics and Vortex-Induced Aeroelastic Instability of a Structural Angle Section," *Journal of Wind Engineering and Industrial Aerodynamics*, **11**, 321–334.

Mulcahy, T. M., H. Halle, and M. W. Wambsganss (1986) "Prediction of Tube Bundle Instabilities: Case Studies," Argonne National Laboratory Report ANL-86-49.

Overvik, T., G. Moe, and E. Hjort-Hansen (1983) "Flow Induced Motions of Multiple Risers," *Journal of Energy Resources Technology*, **105**, 83–89.

Paidoussis, M. P. (1982) "A Review of Flow-Induced Vibration in Reactors and Reactor Components," *Nuclear Engineering and Design*, **74**, 31–60.

—— (1987) "Flow-Induced Instabilities of Cylindrical Structures," *Applied Mechanic Reviews*, **40**, 163–175.

Paidoussis, M. P., and S. J. Price (1988) "The Mechanisms Underlying Flow-Induced Instabilities of Cylinder Arrays in Cross Flow," *Journal of Fluid Mechanics*, **187**, 45–59.

Paidoussis, M. P., D. Mavriplis, and S. J. Price (1984) "A Potential Flow Theory for the Dynamics of Cylinder Arrays in Cross Flow," *Journal of Fluid Mechanics*, **146**, 227–252.

Pettigrew, M. J., Y. Sylvestre, and A. O. Campana (1978) "Vibration Analysis of Heat Exchanger and Steam Generator Designs," *Nuclear Engineering and Design*, **48**, 97–115.

Pettigrew, M. J., J. H. Tromp, and M. Mastaoakos (1984) "Vibration of Tube Bundles Subjected to Two-Phase Cross-Flow," in *Symposium on Flow-Induced Vibration*, Vol. 2, M. P. Paidoussis, ed., ASME, New York, pp. 251–268.

Pettigrew, M. J., et al. (1988) "Vibration of Tube Bundles in Two-Phase Cross Flow: Part 1—Hydrodynamic Mass and Damping," in *1988 Symposium on Flow-Induced Vibration and Noise*, M. P. Paidoussis, ed., Vol. 2, ASME, New York, pp. 79–104.

Price, S. J. (1975) "Wake Induced Flutter of Power Transmission Conductors," *Journal of Sound and Vibration*, **38**, 125–147.

Price, S. J., and M. P. Paidoussis (1984) "The Aerodynamic Forces Acting on Groups of Two and Three Circular Cylinders when Subject to a Cross-Flow," *Journal of Wind Engineering and Industrial Aerodynamics*, **17**, 329–347.

——— (1986) "A Single Flexible Cylinder Analysis for the Fluidelastic Instability of an Array of Flexible Cylinders in Cross-Flow," *Journal of Fluids Engineering*, **108**, 193–199.

Price, S. J., et al. (1987) "The Flow-Induced Vibration of a Single Flexible Tube in a Rotated Square Array," *Journal of Fluids and Structures*, **1**, 359–378.

Price, S. J., and P. Piperni (1986) "An Investigation of the Effects of Mechanical Damping to Alleviate Wake-Induced Flutter of Overhead Power Conductors," in *Flow-Induced Vibration—1986*, PVP—Vol. 104, ASME, New York, pp. 127–137. Also *Journal of Fluids and Structures*, **2**, 53–71 (1988).

Price, S. J., and N. R. Valerio (1989) "A Nonlinear Investigation of Cylinder Arrays in Cross Flow," in *Flow-Induced Vibration—1989*, Vol. 154, American Society of Mechanical Engineers, New York, pp. 1–10.

Roberts, B. W. (1966) "Low Frequency, Aeroelastic Vibrations in a Cascade of Circular Cylinders," *Mechanical Engineering Science Monograph No. 4*.

Simpson, A. (1971a) "Wake Induced Flutter of Circular Cylinders: Mechanical Aspects," *Aeronautical Quarterly*, **22**, 101–118.

——— (1971b) "On the Flutter of a Smooth Circular Cylinder in a Wake," *Aeronautical Quarterly*, **22**, 25–41.

Simpson, A., and J. W. Flower (1977) "An Improved Mathematical Model for the ·Aerodynamic Forces on Tandem Cylinder in Motion with Aeroelastic Application," *Journal of Sound and Vibration*, **51**, 183–217.

Small, W. M., and R. K. Young (1980) "Exchanger Design Cuts Tube Vibration Failures," *Oil and Gas Journal*, 77–80.

Soper, B. M. H. (1983) "The Effect of Tube Layout on the Fluid Elastic Instability of Tube Bundles in Cross Flow," *Journal of Heat Transfer* **105**, 744–750.

Southworth, P. J., and M. M. Zdravkovich (1975) "Effect of Grid-Turbulence on the Fluid-Elastic Vibrations of In-Line Tube Banks in Cross Flow," *Journal of Sound and Vibration*, **39**, 461–469.

Stevens-Guille, P. D. (1974) "Steam Generator Tube Failures: A World Survey of Water-Cooled Nuclear Reactors in Operation During 1972," AECL-4753.

Tanaka, H., and S. Takahara (1981) "Fluid Elastic Vibration of Tube Array in Cross Flow," *Journal of Sound and Vibration*, **77**, 19–37.

Tsui, Y. T. (1977) "On Wake-Induced Flutter of a Circular Cylinder in the Wake of Another," *Journal of Applied Mechanics*, **99**, 194–200.

Tsui, Y. T., and C. C. Tsui (1980) "Two Dimensional Stability Analysis of Two Coupled Conductors with One in the Wake of Another," *Journal of Sound and Vibration*, **69**, 361–394.

Vickery, B. J., and R. D. Watkins (1962) "Flow-Induced Vibration of Cylindrical Structures," in *Proceedings of the First Australian Conference, Held at the University of Western Australia.*

Wardlaw, R. L., K. R. Cooper, and R. H. Scanlan (1973) "Observations on the Problem of Subspan Oscillation of Bundled Power Conductors," in *Proceedings of the International Symposium on Vibration Problems in Industry, Held in Keswick, England, 1973,* U.K. Atomic Energy Authority, England, Paper 323.

Weaver, D. S., and M. El-Kashlan (1981) "The Effects of Damping and Mass Ratio on the Stability of a Tube Bank," *Journal of Sound and Vibration,* **76,** 283–294.

Weaver, D. S., and J. A. Fitzpatrick (1987) "A Review of Flow Induced Vibrations in Heat Exchanger Tube Arrays," *Journal of Fluids and Structures,* **2,** 73–94.

Weaver, D. S., and H. G. D. Goyder (1989) "An Experimental Study of Fluidelastic Instability in a 3 Span Tube Array," in *Flow-Induced Vibration—1989,* Vol. 154, American Society of Mechanical Engineers, New York, pp. 113–124.

Weaver, D. S., and D. Korogannakis (1983) "Flow-Induced Vibrations of Heat Exchanger U-Tubes: A Simulation to Study the Effect of Asymmetric Stiffness," *Journal of Vibration, Acoustics, Stress and Reliability in Design,* **105,** 67–75.

Weaver, D. S., and J. Lever (1977) "Tube Frequency Effects on Cross Flow Induced Vibrations in Tube Arrays," in *Proceedings 5th Biennial Symposium on Turbulence,* University Missouri-Rolla, pp. 323–331.

Weaver, D. S., and W. Schneider (1983) "The Effect of Flat Bar Supports on the Crossflow-Induced Response of Heat Exchanger U Tubes," *Journal of Engineering for Power,* **105,** 775–781.

Weaver, D. S., and H. C. Yeung (1983) "Approach Flow Direction Effects on the Cross-Flow Induced Vibrations of a Square Array of Tubes," *Journal of Sound and Vibration,* **87,** 469–482.

Whiston, G. S., and G. D. Thomas (1982) "Whirling Instabilities in Heat Exchanger Tube Arrays," *Journal of Sound and Vibration,* **81,** 1–31.

Yeung, H. C., and D. S. Weaver (1983) "The Effect of Approach Flow Direction on the Flow-Induced Vibration of a Triangular Tube Array," *Journal of Vibration, Acoustics, Stress and Reliability in Design,* **105,** 76–82.

Zdravkovich, M. M. (1977) "Review of Flow Interference Between Two Circular Cylinders in Various Arrangements," *Journal of Fluids Engineering,* **99,** 618–633.

―――― (1984) "Classification of Flow-Induced Oscillations of Two Parallel Circular Cylinders in Various Arrangements," in Vol. 2 of *Symposium on Flow-Induced Vibrations,* M. P. Paidoussis, et al., eds, ASME, New York, pp. 1–18.

―――― (1985) "Flow Induced Oscillations of Two Interfering Circular Cylinders," *Journal of Sound and Vibration,* **101,** 511–521.

Zdravkovich, M. M., and J. E. Namork (1979) "Structure of Interstitial Flow Between Closely Spaced Tubes in Staggered Array," in *Flow Induced Vibrations,* ASME, New York.

Chapter 6

Vibrations Induced by Oscillating Flow

Ocean waves destroyed Texas Tower No. 4, a platform off the New Jersey Coast, on January 15, 1961 (U.S. Senate, 1961). The floating platform Alexander Kielland capsized in a North Sea storm in March 1980. Floating platforms, guided towers, petroleum pipelines, and tension-leg platforms all respond to sea waves and currents. Offshore structures are now reaching depths of 300 m (1000 ft). Depths five times greater are anticipated. It is impossible to stiffen these deep-water structures out of the range of resonance with waves and vortex shedding. This chapter develops analysis for the dynamic response of structures to the oscillatory flow of ocean waves.

6.1. INLINE FORCES AND THEIR MAXIMUM

Almost all studies of fluid forces that act in line with an oscillating flow utilize the Morison approach (Morison et al., 1950; also see Sarpkaya and Isaacson, 1981): the inline fluid forces are considered to be the sum of an inertial force and a drag force. The inertial force is due to fluid acceleration and the drag force is associated with relative velocity.

The inertial force is produced by two mechanisms. First, if a flow accelerates, a pressure gradient is generated in the flow field. The pressure gradient exerts a buoyancy force on a structure in the fluid just as a buoyancy force is exerted on a structure immersed in a still fluid. The buoyancy force is equal to the mass of the fluid displaced by the structure multiplied by the acceleration of the flow. The second component of the inertial force is added mass, which accounts for fluid entrained by an accelerating structure; see Chapter 2, Section 2.2. The total inertial force per unit length on a stationary structure in an accelerating flow is the sum of the buoyancy and added mass components,

$$F_I = \rho A \dot{U} + C_a \rho A \dot{U}, \tag{6-1}$$

where ρ is the fluid density, A is the cross-sectional area, $U(t)$ is the flow velocity, and C_a is the added mass coefficient. The raised dot ($\dot{}$) denotes derivative with respect to time. The theoretical value of the added mass coefficient of a right circular cylinder in an inviscid flow is $C_a = 1.0$. Added mass coefficients for various structures are given in Chapter 2, Table 2-2, and by Blevins (1979).

The added mass force is proportional to the relative acceleration between the structure and the fluid. Thus, the net inertial force per unit length exerted on an accelerating structure in an accelerating fluid is

$$F_I = \rho A \dot{U} + C_a \rho A (\dot{U} - \ddot{x}), \qquad (6\text{-}2)$$

where $x(t)$ is the displacement of the structure in line with the flow. The coordinate system is shown in Figure 6-1.

The fluid dynamic drag on a structure acts in line with the relative velocity between the structure and the fluid. The drag force per unit length is

$$F_D = \tfrac{1}{2}\rho \,|U - \dot{x}|\,(U - \dot{x})DC_D, \qquad (6\text{-}3)$$

where C_D is the drag coefficient, D is the characteristic width of the structure used in the measurement of C_D, and $U - \dot{x}$ is the relative velocity.

The total inline fluid force per unit length is the sum of the inertial component (Eq. 6-2) and the drag component (Eq. 6-3),

$$F = \rho A \dot{U} + \rho A C_a (\dot{U} - \ddot{x}) + \tfrac{1}{2}\rho\,|U - \dot{x}|\,(U - \dot{x})DC_D. \qquad (6\text{-}4)$$

For a stationary structure, $x = \dot{x} = \ddot{x} = 0$, this expression becomes

$$F(x = 0) = \rho A C_m \dot{U} + \tfrac{1}{2}\rho\,|U|\,UDC_D, \qquad (6\text{-}5)$$

where the inertia coefficient,

$$C_m = 1 + C_a, \qquad (6\text{-}6)$$

is defined as one plus the added mass coefficient.

Equation 6-4, the Morison equation, is exact for (1) acceleration in an inviscid fluid and (2) steady velocity at Reynolds numbers above 1000. It is approximate for all other cases. However, one is more or less free to

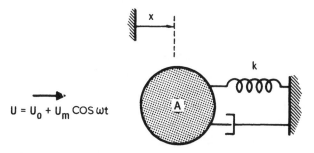

Fig. 6-1 Spring-supported structure in an oscillating flow.

pick C_m and C_D to fit experimental data, and with well-fitted coefficients, the Morison equation can match the time history of inline force over a wide variety of regular and irregular oscillating flows (Dawson, 1985; Sarpkaya and Storm, 1985; Hudspeth and Nath, 1988). Also see Keulegan and Carpenter (1958), Bishop (1985), Starsmore (1981), and Sarpkaya and Isaacson (1981) for discussion of the accuracy of Morison's equation.

It is useful here to note that the circular frequency ω of flow oscillation can be related to the frequency in hertz (cycles per second) f and the period T as follows:

$$\omega = 2\pi f = \frac{2\pi}{T}. \tag{6-7}$$

The parameter $U_m/(fD)$ often arises in ocean engineering. It is called the *Keulegan–Carpenter number*:

$$Kc = \frac{U_m}{fD} = \frac{2\pi U_m}{\omega D} = \frac{U_m T}{D}, \tag{6-8}$$

where U_m is the velocity amplitude of the flow that oscillates with frequency f. (If f is the frequency of structural motion in a steady flow of velocity U_m, then this parameter is called *reduced velocity* (Section 1.1.2).) For circular cylindrical structures, the cross-sectional area A is related to the diameter D by

$$\frac{A}{D^2} = \frac{\pi}{4}, \tag{6-9}$$

so this value can be inserted into various equations that follow.

Consider a stationary structure in a flow with a mean component and a component that oscillates sinusoidally with circular frequency ω,

$$U = U_0 + U_m \cos \omega t. \tag{6-10}$$

U_0 is the mean flow and U_m is the amplitude of the oscillation. The inline force per unit length on a stationary structure is found by substituting this equation into Eq. 6-5:

$$F(t) = -\rho A C_m \omega U_m \sin \omega t$$
$$+ \tfrac{1}{2}\rho \, |U_0 + U_m \cos \omega t| \, (U_0 + U_m \cos \omega t) D C_D. \tag{6-11}$$

The inertial and drag components add to produce the total force, which has a period $T = 2\pi/\omega$ as shown in Figure 6-2. If the mean flow is zero, $U_0 = 0$, then the average force is zero.

If the flow oscillates with circular frequency ω and the mean flow is zero, $U = U_m \cos \omega t$, then by setting the derivative with respect to time to zero it can be shown that the maximum inline force (Eq. 6-11 with $U_0 = 0$) per unit length on a stationary structure is equal to the larger of two

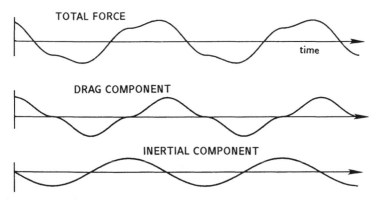

Fig. 6-2 Components of inline fluid force on a stationary structure in an oscillating flow.

values,

$$F_{T_{\max} \atop U_0=0} = \begin{cases} \rho A C_m U_m \omega, & \text{if } \dfrac{U_m}{\omega D} < \dfrac{C_m A}{C_D D^2}, \\[4mm] \frac{1}{2}\rho U_m^2 D C_D + \dfrac{\rho A^2 C_m^2 \omega^2}{2 C_D D}, & \text{if } \dfrac{U_m}{\omega D} > \dfrac{C_m A}{C_D D^2}. \end{cases} \qquad (6\text{-}12)$$

For $U_m/(\omega D) < {\sim}6$, the maximum force is due to the inertial component and the maximum value occurs at a phase angle of 90 or 270 degrees. For $U_m/(\omega D) \gg 6$, the maximum force is increasingly dominated by drag and the maximum occurs at phase angles of 0, 180, and 360 degrees; see Figure 6-2.

If the mean flow is not zero, $U_0 \neq 0$, then the maximum inline force cannot be expressed in closed form, but it can be expressed numerically or graphically. We nondimensionalize the force of Eq. 6-11 by the maximum of the inertial component:

$$F_T = \rho A C_m U_m \omega \left\{ -\sin \omega t + \frac{U_0}{U_m} \frac{U_0}{\omega D} \frac{D^2}{2A} \frac{C_D}{C_m} \right.$$

$$\left. \times \left| 1 + \frac{U_m}{U_0} \cos \omega t \right| \left(1 + \frac{U_m}{U_0} \cos \omega t \right) \right\}. \qquad (6\text{-}13)$$

The maximum of the factor in braces on the right-hand side of this equation is plotted in Figure 6-3. The phase angle at which the maximum occurs is also plotted. Again we see that, for small values of $U_m/(fD)$, the maximum force is dominated by the inertial component, while for large values and large ratios of mean flow to oscillating flow, the maximum force is dominated by drag.

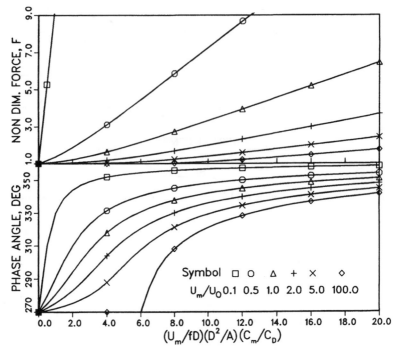

Fig. 6-3 Maximum inline force on a structure in an oscillating flow with a current.

Exercises

1. Derive Eq. 6-12 by setting the time derivative of Eq. 6-11 to zero, solving the result to the value of ωt at which the maximum occurs, and substituting the value(s) back into Eq. 6-11 with $U_0 = 0$. What is the phase angle at which the maximum forces occur? *Hint*: The derivative of the absolute value of a function is either $+$ or $-$ the derivative of that function.

2. In shallow-water wave theory, the horizontal component of velocity is given by $U_m = (H/2)(g/d)^{1/2}$, where H is the trough-to-crest wave height, d is the water depth, and g is the acceleration due to gravity ($9.8 \, \text{m/sec}^2$). In water 10 m deep, 10 sec period waves with heights of 1, 2, and 3 m pass a piling 0.6 m in diameter and 10 m high. What is the maximum total force exerted on the piling for each wave height?

3. For shallow water waves, U_m is given in Exercise 2 and the wavelength (distance between crests) is $L = 2\pi(gd)^{1/2}/\omega$, where g is the acceleration due to gravity and d is the mean water depth. Determine the relationship between cylinder diameter, wavelength,

depth, and wave height for the wave force on a cylinder ($A = \pi D^2/4$. $C_D \approx 1$, $C_m \approx 2$) to be dominated by inertia forces (Eq. 6-12). Are very large cylinders generally dominated by inertia forces? Wave forces on cylinders with diameter comparable to a wavelength are computed by *diffraction theory* (Blevins, 1984b; Sarpkaya and Isaacson, 1981).

6.2. INLINE MOTION

6.2.1. Equations of Motion and Nonlinear Solutions

The equation of motion for the structure shown in Figure 6-1 is found by applying the fluid forces of Eq. 6-4 to the one-degree-of-freedom elastic structure:

$$(m + \rho A C_a)\ddot{x} + 2m\zeta_s\omega_n\dot{x} + kx = \rho A C_m \dot{U} + \tfrac{1}{2}\rho\,|U - \dot{x}|\,(U - \dot{x})DC_D.$$

$$(6\text{-}14)$$

The added mass force term ($\rho A C_a \ddot{x}$) has been brought over to the left side of Eq. 6-14. The inline displacement is $x(t)$, m is the mass per unit length of the structure and any internal fluids, k is the structure spring constant, and ζ_s is the damping coefficient due to energy dissipation within the structure. This is a nonlinear equation owing to the nonlinear drag term on the right-hand side. There are three approaches to its solution: (1) direct numerical solutions obtained by time-history integration, (2) linearization of the equation to make possible approximate steady-state (frequency-domain) solution, and (3) static solutions obtained by neglecting the time-dependent terms dx/dt and d^2x/dt^2. Nonlinear solutions are briefly examined in this section and compared with the static solutions. Linearized solutions are discussed in the following section.

Equation 6-14 can be numerically integrated with time using schemes such as the Runge–Kutta method or the Adams–Bashforth and Euler methods (Ferziger, 1981). Accuracy requires that the time step be a small fraction, say 1/50, of both the wave period and the structural natural period. Some results computed by the author using the Adams–Bashforth second-order integration method are shown in Figure 6-4 for a harmonically oscillating flow, $U_m \cos \omega t$. The integration was started with zero displacement and integrated over 40 wave cycles to achieve steady state. The maximum displacement during the 40th cycle was recorded and nondimensionalized by the maximum static displacement. The maximum static displacement was obtained by determining the maximum wave force on a *stationary* structure (Eq. 6-12), then dividing this force

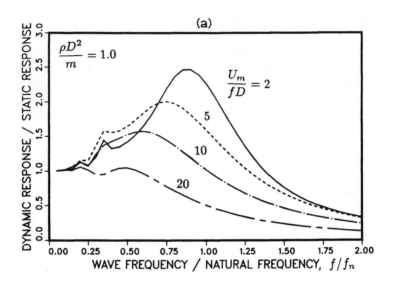

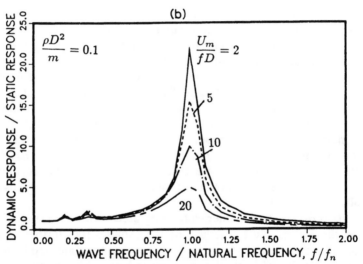

Fig. 6-4 Maximum dynamic inline response of a one-degree-of-freedom structure (Fig. 6-1) to an oscillating flow. $C_D = C_m = 1.5$, $A = \pi D^2/4$, $\zeta_s = 0$.

by the spring constant k:

$$x_{\text{static}} = \frac{F_{\max}(x = 0)}{k}. \tag{6-15}$$

The results represent the ratio between static and dynamic analysis. The horizontal axis shows the wave frequency divided by the natural frequency of the structure,

$$f_n = \frac{\omega_n}{2\pi} = \frac{1}{2\pi} \left(\frac{k}{m_t}\right)^{1/2}, \qquad (6\text{-}16)$$

where the total mass per unit length is

$$m_t = m + \rho A C_a, \qquad (6\text{-}17)$$

the sum of structural and added mass.

Figure 6-4(a) shows results for a structure whose mass is roughly equal to the displaced fluid, i.e., neutrally buoyant. Figure 6-4(b) shows results for a structure whose mass is a factor of 10 greater than the mass of the displaced fluid, for example, a solid steel structure in water. In both cases the structural damping is zero, $\zeta_s = 0$. The dynamic amplification is greatest for small $U_m/(fD)$ because here the wave force is dominated by the inertial term (Eq. 6-1) and this term is independent of structural motion, whereas for drag-dominated cases, the net drag force on the structure is reduced by sympathetic structural motion (Eq. 6-14).

If the structure is stiff, so that its natural frequency is more than five times the wave frequency (Eqs. 6-16, 6-7), then Figure 6-4 implies that the static analysis (Eq. 6-15) is adequate for predicting the structural response. However, as the wave frequency approaches the structural natural frequency, there can be significant dynamic amplification.

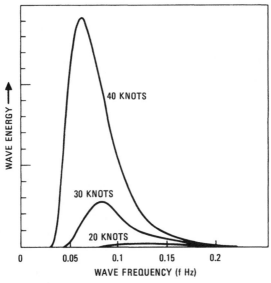

Fig. 6-5 Continuous-wave spectrum for a fully arisen sea at various wind speeds. Note that the maximum spectral energy of waves moves from high to lower frequencies with increasing wind speed. 1 knot = 1.69 ft/sec (Pierson et al., 1967).

The fundamental natural frequencies of (1) fixed bottom, steel-jacketed, tubular offshore platforms (Wirsching and Prasthofer, 1976) and (2) production riser steel piping (author's estimate) are approximately

$$f_1(\text{Hz}) \approx \begin{cases} 35/L, & \text{fixed bottom steel platform,} \\ 15/L, & \text{tensioned production riser, 0.7 m dia.,} \end{cases} \quad (6\text{-}18)$$

where L is the water depth in meters. The energy content of ocean storm wind waves, shown in Figure 6-5, is concentrated at frequencies between 0.1 and 0.05 Hz (periods between 10 and 20 sec). Thus, for the fundamental structure natural frequencies to be five times the wave frequencies, the depth can be no greater than about 40 m for risers and no greater than 100 m for platforms. These depths are shallow by modern standards. Thus, some dynamic amplification is expected in most deepwater platforms and risers and dynamic analysis is required to accurately predict the response.

Exercises

1. Write a computer program that integrates Eq. 6-14 in time. *Hint*: decompose this equation into two first-order equations, with $dx/dt = x_1$, where x_1 is the x velocity. Then use Euler's technique, $x_{i+1} = \Delta t \, (dx/dt)_i$, where i is the time stepping index, to integrate.

2. Compute a result using Exercise 1. If you have a computer, integrate over at least 10 wave periods. If you don't, then integrate by hand over two time steps. Use the parameters of Figure 6-4(a) at $f = f_n$.

3. Using Eq. 6-18, predict the depth at which risers and fixed bottom platforms will resonate, $f_{\text{wave}} = f_{\text{structure}}$, with waves of periods of 10 sec, 15 sec, and 20 sec.

6.2.2. Linearized Solutions

Equation 6-14 does not possess closed-form solutions, because of the nonlinearities in the drag force term. If the wave velocity is generally much larger than the structural velocity, as is often the case ($U \gg dx/dt$), then the nonlinear terms on the right-hand side of Eq. 6-14 can be linearized:

$$|U - \dot{x}| (U - \dot{x}) = |U - \dot{x}| \, U - |U - \dot{x}| \, \dot{x} \simeq |U| \, U - |U| \, \dot{x}$$

$$\sim |U| \, U. \quad (6\text{-}19)$$

Two levels of approximation have been introduced here: (1) Setting

$|U - dx/dt| = |U|$ introduces an error that diminishes as the amplitude of the flow velocity becomes much greater than the structural velocity; (2) the term $|U| \, dx/dt$ induces fluid damping. Neglecting fluid damping can lead to serious overestimation of the dynamic response near resonance.

Substituting the first line of Eq. 6-19 (retaining fluid damping) into Eq. 6-14 and using Eq. 6-17 gives a linearized equation of motion:

$$m_t \ddot{x} + (2m\zeta_s \omega_n + \tfrac{1}{2}\rho DC_D \, |U|)\dot{x} + kx = \rho AC_m \dot{U} + \tfrac{1}{2}\rho \, |U| \, UDC_D. \quad (6\text{-}20)$$

This is a linear equation with time-dependent coefficients, since the fluid damping term $|U| \, dx/dt$ on the left-hand side has the time-dependent coefficient $|U(t)|$. It is possible to solve this equation by expanding both sides in Fourier series and matching the coefficients term by term, but solution is considerably simpler if the time-dependent coefficient can be eliminated.

By expanding $|U|$ in a Fourier series,

$$|U| = |U_m \cos \omega t| = U_m \left[\frac{2}{\pi} + \frac{4}{3\pi} \cos 2\omega t - \frac{4}{15\pi} \cos 4\omega t + \cdots \right], \quad (6\text{-}21)$$

and by retaining only the first term, we see that the variable coefficient becomes the time average (mean) of the absolute value of velocity. With this approximation, Eq. 6-20 is recast in terms of a total damping,

$$m_t \ddot{x} + 2m_t \zeta_t \omega_n \dot{x} + kx = \rho AC_m \dot{U} + \tfrac{1}{2}\rho \, |U| \, UDC_D, \quad (6\text{-}22)$$

where the total damping coefficient ζ_t is the sum of structural and fluid components,

$$\zeta_t = \zeta_s \frac{m}{m_t} + \frac{C_D}{4} \frac{\rho D^2}{m_t} \frac{\overline{|U|}}{\omega_n D}, \quad (6\text{-}23)$$

and where $\overline{|U(t)|}$ is the time average of the absolute value of the inline velocity,

$$\overline{|U|} = \frac{1}{T} \int_0^T |U(t)| \, dt = \begin{cases} (2/\pi)U_m, & \text{if } U = U_m \cos \omega t, \\ U_0, & \text{if } U = U_0 + U_m \cos \omega t \text{ and } U_0 > U_m. \end{cases}$$

$$(6\text{-}24)$$

Equation 6-22 is a linear ordinary differential equation with constant coefficients. It can be solved exactly for the response of the structure. (Other linearization schemes have proved successful. Williamson (1985) introduced a relative velocity variable $\eta = U - dx/dt$, rewrote Eq. 6-14 in terms of this variable, and then developed an equivalent linearization. He found good agreement with experimental data. Tung and Huang (1973), Leira (1987), Langley (1984), Gumestad and Connor (1983), and Chakrabarti and Frampton (1982) discuss various linearization techniques.)

Steady-state solutions to Eq. 6-22 are called frequency-domain solutions because they give the structural response as a function of the wave frequency. Consider a mean flow with an oscillating component, that is, a sea state plus current, where the mean component exceeds the oscillating part. The inline flow velocity is given by Eq. 6-10 with $U_0 > U_m$. Other terms required for the solution of Eq. 6-22 are

$$\frac{dU}{dt} = -\omega U_m \sin \omega t,$$

$$|U| U = U^2 = U_0^2 + \tfrac{1}{2} U_m^2 + 2 U_0 U_m \cos \omega t + \tfrac{1}{2} U_m^2 \cos 2\omega t, \qquad \text{for } U_0 > U_m.$$
(6-25)

Substituting these terms into Eq. 6-22, assuming a steady-state solution at the wave frequency and its harmonics, gives

$$\frac{x(t)}{D} = b_0 + a_1 \sin \omega t + b_1 \cos \omega t + a_2 \sin 2\omega t + b_2 \cos 2\omega t. \qquad (6-26)$$

Substituting this equation into Eq. 6-22, and matching coefficients of each of the five terms, gives the following coefficients in the steady-state solution for $x(t)/D$ as a function of the circular wave frequency ω:

$$b_0 = \frac{\rho D^2}{2m} \left(\frac{U_0^2 + \tfrac{1}{2} U_m^2}{\omega_n^2 D^2} \right) C_D,$$

$$a_1 = \frac{\rho D^2}{m} \frac{U_m}{\omega_n D} \frac{\omega}{\omega_n} \left\{ 2 \zeta_t C_D \frac{U_0}{\omega D} - C_m \frac{A}{D^2} \left[1 - \left(\frac{\omega}{\omega_n} \right)^2 \right] \right\} MF_1,$$

$$b_1 = \frac{\rho D^2}{m} \frac{U_m}{\omega_n D} \frac{\omega}{\omega_n} \left\{ 2 \zeta_t C_m \frac{A}{D^2} + \frac{U_0}{\omega D} \left[1 - \left(\frac{\omega}{\omega_n} \right)^2 \right] C_D \right\} MF_1, \qquad (6-27)$$

$$a_2 = \zeta_t \frac{\omega}{\omega_n} \frac{\rho D^2}{m} \left(\frac{U_m}{\omega_n D} \right)^2 C_D MF_2, \qquad b_2 = \frac{\rho D^2}{4m} \left(\frac{U_m}{\omega_n D} \right)^2 \left[1 - \left(\frac{2\omega}{\omega_n} \right)^2 \right] C_D MF_2,$$

where the magnification factor is a measure of the amplification of the dynamic response,

$$MF_i = \left\{ \left[1 - \left(\frac{i\omega}{\omega_n} \right)^2 \right]^2 + \left(\frac{2 \zeta_t i\omega}{\omega_n} \right)^2 \right\}^{-1}, \qquad i = 1, 2, \ldots . \qquad (6-28)$$

A typical response computed using both the linearized solution and the nonlinear numerical solution is shown in Figure 6-6.

The current U_0 produces a static displacement term (b_0) and amplifies the deformation produced by oscillating components of the flow. There are two resonances associated with the dynamic response: (1) simple resonance when the wave frequency equals the natural frequency of the

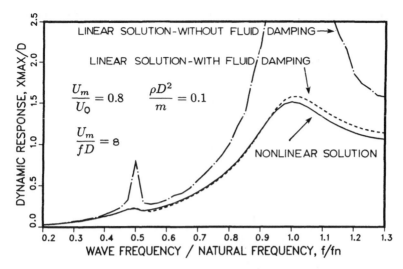

Fig. 6-6 Maximum dynamic inline response in a current plus an oscillating flow. $U_0 = 1.2U_m$, $U_m/fD = 8.0$, $\zeta_s = 0$, $A/D^2 = \pi/4$, $C_D = C_m = 1.5$.

structure, $f \simeq f_n$, and (2) subharmonic resonance when twice the wave frequency equals the natural frequency, $2f \simeq f_n$, that is, when the wave period equals twice the natural period of the structure. In both cases, the amplitude of the resonance is restrained by the fluid damping (Eq. 6-23).

Now consider a purely oscillatory flow, $U = U_m \cos \omega t$. The drag force term on the right-hand side of Eq. 6-14 can be expanded in a Fourier series,

$$U = U_m \cos \omega t,$$

$$|U|\, U = U_m^2\, |\cos \omega t| \cos \omega t = \sum_{i=1,3,5,\ldots}^{\infty} c_i \cos i\omega t, \qquad (6\text{-}29)$$

where

$$c_i = \frac{8}{\pi i (4 - i^2)} \sin\left(\frac{i\pi}{2}\right), \qquad (6\text{-}30)$$

$$c_1 = 0.8488, \qquad c_3 = 0.16977, \qquad c_5 = 0.02425.$$

Similarly, the inline deformation is expanded in a Fourier series with the unknown coefficients a_i and b_i,

$$\frac{x(t)}{D} = \sum_{i=0}^{N} (a_i \sin i\omega t + b_i \cos i\omega t).$$

This equation and Eq. 6-29 are substituted into Eq. 6-22. The coefficients of terms with the same time dependence are equated to determine a_i and b_i. The result is

$$b_0 = 0,$$

$$a_1 = \frac{\omega}{\omega_n} \frac{\rho D^2}{m} \frac{U_m}{\omega_n D} \left\{ \left[1 - \left(\frac{\omega}{\omega_n} \right)^2 \right] \frac{A}{D^2} C_m + \zeta_t c_1 C_D \frac{U_m}{\omega D} \right\} MF_1,$$

$$b_1 = \frac{\omega}{\omega_n} \frac{\rho D^2}{m} \frac{U_m}{\omega_n D} \left\{ c_1 \left[1 - \left(\frac{\omega}{\omega_n} \right)^2 \right] \frac{U_m}{\omega D} \frac{C_D}{2} + 2 \zeta_t C_m \frac{A}{D^2} \right\} MF_1,$$

$$a_i = \frac{\rho D^2}{m} \frac{i\omega}{\omega_n} \left(\frac{U_m}{\omega_{nD}} \right)^2 \zeta_t c_i C_D MF_i, \qquad i = 3, 5, 7, \ldots, \qquad (6\text{-}31)$$

$$b_i = \frac{\rho D^2}{2m} \left[1 - \left(\frac{i\omega}{\omega_n} \right)^2 \right] \left(\frac{U_m}{\omega_{nD}} \right)^2 \zeta_t c_i C_D MF_i, \qquad i = 3, 5, 7, \ldots,$$

$$a_i = b_i = 0, \qquad i = 2, 4, 6, \ldots.$$

With purely oscillatory flow, there is no mean displacement ($b_0 = 0$). However, as in the previous case, the response contains resonances with the inertial and drag components of the fluid force. When the frequency of the wave, ω, approaches the natural frequency of the structure, ω_n, or an odd submultiple of that frequency [$\omega = (1/j)\omega_n$, $j = 1, 3, 5, \ldots$], the structure resonates with a component of the fluid force. The subharmonic resonances can be seen as the bumps at $f/f_n = 1/5$ and $1/3$ in Fig. 6-4.

The existence of subharmonic resonances in the inline response has been confirmed analytically by Laiw (1987) and experimentally by Borthwick and Herbert (1988) who note that the inline response of a flexible cylinder at subharmonics is intertwined with vortex-induced transverse response as discussed Section 6.4.

Exercises

1. Estimate the number of floating-point operations (FLOPS), that is, additions, subtractions, multiplications, or divisions, required to determine the steady-state response of a simple structure to ocean waves using (a) direct time history integration for 40 flow oscillation cycles with 50 integration steps per cycle and (b) closed-form linear solution using Eqs. 6-30 and 6-31.

2. Consider an oscillating flow with $U(t) = U_m \cos 3\omega_n t$. What is the maximum dynamic response relative to that of a flow with $U(t) = U_m \cos \omega_n t$ and $U(t) = U_m \cos [1/100]\omega_n t$? Let $C_D = C_m = 1.5$ and $A/D^2 = \pi/4$, $\zeta_s = 0$.

6.2.3. Inline Response of Continuous Structures

The dynamic flexural deformation of beams, cables, and rods in an oscillatory flow is described by the following partial differential equation of motion,

$$m \frac{\partial^2 X(z, t)}{\partial t^2} + \mathcal{L}[X(z, t)] = F, \qquad (6\text{-}32)$$

with appropriate boundary conditions at the extreme ends of the span; m is the mass per unit length of the beam; $X(z, t)$ is the inline displacement, which is a function of time and the spanwise coordinate z. $\mathcal{L}[X(z, t)]$ is a linear, homogeneous, self-adjoint operator that gives the force–displacement relationship for a differential element of the riser. It is dominated by bending stiffness for relatively small structural members, but for deep-water risers the effects of mean tension and differential buoyancy can dominate. See Chakrabarti and Frampton (1982), Spanos and Chen (1980), and the example in Section 6.6.

A set of natural frequencies ω_i and mode shapes $\tilde{x}_i(z)$ is associated with the free vibration of the undamped structure. If $\mathcal{L}(X)$ is self-adjoint as is the usual case, the mode shapes are orthogonal over the span (Meirovitch, 1971),

$$\int_0^L \tilde{x}_i \tilde{x}_j \, dz = 0, \qquad i \neq j. \qquad (6\text{-}33)$$

The solutions to Eq. 6-32 are the sum of the modal displacements,

$$X(z, t) = \sum_{i=1}^{N} x_i(t)\tilde{x}_i(z). \qquad (6\text{-}34)$$

This equation is substituted into Eq. 6–32 and multiplied through by a mode shape and integrated over the span form $z = 0$ to $z = L$. Then, using Eq. 6-34, this results in a series of uncoupled ordinary differential equations,

$$m_t \ddot{x}_i + 2m_t \zeta_t \omega_i \dot{x}_i + m\omega_i^2 x_i = \rho A C_m \dot{U}\beta_i + \tfrac{1}{2}\rho \, |U| \, UDC_D\beta_i, \qquad (6\text{-}35)$$

if the flow velocity U is uniform over the span. The linearizing approximations of Section 6.2.2 have been used. The equations in this set are identical to Eq. 6-22, which describes a one-degree-of-freedom spring-supported structure, except for the influence coefficient β_i,

$$\beta_i = \int_0^L \tilde{x}(z) \, dx \Big/ \int_0^L \tilde{x}^2(z) \, dz, \qquad (6\text{-}36)$$

on the right-hand side. The influence coefficients β_i for several mode shapes are given in Table 6-1.

Table 6-1 Influence coefficients[a]

Structural element	Mode shape	Natural frequency (ω_N)	Influence factor (β_N)
Rigid cylinder	1	$\sqrt{\dfrac{k}{m}}$	1.0
Uniform pivoted rod	$\dfrac{z}{L}$	$\sqrt{\dfrac{3k_\theta}{mL^3}}$	1.5
Taut string or cable	$\sin\dfrac{n\pi z}{L}$	$n\pi\sqrt{\dfrac{T}{mL^2}}$	$0, n = 2, 4, 6, \ldots$
			$\dfrac{4}{n\pi}, n = 1, 3, 5, \ldots$
Simply supported uniform beam	$\sin\dfrac{n\pi z}{L}$	$n^2\pi^2\sqrt{\dfrac{EI}{mL^4}}$	$0, n = 2, 4, 6, \ldots$
			$\dfrac{4}{n\pi}, n = 1, 3, 5, \ldots$
Cantilevered uniform beam	$\cosh\lambda_n z - \cos\lambda_n z$ $- \sigma_r(\sinh\lambda_n z - \sin\lambda_n z)$	$\omega_1 = 3.52\sqrt{\dfrac{EI}{mL^4}}$	$\beta_1 = 0.783$
	$\sigma_1 = 0.7340$ $\sigma_2 = 1.018$	$\omega_2 = 22.03\sqrt{\dfrac{EI}{mL^4}}$	$\beta = 0.434$
	$\sigma_3 = 0.992$ $\lambda_n^4 = \omega_n^2 m/EI$	$\omega_3 = 61.70\sqrt{\dfrac{EI}{mL^4}}$	$\beta_3 = 0.254$
Platform on piles	$1 - \cos\dfrac{\pi z}{L}$		0.666

[a] m is mass per unit length, including appropriate added fluid mass (Table 2-2), I is moment of inertia, L is length, T is tension, E is modulus of elasticity.

The solution of Eq. 6-35 is discussed in Section 6.2.2 for $\beta_i = 1$; the deformations scale linearly with β_i. The total displacement is the sum of the modal displacements (Eq. 6-34). Ordinarily the majority of the response is contained in the first few modes, since the influence coefficients β_i and the magnification factors MF_i (Eq. 6-28) tend to diminish in the higher modes.

If the structure consists of multiple piles that are interconnected, as is the case for docks and offshore platforms, then there will be a phase shift between the force on one pile and the force on the next owing to the time required for the wave to travel the distance between piles. The phase difference in the force is $\omega d/c$, where ω is the wave frequency, c is the wave speed, and d is the separation distance between piles in the direction of motion. The net force is the sum of these phase-shifted components. The net deformation can be adapted from the solutions presented in this section.

Exercise

1. Often the flow velocity varies over the span of a structure, that is, $U(z, t) = \bar{u}(z)U(t)$. Using this, develop an equation of the general form of Eq. 6-35 with coefficients that take into account the effect of the spanwise variation of the flow on the response of each mode and the modal damping. Borthwick and Herbert (1988) present a solution.

6.3. FLUID FORCE COEFFICIENTS

6.3.1. Dimensional and Theoretical Considerations

The fluid force coefficients for a structure in an oscillating flow are functions of the geometry of the cross section, the maximum Reynolds number (Re), and the Keulegan–Carpenter number,

$$C_m = 1 + C_a = C_m\left[\text{Geometry}, \frac{U_m D}{\nu}, \frac{U_m}{fD} \right],$$

$$C_D = C_D\left[\text{Geometry}, \frac{U_m D}{\nu}, \frac{U_m}{fD} \right], \tag{6-37}$$

where f is the frequency of flow oscillation in hertz and ν is the fluid kinematic viscosity. There can be additional parameters. Geometry ordinarily includes both the aspect ratio (i.e., span L to diameter D ratio), the relative surface roughness, ϵ/D, and the angle of inclination of the span to the flow. For oscillatory flow, a viscous-frequency parameter β was introduced by Sarpkaya (1977),

$$\beta = \frac{D^2}{\nu T} = \frac{D^2 f}{\nu} = \frac{\text{Re}}{K_C}, \tag{6-38}$$

instead of Reynolds number $\text{Re} = U_m D/\nu$, where f is the frequency of flow oscillation in hertz. However, if there is a current, then one generally uses Reynolds number based on maximum velocity, Keulegan–Carpenter number based on the oscillatory component, $Kc = U_m/fD$, and the ratio of amplitude of the oscillatory flow to the current. To further complicate this specification, the oscillatory flow and the current may not be collinear and the orbital motion of ocean waves can influence the forces (Chaplin, 1988; Section 6.3.3). Of course, if the mean flow velocity becomes much greater than the oscillating flow or the wave period becomes very long (large reduced velocity), then the fluid force is dominated by drag, which can be predicted from steady-flow analysis; see Blevins (1984b) or Hoerner (1965).

The theoretical values of the fluid force coefficients for a smooth circular cylinder in an inviscid oscillating flow are $C_a = 1$, $C_m = 2$ (Section 2.2), and $C_D = 0$ based on the diameter D. If the fluid is slightly viscous and the Keulegan–Carpenter number is below 1 so that the flow does not separate, these coefficients are (Sarpkaya, 1986; Bearman et al., 1985)

$$C_m = 1 + C_a = 2 + 4(\pi\beta)^{-1/2} + (\pi\beta)^{-3/2}, \tag{6-39}$$

$$C_D = Kc^{-1}[\tfrac{3}{2}\pi^3(\pi\beta)^{-1/2} + \tfrac{3}{2}\pi^2\beta^{-1} - \tfrac{3}{8}\pi^3(\pi\beta)^{-3/2}], \tag{6-40}$$

where β is given by Eq. 6-38 and $Kc = U_m/(fD)$, where f is the frequency of flow oscillation. For steady flow at Reynolds numbers $\text{Re} = UD/v$ less than about 1, the drag coefficient of a circular cylinder is approximately (Batchelor, 1967)

$$C_D = \frac{8\pi}{\log_e(7.4/\text{Re})}. \tag{6-41}$$

Unfortunately, in many practical problems, the Reynolds numbers and Keulegan–Carpenter numbers are much greater than 1, these solutions do not apply, and experimental data must be used for C_D and C_m as discussed in the following sections.

6.3.2. Influence of Reynolds Number, Reduced Velocity, and Roughness

Sarpkaya (1977, 1981, 1987) has performed a series of elegant experiments in a U-shaped water channel with oscillatory flow and zero mean flow. His results are shown in Figures 6-7 and 6-8 for smooth cylinders as a function of Keulegan–Carpenter number. The experiments of Chaplin (1988) suggest that the drag coefficient for smooth circular cylinder in oscillating flow dips at a Reynolds number of 10^5.

$U_m/(fD) = 10$

UD/v	10^4	2×10^4	4×10^4	6×10^4	8×10^4	10^5	2×10^5	3×10^5
C_D	2.05	1.70	1.1	0.9	0.7	0.7	0.75	0.9
C_m	0.85	0.9	1.3	1.7	1.75	1.75	1.8	1.8

$U_m/(fD) = 20$

UD/v	10^4	2×10^4	4×10^4	6×10^4	8×10^4	10^5	2×10^5	4×10^5
C_D	2.0	1.7	1.2	0.9	0.75	0.7	0.65	0.7
C_m	0.85	1.0	1.3	1.5	1.6	1.7	1.75	1.75

Figure 6–9 gives drag coefficients in steady flow.

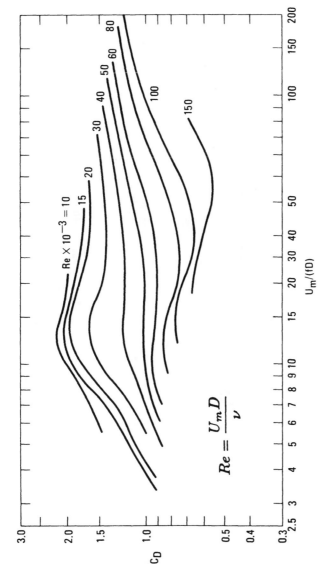

Fig. 6-7 Drag coefficient versus Keulegan–Carpenter number for a smooth circular cylinder in an oscillating flow (Sarpkaya, 1976).

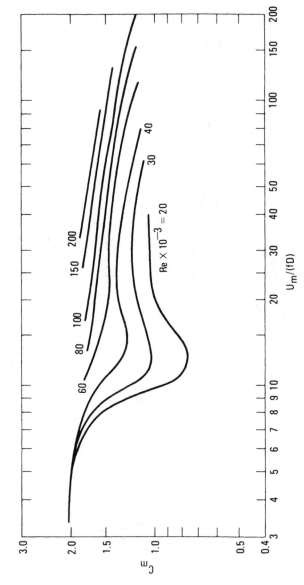

Fig. 6-8 Inertial coefficient versus Keulegan–Carpenter number for a smooth circular cylinder in an oscillating flow (Sarpkaya, 1976).

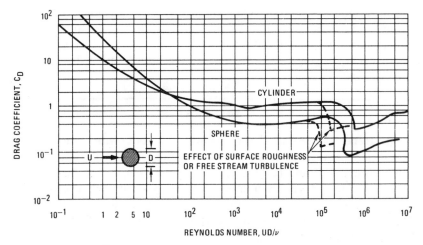

Fig. 6-9 Drag coefficient in a steady flow for smooth circular cylinders. (Massey, 1979.)

Roughness increases the drag coefficient of circular cylinders in steady flow at Reynolds numbers above 10^5 as follows (Norton et al., 1981):

ϵ/D	$<10^{-5}$	10^{-5}	10^{-4}	5×10^{-4}	10^{-3}	2×10^{-3}	10^{-2}	5×10^{-2}	10^{-1}
C_D	0.6	0.63	0.68	0.75	0.85	1.0	1.0	1.1	1.2

Mussels, barnacles, kelp, and sea anemones create surface roughness on offshore structures (Wolfram and Theophanatos, 1985; Nath, 1981). They increase the effective diameter and increase the surface roughness (Nath, 1988). Typically marine roughness ranges from 1 to 100 mm (Norton et al., 1981). In the North Sea, marine growth is monitored, and when it exceeds the design allowance it is scraped off (Wolfram and Theophanatos, 1985). Sarpkaya (1977) and Kasahara (1987) found that roughness as low as $\epsilon/D = 0.005$ would double the drag coefficient and the inertia coefficient can be increased or decreased by roughness, depending on the Keulegan–Carpenter number.

Bishop (1985) reports a series of measurements made on full-scale instrumented piles 0.48 and 2.8 m in diameter in Christchurch Bay at Reynolds numbers of 1.5×10^5 to 2×10^6 for a range of depth, current, and wave conditions. The measured minimum and maximum drag and inertia coefficients are as follows:

$U/(fD)$	<10	12	14	16	18	20	25	30	>40
C_D (min)	1.0	0.90	0.80	0.76	0.70	0.64	0.56	0.52	0.48
C_D (max)	1.0	0.92	0.82	0.81	0.77	0.74	0.70	0.68	0.66

$U/(fD)$	<2	4	6	8	10	12	14	>16
C_m (min)	2.00	2.06	1.98	1.80	1.60	1.47	1.42	1.40
C_m (max)	2.03	2.07	2.04	1.94	1.84	1.80	1.80	1.80

Here U refers to the root-mean-square velocity, including waves and current, and f is the dominant frequency of oscillating components. The coefficients are consistent with Sarpkaya's laboratory measurements for smooth cylinders (Figs. 6-7 and 6-8) at the upper end of his Reynolds number range.

Exercises

1. Repeat Exercise 2 of Section 6.1 but with fluid force coefficients estimated from the data in this section. Assume (1) a smooth cylinder and (2) a roughness $\epsilon = 0.5$ cm due to barnacles. What are the Reynolds number Re, the viscous frequency parameter β, C_m, C_D, and the maximum force per unit length? The kinematic viscosity of seawater at 10°C is $\nu = 1.35 \times 10^{-6}$ m^2/sec.

2. Continuing Exercise 1, what effect do 10% variations in C_D and C_m have on the maximum predicted fluid force for the smooth and rough cylinders? Also see Labbe (1983).

3. An experimental measurement gives the inline force on a stationary cylinder in an oscillatory flow as a function of time. Analysis must be made to break down the time record into the coefficients C_D and C_m. Propose three methods for accomplishing this and develop the mathematics for one of these. For help, see Sarpkaya (1981, 1982), Keulegan and Carpenter (1958), and Bishop (1985).

4. Plot Bishop's and Chaplin's data for C_D and C_m, above, on Figures 6-7 and 6-8. Are the two sets of data mutually consistent?

6.3.3. Dependence on Inclination and Proximity

In general, the flow field surrounding a slender structure is three-dimensional. It is always possible to resolve the fluid velocity components into a velocity vector at an angle α to the axis of the structure. Similarly, the net force can be resolved into components parallel to and perpendicular to the structural axis, as shown in Fig. 6-10. These normal and tangent forces are functions of the angle α.

Steady-flow tests of marine cables, rigid cylinders, and cylindrical pilings show that the drag force normal to the axis of a long cylinder varies as the normal component of flow velocity (Springston, 1967; Norton et al., 1981; Nordell and Meggitt, 1981)

$$F_{DN} = F_D(\alpha = 90 \text{ deg}) \sin^2 \alpha. \tag{6-42}$$

This is called the independence principle or cross-flow principle. Garrison

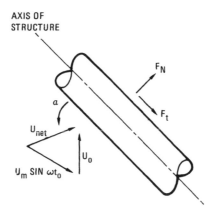

Fig. 6-10 Relative velocity in a three-dimensional flow.

(1985) and Cotter and Chakrabarti (1984) found agreement with the independence principle for cylinders in oscillating flows. However, Stansby et al. (1983), Bishop (1985), Borthwick and Herbert (1988), and Chaplin (1988) have found some dependence on the particle orbit in simulated ocean waves. Sarpkaya's experiments (1982) show that the independence principle tended to underestimate normal forces on inclined cylinders.

The independence principle fails for noncircular sections. For example, the normal loading function for a cable with a streamline fairing [Fig. 3-23(d)] contains multiple harmonic terms (Springston, 1967):

$$\frac{F_{DN}}{F_D} (\alpha = 90 \deg) = -1.064 + 1.263 \cos \alpha + 1.865 \sin \alpha - 0.1993 \cos \alpha$$

$$- 0.6926 \sin 2\alpha. \tag{6-43}$$

Thus, it is likely that the independence principle (Eq. 6–42) is only an approximation even for circular cylinders. It is useful until more complete data become available.

In general, the inertial coefficients increase as a cylinder approaches the seabed or as it approaches one or more cylinders (Dalton, 1980; Heideman and Sarpkaya, 1985; Moretti and Lowery, 1976; Laird, 1966; Chen, 1975). Sarpkaya (1976, 1981) has measured the fluid force coefficients for a circular cylinder approaching a wall parallel to the direction of oscillating flow at Reynolds numbers between 4×10^3 and 25×10^3. The results are shown in Figure 6-11. Note that the coefficients increase as the gap decreases. Also see Section 5.4 and Wright and Yamamoto (1979) and case 7 of Table 2-2.

Large fluid forces perpendicular to the wall have been observed for a circular cylinder approaching a wall in both steady flow (see Section 5.4)

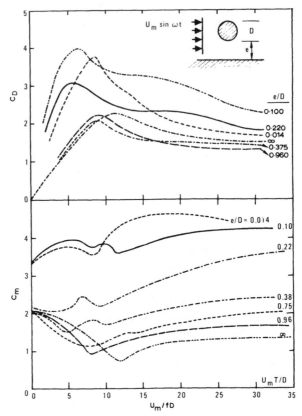

Fig. 6-11 Drag and inertial coefficients for a circular cylinder in an oscillating flow next to a wall (Sarpkaya, 1976).

and oscillating flow. Sarpkaya found that fluid forces per unit length perpendicular to the wall on a circular cylinder in a sinusoidally oscillating flow can vary from $5(\frac{1}{2}\rho U_m^2 D)$ away from the wall to $3(\frac{1}{2}\rho U_m^2 D)$ toward the wall during one cycle of fluid oscillation.

6.4. TRANSVERSE FORCE AND RESPONSE

Circular cylinders shed vortices in oscillating flow. Vortex shedding produces lift forces transverse to the direction of flow that can induce substantial transverse vibration. For Keulegan–Carpenter numbers above about 30, the shedding period $T_s \approx 5D/U_m$ is a small fraction of the flow oscillation period $T = KcD/U_m$ and steady-flow vortex shedding data of Chapter 3 can be applied using the instantaneous flow velocity. For Keulegan–Carpenter numbers less than 30, the shedding period becomes an appreciable fraction of the flow oscillation period and the vortex shedding and flow oscillation interact strongly.

Obasaju et al. (1988), Williamson (1985), Bearman et al. (1984), Bearman and Hall (1987), Sarpkaya (1981, 1975), Dalton and Chantranuvatana (1980), and Zdravkovich and Namork (1977) have experimentally observed the complex patterns of vortices shedding from circular cylinders in oscillating flows. Their results tend to classify the phenomena as follows:

$Kc < 0.4$ The flow does not separate and there are no transverse forces. Equations 6-39 and 6-40 apply.

$0.4 < Kc < 4$ A symmetric pair of vortices is formed in the wake. These vortices reverse during the flow oscillation cycle. Lift forces are minimal.

$4 < Kc < 8$ One of the vortices in the pair becomes stronger and the vortex pair becomes asymmetric. The dominant frequency of lift oscillation (f_s) is twice the frequency of flow oscillation (f).

$8 < Kc < 15$ Vortex pairs are shed alternately into the wake during each half-cycle of flow oscillation. The vortex pairs convect alternately asymmetrically at approximately 45 degrees to the direction of flow oscillation. The dominant frequency of lift oscillation and vortex shedding is twice the flow oscillation frequency, $f_s/f = 2$.

$15 < Kc < 22$ Multiple pairs of vortices are shed in each flow oscillation cycle and the pairs convect at 45 degrees to the direction of flow oscillation. The dominant frequency of vortex shedding and lift oscillation is three times the flow oscillation frequency, $f_s/f = 3$.

$22 < Kc < 30$ Multiple pairs of vortices are shed per cycle. The dominant frequency of vortex shedding and lift oscillation is four times the frequency of flow oscillation, $f_s/f = 4$.

$Kc > 30$ Quasisteady vortex shedding. The frequency of vortex shedding is roughly the nearest multiple of the flow oscillation frequency, corresponding to the Strouhal relationship,

$$f_s/f = 2, 3, 4, 5, \ldots = \text{an integer} \approx 0.2 Kc,$$

$$\text{thus } f_s \approx 0.2 U_m/D. \quad (6\text{-}44)$$

That is, the shedding frequency is given by the Strouhal relationship (Eq. 3-2 with $S = 0.2$) rounded to an integer multiple of the wave frequency f. Subharmonics are often strongly represented. These effects

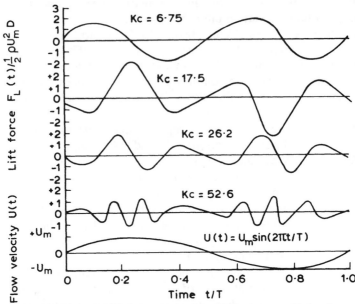

Fig. 6-12 Time-history of lift force on circular cylinder in an oscillating flow over one period (T) of flow oscillation for various Keulegan–Carpenter numbers (Obasaju et al., 1988).

are shown in Figure 6-12, which gives the oscillating lift force during one cycle of flow oscillation for various Keulegan–Carpenter numbers.

Bearman et al. (1984), Bearman and Hall (1987), and Graham (1987) have proposed the following model for the lift force per unit length on a circular-cylinder in harmonically oscillating flow:

$$F_L(t) = \tfrac{1}{2}\rho U_m^2 D C_L \cos\phi \sin^2(2\pi ft) \qquad (6\text{-}45)$$

where ϕ is a function of time

$$\phi = 0.2Kc[1 - \cos(2\pi f)] + \psi. \qquad (6\text{-}46)$$

f is the frequency of flow oscillation, $U(t) = U_m \sin 2\pi ft$, and ψ is a constant. Equations 6-45 and 6-46 are valid over the first half of the flow oscillation cycle. The negative must be used on the second half. The square reproduces the peaking that is evident in Figure 6-12. Graham (1987) has found Eq. 6-45 to be in good agreement with data above $Kc = 20$.

Measurements of the lift coefficient C_L for stationary cylinders in oscillatory flow show that C_L diminishes in the higher Keulegan–Carpenter number range.

$U_m/(fD)$

	5	10	15	20	25	30	35	40	$Re = U_m D/v$
C_L (Bearman et al., 1984)	4.5	3.0	2.5	2.0	1.6	1.5	1.4	1.3	2.2×10^4
C_L (Sarpkaya, 1977)	1.2	3.7	3.4	2.8	2.4	2.0	1.8	1.4	3×10^4
C_L (Sarpkaya, 1977)	—	—	—	—	0.6	0.5	0.45	0.4	10^5
C_L (Kasahara et al., 1987, smooth)	0.4	1.4	1.2	1.0	0.7	0.4	0.4	0.35	10^6
C_L (Kasahara et al., 1987, rough)	0.8	2.2	1.8	1.8	1.4	1.5	1.1	1.0	10^6

Sarpkaya and Isaacson (1981) found that C_L is strongly a function of cylinder motion, as is the case with vortex shedding in a steady flow (Chapter 3, Sections 3.3 and 3.5). Obasaju et al. (1988) and Borthwick and Herbert (1988) similarly found that finite spanwise correlation of vortex shedding [Obasaju et al. (1988) found a spanwise correlation length of 4.6 diameters] influences the net lift force on the cylinder span.

Vortex shedding produces lift forces that can induce substantial transverse vibration of elastic cylinders. The transverse vibration is greatest when the vortex shedding frequency, f_s (Eq. 6-44), or its harmonics coincide with the natural frequency, f_n (Angrilli and Cossalter, 1982; Bearman and Hall, 1987). $f_s \approx f_n$ occurs when $Kc(f/f_n) \approx 5$ (see Chapter 3, Eq. 3-5). Figure 6-13 shows that the peak transverse displacement of an elastically supported circular cylinder in oscillatory flow with a ratio of frequency of flow oscillation to cylinder natural frequency of $f/f_n = 1/4.4$ occurs at $Kc = 18.7$, 23.1, and 33, which

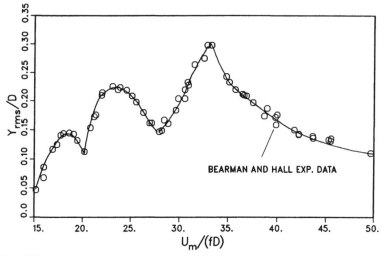

Fig. 6-13 Transverse response of an elastically supported circular cylinder to oscillatory flow; $m/(\rho D^2) = 2.58$, $\zeta_s = 0.08$. (Bearman and Hall, 1987.)

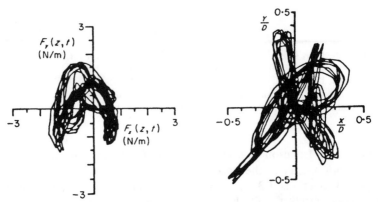

Fig. 6-14 Orbits of force and displacement of an elastically mounted cylinder to inline and vortex forces in oscillatory flow; $Kc = 11.8$, $f/f_n = 0.382$ (Borthwick and Herbert, 1988).

corresponds to $Kc(f/f_n) = 4.25$, 5.25, and 7.5. The combination of inline forces and transverse vortex forces produces orbital motion with roughly equal inline and transverse responses, as shown in Figure 6-14.

In both oscillatory and steady flow, vortex shedding produces a resonant response, a lock-in band, and a self-limiting maximum amplitude of about two to three diameters peak-to-peak (Bearman and Hall, 1987; Sarpkaya and Isaacson, 1981; Zedan et al., 1980; Chapter 3). It is reasonable to apply steady flow results of Chapter 3 to oscillatory flow, at least at higher Keulegan–Carpenter numbers. The principal differences are that, in oscillatory flows, the shedding frequency tunes itself to an integer multiple of the wave frequency (Eq. 6-44), and harmonics are much more in evidence as vortices formed during previous cycles wash back into the current shedding cycle. Because the shedding frequency can vary over the cycle of flow oscillation (Fig. 6-12), resonances can be formed and broken over a single wave period.

Rogers (1983) discussed devices for suppression of vortex shedding in the oscillatory flows of ocean waves; also see Chapter 3, Section 3.6.

Exercises

1. Compare the predictions of Eq. 6-44 with the experimental data of Figure 6-12 by counting the shedding cycles per period. How could the predictions be improved?

2. Compare the predictions of Eq. 6-45 with one of the time histories of Figure 6-12. Comment on the accuracy of the theory.

3. Compare Figure 3-31 with Figure 6-14. Explain the differences and similarities.

6.5. REDUCTION OF VIBRATION INDUCED BY OSCILLATING FLOW

The amplitude of vibrations of a structure in oscillating flow can be reduced by either reducing the amplitude of the fluid oscillation or modifying the structure as follows:

1. *Avoid resonance.* If the fundamental natural frequency of vibration of the structure is greater than five times the frequency of the oscillating flow, then inline resonances through the fifth harmonic are avoided. If the fundamental natural frequency of the structure is raised at least 30% above the vortex shedding frequency (Eq. 6-46), then the possibility of transverse resonance with vortex shedding is avoided. Resonances are ordinarily avoided by stiffening the structure by increasing the diameter of structural members and providing internal bracing.

2. *Increase ratio of structural mass to displaced fluid mass.* The amplitude of vibrations is proportional to $\rho D^2/m$. If the mass m of the structure can be increased without an offsetting decrease in natural frequency, vibration amplitude will be reduced. For tubular structures, thickening the tube wall does not greatly alter their natural frequency, but it does increase structural mass and stiffness with the result that vibrations are reduced.

3. *Modify cross section.* If the cross section is modified to reduce fluid forces, vibrations are reduced. Most offshore structures are fabricated from tubular frames. The circular cross section of tubes is prone to both inline vibration and vortex-induced vibration. By using airfoil-shaped cross sections, drag can be greatly reduced and vortex shedding eliminated, provided the airfoil is stable and always pointed into the flow. In most ocean flows, this requires a pivoting fairing with the associated mechanical complexity and questions of reliability and stability. See Section 3.6, Chapter 4, Gardner and Cole (1982), and Rogers (1983).

6.6. EXAMPLE: OCEAN WAVE-INDUCED VIBRATION OF A RISER

One of the major components in the offshore drilling for petroleum is the marine riser. The marine riser has three functions. First, it guides the drilling string from the vessel to the well; second, it provides a path for the drilling mud; and third, it furnishes the casing for the flow of petroleum. The riser casing is a steel pipe, 12 to 24 in (0.5 to 0.7 m) in diameter and typically 1 in (2 cm) thick and 150 to 3000 ft (50 to 1000 m)

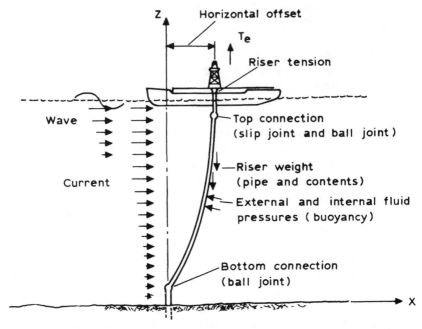

Fig. 6-15 Typical ship and riser configuration (Chakrabarti and Frampton, 1982).

or more long. A typical cross section is shown in Chapter 3, Fig. 3-27(b). The riser section is supported at the top by a platform as shown in Figure 6-15. The top tension is typically 20% greater than that required to support the weight of the riser. This weight can range from 200 000 to 2 million pounds (100 000 to 1 000 000 kg).

Because the riser is a slender member under high tension, it is often possible to approximate the riser natural frequencies from the frequencies of a tensioned string with no bending stiffness. These frequencies are

$$f_{\text{string}} = \frac{1}{T} = \frac{i}{2L} \left(\frac{T_e}{m} \right)^{1/2}, \qquad i = 1, 2, 3 \quad \text{Hz.} \tag{6-47}$$

L is the span, T_e is the tension, m is the mass per unit length, and i is the mode number. Consider the riser described by Spanos and Chen (1980). This riser is 500 ft (152 m) in length, with a mean tension of $T_e = 393\,000$ lb (1.7×10^6 N), stiffness $EI = 6.53 \times 10^8$ lb-ft^2 (270×10^6 N-m^2), a diameter of 2 ft (0.6 m), and a mass including added mass of 669.8 lb/ft (20.80 lbm-sec^2/ft^2, 999 kg/m). The computed periods are listed below in comparison with those obtained by Spanos and Chen using the finite-element method with both tension and bending stiffness:

| | Period (sec) | | | | |
Mode	1	2	3	4	5
Spanos and Chen	7.56	3.43	2.01	1.32	0.919
Tensioned string	7.27	3.64	2.42	1.82	1.455

These results imply that that bending stiffness is relatively unimportant in the lower modes of a tensioned riser.

The riser is exposed to a wave with peak-to-trough height of 13 ft (4 m) and a period of 6.7 sec. The maximum horizontal velocity of 6.09 ft/sec (1.9 m/sec) at the surface decays with depth as $\exp(-2\pi z/L_w)$, where z is water depth and $L_w = 230$ ft (70 m) is the wavelength. The relevant nondimensional parameters for this system are

$$\frac{U_m}{fD} = 20, \qquad \frac{m}{\rho D^2} = 2.62, \qquad \frac{A}{D^2} = 0.785, \qquad \frac{U_m D}{\nu} = 810\,000$$

for a seawater density of $\rho = 64$ lb/ft^3 (1025 kg/m^3) and a kinematic viscosity of $\nu = 1.5 \times 10^{-5}$ ft^2/sec (1.4×10^{-6} m^2/sec). With the data of Section 6.3.2, these imply $C_D = 0.7$ and $C_m = 1.6$. Equation 6-12 implies that the wave forces will be dominated by drag.

Calculation of the total response is complex because the wave force varies continuously over the riser and multiple modes are present. Here we will make some approximations to illustrate the major response. Because the wave period is nearly equal to the natural period of the fundamental riser mode, and there is some uncertainty in both quantities, it is reasonable that the wave frequency may equal the fundamental natural frequency and a resonance will result. The resonant response can be obtained from Eqs. 6-26 and 6-31 for a single-degree-of-freedom system. Limiting attention to a single mode at resonance, $f = f_n$, neglecting any contribution from the inertial component of force by setting $C_m = 0$, and neglecting the contribution of higher harmonics, we find that Eq. 6-31 gives

$$\frac{x(t)}{D} = \frac{c_1 C_D \rho D^2}{4\zeta_t \, m} \left(\frac{U_m}{\omega_n D}\right)^2 \sin \omega t. \tag{6-48}$$

The damping factor ζ_t is given by Eq. 6-23. Neglecting structural damping ζ_s and substituting this equation into the previous equation, we have a simple expression for the resonant response due to drag,

$$\frac{x(t)}{D} = \frac{c_1 \pi}{2} \frac{U_m}{\omega_n D} \sin \omega t = 4.85 \sin \omega t,$$

where $\omega_n = 2\pi/T = 0.838$ rad/sec. Thus, the response amplitude is about 4.9 diameters or about 20 ft (6.1 m), peak to peak. This is, of course, a one-degree-of-freedom calculation. Table 6-1 implies that this response should be multiplied by $4/\pi$ to obtain the continuous structure response at midspan of the fundamental mode.

There is another complication. The wave-induced horizontal velocity diminishes rapidly with depth. At a depth of 50 ft (15 m), the wave amplitude has decayed to 25% of its surface amplitude. This reduces the amplitude of vibration to the order of 0.5 diameter rather than 4.9 diameters as computed above. The details of this calculation are left to the student, with Section 6.2.3 as a guide.

Equation 6-44 implies that the wave-induced vortex shedding frequency for this system is 0.61 Hz, that is, a period one-eleventh of the wave period and the fundamental natural frequency of the riser. This frequency will vary continuously over the riser span with depth as follows:

	Depth, ft						
	0	10	20	30	50	70	100
Wave velocity, ft/sec	6.1	4.6	3.5	2.7	1.55	0.9	0.40
Shedding freq, Hz	0.61	0.46	0.35	0.27	0.16	0.09	0.04
Shedding period, sec	1.64	2.17	2.86	3.70	6.25	11.0	25.0

Because of the rapid variation of shedding frequency, shedding will be correlated only over a small fraction of the riser span at any one frequency. Hence, the potential for causing significant vortex-induced vibration of the lower riser modes is minimal (see Section 3.8). However, it is possible that the current could excite significant overall vortex-induced motion as discussed in Chapter 3, Section 3.4.

Exercises

1. Calculate the influence factors for the fundamental mode of the riser exposed to deep-water waves,

$$\bar{x}(z) = \sin\left(\frac{z\pi}{L}\right), \qquad U_m(z) = U_m \exp\left(\frac{-2\pi z}{L_w}\right),$$

using Eqs. 6-34 and 6-35 as a guide. Neglect the inertial forces $(C_m = 0)$. The result will be a factor similar to Eq. 6-36 but including the influence of depth-dependent velocity. Evaluate for the parameters of Section 6.6.

2. The horizontal displacement of fluid particles in harmonically oscillating flow is $(U_m/\omega) \sin \omega t$. Is it possible for the response of a cylindrical elastic structure in oscillating flow to exceed the fluid displacement? Discuss.

3. The fundamental natural frequency of a uniform simply supported beam in bending is given by $f_b = (\pi/2L^2)\sqrt{EI/m}$ Hz, where EI is the bending stiffness and L is the length. Compare this equation with Eq. 6-47. For what range of length and stiffness will the natural frequency of a tensioned beam be dominated by bending stiffness, and for what range will it be dominated by tension? What about higher modes?

6.7. SHIP MOTION IN A SEAWAY

6.7.1. Description of Ship Motions

Motions of ships in a seaway have been studied since ancient times, yet most ships are designed with little dynamic analysis. There are two reasons for this. First, the motion of the surface of the sea does not obey a deterministic law, and second, the interaction of a ship and a wave is itself a very difficult problem. The first step in the analysis of ship motions was to unravel the complex motion of the surface of the sea. St. Dennis and Pierson (1953) recognized that the random nature of the sea is the result of the superposition of many sinusoidal waves traveling with various amplitudes and frequencies in different directions. Thus, within the limits of linearity, the response of a ship to a complex sea can be found by summing the responses of a ship to a series of regular, sinusoidal, waves (Price and Bishop, 1974; Lloyd, 1989). It is possible to simplify the calculation of fluid forces on the ship by neglecting the forward speed of the ship and computing the pressure on the hull from hydrostatic principles. This is the approach followed in this section. More advanced theories incorporate the effect of forward speed by solving the inviscid problem of wave–ship interaction (Newman, 1977; Salvesen et al., 1970; Meyers et al., 1975; Lloyd, 1989).

The fundamental difference between wave-induced ship motions and the flow-induced vibrations discussed previously is that the ship motions are primarily rigid-body motions. The deformation of the ship is ordinarily very small compared with the rigid-body translations and rotations of the ship (Bishop and Price, 1979). The ship motion is specified by three translations and three rotations, as shown in Figure 6-16. The right-hand coordinate system (x, y, z) is fixed with respect to the mean position of the ship, with z traveling vertically upward through the center of gravity of the ship, x traveling in the direction of forward motion, and the origin lying in the plane of the undisturbed free surface

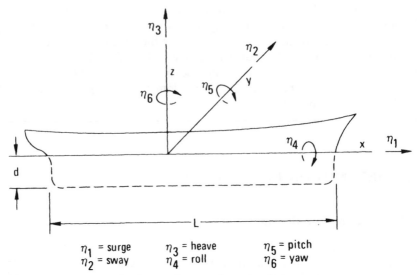

η_1 = surge $\quad\quad$ η_3 = heave $\quad\quad$ η_5 = pitch
η_2 = sway $\quad\quad\quad$ η_4 = roll $\quad\quad\quad$ η_6 = yaw

Fig. 6-16 Sign convention for translatory and angular displacements.

of the sea. The three translations are:

η_1 = surge; that is, translatory oscillation in the direction of headway,

η_2 = sway; that is, horizontal, translatory oscillation perpendicular to the direction of headway,

η_3 = heave; that is, translatory oscillation in the vertical direction.

The three rotations are:

η_4 = roll; that is, rotational oscillation about the longitudinal axis,

η_5 = pitch; that is, rotational oscillation about the transverse axis,

η_3 = yaw; that is, rotational oscillation about the vertical axis.

Of these six motions, only three—pitch, roll, and heave—are opposed by hydrostatic pressure during ship motion at zero forward speed. If the ship has a long, slender hull form, then surge motions are generally negligible in comparison with sway and heave.

6.7.2. Stability and Natural Frequency

A freely floating ship, at equilibrium in an undisturbed sea, is acted upon by two opposite equal forces: the upward force of buoyancy and the downward force of gravity (Fig. 6-17). The coordinates of the center of gravity are

$$\bar{x} = \int_V x \frac{dM}{M}, \qquad \bar{y} = \int_V y \frac{dM}{M}, \qquad \bar{z} = \int_V z \frac{dM}{M}, \qquad (6\text{-}49)$$

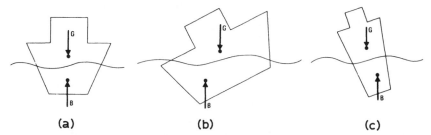

Fig. 6-17 Stability of a ship in roll: (a) equilibrium; (b) stable ship (moment tends to return ship to equilibrium); (c) unstable ship (moment tends to increase roll).

where the x, y, z coordinates are considered to be fixed with respect to the ship, and dM is an element of the mass of the ship that occupies the volume V and has total mass M.

The hydrostatic components of fluid pressure can be resolved into a net buoyancy force that acts upward on a point called the center of buoyancy, which is the centroid of the portion of the ship below the waterline. As the ship oscillates with respect to the sea, the position of the center of buoyancy changes with respect to the hull, because the portion of the ship hull below the water line changes.

The center of gravity lies above the center of buoyancy in nearly all surface ships. At equilibrium, the center of buoyancy is directly below the center of gravity (Fig. 6-17(a)). If the ship rolls slightly and the center of buoyancy moves in the direction of roll (Fig. 6-17(b)), then a righting moment $F_4 = F_4(\eta_4)$ tends to restore the ship to equilibrium and the ship is stable. That is, the ship is stable if $\eta_4 > 0$ implies $F_4(\eta_4) < 0$, or equivalently $dF_4/d\eta_4 < 0$ so that a hydrodynamic moment counters a perturbation in roll. In the linear range, the roll righting moment can be approximated by a linear equation,

$$F_4(\eta_4) = -\rho g V_d \overline{GM} \eta_4, \qquad (6\text{-}50)$$

where ρ is the density of seawater, V_d is the volume of water displaced by the hull, that is, the mass of the ship is $\rho V_d g$, and g is the acceleration due to gravity. The proportionality constant $\overline{GM}$ is called the metacentric height and it is a positive length of approximately 2% to 4% of the beam, 4 to 40 in (0.1 to 1 m) for many ships (Bascom, 1980; Muckle and Taylor, 1987). Rawson and Tupper (1976) present manual methods and Patel (1989) presents numerical methods for the calculation of $\overline{GM}$ and stability.

If the metacentric height is negative, that is, the center of buoyancy moves away from the direction of roll (Fig. 6-17(c)), then the resulting moment tends to increase the roll angle. The ship is unstable and continues to roll until it has turned over. Nearly all surface ships become unstable if they are rolled to sufficiently large angles because the stabilizing qualities of the hull are negated if a large portion of the hull is rolled out of the water. Submarines and buoys achieve stability by having the center of buoyancy above the center of gravity, a condition called pendulum stability, rather than by obtaining sufficient metacentric height. Submarines achieve pendulum stability by using buoyancy tanks in the upper portions of the hull. Buoys achieve pendulum stability by the addition of heavy weights to a tether at the submerged end of the buoy.

Neglecting viscous resistance, the linear equation of motion of a stable ship in roll is

$$(I_{44} + A_{44})\ddot{\eta}_4 + \rho g V_d \overline{GM} \eta_4 = 0, \qquad (6\text{-}51)$$

where $I_{44} = \int_{V_d} (z^2 + y^2) \, dM$ is the polar mass moment of inertia of the ship for roll about the longitudinal axis and A_{44} is the polar mass moment of inertia of the added mass of water entrained by the hull. Typically, A_{44} is about 30% of I_{44} (Blagoveshchensky, 1962).

The solution of Eq. 6-51 oscillates about vertical,

$$\eta_4 = A \sin \omega_4 t. \qquad (6\text{-}52)$$

By substituting this solution into Eq. 6-51, the natural frequency of roll, in hertz, is found to be

$$f_4 = \frac{\omega_4}{2\pi} = \frac{1}{2\pi} \left(\frac{\rho g V_d \overline{GM}}{I_{44} + A_{44}} \right)^{1/2}. \qquad (6\text{-}53)$$

This formula can be simplified by noting that the polar mass moment of inertia of the ship about the longitudinal axis is proportional to the mass of the ship multiplied by the square of the beam: $I_{44} + A_{44} \sim \rho V_d b^2$, where b is the beam of the ship. This gives a natural frequency, in hertz, of the ship in roll:

$$f_4 = \frac{0.35(g\overline{GM})^{1/2}}{b} \approx 0.08(g/b)^{1/2}, \qquad (6\text{-}54)$$

where $\overline{GM}$ is roughly $0.05b$ for large ships. The larger the ship, the lower the roll frequency. The typical frequency of roll for large ships is between one cycle every 4 sec and one cycle every 30 sec.

Heave is the simplest ship motion to analyze. If a ship is submerged below its equilibrium by a small displacement $-\eta_3$, an excess buoyancy force equal to $\rho g S \eta_3$, where S is the plane area enclosed by the waterline, tends to raise the ship. Thus, the linearized equation for pure heave, neglecting dissipative terms, is

$$(M + A_{33})\ddot{\eta}_3 + \rho g S \eta_3 = 0, \qquad (6\text{-}55)$$

where η_3 is the heave displacement, which is positive for upward motion. M is the mass of the ship and A_{33} is the added mass of the water entrained by the heaving hull. The natural frequency of oscillation of the heaving ship is easily found to be

$$f_3 = \frac{1}{2\pi} \left(\frac{\rho g S}{M + A_{33}} \right)^{1/2}. \tag{6-56}$$

Since the added mass of water A_{33} is usually on the same order as the mass of the ship, which in turn is proportional to the volume of the hull below the waterline, $M + A_{33} \sim \rho S d$, where d is the depth of the hull. If we substitute this relationship into the previous equation, the natural frequency of heave oscillations is

$$f_3 = 0.12 \left(\frac{g}{d} \right)^{1/2}, \tag{6-57}$$

where the factor 0.12 is intermediate between the value 0.11 suggested by Muckle and Taylor (1987) and the value 0.13 suggested from experience by Blagoveschensky (1962). Interestingly, the natural frequencies of pitch and heave are very similar for most ships, and these motions are coupled if the ship does not have fore–aft symmetry, that is, the stern is not the same shape as the bow.

Exercise

1. Consider the roll stability of the boxy ship shown in Figure 6-18. The ship is assumed to have uniform density that is equal to one-half the density of the water in which the ship floats. At each roll angle η_4, a trapezoidal section of the ship (Fig. 6-19) is submerged below the waterline.

(a) Show that the centroid of this submerged trapezoid is given by

$$\overline{\xi_1} = -\frac{h}{4}, \qquad \overline{\xi_2} = \frac{b^2}{6h} \tan \eta_4.$$

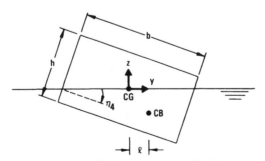

Fig. 6-18 Boxy ship in roll. The righting moment on the ship is equal to the weight of the ship times the distance l between the center of buoyancy and the center of gravity.

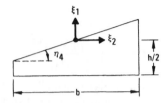

Fig. 6-19 Coordinate system for a trapezoidal area.

(b) Show that these equations imply that the lateral distance between the center of mass and the center of buoyancy is given by the following expression:

$$l = \frac{b^2}{6h} \sin \eta_4 - \frac{h}{4} \sin \eta_4.$$

(c) Derive an expression for the metacentric height for this ship, $\overline{GM} = \partial l / \partial \eta_4$.

(d) Consider ships of various widths $b = 0.5, 1, 2, 4$ for $h = 1$. Plot the lateral distance l for these ships for angles between 0 and 90 degrees. For what angles are these ships stable?

(e) If time permits, test these theories by putting lengths of wood with approximately square and approximately 1 by 2 cross sections in your sink and examine their stability at various angles.

6.7.3. Ship Motion Induced by Waves

Wave forces on a ship hull are generated by three mechanisms: (1) hydrostatic forces due to the increase in water pressure with increasing depth, (2) inviscid forces associated with the generation and diffraction of waves by the hull, and (3) viscous forces generated by the shearing of water against the moving hull. In general, the hydrostatic forces dominate the motions of slow ships such as tankers, the importance of other inviscid forces increases with speed, and viscous forces are most important in computing drag and damping.

The degree to which a ship responds to a wave depends on the magnitude, direction, and wavelength of the sea wave, the geometry of the hull, and the dynamic characteristics of the ship such as natural frequency in pitch, heave, and roll. Figure 6-20 shows nomenclature for incident wave directions. In this section, we will be primarily concerned with roll response to a beam sea.

The speed of deep-water sea waves is $gT/(2\pi)$, where T is the wave period. The wavelength of the sea wave is $gT^2/(2\pi)$. For typical wave

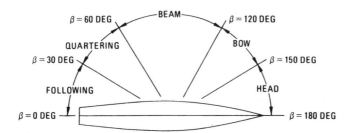

Fig. 6-20 Definition of incident wave directions.

periods of 5 to 15 sec, the wavelengths are greater than 200 ft (60 m), so the wavelength is much greater than the beam of a ship. This allows the surface of the sea to be modeled as a tilting plane as shown in Figure 6-21. Within the limits of linearity, the rolling moment induced on the ship by the wave is

$$F_4 = \rho g V_d \overline{GM}(\theta - \eta_4). \qquad (6\text{-}58)$$

$\rho g V_d$ is the weight of the ship and $\theta - \eta_4$ is the difference between the local angle of the sea surface θ and the roll angle of the ship η_4. The angle of the sea surface is simply the slope of the wave. The surface of the moving wave with amplitude a is $\eta(y, t) = a \sin(ky + \omega t)$, where $k = (2\pi)/\lambda$ is the wave number. The slope of the sea surface is

$$\theta = \frac{\partial \eta}{\partial y}(y, t)\big|_{y=0} = ak \cos \omega t. \qquad (6\text{-}59)$$

If damping terms are neglected, the linear equation of motion for the roll response of the ship to the tilting sea is

$$(I_{44} + A_{44})\ddot{\eta}_4 + \rho g V_d \overline{GM}(\eta_4 - \theta) = 0. \qquad (6\text{-}60)$$

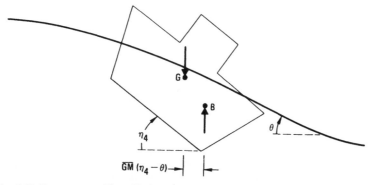

Fig. 6-21 Forces on a rolling ship in a seaway.

This equation can be rewritten using Eq. 6-59 for a sea wave of wavelength λ and amplitude a:

$$\ddot{\eta}_4 + \omega_4^2 \eta_4 = \frac{\rho V_d g \overline{GM}}{I_{44} + A_{44}} \frac{2\pi a}{\lambda} \cos \omega t. \qquad (6\text{-}61)$$

This equation is a classic forced linear oscillator. The right side is a harmonic forcing function whose amplitude is proportional to the wave amplitude and the metacentric height. The metacentric height, which provides the ship with a stabilizing moment, also provides the waves with a handle to rock the boat.

Using the trial solution

$$\eta_4 = \overline{\eta}_4 \cos \omega t, \qquad (6\text{-}62)$$

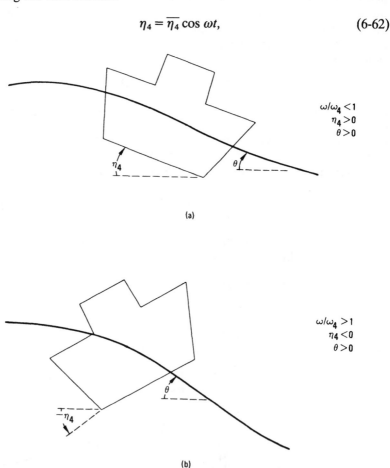

(a)

(b)

Fig. 6-22 The ship rolls (a) with the wave when the wave frequency is less than the natural frequency and (b) against the wave when the wave frequency is greater than the natural frequency.

the solution to Eq. 6-61 is found,

$$\eta_4 = \left(\frac{\rho g V_d \overline{GM}}{I_{44} + A_{44}}\right) \frac{2\pi a}{\lambda} \frac{1}{\omega_4^2 - \omega^2} \cos \omega t, \qquad (6\text{-}63)$$

where ω is the circular wave frequency and ω_4 is the natural frequency of roll.

As the natural frequency of the wave approaches the natural frequency of roll, the amplitude of roll increases sharply. In the frequency range $0.7 < \omega/\omega_4 < 1.2$, the amplitude of roll is at least twice that produced by lower-frequency waves. As the ship passes through resonance, $\omega = \omega_4$, the sign of η_4 changes as Eq. 6-63 and Figure 6-22 show. If the wave frequency is less than the natural frequency of the ship, the ship rolls with the wave; but if the wave frequency is greater than the natural frequency of the ship, the ship rolls against the wave slope, and the chance of water coming on board is greatly increased. Figure 6-23 shows theory in comparison with experiment for roll of a rectangular ship. The amplitude at resonance is limited by viscous effects that are not

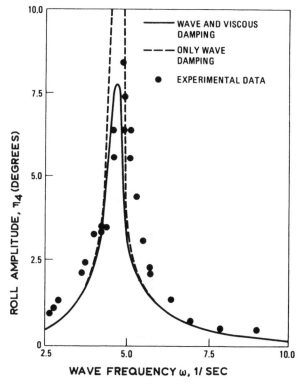

Fig. 6-23 Theoretical and experimental roll amplitudes for a rectangular section in beam waves (Salvesen et al., 1970).

included in inviscid theory (Roberts, 1985). Ships are often fitted with bilge keels to increase viscous resistance or with active hydraulic fins to limit the roll response and minimize discomfort of passengers (Lloyd, 1989; Downie, 1988; Muckle and Taylor, 1987).

The wavelength of a sea wave in pitch and heave is often comparable to the length of a ship. If we assume that the wave is traveling along the longitudinal axis of the ship, then the change in hydrostatic pressure on the ship hull is

$$p(a, t) = \rho g[\eta(x, t) + x\eta_5 - \eta_3], \qquad (6\text{-}64)$$

where $\eta(x, t)$ is the height of the sea wave traveling in the x direction, η_3 is the heave displacement, and $x\eta_5$ is the displacement of each point x during pitching of the ship to a small angle η_5; see Figure 6-24. The change in buoyancy provides a net vertical component on forces on the ship, $dF_3 = pb(x)\,dx$, where $b(x)$ is the beam at position x. This force and the associated pitching moment are functions both of the shape of the ship, $b(x)$, and the wavelength of the sea wave in comparison with the length of the ship. The resultant equation of motion for heave is coupled to the pitch motions,

$$(M + A_{33})\ddot{\eta}_3 + C_{35}\eta_5 + C_{33}\eta_3 = N(t), \qquad (6\text{-}65)$$

where

$$C_{35} = -\rho g \int_L xb(x)\,dx, \qquad C_{33} = \rho g \int_L b(x)\,dx = \rho g S,$$

$$N(t) = \rho g \int_L b(x)\eta(x, t)\,dx. \qquad (6\text{-}66)$$

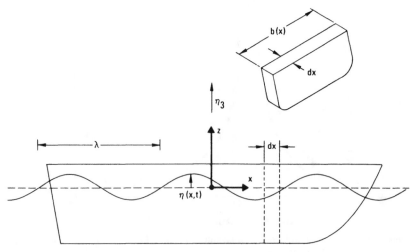

Fig. 6-24 Wave traveling in x direction and passing under ship.

For a fore-and-aft traveling wave $\eta(x, t) = a \sin(kx - \omega t)$,

$$N(t) = -\rho g a \sin \omega t \int_L b(x) \cos kx \, dx + \rho g a \cos \omega t \int_L b(x) \sin kx \, dx.$$

$$(6\text{-}67)$$

This equation implies that pitch and heave are generally coupled; that is, pitch does not generally occur without heave except for ships that are symmetric fore and aft, that is, $b(x) = b(-x)$. Since the two natural frequencies are generally close, the coupling evolves motions of the same order in both degrees of freedom (Timman and Newman, 1962). There can be nonlinear coupling as well (Nayfeh, 1988; Dowie, 1988).

Equation 6-67 also implies that a sea wave can only excite heaving motions of a ship if the length of the wave is comparable to the length of the hull. If the wavelength is much less than the length of the hull, then the forces on the hull are self-canceling when summed over the hull.

Exercise

1. Noting that the increment of pitch moment on the ship is $dF_5 = -xpb(x) \, dx$, develop an equation for the net pitching moment on the ship and develop the associated equation of motion. What is the amplitude of the pitching moment on a boxy ship, $b = $ constant, as a function of the wavelength of the sea wave?

REFERENCES

Angrilli, F., and V. Cossalter (1982) "Transverse Oscillations of a Vertical Pile in Waves," *Journal of Fluids Engineering,* **104,** 46–53.

Bascom, W. (1980) *Waves and Beaches,* Anchor Books, New York, pp. 148–149.

Batchelor, G. K. (1967) *An Introduction to Fluid Mechanics,* Cambridge University Press, London, p. 246.

Bearman, P. W., and P. F. Hall (1987) "Dynamic Response of Circular Cylinders in Oscillatory Flow," Paper D6, *International Conference on Flow Induced Vibrations, Bowness-on-Windermere, England, May 1987,* BHRA, Cranfield, England.

Bearman, P. W., et al. (1985) "Forces on Cylinders in Viscous Oscillatory Flow at Low Keulegan-Carpenter Numbers," *Journal of Fluid Mechanics,* **154,** 337–356.

Bearman, P. W., J. M. R. Graham, and E. D. Obasaju (1984) "A Model Equation for the Transverse Forces on Cylinders in Oscillatory Flows," *Applied Ocean Research,* **6,** 166–172.

Bishop, J. R. (1985) "Wave Force Data from the Second Christchurch Bay Tower," Paper 4953, 17th Annual Offshore Technology Conference, Houston, Tex.

Bishop, R. E. D., and W. G. Price (1979) *Hydroelasticity of Ships,* Cambridge University Press, Cambridge.

Blagoveschensky, S. N. (1962) *Theory of Ship Motions,* Vol. 1, Dover, New York.

Blevins, R. D. (1984a) *Formulas for Natural Frequency and Mode Shape*, Van Nostrand Reinhold, New York, 1979. Reprint by Robert Kreiger, Malabar, Fla.

—— (1984b) *Applied Fluid Dynamics Handbook*, Van Nostrand Reinhold, New York.

Borthwick, A. G. L., and D. M. Herbert (1988) "Loading and Response of a Small Diameter Flexibly Mounted Cylinder in Waves," *Journal of Fluids and Structures*, **2**, 479–501.

Chakrabarti, S. K., and R. E. Frampton (1982) "Review of Riser Analysis Techniques," *Applied Ocean Research*, **4**, 73–90.

Chaplin, J. R. (1988) "Loading on a Cylinder in Uniform Oscillatory Flow," *Applied Ocean Research*, **10**, 120–128, 199–206.

Chen, S. S. (1975) "Vibration of Nuclear Fuel Rod Bundles," *Nuclear Engineering and Design*, **35**, 399–422.

Cotter, D. C., and S. K. Chakrabarti (1984) "Wave Force Tests on Vertical and Inclined Cylinders," *Journal of the Waterway, Port, Coastal and Ocean Engineering Division*, **110**, 1–14.

Dalton, C. (1980) "Inertia Coefficients for Riser Configurations," *Journal of Energy Resources Technology*, **102**, 197–202.

Dalton, C., and B. Chantranuvatana (1980) "Pressure Distributions Around Circular Cylinders in Oscillating Flow," *Journal of Fluids Engineering*, **102**, 191–195.

Dawson, T. H. (1985) "In-Line Forces on Vertical Cylinders in Deepwater Waves," *Journal of Energy Resources Technology*, **107**, 18–23.

Downie, M. J. (1988) "Effect of Vortex Shedding on the Coupled Roll Response of Bodies in Waves," *Journal of Fluid Mechanics*, **189**, 243–264.

Ferziger, J. H. (1981) *Numerical Methods for Engineering Application*, Wiley, New York.

Gardner, T. N., and N. W. Cole (1982) "Drilling in Strong Current, Deep Water," *Ocean Industry*, **17**, 45–48.

Garrison, C. J. (1985) "Comments on Cross-Flow Principle and Morison's Equation," *Journal of the Waterway, Port, Coastal and Ocean Engineering Division*, **111**, 1075–1079.

Graham, J. M. R. (1987) "Transverse Forces on Cylinders in Random Seas," Paper D7, *International Conference on Flow Induced Vibrations, Bowness-on-Windermere, England, May, 1987*, BHRA, Cranfield, England.

Gudmestad, O. T., and J. J. Connor (1983) "Linearization Methods and the Influence of Current on the Nonlinear Hydrodynamic Drag Force," *Applied Ocean Research*, **5**, 184–194.

Heideman, J. C., and T. Sarpkaya (1985) "Hydrodynamic Forces on Dense Arrays of Cylinders," Paper 5008, 17th Annual Offshore Technology Conference, Houston, Tex.

Hoerner, S. F. (1965) *Fluid Dynamic Drag*, published by the author, New Jersey.

Hudspeth, R. T., and J. H. Nath (1988) "Wave Phase/Amplitude Effects on Force Coefficients," *ASCE Journal of Waterway, Port, Coastal and Ocean Engineering*, **114**, 34–49.

Kasahara, Y., et al. (1987) "Wave Forces Acting on Rough Circular Cylinders at High Reynolds Numbers," Paper OTC 5372, 1987 Offshore Technology Conference, Houston, Tex.

Keulegan, G. H., and L. H. Carpenter (1958) "Forces on Cylinders and Plates in an Oscillating Fluid," *Journal of Research of the National Bureau of Standards*, **60**, 423–440.

Labbe, J. R. (1983) "Sensitivity of Marine Riser Response to the Choice of Hydrodynamic Coefficients," Paper OTC 4592, 15th Annual Offshore Technology Conference, Houston, Tex.

Laird, A. D. (1966) "Flexibility in Cylinder Groups Oscillated in Water," *ASCE Journal of the Waterways and Harbors Division*, **92**, 69–84.

Laiw, C.-Y. (1987) "Subharmonic Response of Offshore Structures," *ASCE Journal of Engineering Mechanics*, **113**, 366–377.

Langley, R. S. (1984) "The Linearization of Three Dimensional Drag Force in Random Sea Current," *Applied Ocean Research*, **6**, 126–131.

Leira, B. J. (1987) "Multidimensional Stochastic Linearization of Drag Forces," *Applied Ocean Research*, **9**, 150–162.

Lloyd, A. R. J. M. (1989) *Seakeeping: Ship Behaviour in Rough Weather*, John Wiley, New York.

Massey, B. S. (1979) *Mechanics of Fluids*, 4th ed., Van Nostrand Reinhold, New York.

Meirovitch, L. (1971) *Analytical Methods in Vibrations*, Macmillan, New York.

Meyers, W. G., et al. (1975) "Manual—NSRDC Ship Motion and Sea-Load Computer Program," Naval Ship Research and Development Report 3376.

Moretti, P. M., and R. L. Lowery (1976) "Hydrodynamic Inertia Coefficients for a Tube Surrounded by Rigid Tubes," *Journal of Pressure Vessel Technology*, **98**, 190–193.

Morison, J. R., et al. (1950) "The Force Exerted by Surface Waves on Piles," *AIME Petroleum Transactions*, **189**, 149–154.

Muckle, W., and D. A. Taylor (1987) *Muckle's Naval Architecture*, 2d ed., Butterworths, London.

Nath, J. H. (1981) "Hydrodynamic Coefficients for Macro-Roughness," Paper 3989, 13th Annual Offshore Technology Conference, Houston, Tex.

Nath, J. H. (1988) "Bifouling and Morison Equation Coefficients," in *Proceedings of the Seventh International Conference on Offshore Mechanics and Arctic Engineering*, J. S. Chung (ed.), American Society of Mechanical Engineers, New York, pp. 55–64.

Nayfeh, A. H. (1988) "On the Undesirable Roll Characters of Ships in Regular Seas," *Journal of Ship Research*, **32**, 92–100.

Newman, J. N. (1977) *Marine Hydrodynamics*, The MIT Press, Cambridge, Mass.

Nordell, N. J., and D. J. Meggitt (1981) "Under Sea Suspended Cable Structures," *ASCE Journal of Structures Division*, **107**, 1025–1040.

Norton, D. J., J. C. Heideman, and W. W. Mallard (1981) "Wind Tunnel Tests of Inclined Circular Cylinders," Paper 4122, 13th Annual Offshore Technology Conference, Houston, Tex.

Obasaju, E. D., P. W. Bearman, and J. M. R. Graham (1988) "A Study of Forces, Circulation and Vortex Patterns Around a Circular Cylinder in Oscillating Flow," *Journal of Fluid Mechanics*, **196**, 467–494.

Patel, M. H. (1989) *Dynamics of Offshore Structures*, Butterworths, London.

Pierson, W. J., G. Neumann, and R. James (1985) *Observing and Forecasting Ocean Waves*, U.S. Naval Oceanographic Office, p. 34 (H.O. Pub. No. 603). Reprinted 1967.

Price, W. G., and R. E. D. Bishop, *Probabilistic Theory of Ship Dynamics*, Wiley, New York, 1974.

Rawson, K. J., and E. C. Tupper (1976) *Basic Ship Theory*, Vol. 1, Longman, London.

Roberts, J. B. (1985) "Estimation of Nonlinear Roll Damping from Free-Decay Data," *Journal of Ship Research*, **29**, 127–138.

Rogers, A. C. (1983) "An Assessment of Vortex Suppression Devices for Production Risers and Towed Deep Ocean Pipe Strings," Paper 4594, 15th Annual Offshore Technology Conference, Houston, Tex.

Salvesen, N., E. O. Tuck, and O. Faltinsen (1970) "Ship Motions and Sea Loads," *Transactions of the Society of Naval Architects and Marine Engineers*, **78**, 250–287.

Sarpkaya, T. (1975) "Forces on Cylinders and Spheres in a Sinusoidally Oscillating Fluid," *Journal of Applied Mechanics*, **42**, 32–37.

——— (1976) "Forces on Cylinders Near a Plane Boundary in Sinusoidally Oscillating Fluid," *Journal of Fluids Engineering*, **98**, 499–505.

——— (1977) "Inline and Transverse Forces on Cylinders in Oscillatory Flow with High Reynolds Numbers," *Journal of Ship Research*, **21**, 200–216.

——— (1982) "Wave Forces on Inclined Smooth and Rough Circular Cylinders," Paper 4227, 14th Annual Offshore Technology Conference, Houston, Tex.

———— (1986) "Force on a Circular Cylinder in Viscous Oscillatory Flow at Low Keulegan–Carpenter Number," *Journal of Fluid Mechanics* **165,** 61–71.

———— (1987) "Oscillating Flow about Smooth and Rough Cylinders," in *Proceedings of the Sixth (1987) International Offshore Mechanics and Arctic Engineering Symposium,* Vol. II, ASME, N.Y.

Sarpkaya, T., and M. Isaacson (1981) *Mechanics of Wave Forces on Offshore Structures,* Van Nostrand Reinhold, New York.

Sarpkaya, T., and M. Storm (1985) "In-Line Force on a Cylinder Translating in Oscillatory Flow," *Applied Ocean Research,* **7,** 188–196.

Sarpkaya, T., F. Rajabi, and M. F. Zedan (1981) "Hydroelastic Response of Cylinders in Harmonic and Wave Flow," Paper 3992, 13th Annual Offshore Technology Conference, Houston, Tex.

Spanos, P. D., and T. W. Chen (1980) "Vibrations of Marine Riser Systems," *Journal of Energy Resources Technology,* **102,** 203–213.

Springston, G. B. (1967) "Generalized Hydrodynamic Loading Function for Bare and Faired Cables in Two-Dimensional Steady-State Cable Configurations," Naval Ship Research and Development Center Report 2424.

St. Dennis, M., and W. J. Pierson (1953) "On the Motion of Ships in Confused Seas," *Transactions of the Society of Naval Architects and Marine Engineers,* **61,** 280–332.

Stansby, P. K., G. N. Bullock, and I. Short (1983) Quasi-2-D Forces on Vertical Cylinders in Waves," *Journal of Waterway, Port, Coastal and Ocean Engineering Division,* **109,** 128–132.

Starsmore, N. (1981) "Consistent Drag and Added Mass Coefficients from Full-Scale Data," Paper OTC 3990, 13th Annual Offshore Technology Conference, Houston, Tex.

Timman, R., and J. N. Newman (1962) "The Coupled Damping Coefficient of Symmetric Ships," *Journal of Ship Research,* **5,** 1–7.

Tung, C. C., and N. E. Huang (1973) "Combined Effects of Current and Waves on Fluid Force," *Ocean Engineering,* **2,** 183.

U.S. Senate Committee on Armed Services Preparedness Investigating Subcommittee, "Enquiry into the Collapse of Texas Tower No. 4," 87th Congress, First Session.

Williamson, C. H. K. (1985) "In-line Response of a Cylinder in Oscillatory Flow," *Applied Ocean Research,* **7,** 97–106.

———— (1985) "Sinusoidal Flow Relative to Circular Cylinders," *Journal of Fluid Mechanics,* **155,** 141–174.

Wirsching, P. H., and P. H. Prasthofer (1976) "Preliminary Assessment of Deepwater Platforms," *ASCE Journal of the Structural Division,* **102,** 1447–1462.

Wolfram, J., and A. Theophanatos (1985) "The Effects of Marine Fouling on the Fluid Loading of Cylinders: Some Experimental Results," Paper 4954, 17th Annual Offshore Technology Conference, Houston, Tex.

Wright, J. C., and T. Yamamoto (1979) "Wave Forces on Cylinders Near Plane Boundaries," *ASCE Journal of the Waterway, Port, Coastal and Ocean Engineering Division,* **103,** 378–383.

Zdravkovich, M. M., and J. E. Namork (1977) "Formation and Reversal of Vortices around Circular Cylinders Subjected to Water Waves," *ASCE Journal of the Waterway, Port, Coastal and Ocean Engineering Division,* **105,** 1–13.

Zedan, M. F., et al. (1980) "Dynamic Response of a Cantilever Pile to Vortex Shedding in Regular Waves," Paper 3799, 12th Annual Offshore Technology Conference, Houston, Tex.

Chapter 7

Vibration Induced by
Turbulence and Sound

Gusts of wind sway trees and tall buildings. A ship bobs and rolls in a complex, turbulent sea. Aircraft skin panels respond to the sound of a turbojet engine. In each of these cases, a structure responds to random surface pressures imposed by the surrounding fluid. Analysis of these random vibrations is developed in this chapter. The results are applied to turbulence excitation of rods and tubes, sonic fatigue of plates, wind-induced vibration of buildings, and buffeting of aircraft by gusts.

7.1. ELEMENTS OF THE THEORY OF RANDOM VIBRATIONS

Turbulent flows are the sum of components that oscillate with many frequencies. Deterministic analysis for each oscillating component becomes very tedious. It is more practical to treat the flow and structural response statistically and deal only with time-averaged quantities. The theory of random vibrations provides a framework for this analysis.

In many practical cases, time-averaged measurements of random turbulence are independent of the start of the time-averaging interval and nearly independent of the position of the measurement. Such phenomena are called *stationary*, that is, steady, *homogeneous random processes*. Crandall and Mark (1963), Lin (1967), Bolotin (1984), Yang (1986), Dowell (1978), and Simiu and Scanlan (1986) review the response of structures to stationary random processes. This chapter develops the response of stable structures in a single mode to stationary, homogeneous random turbulence and acoustic pressure fields.

The *auto spectral density*, also called the *power spectral density, $S_y(f)$* of a stationary random process $y(t)$ is defined so that the integral of the auto spectral density over the frequency range is its overall mean square value,

$$\overline{y(t)^2} = \int_{f_1}^{f_2} S_y(f)\, df, \qquad (7-1)$$

239

for frequencies $f_1 = 0$ and $f_2 = \infty$. For other frequency limits, the integral of the auto spectral density gives the mean square of components within those limits. As the frequency limits become close together so that $f_2 - f_1 = \delta f$, the auto spectral density becomes

$$S_y(f) = \frac{\overline{y_f(t)^2}}{\delta f}, \tag{7-2}$$

where $y_f(t)$ are those components of $y(t)$ that have frequencies near the frequency f. Thus, one method of determining the auto spectral density at a frequency is to filter the time history signal at that frequency, square the output, average it, and divide by the bandwidth of the filter. The units are (units of $y)^2$/Hz. The reader should be warned that other definitions of auto spectral density are sometimes used. For example, in theoretical studies a frequency spectrum is often defined symmetrically over the positive and negative frequencies, $-\infty < f < \infty$ with frequency in radians per second instead of hertz (Bendat and Piersol, 1986). These spectra are a factor of $1/(4\pi)$ smaller than the present spectra. Digital computation of auto spectral density is discussed in Appendix D.

The mean square of a stationary process is defined as the average of its square over a time period T that includes many cycles of oscillation:

$$\overline{y^2(t)} = \frac{1}{T} \int_0^T y^2(t) \, dt. \tag{7-3}$$

The bar ($^-$) denotes time average. Consider a sinusoidal process with amplitude Y_0,

$$y(t) = Y_0 \sin (2\pi f t). \tag{7-4}$$

The mean (average), mean square, and root mean square (rms) values for the sinusoidal process are

$$\overline{y(t)} = 0, \qquad \overline{y^2(t)} = \frac{Y_0^2}{2}, \qquad y_{\text{rms}} = \sqrt{\overline{y^2(t)}} = \frac{Y_0}{\sqrt{2}}. \tag{7-5}$$

For a sinusoidal process, the ratio of the peak value (Y_0) to the rms value is $\sqrt{2}$. If a process has zero mean, as is the case for the sinusoidal process, the root mean square is also equal to the standard deviation.

The *probability density function* $p(y)$ is the fraction of time that the process spends in a small interval $y \pm (dy/2)$, divided by the width of the interval. In other words, $p(y_1) \, \delta y$ is the probability that y will have values in the range $y_1 - \delta y/2 < y < y_1 + \delta y/2$. The probability densities of a sinusoidal process and a Gaussian random process are (Bendat and Piersol, 1986)

$$p(y) = \begin{cases} 1/[\pi(Y_0^2 - y^2)^{1/2}], & |y| \le Y_0, \text{ otherwise } 0; \text{ sinusoidal;} \\ [y_{\text{rms}}(2\pi)^{1/2}]^{-1} e^{-y^2/2y_{\text{rms}}^2} & \text{Gaussian.} \end{cases}$$

$$\tag{7-6}$$

The units of probability density are probability/unit y. The probability density of a sinusoidal process peaks near $y(t) = Y_0$ because a large fraction of the sinusoid occurs near its peaks.

The Central Limit Theorem of statistics (Loeve, 1977; Bendat and Piersol, 1986) states that the probability distribution of the sum of a number of independent random events will tend to the Gaussian distribution (also called the *Normal distribution*). The probability density of elastic structures responding to turbulent flows is reasonably well represented by the Gaussian distribution (Wambsganss and Chen, 1971; Basile et al., 1968; Ballentine et al., 1968). For example, Figure 7-1(a) shows that the probability distribution of an aircraft thrust reverser responding to the jet of a turbojet engine is near Gaussian. The Central Limit Theorem also states that the mean square of the sum of independent random processes is the sum of the mean squares of the individual processes. This implies that the overall mean square response of an elastic structure to turbulence will be the sum of the mean square of the individual modes,

$$\overline{y^2(t)} = \sum_{i=1}^{N} \overline{y_i^2(t)}, \tag{7-7}$$

where $y_i(t)$ are the component modes, provided each mode responds independently. Similarly, the overall mean square stress and strain are

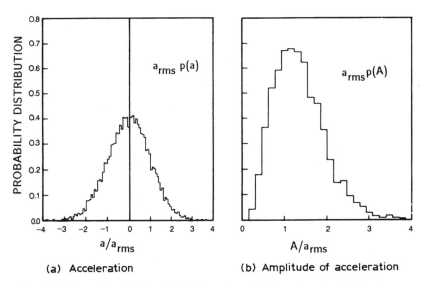

(a) Acceleration   (b) Amplitude of acceleration

Fig. 7-1 Measured probability distributions of acceleration of an aircraft thrust reverser. Compare with Eqs. (7-6) and (7-8). Courtesy J Mekus, Rohr Industries.

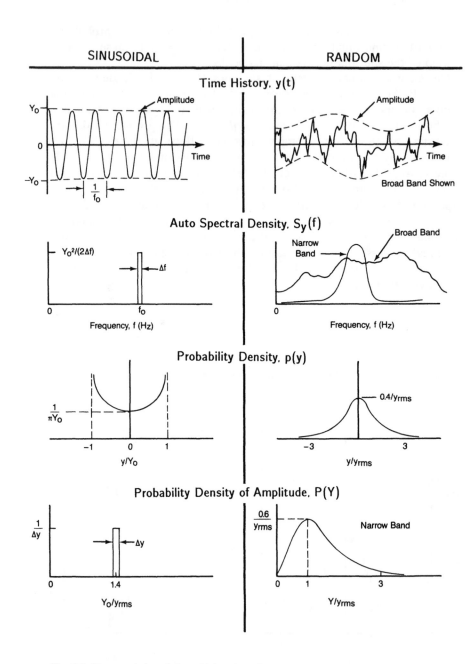

Fig. 7-2 Characteristics of sinusoidal and random processes.

242

determined by the sum of the squares of the stress and strain in each mode.

The difference between the Gaussian and the sinusoidal probability distributions (Eq. 7-6, Fig. 7-2) can be used to differentiate between random and deterministic vibration of a structure in a turbulent flow. For example, if the probability distribution of the displacement of a structure is Gaussian, then it is reasonable to assume that the structure is responding to the random tubulence-induced pressures on its surface. However, if the probability density dips in the middle with increasing flow velocity, we would suspect that the structure is responding sinusoidally at its natural frequency, perhaps to onset of an instability.

A random process is *narrow band* if a small finite band of frequencies dominates the process. An elastic structure responding in one mode to turbulent pressures is usually narrow band because the majority of the response occurs at a narrow band of frequencies centered about the natural frequency (Figs. 7-2, 7-3). It is possible to identify an amplitude with individual cycles of a narrow-band process as the magnitude at points where the velocity changes sign. The probability distribution of the amplitudes of a narrow-band Gaussian random process is approximately the Rayleigh distribution (Crandall and Mark, 1963; Cartwright and Longuet-Higgins, 1956),

$$p(Y) = \frac{Y}{y_{rms}^2} e^{-Y^2/2y_{rms}^2}, \qquad \text{for } Y \geq 0. \qquad (7-8)$$

Y is the amplitude of the cycles. The *cumulative probability distribution* is the probability the process will exceed a given value. The cumulative probability distribution $P(Y > Y_1)$ of the Rayleigh distribution is

$$P(Y > Y_1) = \int_{Y_1}^{\infty} p(Y)\, dY = e^{-Y_1^2/2y_{rms}^2}. \qquad (7-9)$$

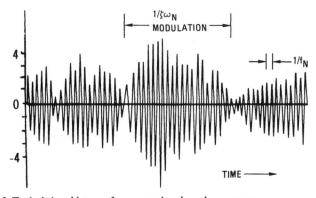

Fig. 7-3 Typical time-history of a narrow-band random process.

The cumulative probability that a sinusoidal process will exceed the amplitude of the sinusoid is zero (Eq. 7-6). The cumulative probability that the amplitude will exceed three standard deviations, $Y > 3y_{rms}$, in a Rayleigh distribution is 0.0111 (Eq. 7-9).

The Rayleigh distribution (Eq. 7-8) implies that extreme amplitudes occur from time to time. In practice, values of stress, acceleration, or displacement greater than six times the root mean square are virtually never observed. Finite energy and structural nonlinearities curtail extreme amplitudes. Moreover, an exceptional peak is unlikely to occur in a finite sampling interval. Davenport (1964; also see Solari, 1982) showed that the ratio of the most likely largest amplitude to the rms value for a narrow-band random process during sampling time T is

$$\bar{g} = \frac{\text{Typical peak value}}{\text{rms value}} = \sqrt{1.175 + 2 \log_e (fT)}. \qquad (7\text{-}10)$$

Some values of this peak response factor are:

Cycles in interval, fT	100	1000	10000	10^5	10^6
Peak factor, $\bar{g}$	3.22	3.87	4.42	4.92	5.36

For example, an aircraft thrust reverser experiences about 500 to 10 000 vibration cycles during a landing. Equation 7-10 predicts a peak response between 3.5 and 5.0 times the rms response and the data of Figure 7-1(b) show that the measured peak response during a landing is 3.8 times the rms value. Peak values between 2.5 and 4.5 times the rms are generally used in design (see Section 7.4.2).

Exercises

1. If the units of a random time history of acceleration $\ddot{y}(t)$ are meters/(sec)^2, what are the units of the auto spectral density of acceleration $S_{\ddot{y}}(f)$, probability distribution $p(\ddot{y})$, and cumulative probability distribution of acceleration $P(\ddot{y})$? Hint: See Appendix D.

2. Plot the Gaussian probability density of Eq. 7-6 and the Rayleigh probability density of amplitude of Eq. 7-8 on Figure 7-1. How well do the data follow these equations?

3. Make a table of the amplitudes of each of the first 20 cycles of the time history of Figure 7-3. Note that one cycle extends from a positive peak of the time history to the next positive peak. The mean square of an individual cycle is the square of the peak divided by 2 (Eq. 7-5). Compute the overall root mean square of the peak divided by 2 (Eq. 7-5). Compute the overall root mean square for

the first 20 cycles by summing the mean squares of the cycles, averaging the result, and taking the square root. What is the peak-to-rms ratio? Compare it with the prediction of Eq. 7-10.

4. Continuing the previous exercise, sort the amplitudes of these cycles by ranges 0–0.5, 0.5–1.0, 1.0–1.5, and so on. The number of peaks in each range divided by the width of the range and the total number of cycles is the probability density of amplitude. Nondimensionalize by dividing by the rms value and plot this on Figure 7-1(b) and compare with the Rayleigh distribution (Eq. 7-8).

7.2. SOUND- AND TURBULENCE-INDUCED VIBRATION OF PANELS

7.2.1. Analytical Formulation

Table 7-1 gives estimates of the magnitude of oscillating near-field surface pressures generated by turbulent jets and aircraft engines. Aircraft turbojet engines can produce near-field sound levels of 140 to 170 decibels (dB). Sound levels over 185 dB have been estimated for the ram and scram jet engines of hypersonic aircraft. In-flight, turbulent boundary layers induce oscillating pressures on aircraft skin panels. Interaction with a shock wave can augment these oscillating boundary layer pressures by 10 to 45 dB (Ungar et al., 1977; Zorumski, 1987; Raghunathan, 1987). Such high sound pressure levels can induce fatigue failure of aircraft skin panels. This is called *sonic fatigue*. In most cases of sonic fatigue, only a relatively few structural modes of the panel contribute to the failures. Using one mode, Miles (1954) and Clarkson and Fahy (1968) have developed a method for sonic fatigue of plates which is the basis for AGARD design methods (Thompson and Lambert, 1972). Blevins (1989) has extended their approach to complex shapes and higher modes.

Acoustic pressures on the surface of a panel that responds in a single mode are shown in Figure 7-4. The dynamic equation of motion of the panel shown is

$$m \frac{\partial^2 W}{\partial t^2} + \mathscr{L}[W] = P. \qquad (7\text{-}11)$$

$W(x, y, z, t)$ is the displacement; x, y, z are the spatial coordinates; $P(x, y, z, t)$ is the pressure on the exposed surface; and t is time. $m(x, y, z, t)$ is the mass per unit area, and $\mathscr{L}$ is a linear operator representing the load–deflection relationship of the panel. For the flat

Table 7-1 Oscillating surface pressures[a]

Case	Root-mean-square pressure, P_{rms}, at P	Spectral shape, typical levels, and comments
1. Impinging turbulent jet 	αq, where $0.1 \leqslant \alpha \leqslant 0.2$	Broad-band spectrum, declining in higher frequencies. For transonic jet, a typical 1/3 octave sound pressure level at 500 Hz is 160 dB
2. Turbulent boundary layer 	$\dfrac{0.006\rho U^2}{1 + 0.14M^2}$	Broad-band spectrum, declining in frequencies above U/δ. See Laganelli et al. (1983) The oscillating surface pressures are zero for laminar boundary layers
3. Separated turbulent boundary layer 	αq, where $0.04 \leqslant \alpha \leqslant 1$	Broad-band spectrum, declining in higher frequencies

4. Transonic shock in turbulent boundary layer

$$\frac{\Delta p_{shock}}{2}$$

Broad-band spectrum. Oscillating pressures are result of shock motion due to its interaction with boundary layer

5. Ducted fan or propeller

αq, where $0.02 \leq \alpha \leq 0.1$ for either P_1 or P_2

Spectrum is series of distinct peaks at harmonics of rotation frequency f. Largest peak occurs at nf, where n = number of blades. Typical *SPL* of individual peak is 145 to 160 dB for sonic blade tip speed. Harris (1979) discusses far field radiated noise.

6. Turbojet exhaust

αq, where $0.02 \leq \alpha \leq 0.1$

Broad-band spectrum. Typical 1/3 octave sound pressure levels are 140 to 150 dB and 161 dB overall without afterburning and 152 to 160 dB and 170 dB overall with afterburning.

247

Table 7-1 Oscillating surface pressures[a] (*continued*)

Case	Root-mean-square pressure, p_{rms}, at P	Spectral shape, typical levels, and comments
7. Rocket Engine	Overall sound pressure level radiated to far field $SPL = 100 + 10 \log_{10} W$, where $W = \frac{1}{2}$ thrust $\times U_e$ (watts)	Broad-band spectrum peaking at $0.01 U_e/d_e$ See NASA SP-8072, "Acoustic Loads Generated by Propulsion System", 1971

Sources: Brase (1988), Laganelli et al. (1983), Lowson (1968), Unger et al. (1977).

[a]
f	Frequency, Hz
$M = U/c$	Mach number
p	Pressure on surface
p_{rms}	Overall root-mean-square pressure on surface
q	Dynamic pressure, $\frac{1}{2}\rho U^2$
SPL	Sound pressure level, Eq. 9-38
U	Velocity
ρ	Fluid density
δ	Boundary layer thickness

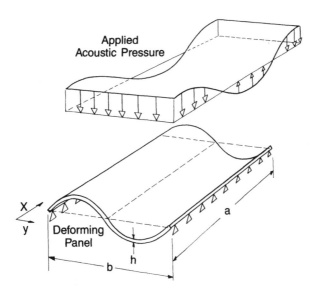

Fig. 7-4 A rectangular panel responding to acoustic pressures on its surface.

plate of Figure 7-4, $\mathscr{L}[W]$ is

$$\mathscr{L}[W] = \frac{Eh^3}{12(1-v)}\left(\frac{\partial^4 W}{\partial x^4} + 2\frac{\partial^4 W}{\partial x^2 \, \partial y^2} + \frac{\partial^4 W}{\partial y^4}\right). \tag{7-12}$$

W is the displacement normal to the plane of the plate. The plate thickness is h, E is the modulus of elasticity, v is Poisson's ratio, and x and y are rectangular coordinates in the plane of the plate.

The following eigenvalue problem is associated with Eq. 7-11 (Appendix A):

$$\omega_i^2 m\bar{w}_i - \mathscr{L}[\bar{w}_i] = 0, \qquad i = 1, 2, 3, \ldots, \tag{7-13}$$

where ω_i are natural frequencies of the ith mode. The mode shapes $\bar{w}_i$ are orthogonal over the structure (Meirovitch, 1967),

$$\int_D \bar{w}_i m\bar{w}_j \, ds = 0, \qquad i \neq j, \tag{7-14}$$

$$\int_D \bar{w}_i \mathscr{L}[\bar{w}_j] \, ds = 0, \qquad i \neq j, \tag{7-15}$$

provided that the natural frequencies are not repeated, $\omega_i \neq \omega_j$. D is the domain of the panel and ds is an element of that domain. The natural frequencies and fundamental mode shapes of flat plates with simply supported and clamped edges are given in Table 7-2.

The displacement solution to Eq. 7-11 is the sum of the modal

Table 7-2 Response of rectangular plates with simply supported and clamped edges to broad-band acoustic loads[a]

Quantity	Simply supported edges	Clamped edges
Fundamental natural frequency, Hz	$\dfrac{\pi[1/a^2+1/b^2]}{2}\sqrt{\dfrac{Eh^2}{12\rho(1-\nu^2)}}$	$\dfrac{2\pi[3(a/b)^2+3(b/a)^2+2]^{1/2}}{3ab}\sqrt{\dfrac{Eh^2}{12\rho(1-\nu^2)}}$
Mode shape of fundamental mode, $\bar{w}_1(x,y)$	$\sin\dfrac{\pi x}{a}\sin\dfrac{\pi y}{b}$	$\left[1-\cos\dfrac{2\pi x}{a}\right]\left[1-\cos\dfrac{2\pi y}{b}\right]$
Maximum modal deformation, $\bar{w}_1(x,y)_{max}$	1.0	4.0
Maximum modal stress, $\bar{\sigma}_1(x,y)_{max}$ [b]	$\dfrac{6\pi^2}{h^2}\left[\dfrac{1}{a^2}+\dfrac{\nu}{b^2}\right]\dfrac{Eh^3}{12(1-\nu^2)}$	$\dfrac{48\pi^2}{a^2h^2}\dfrac{Eh^3}{12(1-\nu^2)}$
Maximum stress due to uniform pressure, σ_0	$\dfrac{96}{\pi^4h^2}\dfrac{(1/a^2+\nu/b^2)}{(1/a^2+1/b^2)^2}$	$\dfrac{24}{\pi^2}\left(\dfrac{b}{h}\right)^2\dfrac{1}{[3(a/b)^2+3(b/a)^2+2]}$
Miles equation maximum stress, Eq. 7-32	$\dfrac{96}{\pi^4h^2}\dfrac{(1/a^2+\nu/b^2)}{(1/a^2+1/b^2)^2}\sqrt{\dfrac{\pi f_1 S_p(f_1)}{4\zeta_1}}$	$\dfrac{24}{\pi^2}\left(\dfrac{b}{h}\right)^2\dfrac{1}{[3(a/b)^2+3(b/a)^2+2]}\sqrt{\dfrac{\pi f_1 S_p(f_1)}{4\zeta_1}}$
Eq. 7-31 maximum stress	$\dfrac{6}{\pi^2h^2}\dfrac{(1/a^2+\nu/b^2)}{(1/a^2+1/b^2)^2}\sqrt{\dfrac{\pi f_1 S_p(f_1)}{4\zeta_1}}$	$\dfrac{54}{4\pi^2}\left(\dfrac{b}{h}\right)^2\dfrac{1}{[3(a/b)^2+3(b/a)^2+2]}\sqrt{\dfrac{\pi f_1 S_p(f_1)}{4\zeta_1}}$

Source: Adapted in part from Ballentine et al. (1968).
[a] a = width, b = length, h = thickness, E = modulus of elasticity, ρ = density, ν = Poisson's ratio.
[b] Modal stress is the stress produced by a unit modal deformation. Maximum stress is at center of long edge for clamped edges and at center of plate for simply supported edges. Stress results are valid for $b \geq a$.

250

responses (Appendix A),

$$W(x, y, z, t) = \sum_{i=1}^{N} \bar{w}_i(x, y, z) w_i(t), \qquad (7\text{-}16)$$

where $w_i(t)$ are functions of time and the mode shapes $\bar{w}_i(x, y, z)$ are only functions of space. Likewise, it is reasonable to believe that most sound fields can be decomposed into functions of space and time,

$$P(x, y, z, t) \approx \bar{p}(x, y, z) p_i(t). \qquad (7\text{-}17)$$

For example, a traveling acoustic wave in the x direction can be decomposed into two standing waves,

$$P(x, t) = P_0 \sin (kx - \omega t) = P_0 \sin (kx) \cos (\omega t) - P_0 \cos (kx) \sin (\omega t). \qquad (7\text{-}18)$$

Equations 7-17 and 7-18 can be considered modal expansions for the acoustic field. However, the majority of the structural response is contained at frequencies near the structural natural frequencies, and it is convenient to consider that $p_i(t)$ is the component of acoustic pressure at frequencies near the ith structural natural frequency.

By substituting Eqs. 7-16 and 7-17 into Eq. 7-11, multiplying through by $\bar{w}_j$, integrating over the domain of the structure, utilizing the orthogonality conditions (Eqs. 7-14 and 7-15), and introducing the modal damping factor ζ_i, the following equation is produced, which describes the response of each mode of the panel to the applied surface pressure:

$$\frac{1}{\omega_i^2} \ddot{w}_i + \frac{2\zeta_i}{\omega_i} \dot{w}_i + w_i = J_i p_i(t). \qquad (7\text{-}19)$$

This equation is solved in the following section. The results are summed (Eq. 7-16 or Eq. 7-7) for the total displacement of the panel.

The joint acceptance for the panel is defined by

$$J_i = \frac{\displaystyle\int_D \bar{p}_i(x, y, z)\bar{w}_i(x, y, z) \, ds}{\omega_i^2 \displaystyle\int_D m\bar{w}_i^2(x, y, z) \, ds}. \qquad (7\text{-}20)$$

The joint acceptance of a mode is a measure of the degree to which the spatial distribution of pressure on the surface of the panel $\bar{p}_i(x, y, z)$ is compatible with the structural mode shape $\bar{w}_i(x, y, z)$. J_i is determined by spatial integrations. The units of J_i depend on the units of $\bar{p}_i$ and $\bar{w}_i$. There is some arbitrariness in this because the mode shapes can be normalized in various ways. However, the present method ensures that the physical displacements and stress are independent of the manner in which the modes are normalized.

Exercise

1. The wavelengths of acoustic waves propagating in fluids and flexural waves propagating in large flat plates are as follows (Junger and Feit, 1986):

$$\lambda_{\text{acoustic}} = \frac{c}{f}, \qquad \lambda_{\text{flex-plate}} = \left(\frac{Eh^2}{12\rho}\right)^{1/4}\left(\frac{2\pi}{f}\right)^{1/2},$$

where c is the speed of sound in the fluid, ρ is the plate density, h is its thickness, and f is frequency in hertz. At what frequency do the two wavelengths coincide? For lower frequencies, is the flexural wavelength greater than or less than the acoustic wavelength? How will coincidence or lack of coincidence influence the joint acceptance?

7.2.2. Temporal Solutions for the Modal Response

Consider that the joint acceptance (Eq. 7-20) is unity, $J_i = 1$, which will be the case if the shape of the pressure field is identical to the mass-weighted mode shape,

$$\bar{p}_i(x, y, z) = m\omega_i^2 \bar{w}_i(x, y, z). \tag{7-21}$$

This is often a conservative assumption; rarely will the shape of the pressure field line up exactly with the mode shape. With this assumption, the equation describing the response of each mode (Eq. 7-19) becomes the classic single-degree-of-freedom forced oscillator equation,

$$\frac{1}{\omega_i^2}\ddot{w}_i(t) + \frac{2\zeta_i}{\omega_i}\dot{w}_i(t) + w_i(t) = p_i(t). \tag{7-22}$$

Cases for nonunity joint acceptance can be considered to scale the right-hand side of this equation.

If the pressure is sinusoidal at frequency f,

$$p_i(t) = P_0 \sin(2\pi f t), \tag{7-23}$$

then Eq. 7-22 has a classical solution (Thomson, 1988),

$$w_i(t) = \frac{P_0 \sin(2\pi f t - \phi)}{\sqrt{[1 - (f/f_i)^2]^2 + (2\zeta_i f/f_i)^2}}, \tag{7-24}$$

where the phase angle is $\tan\phi = 2\zeta_i(f/f_i)/[1 - (f/f_i)^2]$. If the forcing frequency coincides with the natural frequency $f = f_i = \omega_i/(2\pi)$, creating a resonance, then the response is

$$w_i(t) = -\frac{P_0}{2\zeta_i}\cos(2\pi f t), \qquad \text{or equivalently,} \qquad w_{i,\text{rms}} = \frac{P_{\text{rms}}}{2\zeta_i}, \tag{7-25}$$

where the subscript rms denotes root mean square (Section 7.1). The rms response to a sinusoidal pressure distribution is identical to the rms response to narrow-band random forcing provided that the bandwidth of the random pressures is a small fraction of the bandwidth of the structural response ($BW = 2\zeta_i f_i$, Hz; Chapter 8, Sections 8.3.1 and 8.4.5).

The auto spectral density of the response $S_w(f)$ of an oscillator (Eq. 7-22) to random pressures is found by utilizing Eq. 7-24 and computing the mean square response at each frequency component (Eq. 7-2) to the auto spectral density $S_p(f)$ of the applied pressure,

$$S_{wi}(f) = \frac{S_p(f)}{[1 - (f/f_i)^2]^2 + (2\zeta_i f/f_i)^2}. \tag{7-26}$$

The overall mean square response is found by integrating over the frequency range (Eq. 7-1). The result depends on the form of $S_p(f)$. If the bandwidth of $S_p(f)$ is greater than the bandwidth of the structure, it is called intermediate band or *broad band*. Most turbulence-induced pressures are broad band, as shown in Figure 7-5. The response to broad-band pressure is the integral of the power spectral density of the response. This integral has an exact solution (Blevins, 1989) and a well-known approximate solution (Crandall and Mark, 1963),

$$w^2_{i,\mathrm{rms}} = \int_{f_1}^{f_2} \frac{S_p(f)\,df}{[1 - (f/f_i)^2]^2 + (2\zeta_i f/f_i)^2},$$

$$\approx \frac{\pi}{4\zeta_i} f_i S_p(f_i), \qquad \text{if } S_p(f) = \text{constant and } f_1 \ll f_i \ll f_2. \tag{7-27}$$

Comparing this equation with Eq. 7-25, we see that the rms resonant response to sinusoidal loading varies inversely with damping, while the rms response to broad-band random forcing varies inversely as the square root of damping. The contribution of the nonresonant 'tails' of the response curve decreases the sensitivity of broad-band response to damping.

7.2.3. Approximate Solutions

The procedure for calculating the dynamic response of panels to surface pressures is assembled from Sections 7.2.1 and 7.2.2 on a mode-by-mode basis using the single mode expansion, $W_i(x, y, z, t)/\bar{w}_i(x, y, z, t) = w_i(t)$ (Eq. 7-16), where $W_i(x, y, z, t)$ is the physical deflection in the ith mode. If the pressure distribution over the plate is known, the joint acceptance (Eq. 7-20) can be computed and the result is exact. Here the pressure

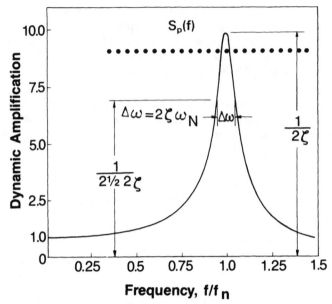

Fig. 7-5 Random forcing and response in the frequency domain.

distribution is assumed to match the mass-weighted mode shape (Eq. 7-21, and Blevins, 1989). Other approximations are considered in Section 7.2.4.

Step 1. Identify the rms acoustic pressure applied to the surface of the structure in the frequency range of interest. Sound pressure levels are expressed in decibels. The conversion to rms pressure is

$$P_{rms} = p_{ref}10^{SPL/20}, \tag{7-28}$$

where SPL is sound pressure level in decibels and $P_{ref} = 2.9 \times 10^{-9}$ psi $(2 \times 10^{-5}$ Pa$)$. The pressure spectrum can be calculated from time history measurements as discussed in Appendix D. If the overall rms pressure in a frequency band is known, then the pressure auto spectral density is $S_p(f_1 < f < f_2) = P_{rms}^2/(f_2 - f_1)$, where P_{rms} is the rms pressure over the frequency band $f_1 < f < f_2$. For one-third octave pressures, $f_2 - f_1 = 0.232f_c$, where f_c is the band center pressure. Note that $S_p(f)$ has units of (pressure)^2/Hz.

Step 2. Perform a modal analysis to determine the natural frequencies f_i, the mode shapes $\bar{w}_i$, and, if stress information is desired, the stresses $\bar{\sigma}_i$ associated with the deformation in each mode. The modal stresses are the stresses that result from a unit modal deformation. The modal stresses vary over the panel and they have a maximum value (Table 7-2). The finite-element method is often utilized to compute natural frequency, mode shape, and modal stresses for complex panels.

Step 3. Choose a characteristic point on the structure that will be matched to the applied acoustic pressure. Ordinarily this reference point is the point of maximum modal displacement. Calculate the characteristic modal pressure for the reference point,

$$\bar{P}_{ic} = \rho h (2\pi f_i)^2 |\bar{w}_i(x_c, y_c, z_c)|, \qquad (7\text{-}29)$$

where f_i is the natural frequency of the ith mode in hertz; ρh is the mass per unit area at the reference point. Specifying the magnitude of the acoustic pressure at the reference point determines the pressure at all other points under the assumption that the shape of the pressure fields matches the mass-weighted mode shape (Eq. 7-21).

Step 4. Calculate the stress and deformation in the mode of interest. The response is scaled relative to the modal response. For sinusoidal excitation with pressure amplitude P_0 at frequency f in hertz, the ratio of the physical deformation W_i and stress σ_i in the ith mode to the corresponding modal stress and deformation is

$$\frac{W_i(x, y, z, t)}{\bar{w}_i(x, y, z)} = \frac{\sigma_i(x, y, z, t)}{\bar{\sigma}_i(x, y, z)}$$

$$= \begin{cases} \dfrac{1}{2\zeta_i} \dfrac{P_0}{\bar{P}_{ic}} \cos(2\pi f_i t), & \text{if } f = f_i; \\[4mm] \dfrac{1}{[[1-(f/f_i)^2]^2 + (2\zeta_i f/f_i)^2]^{1/2}} \dfrac{P_0}{\bar{P}_{ic}} \sin(2\pi f t - \phi), & \text{if } f \neq f_i. \end{cases}$$

$$\qquad (7\text{-}30)$$

If the pressures have a random broad-band spectrum, then

$$\frac{W_i(x, y, z, t)_{\text{rms}}}{\bar{w}_i(x, y, z)} = \frac{\sigma_i(x, y, z, t)_{\text{rms}}}{\bar{\sigma}_i(x, y, z)} = \sqrt{\frac{\pi f_i}{4\zeta_i} \frac{S_p(f_i)}{\bar{P}_{ic}^2}}. \qquad (7\text{-}31)$$

7.2.4. Application and Extensions

As an example, consider the first mode ($i = j = 1$) of a flat rectangular plate as shown in Figure 7-4 with two sets of boundary conditions: (1) all edges simply supported, as shown in Figure 7-4, and (2) all edges clamped. The natural frequencies, mode shapes, and maximum modal stresses are given in Table 7-2. In the fundamental mode, the stress is maximum in the center of a simply supported plate and maximum at the center of the long edges for a clamped plate. The response for broad-band pressure loading is predicted by Eq. 7-31.

The 'Miles equation' for estimating stress induced by broad-band acoustic pressure is based on the assumption that the pressure is uniform over the surface (Richards and Meade, 1968; Ballentine et al., 1968),

$$\sigma_{\text{Miles}} = \left[\frac{\pi f_1 S_p(f_1)}{4 \zeta_1} \right]^{1/2} \sigma_0, \qquad (7\text{-}32)$$

where σ_0 is the stress induced by a uniform unit surface pressure. As can be seen from Table 7-2, the Miles equation predictions for the plate with simply supported edges are a factor of $16/\pi^2$ higher than those predicted by Eq. 7-31 for the simply supported plate. The reason for this difference is that the Miles equation assumes that the pressures are uniform over the surface, whereas the present method assumes that the pressures follow the mass-weighted mode shape (Eq. 7-21), which falls to zero at the edges. Thus, we could apply a correction factor of $\pi^2/16 = 1.62$ to the present method for uniform loading.

In general, the acoustic loading can be either uniform or variable over the surface of the plate according to the wavelength of the acoustic wave. The present method can be adapted to this situation. The mode shapes of panels can be approximated by sine waves of the form $\bar{w}_i(x) \sim \sin(ix/L)$, where $i = L/(\lambda_{\text{flex}}/2)$ is the number of half waves, that is, the number of times the mode shape passes through zero in the span L of the plate. The shape of a propagating acoustic wave (Eq. 7-18 with $k = 2\pi/\lambda_{\text{acoustic}}$) is also sinusoidal. These two shapes can be integrated to calculate a one-dimensional joint acceptance (Eq. 7-20) for a propagating acoustic wave,

$$J_{1D\text{-wave}} = \sqrt{J_{1D}^2(\phi = 0) + J_{1D}^2(\phi = 90°)}, \qquad (7\text{-}33)$$

where the joint acceptance for a phase-shifted wave $\bar{p}(x) = m\omega_i^2 \sin(kx + \phi)$ is

$$J_{1D} = \frac{\sin(kL - i\pi + \phi)}{kL - i\pi} - \frac{\sin(kL + i\pi + \phi) - \sin\phi}{kL + i\pi}. \qquad (7\text{-}34)$$

J is a function of the ratio of the acoustic wavelength to the structural wavelength (λ_f/λ_a) and the number of waves in the span of the panel $i = L/(\lambda_f/2)$. $J = 1$ if the acoustic wavelength and the panel wavelength are equal. Other values are given in Table 7-3. For an acoustic wavelength much greater than the span of the structure, the joint acceptance for the fundamental mode, $L/(\lambda_f/2) = 1$, is $J = 4/\pi = 1.27$. This is only one dimension. The two-dimensional joint acceptance is the

Table 7-3 One-dimensional joint acceptance (Eq. 7-34)

$\dfrac{L}{(\lambda_f/2)}$	λ_f/λ_a					
	0.0001	0.010	0.100	0.200	0.300	0.500
1	1.273	1.273	1.270	1.261	1.247	1.200
2	0.000	0.020	0.199	0.390	0.566	0.849
3	0.424	0.424	0.382	0.260	0.073	0.400
4	0.000	0.020	0.189	0.315	0.333	0.000
5	0.255	0.254	0.182	0.000	0.198	0.240
6	0.000	0.020	0.173	0.210	0.072	0.283
7	0.182	0.181	0.083	0.111	0.197	0.171
8	0.000	0.020	0.153	0.097	0.103	0.000
9	0.141	0.140	0.022	0.140	0.071	0.133
10	0.000	0.020	0.129	0.000	0.140	0.170
	0.700	0.900	1.000	1200	1.500	2.000
1	1.133	1.048	1.000	0.894	0.720	0.424
2	1.010	1.035	1.000	0.850	0.509	0.000
3	0.822	1.014	1.000	0.780	0.240	0.141
4	0.594	0.985	1.000	0.688	0.000	0.000
5	0.353	0.948	1.000	0.579	0.144	0.085
6	0.129	0.904	1.000	0.459	0.170	0.000
7	0.056	0.853	1.002	0.334	0.103	0.061
8	0.183	0.797	1.000	0.213	0.000	0.000
9	0.247	0.735	0.999	0.099	0.080	0.047
10	0.250	0.670	1.000	0.000	0.102	0.000

product of the joint acceptances along each coordinate. Thus, $J_{panel} = J_x J_y$, where J_x and J_y are found from Eq. 7-31 or Table 7-3. Therefore, for the case where the acoustic wavelength is much greater than the panel dimensions in both coordinate dimensions, $J_x = J_y = 1.27$ and $J_{panel} = 1.62 = 1.27^2 = 16/\pi^2$. The results of Eqs. 7-30 and 7-31 can be increased by this factor to approximate uniform loading as discussed previously.

Consider the situations shown in Figure 7-4. Here the acoustic wave is propagating in the x direction. In the y direction, the pressure is uniform. Effectively, the acoustic wavelength is infinite in the y direction, $\lambda_f/\lambda_a = 0$. The panel is deforming in its second transverse mode, so $b/(\lambda_f/2) = 2$. For this case, Table 7-3 shows $J_y = 0$, because the panel mode is antisymmetric, whereas the acoustic mode is symmetric from left to right. Symmetric pressures will not excite an antisymmetric panel mode.

One consequence of Eqs. 7-30 is that the sound-induced stress caused by sinusoidal pressure–time histories scales approximately as $\sigma \sim (b/h)^2$, so that doubling the thickness or having the unsupported span decreases the sound-induced stress by a factor of 4. The sound-induced stress for broad-band pressures (Eq. 7-31) scales as $\sigma \sim b/h^{1.5}$ (Table 7-2). Increasing thickness is not as effective in reducing stress for broad-band forcing because increasing frequency also increases the bandwidth of the response, admitting more broad-band energy.

Exercises

1. Consider a simply supported plate with the following properties: $E = 10 \times 10^6$ psi $(6.9 \times 10^{10}$ Pa), $\nu = 0.33$, $\rho = 0.1$ lb/in^3 $(2.7$ gm/cm^3), $h = 0.050$ in $(1.25$ mm), $b = 8$ in $(0.2$ m), $a = 39$ in $(1$ m). It is exposed to random pressures, uniform over the surface, of 150 db overall. The random pressures have constant auto spectral density over the frequency range from 0 to 1000 Hz. What is the fundamental natural frequency of the plate? If the damping is $\zeta = 0.015$, what is the maximum stress in the plate in the fundamental mode? What is the maximum displacement at the center of the plate?

2. Develop an expression for the maximum stress in a flat rectangular plate with clamped edges responding to uniform pressure with sinusoidal time history.

3. A traveling sound wave passes over the plate parallel to the long edge. The wavelength of the sound is 40 ft (10 m). How does this influence the calculation of Exercise 1? What if the wavelength is 8 in (0.2 m)?

7.2.5. Example: Acoustic Excitation of a Panel

An integrally stiffened, flat, rectangular titanium panel was exposed to acoustic waves traveling along its length on one side. The panel was fabricated from titanium with a triangular grid of stiffeners integrally bonded onto a thin skin. The skin was 0.018 in (0.46 mm) thick and the stiffening ribs were 0.17 in (4.32 mm) high. The overall panel dimensions were 23 by 33 in (58.4 by 83.8 cm). The edges of the panel were clamped and the average mass of the panel was 0.00553 lb/in^2 (3.88 kg/m^2). The panel was modeled using the finite-element technique. Figure 7-6 shows the first four natural modes of vibration and their natural frequencies. These calculated modes and frequencies agreed within 15% of experimental measurements. The measured damping was $\zeta_1 = 0.015$ in the first mode and $\zeta_3 = 0.009$ in the third mode. A displacement transducer was placed in the center of the panel. The response to broad-band acoustic pressures with a power spectral density of $S_p(f) = 8 \times 10^{-6}$ psi^2/Hz was observed.

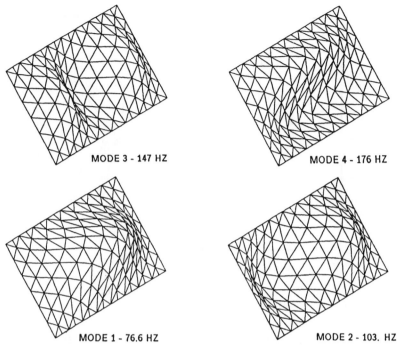

MODE 3 - 147 HZ

MODE 4 - 176 HZ

MODE 1 - 76.6 HZ

MODE 2 - 103. HZ

Fig. 7-6 Mode shapes of an integrally reinforced titanium panel (Blevins, 1989).

The theory of the previous section was used to predict the following power spectral density of displacement at the center of the panel in each mode at the center of the panel:

$$S_d(f) = \frac{S_p(f)}{[1 - (f/f_i)^2]^2 + (2\zeta_i f/f_i)^2} \frac{(J_x J_y)^2 \bar{w}_i^2}{\bar{P}_{ic}^2}, \qquad (7\text{-}35)$$

where $\bar{w}_i$ corresponds to the modal displacement at the center of the panel, the location of the transducer. Since the acoustic wave travels two-dimensionally (Fig. 7-4) down the length of the panel $\lambda_y = \infty$, $J_y = 1.27$. In the direction of propagation (x direction), the acoustic wavelengths were calculated: $J_x = 1.239$ for the first mode, $J_x = 0.442$ for the second mode, and $J_x = 0.232$ for the third mode, corresponding to $\lambda_i = c/f_i = 157$, 10.9, and 7.6 ft (4.8, 3.3, and 2.33 m) for the first three modes, where $c = 1100$ ft/sec (343 m/sec). Since the second mode has zero displacement at the center of the panel ($\bar{w}_2(x_c, y_c) = 0$, Fig. 7-6), no response was predicted in the second mode at the transducer.

The theoretical and measured displacements at the center of the panel are shown in Figure 7-7. The general agreement between theory and experiment is good. The errors are associated with the finite-element method, which overpredicted the natural frequencies by about 10%,

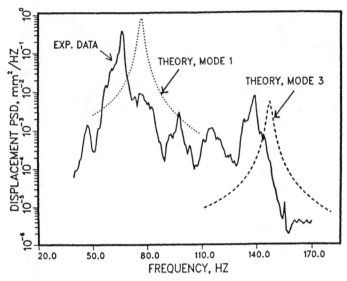

Fig. 7-7 Auto spectral density of displacement at the center of the panel when exposed to traveling acoustic waves (Blevins, 1989).

perhaps owing to motion of the panel edges, and approximately 30% uncertainty in the applied acoustic levels, which could not be controlled precisely since the panel motion and acoustic radiation from the vibrating panel affected the incoming acoustic wave.

The agreement of the data with the third-mode response is gratifying. The wavelength of the sound in the third mode is comparable to the panel dimensions. This and the complexity of the third-mode shape considerably diminish its response as compared with the first mode. Nonlinear effects in panels responding with large amplitudes increase both stiffness and damping (Mei and Prasad, 1987).

7.3. TURBULENCE-INDUCED VIBRATION OF TUBES AND RODS

7.3.1. Analytical Formulation

Heat-exchanger tubes are buffeted by the surrounding flow. The resultant vibrations can induce wear and, eventually, tube failure. Here we will investigate methods of predicting the vibrations of stable (Chapters 5 and 10) tubes and rods in response to external turbulent flows.

Surface pressures are shown in Figure 7-8 for flow over a rod parallel to a mean flow. At any instant of time, these surface pressures will exert

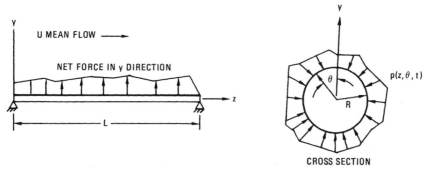

Fig. 7-8 Pressures acting on the surface of a rod in a turbulent flow.

a net lateral force per unit length on the rod,

$$F_y = -\int_0^{2\pi} p(x,\,\theta,\,t) \cos\theta R\,d\theta\,dx. \qquad (7\text{-}36)$$

F_y is the net lateral force per unit length resolved into the vertical (y) direction. The partial differential equation of motion of a uniform rod responding to this force is developed in Appendix A,

$$EI\frac{\partial^4 Y(z,\,t)}{\partial z^4} + m\frac{\partial^2 Y(z,\,t)}{\partial t^2} = F_y(z,\,t). \qquad (7\text{-}37)$$

$Y(z,\,t)$ is the displacement of the rod in the vertical direction. Solutions are sought in terms of a modal series,

$$Y(z,\,t) = \sum_{i=1}^{N} \bar{y}_i(z)y_i(t). \qquad (7\text{-}38)$$

For example, the natural frequencies and mode shapes of a rod or tube with pinned ends, that is, simply supported, are (Blevins, 1979a)

$$f_i = \frac{\omega_i}{2\pi} = \frac{(i\pi)^2}{2\pi L^2}\left(\frac{EI}{m}\right)^{1/2},\text{Hz} \qquad (7\text{-}39)$$

$$\bar{y}_i(z) = \sin\left(i\pi z/L\right), \qquad i = 1,\,2,\,3,\,\dots\,. \qquad (7\text{-}40)$$

E is the modulus of elasticity, m is the mass per unit length including added mass, and I is the moment of inertia of the cross section about the neutral axis.

Substituting Eq. 7-38 into Eq. 7-37, multiplying by a mode shape $\bar{y}_j(z)$, integrating over the span from $z = 0$ to $z = L$, utilizing the orthogonality conditions (Eq. 7-15), introducing a modal damping factor ζ_i, and assuming that any force-induced cross coupling between modes is negligible produces the following equation describing the dynamic response of each mode to the turbulent pressures (Appendix A):

$$\frac{1}{\omega_i^2}\ddot{y}_i(t) + \frac{2\zeta_i}{\omega_i}\dot{y}_i(t) + y_i(t) = \frac{\displaystyle\int_0^L F_y(z, t)\bar{y}_i(z)\,dz}{\displaystyle m\omega_i^2 \int_0^L \bar{y}_i^2(z)\,dz}.$$ (7-41)

The analysis is made mode by mode following the technique used in Section 7.2.1. We consider that the components of the lateral force contain a broad range of frequencies. Those components $F_i(t)$ with frequencies near the ith mode natural frequency will produce the majority of the response. Furthermore, if the force is fully correlated along the rod, it may be possible to separate the lateral force into functions of space and time,

$$F_y(z, t) \approx m\omega_i^2 \bar{g}_i(z)F_i(t),$$ (7-42)

where $\bar{g}_i(z)$ is the shape of the distribution of the lateral force along the span of the rod, normalized to a maximum value of 1.0. The scale factor $m\omega_i^2$ has been added to simplify the notation required in the subsequent equations. Substituting Eq. 7-42 into Eq. 7-41 produces the equation of motion of each mode,

$$\frac{1}{\omega_i^2}\ddot{y}_i + \frac{2\zeta_i}{\omega_i}\dot{y}_i + y_i = J_i F_i(t),$$ (7-43)

where the joint acceptance J_i is a measure of the compatibility of the distribution of the force along the span of the rod and the mode shape of the rod,

$$J_i = \frac{\displaystyle\int_0^L \bar{g}_i(z)\bar{y}_i(z)\,dz}{\displaystyle\int_0^L \bar{y}_i^2(z)\,dz}.$$ (7-44)

$J_i = 1$ if the shape of the force distribution is identical to mode shape $\bar{g}_i(z) = \bar{y}_i(z)$. Solutions to Eq. 7-43 are presented in Section 7.2.2; solutions for nonunity J_i can be found by scaling these results by J_i. The rms displacement of a rod responding in the ith mode to broad-band random pressures is obtained from Eq. 7-27,

$$\sqrt{\overline{Y^2(z, t)}} = \frac{J_i \bar{y}_i(z)}{m\omega_i^2}\sqrt{\frac{\pi f_i}{4\zeta_i}\,S_{F_y}(f_i)}.$$ (7-45)

$S_{F_y}(f)$ is the auto spectral density of the lateral force per unit length on the rod; see Section 7.1. The bar ($^-$) denotes time average over many cycles.

Turbulence is not generally fully correlated. Consider that the turbulence is *homogeneous*; that is, the time-averaged surface pressures and lateral force are independent of the coordinate z so that $\overline{F_y^2(z_1, t)} = \overline{F_y^2(z_2, t)}$. Then the spanwise correlation function between any two points on the rod is only a function of the separation between the points,

$$r(z_1, z_2) = r(z_1 - z_2) = \frac{\overline{F_y(z_1, t)F_y(z_2, t)}}{\overline{F_y^2}}. \tag{7-46}$$

Note that $r(z_1, z_2) \equiv 1$ when $z_1 = z_2$. For example, the correlation functions developed by Blakewell (1968) and Corcos (1964) for boundary layer turbulence decay exponentially with increasing separation,

$$r(z_1, z_2) = e^{-2|z_1 - z_2|/l_c} \cos\left[\omega(z_1 - z_2)\cos\theta/U_c\right], \tag{7-47}$$

where U_c is the convection velocity, θ is the angle between the axis and the flow direction, ω is the frequency, and l_c is the correlation length. The maximum value of the correlation function for homogeneous turbulence is unity; its minimum value is -1; and if the two points are very widely separated, then the correlation falls to zero.

Consider a spring-supported rigid rod that responds to turbulence. The mode shape is $\bar{y} = 1$. The mean square joint acceptance for homogeneous turbulence (Eqs. 7-41 and 7-44) is

$$J_i^2 = \frac{1}{L^2} \int_0^L \int_0^L r(z_1 - z_2) \, dz_1 \, dz_2. \tag{7-48}$$

The evaluation of this integral is a bit tricky. Following Frenkiel (1953), we make a change in variables $\xi = z_1 - z_2$ and $z = z_2$,

$$\int_0^L \int_0^L r(z_1 - z_2) \, dz_1 \, dz_2 = \int_0^L \int_{-z}^{L-z} r(\xi) \, d\xi \, dz$$

$$= \int_0^L \int_0^{L-z} r(\xi) \, d\xi \, dz + \int_0^L \int_{-z}^0 r(\xi) \, d\xi \, dz. \tag{7-49}$$

The two double integrals on the right-hand side of this equation can be evaluated by inverting the order of integration. The cross-hatched area of Figure 7-9 shows the area of integration in the first double integral on the right-hand side of Eq. 7-49. The integration may be made either by first integrating ξ from 0 to $L - z$ (vertical arrows) and then integrating z from 0 to L (horizontal arrows) or by inverting the order of integration,

$$\int_0^L \int_0^{L-z} r(\xi) \, d\xi \, dz = \int_0^L \int_0^{L-\xi} r(\xi) \, dz \, d\xi = \int_0^L (L - \xi) r(\xi) \, d\xi. \tag{7-50}$$

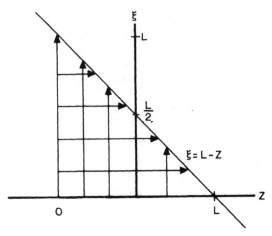

Fig. 7-9 Area of integration for correlation.

The last double integral on the right-hand side of Eq. 7-50 gives the same result. Thus,

$$J_i^2 = \frac{l_c}{L}\left(1 - \frac{\gamma}{L}\right),$$

$$\approx \frac{l_c}{L} \qquad \text{for } l_c \ll L. \qquad (7\text{-}51)$$

The correlation length l_c is defined as the area under both sides of the correlation function, and γ is the position of the centroid of the correlation area,

$$l_c = 2\int_0^L r(\xi)\,d\xi, \qquad \gamma = \int_0^L \xi r(\xi)\,d\xi \Big/ \int_0^L r(\xi)\,d\xi. \qquad (7\text{-}52)$$

The correlation length is of the order of 2 to 10 diameters for turbulent flow over rods; see Chapter 3, Section 3.2.

A small joint acceptance greatly reduces structural response. For example, suppose 100 wrens stand on a drum head covered with birdseed. If every wren pecked the drum head without regard for its neighbors, the drum would emit only a low random rumble because the energy put into the drum would most likely be counteracted by another bird pecking at a different time. However, if the birds banded together in groups of 10 and every wren in a group hit the drum at the same time, but different groups ignored each other, the sound would be louder because the correlation length (number of wrens in a group) would be larger with respect to the drum head. The loudest sound would be obtained when all the wrens pecked the drum at the same time. This is the fully correlated case. Thus, as the correlation length increases, the net force on the structure increases.

Exercises

1. Plot the correlation function of Eq. 7-47 as a function of $(z_1 - z_2)/l_c$ over the range from -8 to $+8$ for $\omega \cos \theta / U_c = 1/l_c$.
2. For the correlation function of Eq. 7-47 with $\theta = 0$, calculate the joint acceptance using Eq. 7-51.
3. Show that for fully correlated flow, the maximum values of correlation length and centroid of the correlation area (Eq. 7-52) are $l_{c,max} = 2L$, $\gamma_{max} = L/2$. What is the corresponding joint acceptance (Eq. 7-51)?
4. Calculate the correlation length corresponding to the correlation functions given in Chapter 3, Figure 3-9.

7.3.2. Tubes and Rods in Cross Flow

Fluid flows over cylinders and arrays of cylinders arise in industrial heat exchangers, steam generators, boilers, and condensers; in nuclear reactor fuel rod clusters; in power transmission line bundles and chimneys exposed to the wind; and in offshore oil rigs. The component of flow normal to the cylinder axis, cross flow, causes turbulence-induced vibration and, at sufficiently high velocity, instability (Chapter 5).

It is possible to express the auto spectral density of turbulence-induced force normal to the axis of a cylinder in cross flow in nondimensional form (Blevins et al., 1981),

$$S_{F_y} = (\tfrac{1}{2}\rho U^2 D)^2 \frac{D}{U} \Phi(fD/U), \qquad (7-53)$$

where ρ is the fluid density, U is the average cross-flow velocity through the minimum gap between tubes, D is the tube outside diameter, f is the frequency in hertz, and $\Phi(fD/U)$ is a dimensionless spectral shape function. Figure 7-10 shows that the nondimensionalization of Eq. 7-53 allows the data taken at various velocities to collapse onto a single curve. (The rms force on the tubes is proportional to the square of the dynamic head (U^4), as postulated by Pettigrew and Gorman (1978), but this force is spread over an expanding range of frequencies as the velocity increases with the net result that the auto spectral density of lateral force increases as U^3.)

$\Phi(fD/U)$ is a function of reduced frequency, Reynolds number, position in the tube array, tube array pattern, and upstream turbulence generators. In tube arrays with smooth entrance flow, the level of turbulence increases from the first row. By about the fifth row, the fully

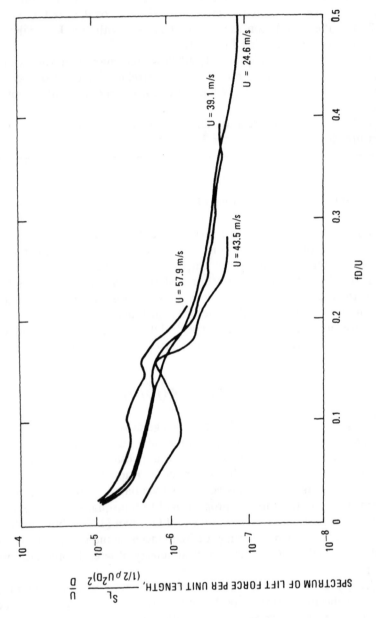

Fig. 7-10 Auto spectral density of lateral force in cross flow at various velocities for a tube in front row of a tube array (Blevins et al., 1981).

developed turbulence is independent of upstream conditions (Blevins et al., 1981; Sandifer and Bailey, 1984; Chen and Jendrzejczyk, 1987). Figure 7-11 shows the auto spectral density measured in water flow by Chen and Jendrzejczyk (1987) on an inline rectangular tube array with $T/D = L/D = 1.75$ for $15 \times 10^3 < \mathrm{Re} < 3.3 \times 10^5$ and by Taylor et al. (1986) on tube rows with $P/D = 1.5$ and 3. Axisa et al. (1988) recommend the following formula for the dimensionless auto spectral density of lateral force on a tube in a tube array:

$$\Phi(fD/U) = \begin{cases} 4 \times 10^{-4}(fD/U)^{-0.5}, & 0.01 \le fD/U \le 0.2, \\ 3 \times 10^{-6}(fD/U)^{-3.5}, & 0.2 \le fD/U \le 3. \end{cases} \tag{7-54}$$

U is the average velocity through the minimum gap between tubes. As observed by Savkar (1984) and these authors, the lift and drag forces have the same frequency in closely spaced tube arrays, which suggests that a regular vortex street does not form in these arrays, but instead large individual vortices are created in the spaces between tubes that are swept aft to impinge on downstream tubes. See Fig. 9-14.

Substituting the lateral force auto spectral density (Eq. 7-54) into the equation for broad-band response (Eq. 7-45) produces an equation for the characteristic rms response of the tube in a single mode,

$$\frac{Y_{i,\mathrm{rms}}}{D} = \frac{1}{16\pi^{3/2}} \frac{\rho D^2}{m} \left(\frac{U}{f_i D} \right)^{1.5} \frac{J_i}{\zeta^{1/2}} [\Phi(f_i D/U)]^{1/2} \bar{y}_i(z). \tag{7-55}$$

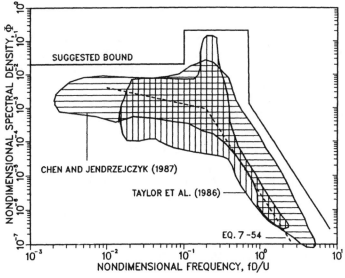

Fig. 7-11 Auto spectral density of turbulence-induced force on a cylinder array due to cross flow.

f_i is the natural frequency of the ith mode in hertz (Eq. 7-39), $\bar{y}_i(z)$ is the mode shape which has been normalized to a maximum value of 1.0, and J_i is the joint acceptance. Two limiting cases for the joint acceptance are $J_i = 1$ if the flow is completely correlated and $J_i = (l_c/L)^{1/2}$ for partially correlated flow with $l_c \ll L$ (Eq. 7-51). Blevins et al. (1981) found $l_c \approx 6.8D$ in a tube array. Axisa et al. (1988) suggest $L_c = 8D$.

Equation 7-55 is compared with experimental results of Weaver and Yeung (1984) in Figure 7-12. The experimental data are for water flow through an array of 0.5 in (1.27 cm) diameter brass tubes in a triangular pattern with a pitch-to-diameter ratio of 1.5. Note that the dominant frequency of the response, shown in the upper portion of the figure, is within a few percent of the natural frequency in still water. $\Phi = 0.006$ and $J = 1$ gives good agreement with the data except near the vortex resonance at $U/fD = 1.7$ and the onset of instability at $U/(fD) = 3.6$. Ordinarily, vortex-induced vibration (Chapter 3) and fluid elastic instability (Chapter 5), rather than turbulence, dominate the response of tubes and rods in cross flow.

Taylor et al (1988) discuss two-phase turbulence excitation in cross flow.

Exercises

1. Equations 7-1 and 7-53 imply that the overall rms lateral force on the rod is given by

$$F_{\text{rms}} = \tfrac{1}{2}\rho U^2 D C_{\text{rms}} = \tfrac{1}{2}\rho U^2 D \left[\int_0^\infty \Phi(fD/U)\, d(fD/U) \right]^{1/2}.$$

Perform this integration approximately using the nondimensional spectral density function of Eq. 7-54 to determine C_{rms}. Compare the results with measurements of lift coefficient and drag coefficient on single tubes (Sections 3.5.1 and 6.3.2).

2. Predict the onset of fluid elastic instability for the tube array of Figure 7-12 using Chapter 5, Eq. 5-2. How well does theory agree with experiment?

3. The peak in Figure 7-11 near $fD/U = 0.2$ may be associated with vortex shedding. Using $\Phi = 0.2$, compare the response of this peak predicted by Eq. 7-85 with that predicted by Eq. 3-11 using $C_L = 0.5$.

7.3.3. Tubes and Rods in Parallel Flow

Parallel flows does not ordinarily cause failure of tubes and rods because tubes of heat exchangers are generally too stiff to experience a parallel

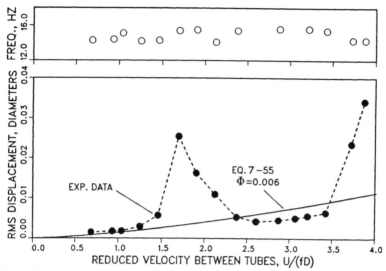

Fig. 7-12 Response of a $D = 1.27$ cm diameter brass tube with a natural frequency of $f = 14.6$ Hz in a tube array to water cross flow. $m/(\rho D^2) = 8.16$, $\zeta = 0.0057$ (Weaver and Yeung, 1984).

flow-induced instability (Paidoussis, 1982; Chapter 10, Section 10.2). Nevertheless, a turbulent parallel flow (Fig. 7-8) will excite vibrations and induce wear. Semiempirical correlations fit measurements of the characteristic maximum amplitude Y of the vibration to an expression of the form (Paidoussis, 1982, 1981; Chen, 1970; and Blevins, 1979b)

$$Y = K\rho^a U^b d^c D^e m^g f^h \zeta^j, \qquad (7\text{-}56)$$

where a through j are dimensionless exponents whose values are given in Table 7-4 for metallic rods and tubes between 0.4 and 1.5 in (1 and 3 cm) in diameter and between 18 and 40 in (0.5 and 2 m) in span and for water flows up to 40 ft/sec (10 m/sec). An order of magnitude scatter of the data about the predictions, as shown in Figure 7-13, is due in part to the influence of upstream turbulence generators.

The Paidoussis (1981) correlation for tube and rod vibration induced by parallel flow is

$$\frac{Y^*}{D} = \frac{5 \times 10^{-4} K}{\alpha^4} \left[\frac{u^{1.6}(L/D)^{1.8} \mathrm{Re}^{0.25}}{1 + u^2} \right] \left(\frac{D_h}{D} \right)^{0.4} \left[\frac{\beta^{3/2}}{1 + 4\beta} \right]. \qquad (7\text{-}57)$$

Y^* is the characteristic vibration amplitude; α is the dimensionless first-mode eigenvalue of the cylinder, equal to $\alpha = \pi$ for pinned (simply supported) cylinders and $\alpha = 4.73$ for clamped ends; D is the cylinder diameter; u is the dimensionless flow velocity, defined as $u = (\rho A/EI)^{1/2}UL$, where ρ is the fluid density; $A = \pi D^2/4$ is the

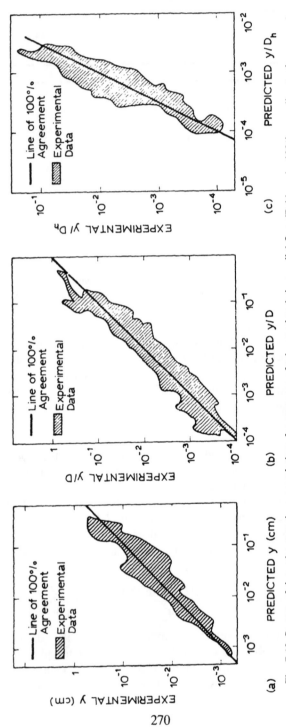

Fig. 7-13 Scatter of data about various correlations for response of tubes and rods in parallel flow (Paidoussis, 1981) according to the expressions of (a) Reavis (1969), (b) Paidoussis (1981), (c) Y. N. Chen (1970).

Table 7-4 Comparisons of exponents in theories of parallel flow-induced vibration of cylindrical structures (Eq. 7-56)[a]

Correlation	Exponent of						
	ρ	U	d	D	m	f	ζ
Basile et al. (1968)	0.25	1.5	0.5	−0.5	−0.25	−1	0
Burgreen et al. (1958)	0.385	2.3	1	0.77	−0.65	−2.6	0
Chen (1970)	1	2	0	1	0	−2	0
Paidoussis (1981)	0.8	2.6	0.8	2.2	−0.66	−1.6	0
Reavis (1969)	1	1.5	0.4	1.5	−1	−1.5	−0.5
Wambsganss (1971)	0	2	1.5	1.5	−1	−1.5	−0.5
Eq. 7-55	1	1.5	0	1.5	−1	−0.5	−0.5

Source: Blevins (1979b).
[a] d Hydraulic diameter
D Cylinder diameter
f Natural frequency, Hz
m Mass per unit length including added mass
U Velocity of parallel flow
ρ Fluid density
ζ Damping, including fluid damping (Chapter 8)

cross-sectional area of the cylinder; E is its modulus of elasticity; U is the parallel flow velocity; L is the span of the cylinder; $\mathrm{Re} = UD/\nu$ is the Reynolds number; $D_h = 4(\text{flow area})/(\text{wetted perimeter})$ is the hydraulic diameter of the flow channel containing the cylinder; $\beta = \rho A/(\rho A + m)$ is the mass ratio, where m is the cylinder mass per unit length; and K is a parameter that is equal to 1 for very "quiet" circulating systems such as experimental water tunnels and equal to 5 for more turbulent industrial environments.

Other correlations for parallel flow-induced vibrations of tubes and rods are given by Wambsganss and Chen (1971), Basile et al. (1968), Burgreen et al. (1958), Y. N. Chen (1970), and Reavis (1969).

7.4. VIBRATION INDUCED BY WINDS

The study of the random patterns of winds that envelop the earth is the most ancient form of fluid mechanics. Homer's Odysseus was tormented by the random occurrences of calm and gale inflicted by Aeolus, the god of the winds. The destructive effect of extreme wind on civil structures is well known. Codes of Practice provide standards of building design for wind loading. Five of these are:

Great Britain   Code of Basic Data for the Design of Buildings, Chapter V, Part 2, Wind Loads, British Standards Institution CP3, 1972.

Canada	National Building Code of Canada, National Research Council, Ottawa, Canada, NRCC No. 17303, 1980.
United States	Building Code Requirements for Minimum Design Loads in Buildings and Other Structures, American National Standard A58.1-1982, American National Standards Institute, New York, 1982.
Australia	Minimum Design Loads on Structures, Australian Standard AS 1170, Part 2, Standards Association of Australia, Sydney, 1981.
Switzerland	Normen fur die Belastungsannahmen, die Inbetriebnahme und die Uberwachung der Bauten, *Schweizerischer Ingenieur-und Architekten-Verein,* No. 160, 1956.

Books on wind engineering include those of Simiu and Scanlan (1986), Sachs (1978), Panofsky and Dutton (1984), Kolousek et al. (1984), Houghton and Carruthers (1976), Lawson (1980), Plate (1982), and Geissler (1970). Excellent review articles have been written by Cermak (1975), Scruton (1981), Davenport (1982) and the ASCE (1980). Many papers in this field are published in the *Journal of Wind Engineering and Industrial Aerodynamics,* Elsevier Science Publishers. This section emphasizes the development of an understanding of the turbulent boundary layer of the wind and wind-induced vibration of buildings.

7.4.1. Turbulent Boundary Layer of the Earth

As air, or any viscous fluid, flows over a solid surface, the fluid at the surface is slowed as fluid shears against the surface. The viscous shearing of layers of fluid adjacent to the surface triggers rolling eddies higher in a turbulent boundary layer that exchange momentum between the free stream flow and the viscous sublayer. The change in wind velocity with height above the surface of the earth in this boundary layer can be seen in Figure 7-14. The boundary layer thickness increases with roughness of the surface and the fetch over which the flow (wind) acts, as shown in Figure 7-15.

Although theoretical results are available for predicting the wind boundary layer profile, these predictions have proven no more useful in wind engineering than the simple power-law profile:

$$\frac{\bar{U}(z)}{\bar{U}_G} = \left(\frac{z}{z_G}\right)^{\alpha}. \tag{7-58}$$

$\bar{U}(z)$ is the mean reference velocity averaged over a time interval (ordinarily 1 to 10 minutes) at a height z above the surface. At the top of the boundary layer, the wind approaches the velocity $\bar{U}_G$, which is called

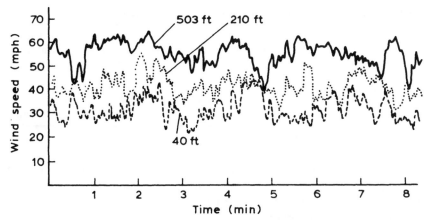

Fig. 7-14 Time-history of wind speed at three heights on a 500 ft mast (Deacon, 1955).

the gradient velocity since it can be predicted from the pressure gradients in the earth's atmosphere. The gradient height z_G is the height at which the gradient velocity is matched with this profile, typically between 700 and 2000 ft (200 and 600 m). Table 7-5 gives values of the gradient height and exponent α over different surface covers. These are average results, and large deviations can be expected in a city, where upstream buildings can block the flow and generate large turbulent eddies.

Winds are not steady. As shown in Figure 7-16, wind ranges in frequency from the 1 year period of the annual weather cycle and the 4 day period typical of large weather systems to gusts with periods of 10

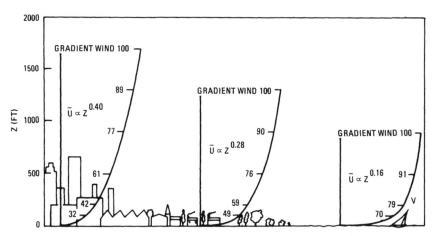

Fig. 7-15 Profiles of mean wind velocity over level terrains of differing roughness (Davenport, 1963).

Table 7-5 Parameters of mean wind speed profiles over different surfaces

Surface	Roughness length, z_0(m)	Exponent, α	Gradient height, z_G(m)	Surface drag coefficient, κ
Rough sea	0.003	0.11	250	0.002
Prairie, farmland	0.03	0.16	300	0.005
Forest, suburbs	0.3	0.28	400	0.015
City center	3.0	0.40	500	0.05

Sources: Davenport (1963), Chamberlain (1983).

seconds or less. The measurement of wind velocity thus depends on the duration of the time-averaging interval. Durst (1960) gives the ratio of the maximum wind speed averaged over a period t to the wind speed averaged over 1 hour as follows:

Averaging time t (sec)	2	5	10	20	50	100	200	1000	3600
U_t/U_{3600}	1.53	1.48	1.43	1.37	1.26	1.19	1.13	1.08	1.00

A cup anemometer averages wind speeds over two to three seconds, and so the maximum gust velocity measured by a cup anemometer would be expected to exceed the hourly average by about 1.5. Similarly, the National Building Code of Canada suggests that the maximum gust speeds are $\sqrt{2}$ times the highest hourly mean. These 1.4–1.5 ratios are apparent in Figure 7-17, which shows the gust maximum and hourly mean winds over the United Kingdom. Also see Cheng and Chiu (1985).

Extreme winds are presented by Changery et al. (1984), Thom (1968), and Batts (1982), and by the U.S. Atomic Energy Commission (1974), as well as in the National Standards and in Figures 7-17 and 7-18. Simple

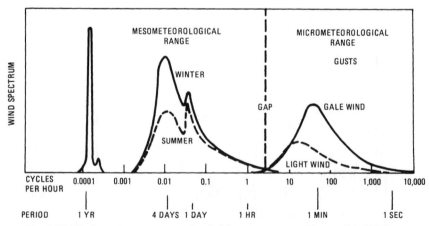

Fig. 7-16 Wind spectrum over an extended frequency range (Davenport, 1970).

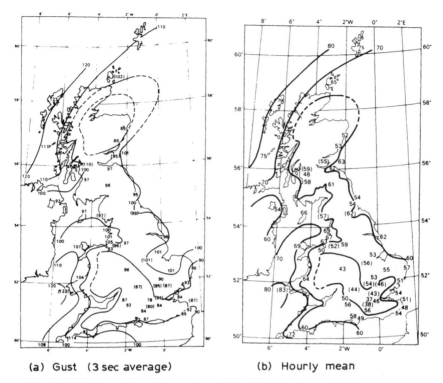

(a) Gust (3 sec average)   (b) Hourly mean

Fig. 7-17 Wind speed in miles per hour. United Kingdom annual highest wind speeds, 50 year recurrence interval, miles per hour 33 ft above surface. "Wind Loads," British Standards Institution, 1972.

design-limit wind speeds can be useful. The American Association of Highway and Transportation Officials (1986) recommends a design velocity of 100 mph (160 km/hr) for all highway bridges.

The cumulative distribution function of the wind speed $P(U)$ is defined as the probability that the hourly average wind speed in any one year will not exceed the speed U (see Eq. 7-9). Davenport (1982) has found, at least for the gradient wind, that the Rayleigh distribution

$$P(U > U_1) = e^{-U_1^2/2U_{rms}^2} \qquad (7\text{-}59)$$

gives a good approximation to the cumulative distribution function. A wide range of probabilistic methods is being used to estimate extreme winds. See Simiu and Scanlan (1986), Benjamin and Cornell (1970), and Panofsky and Dutton (1984).

$1 - P(U)$ is the probability that the hourly average wind speed will exceed U in any year and, on the average, a period of N years

$$N = 1/(1 - P) \qquad (7\text{-}60)$$

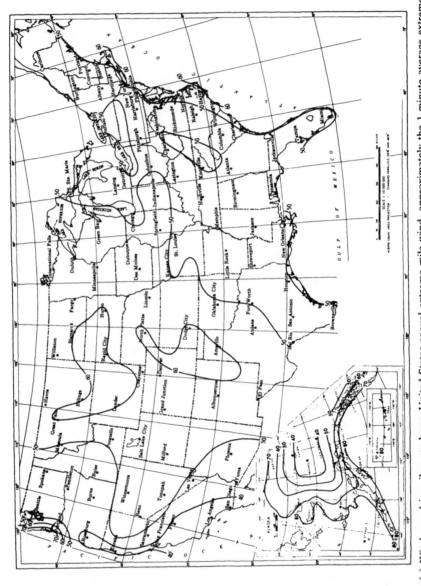

Fig. 7-18 (a) Wind speed in miles per hour. United States annual extreme mile wind, approximately the 1 minute average extreme wind, 30 ft above ground; 2 year mean recurrence interval (Thom, 1968).

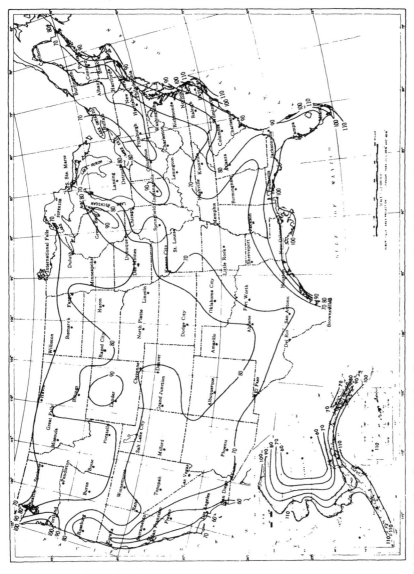

Fig. 7-18 (b) Wind speed in miles per hour. United States annual extreme mile wind, approximately the 1 minute average extreme wind, 30 ft above ground; 50 year mean recurrence interval (Thom, 1968).

must elapse for U to be exceeded. N is called the recurrence interval or, more commonly, the return period. In L years, the probability that U will not be exceeded is P^L. The probability that it will be exceeded is $r = 1 - P^L$. In terms of the return period (Eq. 7-60), the probability that wind with a return period of N years will be exceeded at least once in L years is (Mehta, 1984)

$$r = 1 - \left(1 - \frac{1}{N}\right)^L.$$ (7-61)

For example, if the design wind speed is based on an $N = 50$ year recurrence interval wind, there exists a probability of 0.64 that the design wind will be exceeded in the $L = 50$ year life of the structure. Equation 7-61 can be solved for the return period N associated with the risk of exceedance r in L years (Holand et al., 1978):

$$N = \frac{1}{1 - (1 - r)^{1/L}} \approx \frac{L}{r}, \qquad \text{for small } r.$$ (7-62)

If there is to be no greater than a 10% chance that a design wind speed will be exceeded in a 50 year lifetime, then the return period of that wind is $N \approx 50/0.1 = 500$ yr. Design winds of this high return period are rarely used. One-hundred or fifty year design winds are commonly used, although it is probable that these winds will be exceeded in the life of most civil engineering structures. Factors of safety in the design compensate for this to a degree, as does the fact that some damage from extreme winds is often acceptable.

The time history of velocity fluctuations in the direction of the mean wind is characterized by auto spectral density (Section 7.1). Two expressions for the auto spectral density of horizontal gustiness $S_u(f)$ are

$$\frac{fS_u(f)}{\kappa \bar{U}^2} = \begin{cases} 4\gamma^2/(1 + \gamma^2)^{4/3} & \text{Davenport (1963);} \\ 200\gamma/(1 + 50\gamma)^{5/3} & \text{Kaimal (1972).} \end{cases}$$ (7-63)

The Davenport spectrum has been incorporated in the Canadian National Building Code. f is gust frequency in hertz. $\gamma = fL_t/\bar{U}(10)$ for Davenport's spectrum, where $U(10)$ is the mean wind velocity at 10 meters above the surface, κ is the surface drag coefficient, given in Table 7-5, and L_t is the length scale of turbulence, equal to $\sqrt{3}$ times the wavelength at which the function $fS_u(F)/\kappa\bar{U}^2$ is maximum. Davenport has found that $L_t = 1200 \text{ m} \times \bar{U}/\bar{U}_{10}$ gives good agreement with experimental data. $\gamma = fz/U(z)$ for Kaimal's spectrum, where z is the height above the surface. The Davenport spectrum is shown in Figure 7-19. The peak in the spectrum at $\bar{U}/f = 700$ m corresponds to the peak in the spectrum of Figure 7-16 near the 1 minute period. These spectra cover only the higher frequency ranges of the gusts. For a mean velocity of

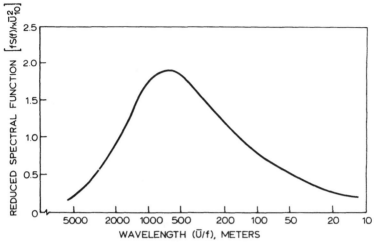

Fig. 7-19 Spectrum of horizontal gustiness of high winds (Davenport, 1961).

33 ft/sec (10 m/sec), the frequency over which the spectra are valid ranges from about 0.02 to 1 Hz.

The mean square component of the horizontal component of the gust velocity is the integral of the auto spectral density (Eq. 7-1). Using the Davenport spectrum, this is

$$\overline{u^2} = \int_0^\infty S_u(f)\, df = 2.35^2 \kappa \bar{U}^2. \tag{7-64}$$

Since κ ranges from about 0.0005 to 0.05, the magnitude of the root mean square of the horizontal component of wind turbulence ranges between 5% and 50% of the mean wind. Experimental measurements of wind turbulence, such as those shown in Figure 7-20, suggest that the unsteady component of the velocity in line with the mean wind decays exponentially with increasing separation,

$$r(z_1 - z_2) = e^{-C_z f |z_1 - z_2|/\bar{U}}, \tag{7-65}$$

where $z_1 - z_2$ is the vertical separation of points. The data of Figure 7-20 suggest that $C_z \approx 7$ and is dependent on the roughness of the surface. Atmospheric tests suggest that $C_z \approx 10$ and that the coefficient of correlation for horizontal separation is $C_y \approx 16$ (Vickery, 1969).

7.4.2. Prediction of Alongwind Response

Techniques for estimating alongwind response have been developed by Davenport (1967), Vickery (1969), Vellozzi and Cohen (1968), Simiu

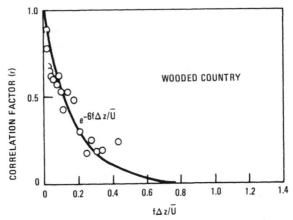

Fig. 7-20 Correlation factor of wind speed in the vertical direction as a function of the ratio of separation to $\bar{U}/f$ (Davenport, 1963; Singer, 1960).

(1980), Solari (1982), and Simiu and Lozier (1979). Simiu and Hendrickson (1987) review methods for predicting failure of windows and glass cladding due to wind. These analyses are quasisteady; they are most applicable to slender, stable structures, outside the range of vortex-induced resonance (Chapters 3 and 4). The analysis presented here follows Vickery's technique.

The response of elastic structures to wind is the sum of mean and fluctuating components. The mean drag force per unit height on a tall, slender structure is

$$F_x = \tfrac{1}{2}\rho \bar{U}^2(z)DC_D, \qquad (7\text{-}66)$$

where $\bar{U}$ is the average wind speed and ρ is the air density. The drag coefficient of rectangular sections at high Reynolds numbers is a function of the ratio of the breadth in the direction of flow B to the lateral width D (Blevins, 1984; Nagano et al., 1983):

B/D	0.2	0.4	0.5	0.65	0.8	1.0	1.2	1.5	2.0	3.0
C_D	2.1	2.35	2.5	2.9	2.3	2.2	2.1	1.8	1.6	1.3

The mean inline response of a building to the drag force of the wind can be obtained from Eqs. 7-41, 7-58, and 7-66 by neglecting dynamic effects and considering only a single mode,

$$\bar{X}(z) = \tfrac{1}{2}\rho \bar{U}^2(L)DC_D \bar{x}(z) \int_0^L z^{2\alpha}\bar{x}(z)\,dz \Big/ \left[m\omega_n^2 L^{2\alpha} \int_0^L \bar{x}^2(z)\,dz \right]$$

$$= \frac{2\beta+1}{2(2\alpha+\beta+1)} \frac{\rho \bar{U}^2(L)DC_D}{m\omega_n^2} \left(\frac{z}{L}\right)^{\beta}, \qquad \text{if } \bar{x}(z) = \left(\frac{z}{L}\right)^{\beta}. \quad (7\text{-}67)$$

$\bar{X}$ is the mean deflection in the direction of wind of the fundamental mode with generalized mass m per unit length. The fundamental mode of most buildings is a cantilever mode, which is well approximated by $\bar{x}(z) = (z/L)^\beta$, where L is the height of the building. $\beta = 1.86$ gives a good approximation to the mode shape of a bending cantilever. Equation 7-67 can easily be generalized to structures whose cross section varies with height if the drag coefficient on each section is known. End effects can have a profound effect on the drag of relatively low structures, as shown in Figure 7-21. Note the sharp drop in wind pressure near the top of the building due to the spill of pressure over the roof.

A building is buffeted by gusts of wind. Consider that the wind velocity is the sum of a steady component and a sinusoidally oscillating component,

$$U = \bar{U} + u_m \sin \omega t. \qquad (7\text{-}68)$$

If the oscillating component u_m is a small fraction of the mean component $\bar{U}$ and buoyancy effects are negligible (see Chapter 6, Section 6.1), then terms proportional to u_m^2 and ωu_m can be neglected, and the stagnation pressure becomes

$$p = \tfrac{1}{2}\rho U^2 \approx \tfrac{1}{2}\rho \bar{U}^2 + \rho u_m \bar{U} \sin \omega t. \qquad (7\text{-}69)$$

The mean component of stagnation pressure p is denoted by $\bar{p}$, and the

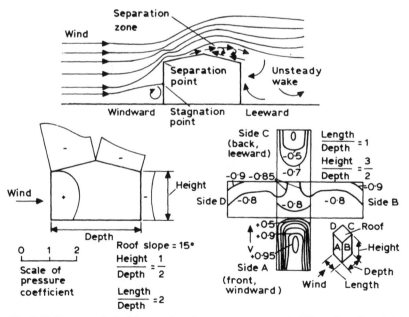

Fig. 7-21 Patterns of wind flow and surface pressure over a building of medium height (Liu, 1979).

fluctuating component is $p - \bar{p}$. The ratio of the mean square fluctuating component to the mean component is

$$\frac{\overline{(p - \bar{p})^2}}{\bar{p}^2} = \frac{4\overline{u_m^2}}{\bar{U}^2}, \quad \text{which implies} \quad \frac{S_p(f, z)}{\bar{p}^2} = \frac{4S_u(f, z)}{\bar{U}^2}. \quad (7\text{-}70)$$

Solving for the spectral density of pressure on the surface gives

$$S_p(f, z) = [\tfrac{1}{2}\rho \bar{U}^2(L)C_D]^2 \left(\frac{\bar{U}(z)}{\bar{U}(L)}\right)^2 \left(\frac{4\overline{u^2}(L)}{\bar{U}^2(L)}\right) \frac{S_u(f, L)}{\overline{u^2}(L)}. \quad (7\text{-}71)$$

The wind spectrum at the roof, $z = L$, has been used to characterize the wind spectra over the entire building surface. A related spectrum may be defined as

$$S_p'(f) = S_p(f, z)\left[\frac{\bar{U}(L)}{\bar{U}(z)}\right]^2, \quad (7\text{-}72)$$

which has no height dependence, and the height dependence factor $[\bar{U}(z)/\bar{U}(L)]^2$ is shifted into the joint acceptance.

The correlation of wind over the building surface decays exponentially in the lateral (y) and vertical (z) directions,

$$r(z_1, z_2, y_1, y_2) = \exp\left(\frac{-2f}{\bar{U}(z_1) + \bar{U}(z_2)}[C_z^2(z_1 - z_2)^2 + C_y^2(y_1 - y_2)^2]^{1/2}\right)$$

$$= \exp\left(\frac{-2a[(z_1 - z_2)^2/L^2 + \lambda(y_1 - y_2)^2/D^2]^{1/2}}{(z_1/L)^\alpha + (z_2/L)^\alpha}\right) \quad (7\text{-}73)$$

for the power-law wind profile (Eq. 7-58) and where $a = C_z f L/\bar{U}(L)$ and $\lambda = C_y D/(C_z L)$.

The joint acceptance of turbulence can be found from the mode shape and the correlation function for surface pressure of the wind. Since wind turbulence is not homogeneous (it changes with height), it is necessary to include the height dependence of the wind spectrum within the spatial integrals of the joint acceptance. A joint acceptance is defined as

$$J^2(f) = \int_0^A \int_0^A \frac{\bar{U}(z_1)\bar{U}(z_2)}{\bar{U}^2(L)} \tilde{x}(z_1)\tilde{x}(z_2)r(z_1, z_2, y_1, y_2) \frac{dA_1}{A} \frac{dA_2}{A}$$

$$= \int_0^1 \int_0^1 \int_0^1 \int_0^1 \left(\frac{z_1}{l}\right)^{\alpha+\beta} \left(\frac{z_2}{l}\right)^{\alpha+\beta}$$

$$\times r\left(\frac{z_1}{L}, \frac{z_2}{L}, \frac{y_1}{L}, \frac{y_2}{L}\right) d\left(\frac{z_1}{L}\right) d\left(\frac{z_2}{L}\right) d\left(\frac{y_1}{L}\right) d\left(\frac{y_2}{L}\right) \quad (7\text{-}74)$$

for the power-law wind profile and a power-law approximation of the mode shape. A is the frontal area of the structure, $A = DL$. The auto spectral density of the generalized force on the structure is assembled from the auto spectral density of the surface pressure times the frontal area times the square of the joint acceptance,

$$S_F(f) = [\tfrac{1}{2}\rho \bar{U}^2(L) C_D DL]^2 \left(\frac{4\overline{u^2}(L)}{\bar{U}^2(L)}\right) J^2(f) \frac{S_u(f)}{\overline{u^2}(L)}. \tag{7-75}$$

The spectrum of horizontal gustiness, $S_u(f)$, is given by Eq. 7-63.

The root mean square of the dynamic response of the structure to wind gusts is found by integration of the generalized force spectrum divided by the structural impedance over the frequency range of the gusts. The mean square of the alongwind response of the top of the building is obtained by integrating the generalized force times the transfer function over the frequency range (Eq. 7-27),

$$\overline{(X - \bar{X})^2} = \int_{f_1}^{\infty} \frac{S_F(f)\, df}{m^2 (2\pi f_n)^2 [(1 - (f/f_n)^2)^2 + (2\zeta f/f_n)^2]^2} \tag{7-76}$$

where f_1 is the frequency of the beginning of the gust portion of the spectrum and $\bar{X}$ is the mean response at the top of the building (Eq. 7-68). This integration can be considerably simplified by dividing the response into resonant and nonresonant components. For the portion of the gust spectrum with frequencies much less than the natural frequency of the building, $f \ll f_n$, $[(1 - (f/f_n)^2]^2 + (2\zeta f/f_n)^2 \approx 1$, which greatly simplifies the integration. For the portion of the gust spectrum with frequencies near the building natural frequency f_n, a closed-form solution is given by Eq. 7-27. The root mean square dynamic response is the square root of the sum of the nonresonant and resonant components,

$$\frac{\overline{[(X - \bar{X})^2]}^{1/2}}{\bar{X}} = \left(\frac{1 + 2\alpha + \beta}{1 + \alpha + \beta}\right) \frac{2[\overline{u^2}(L)]^{1/2}}{\bar{U}(L)} \left(\frac{2(1 + \alpha + \beta)^2}{3} \int_0^{\infty} \frac{\gamma J^2\, d\gamma}{(1 + \gamma^2)^{4/3}} \right.$$

$$\left. + \frac{\pi}{6} \frac{\gamma^2(f_n)}{[1 + \gamma^2(f_n)]^{4/3}}\right)^{1/2}, \tag{7-77}$$

where f_n is the natural frequency of the structure. In more compact form, the ratio of the root mean square dynamic response to the mean response is

$$\frac{\overline{[(X - \bar{X})^2]}^{1/2}}{\bar{X}} = R\left(B + \frac{SF}{\zeta}\right)^{1/2}. \tag{7-78}$$

The building response to the gust spectrum is thus the sum of a nonresonant buffeting response, $RB^{1/2}$, and the resonant response at frequencies about the natural frequency, $R(SF/\zeta)^{1/2}$.

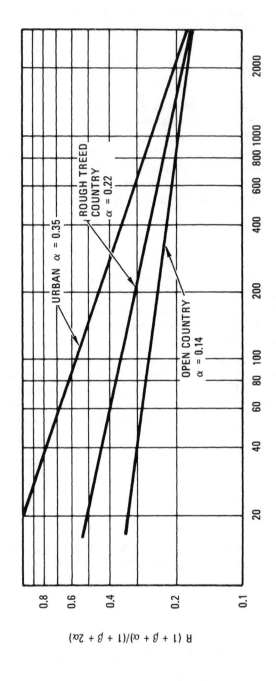

Fig. 7-22 Suggested values of $2\sqrt{\bar{u}^2}/\bar{U}(L)$, which is equal to $R(1 + \beta + \alpha)/(1 + \beta + 2\alpha)$. (Vickery, 1969.)

284

The surface roughness factor R and the dynamic response are directly proportional to the rms wind turbulence,

$$R = \frac{2[\overline{u^2}(L)]^{1/2}}{\bar{U}(L)} \frac{(1 + 2\alpha + \beta)}{(1 + \alpha + \beta)} = R(L, \text{ surface roughness, } \alpha, \beta). \quad (7\text{-}79)$$

R is plotted in Figure 7-22. R decreases with the ratio of surface roughness to building height. The background excitation factor B, plotted in Figure 7-23,

$$B = B\left(\frac{C_z L}{L_t}, \frac{C_y D}{C_z L}\right), \qquad S = S\left(\frac{C_z f_n L}{\bar{U}(L)}, \frac{C_y D}{C_z L}\right), \quad (7\text{-}80)$$

decreases with increasing fineness of the building. Davenport (1961) has estimated $L_t(L)$ to be $l\bar{U}(L)/\bar{U}(z_{10})$, where $l = 4000$ ft (1200 m) and $z_{10} = 33$ ft (10 m). The size reduction factor S, plotted in Figure 7-24, decreases with increasing reduced velocity. f_n is the natural frequency of the structure. The gust energy factor F, plotted in Figure 7-25,

$$F = \frac{\pi}{6} \frac{\gamma^2(f_n)}{[1 + \gamma^2(f_n)]^{4/3}}, \quad (7\text{-}81)$$

decreases with increasing frequency of the structure, since the gust energy decreases in the high frequency ranges $\gamma(f_n) = f_n L_t / \bar{U}(L)$.

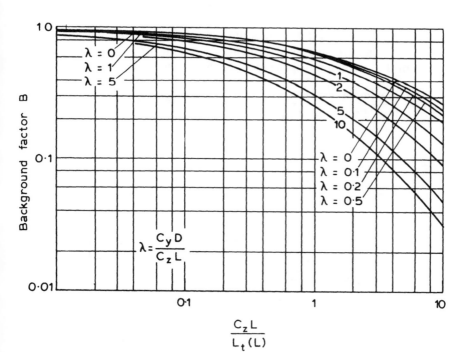

Fig. 7-23 Background excitation factor B (Vickery, 1969).

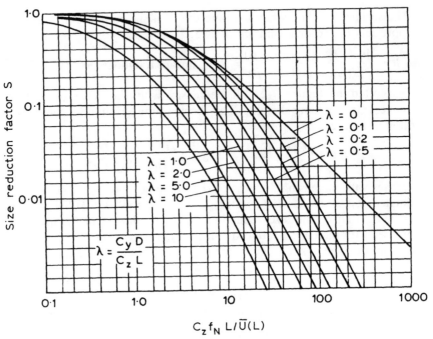

Fig. 7-24 Size reduction factor S (Vickery, 1969).

The maximum displacement is the sum of the mean displacement plus the maximum dynamic displacement. A gust factor G is defined as the ratio of the maximum displacement to the mean displacement,

$$G = \frac{\text{Maximum displacement}}{\text{Mean displacement}}_{\text{(Eq. 7-67)}} = \frac{\bar{X} + \bar{g}\overline{(X - \bar{X})^2}}{\bar{X}} = 1 + \bar{g}R\left(B + \frac{SF}{\zeta}\right)^{1/2}.$$

(7-82)

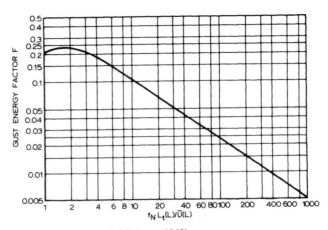

Fig. 7-25 Gust energy factor F (Vickery, 1969).

The peak factor $\bar{g}$ is equal to the ratio of the maximum dynamic response to the rms dynamic response (Section 7.1, Eq. 7-10). For time intervals between 10 and 30 min, $\bar{g}$ is between 3.0 and 3.7 (Eq. 7-10). Vickery (1969) found that the value $\bar{g} = 3.5$ has given good agreement with experimental data. Experience shows that the gust factor G is ordinarily between 2.0 and 3.0, with $G = 2.5$ being typical for most buildings. With this factor the maximum response is estimated from Eq. 7-82. Because the mean response is proportional to $\bar{U}^2$, the dynamic and total responses are also proportional to the velocity squared, as shown in Figure 7-26.

The maximum acceleration is equal to the peak factor times the rms acceleration,

$$\text{Peak acceleration} = \bar{g}(2\pi f_n)^2 \bar{X}(L) R(SF/\zeta)^{1/2}. \qquad (7\text{-}83)$$

Human tolerance to acceleration is as follows for the low frequencies associated with building motion (Chang, 1973):

Peak acceleration, g	less than 0.005	0.005 to 0.015	0.015 to 0.05	0.05 to 0.15	Greater than 0.15
Human discomfort	Imperceptible	Perceptible	Annoying	Very annoying	Intolerable

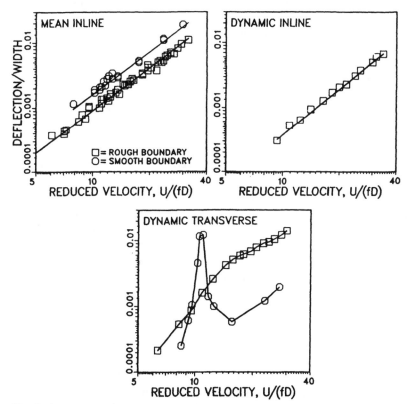

Fig. 7-26 Response of a rectangular structure in the direction of flow and perpendicular to the direction of flow. Damping factor, $\zeta = 0.01$ (Davenport, 1982).

where g is the acceleration due to gravity. Parsons and Griffin (1988) found that the threshold of human perception of sinusoidal vibration at frequencies between 2 and 100 Hz ranges from 0.001 to 0.01g rms, which is consistent with the results of Chang.

The previous procedure for estimating alongwind response is relatively insensitive to α and β; $\beta = 1.86$ well approximates a cantilever mode and $\alpha = 1/7$ approximates a relatively open environment. By substituting these values into Eq. 7-67, the mean deflection at the top of a building is estimated from Eq. 7-67 to be

$$\bar{X} \approx 0.02 \frac{\rho \bar{U}^2 D C_D}{m f_n^2}. \tag{7-84}$$

D is the width of the building perpendicular to the wind; $\bar{U}$ is the wind speed at the top of the building; $m = \rho_{\text{Bldg}} DB$, where DB is the cross-sectional area of the building and the density of most conventional buildings lies between 6.2 lb/ft^3 (100 kg/m^3) $< \rho_{\text{Bldg}} <$ 12.4 lb/ft^3 (200 kg/m^3) with $\rho_{\text{Bldg}} = 9.3$ lb/ft^3 (150 kg/m^3) being typical. α is estimated from Table 7-5. $C_D \approx 2$ is typical of most rectangular buildings [see values given earlier in this section; Simiu and Scanlan (1986) and Hoerner (1965)]. Two approximate formulas for the fundamental natural frequency in hertz of conventional buildings are (Housner and Brody, 1963; Rinne, 1952; Ellis, 1980)

$$f_n = \frac{\omega_n}{2\pi} = \begin{cases} 46/L, & L = \text{height in meters}, \\ c\sqrt{B}/L, & B = \text{width in direction of vibration}, \end{cases} \tag{7-85}$$

where $c = 20$ ft$^{1/2}$/sec (11 m$^{1/2}$/sec).

The quasisteady analysis presented in this section is approximate. The building aerodynamics have been reduced to a flat plate facing into the wind. Various theories for alongwind response can differ from one another by 50% (Solari, 1982; Simiu et al., 1977). The variation from full-scale measurements can be ±50% (Davenport, 1982), in part because effects of an unsteady wake and lateral reponse are not included in the theory (see Chapters 3 and 4). Figure 7-26 shows that cross-wind response can be as large as or larger than alongwind response. Improvement in the accuracy of prediction of wind-induced vibration requires test data.

7.4.3. Wind-Tunnel Simulation

With careful scale modeling of the structure and its dynamics (Chapter 1), it is possible to accurately simulate the dynamic response of a structure to the natural wind in aerodynamic wind tunnels with a long, roughened

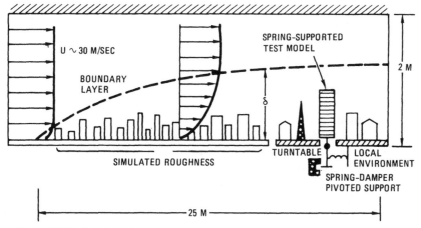

Fig. 7-27 Typical boundary layer wind tunnel for dynamic modeling of building response in the wind.

fetch upstream of the model, as shown in Figure 7-27. The turbulence intensity and the boundary layer thickness are adjusted by varying the upstream roughness. Boundary layers up to 3 ft (1 m) thick have been developed to permit scaling of buildings for geometric scale factors between 1/100 and 1/600. Davenport and Isyumov (1967), Cermak (1977, 1987), and Hansen and Sorensen (1985) describe boundary wind tunnels of this type at the University of Western Ontario, Colorado State University, and the Danish Maritime Institute. Wind-tunnel modeling techniques are reviewed by Plate (1982), Simiu and Scanlan (1986), and Reinhold (1982).

The principal scale factors that are preserved in modeling describe the turbulent boundary layer, the dynamics of the structure, and their interaction. The principal scale factors are

L/D, δ/D = model geometry, boundary layer,

$\bar{U}/f_n D$ = reduced velocity; f_n = natural frequency of structure,

$m/\rho D^2$ = model mass/mass of displaced fluid,

ζ = damping factor,

$fS_u(f)/\bar{U}^2$ = nondimensional gust spectrum.

These parameters are evaluated for the full-scale building. Models are built which achieve the same values at model scale.

For example, if a full-scale structure 300 ft (100 m) high is exposed to a turbulent boundary layer of wind 1200 ft (400 m) thick, and a boundary layer 3 ft (1 m) thick can be developed in a wind tunnel, then a geometric scale factor of 1/400 would be appropriate. A 300 ft (100 m)/400 = 1.2 ft

(0.4 m) high scale model would be used to represent the structure. The roughness of the fetch of the wind tunnel is adjusted to achieve modeling of the nondimensional turbulence spectra over the maximum possible range of nondimensional frequency, $fD/\bar{U}$, where f refers to the frequency of each component of the turbulence. Structures and terrain adjacent to the structure of interest are modeled so that the local flow distribution is simulated. The entire local region is mounted on a turnable so that the wind direction can be varied relative to the model.

Considerable skill and patience are required to create an accurate aeroelastic model of a complex structure (Plate, 1982; Reinhold, 1982). Preserving mass ratio implies that the model density is equal to the full-scale density. A building model 14 in (0.4 m) high with a 7 in by 3.5 in (0.2 m by 0.1 m) cross section that has a typical full-scale density of 9.3 lb/ft³ (150 kg/m³) would weigh 2.6 lb (1.2 kg). Lightweight models are constructed with balsa wood or plastic skins. The flexibility of the model can be achieved by using a rigid model that is attached to a spring mount as shown in Figure 7-27, by building the model about a flexible central spine, or by creating a flexible steel wire skeleton that carries the external skins. The model frequency is adjusted to achieve reduced velocity scaling. For example, if the fundamental natural frequency of the full-scale structure is 0.3 Hz and the mean wind-tunnel velocity is one-quarter of the atmospheric wind speed, then for equal reduced velocities, the natural frequency of a 1/400 scale model must be 30 Hz. Damping factor is modeled on a one-to-one scale. The measured model displacements,

$$\frac{X(t)}{D} = \text{model displacement/model width,}$$

provide a measure of the full-scale displacements and can be used to predict stresses, accelerations, and their effect on the structure and its occupants (Chang, 1973; Reed, 1971).

Reynolds number is not preserved in wind modeling. Modeling Reynolds number requires very high flow velocities, which result in unworkable dynamic pressures and Mach numbers. Model tests are generally conducted at Reynolds numbers above 1000 where the boundary layer is turbulent and at Mach numbers below 0.3 so that compressibility does not influence the results. Reynolds number modeling is closely tied to the turbulence of wind, surface roughness of the model, and separation of flow from the model. For structures with rounded members, the aerodynamic force coefficients are Reynolds number dependent and it may be possible to roughen the model surface to encourage early laminar/turbulent boundary layer transition that would occur at higher Reynolds numbers (Armitt, 1968). The aerodynamic forces on sharp-edged sections are largely independent of Reynolds number.

Full-scale and model-scale tests by Dalgliesh (1982) suggest that it may be possible to predict full-scale response to natural wind within 20% from scaled model tests. There may be little hope of predicting full-scale response with better accuracy because the natural environment is subject to many uncontrolled variables.

Exercises

1. A 1/400 scale model is tested in a wind tunnel. The full-scale natural frequency is 0.3 Hz. The model natural frequency is 30 Hz. What is the relationship between model accelerations and full-scale accelerations?

2. The initial impulse of most aerodynamics engineers in wind-tunnel testing is to match Reynolds number. Consider an atmospheric wind speed of 100 ft/sec (30 m/sec) and a 1/400 scale building model. What wind tunnel velocity is required for one-to-one Reynolds number modeling? What Mach numbers and dynamic pressures does this imply? If the wind tunnel velocity is limited to 65 ft/sec (20 m/sec), what Reynolds numbers are achieved in the model and at full scale? What influence could this have on drag and vortex shedding from the model (see Sections 3.1, 3.2, and 6.3)?

7.5. RESPONSE OF AIRCRAFT TO GUSTS

Gust loading is a critical design condition for aircraft. Gusts can induce vertical aircraft accelerations exceeding the acceleration due to gravity. Gusts are associated with thermal and humidity gradients in the atmosphere, such as in clouds, storms, and thermal currents (Houbolt, 1973). Figure 7-28 shows a vertical gust measured by an airplane entering a cloud. Gusts occur in all directions, but those normal to the flight path produce the greatest loading on aircraft.

Agencies regulating the design of aircraft have taken the approach that the durability of aircraft under gust loading is a safety issue and they therefore require that aircraft withstand gusts of a certain magnitude. The maximum positive (up) or negative (down) discrete gust specified by the United States Federal Aviation Administration (1988) is 66 ft/sec (20 m/sec) at altitudes between sea level and 20 000 ft (6000 m), reduced linearly from 66 to 38 ft/sec (20 to 10 m/sec) at 50 000 ft (15 000 m). The Federal Aviation Administration also permits certification using continuous turbulence as an alternative to discrete gusts.

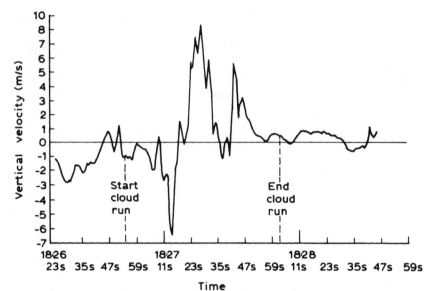

Fig. 7-28 Profile of vertical gust measured by an aircraft penetrating a cloud (Carlson and Sheets, 1971).

Certain simplifying assumptions are useful in computing the response of aircraft to gusts: (1) the aircraft is a rigid body; (2) the aircraft horizontal velocity is constant; (3) the gust is normal to flight; (4) the aicrcraft does not pitch; and (5) quasisteady aerodynamic analysis can be used. In addition, we will use two-dimensional aerodynamics.

An aircraft encountering a discrete sharp-edged gust is shown in Figure 7-29. The equation of rigid body motion of the aircraft in the vertical direction is

$$m\ddot{y} - mg = -F_y = -(F_L \cos \alpha + F_D \sin \alpha),$$

$$\approx -F_L = -\tfrac{1}{2}\rho U^2 c C_L(\alpha),$$ (7-86)

where y is the vertical position of the aircraft, positive downward; m is the mass per unit span of the aircraft; c is the chord of the wing; U is the forward speed; g is the acceleration due to gravity; F_y is the vertical force per unit span, positive upward; F_L and F_D are the aerodynamic lift and

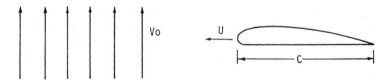

Fig. 7-29 Aircraft encountering an abrupt gust.

drag forces per unit span; and α is the angle of attack of the air flow with respect to the wing. Under normal conditions, the lift on an aircraft greatly exceeds the drag, $F_L \gg F_D$, and angle of attack and the changes in angle of attack, in radians, are much less than unity, $\alpha_0 \ll 1$ and $(\alpha - \alpha_0) \ll 1$. (Note that the first assumption is not valid at stall or for bluff sections; see Chapter 4, Section 4.2.) For small changes in angle of attack, the angle of attack and the lift coefficient can be expanded in power series,

$$\alpha - \alpha_0 = \arctan\left(\frac{\dot{y}}{U} + \frac{v}{U}\right) \approx \frac{\dot{y}}{U} + \frac{v}{U}, \tag{7-87}$$

$$C_L(\alpha) \approx C_L(\alpha_0) + \frac{\partial C_L}{\partial \alpha}\bigg|_{\alpha_0} (\alpha - \alpha_0) \approx C_L(\alpha_0) + \frac{\partial C_L}{\partial \alpha}\bigg|_{\alpha_0} \left(\frac{\dot{y}}{U} + \frac{v}{U}\right), \tag{7-88}$$

where v is the vertical gust velocity, positive upward. The lift coefficient C_L is the lift coefficient of the entire aircraft; it includes the wing, the effects of the fuselage, and the use of flaps and other high lift devices; c is the chord of the wing.

Before encountering the gust, the aircraft is in vertical equilibrium, $mg = \frac{1}{2}\rho U^2 c C_L(\alpha_0)$. Substituting the expansions of Eqs. 7-87 and 7-88 into Eq. 7-86 and subtracting out the steady solution gives the dynamic equation of vertical motion of a rigid aircraft encountering a vertical gust:

$$m\ddot{y} = -\frac{1}{2}\rho U^2 c \frac{\partial C_L}{\partial \alpha}\bigg|_{\alpha_0} \left(\frac{\dot{y}}{U} + \frac{v}{U}\right). \tag{7-89}$$

Following Fung (1969, p. 282), we define a parameter η with units of frequency that characterizes the response frequency of the aircraft

$$\eta = \frac{1}{2}\frac{\rho U c}{m}\frac{\partial C_L}{\partial \alpha}\bigg|_{\alpha_0}. \tag{7-90}$$

With this substitution, Eq. 7-89 becomes

$$\ddot{y}(t) + \eta\dot{y}(t) = -\eta v(t). \tag{7-91}$$

This equation of motion with initial condition $y(0) = 0$ has the following solution for an impulsive gust $[\int_0^\epsilon v(t)\, dt = 1, \ v(t > \epsilon) = 0]$:

$$y(t) = e^{-\eta t} - 1. \tag{7-92}$$

The general solution to Eq. 7-91 is found from a *convolution integral* also called *Duhamel's integral* in terms of the impulse solution (Meirovitch, 1967; Fung, 1969),

$$y(t) = \int_0^t v(\xi)[e^{-\eta(t-\xi)} - 1]\, d\xi, \tag{7-93}$$

which is valid for any gust time-history.

Two discrete gust time-histories of practical interest are the abrupt gust and the FAA gust,

$$v(t > 0) = \begin{cases} V_0, & \text{abrupt gust,} \\ \dfrac{V_g}{2}\left(1 - \cos \dfrac{2\pi Ut}{25c}\right), & \text{Federal Aviation} \\ & \text{Administration gust (1988).} \end{cases} \quad (7\text{-}94)$$

Both of these gusts are zero for $t \le 0$. The time-history vertical displacement in response to the abrupt gust is easily found from Eq. 7-93,

$$y(t)_{\text{abrupt gust}} = \frac{V_0}{\eta}(1 - e^{-\eta t}) - V_0 t. \quad (7\text{-}95)$$

Vertical acceleration of the aircraft loads the aircraft structure. The maximum vertical acceleration in response to the abrupt gust occurs at $t = 0$. The ratio of this acceleration to the acceleration due to gravity is (Fung, 1969, p. 282)

$$\frac{|\ddot{y}(0)|}{g} = \frac{\eta V_0}{g} = \frac{\rho U}{2} \frac{c V_0}{mg} \frac{\partial C_L}{\partial \alpha}\bigg|_{\alpha_0}. \quad (7\text{-}96)$$

The aircraft acceleration increases with the product of the gust velocity times the aircraft forward velocity times the slope of the lift curve divided by the mass. As a result, small aircraft have greater response to gusts than large aircraft (Houbolt, 1973).

Gustiness can also be modeled as a continuous process. Consider a "tuned" sinusoidal, vertical gust

$$v(t) = V_0 \sin \omega t, \qquad -\infty < t < \infty \quad (7\text{-}97)$$

with circular frequency ω and amplitude V_0. With a trial steady-state solution to Eq. 7-91

$$y(t) = A_y \sin(\omega t + \phi), \quad (7\text{-}98)$$

it can easily be shown by substituting Eqs. 7-97 and 7-98 into Eq. 7-91 that the amplitude and phase of the steady-state response are

$$A_y = \frac{\eta}{\omega} \frac{V_0}{(\omega^2 + \eta^2)^{1/2}}, \qquad \phi = \arctan^{-1} \frac{\eta}{\omega}. \quad (7\text{-}99)$$

The ratio of the aircraft acceleration amplitude to the acceleration of gravity,

$$\frac{|\ddot{y}(\omega)|}{g} = \frac{\eta V_0}{g} \frac{1}{(1 + \eta^2/\omega^2)^{1/2}}, \quad (7\text{-}100)$$

is less than that of a discrete gust of the same magnitude (Eq. 7-96), a fact that leads aircraft designers to exploit regulations that permit design by continuous gusts rather than discrete gusts. The effects of continuous gusts can also be alleviated by means of feedback control (Houbolt, 1973).

Consider continuous random gusts. Using the transfer function between gust velocity and aircraft acceleration established by Eq. 7-99, the auto spectral density of acceleration $S_{\ddot{y}}$ to random gusts with auto spectral density $S_v(f)$ is

$$S_{\ddot{y}}(f) = \frac{\eta^2}{1 + \eta^2/\omega^2} S_v(f). \tag{7-101}$$

The von Karman vertical gust spectrum (1961) is widely used for aircraft design (Etkin, 1972, p. 539),

$$S_v(f) = v_{rms}^2 \frac{2L}{U} \frac{1 + \frac{8}{3}(2\pi a f L/U)^2}{[1 + (2\pi a f L/U)^2]^{11/6}}, \tag{7-102}$$

where v_{rms} is the root-mean-square vertical velocity, equal to 85 ft/sec (26 m/sec) between 0 and 30 000 ft (9000 m) altitude and then decreasing linearly to 30 ft/sec (9.1 m/sec) at 80 000 ft (24 300 m); $L = 2500$ ft (760 m) is an integral length scale of the turbulent eddies (FAA, 1988); $a = 1.339$ is the von Karman constant; and f is the frequency in hertz. Note that this is a single-sided spectrum defined with respect to f in hertz, $\int_0^\infty S_v(f)\, df = v_{rms}^2$. U is the aircraft forward speed. Etkin (1972) discusses multiple-degree-of-freedom aircraft response to turbulence. Other statistical models for atmospheric turbulence are presented by Justus (1989), Campbell (1986), Houbolt (1973), Panofsky and Dutton (1984), and Lappe (1966), and by Eq. 7-63.

Exercises

1. Compare the von Karman spectrum (Eq. 7-101) with the Davenport spectrum (Eq. 7-63) for gustiness of winds by plotting both spectra on Figure 7-19.

2. What is the response of an aircraft to the FAA gust (Eq. 7-94)?

3. Consider an aircraft weighing 670 lb/ft (1000 kg/m) of span with a 20 ft (6 m) chord wing, $\partial C_L/\partial \alpha = 2\pi$, flying at 320 ft/sec (100 m/sec) and encountering an abrupt gust with maximum velocity of 65 ft/sec (20 m/sec) and air density of 0.075 lb/ft^3 (1.2 kg/m^3). What is the acceleration, relative to the acceleration due to gravity, resulting from an abrupt gust and an FAA gust?

7.6. REDUCTION OF VIBRATION INDUCED BY TURBULENCE

Vibrations induced by turbulence can be reduced by either reducing the intensity of the turbulence or modifying the structure as follows:

1. *Change the structural shape to reduce aerodynamic loads.* The mean and unsteady aerodynamic forces can be reduced by decreasing the height of the structure or by orienting the structure so that the minimum cross section is exposed to the direction of maximum velocity or by using an open lattice "flow through" construction rather than a sheet-metal box beam.

2. *Increase stiffness of the structure.* The mean deflection, the nonresonant vibration amplitude, and the resonant vibration are inversely proportional to the structural stiffness. Structures can be stiffened by increasing the cross section, increasing the gauges of load-bearing structure, stiffening joints, using diagonal bracing, or replacing low-modulus materials such as wood, plastics, or aluminum with higher-modulus materials such as steel.

3. *Increase structural mass.* Increasing structural mass increases mass ratio and generally has a beneficial effect on vibration provided that natural frequency and damping are maintained. In practical structures, mass can be increased by increasing tube wall thickness and use of heavy construction.

4. *Increase damping.* Increasing damping reduces resonant vibration, but it has little effect on nonresonant vibration. Damping can be increased by incorporating energy-absorbing materials such as rubber, wood, and soft polymers in the design, by permitting small local motions at joints by using built-up rather than welded construction, and by using specially designed dampers; see Chapter 8, Section 8.5.

7.7. EXAMPLE: WIND-INDUCED VIBRATION OF A BUILDING

The Alcoa Building in San Francisco is 27 stories (116 m) high. In plan the building is rectangular (64 m by 33 m), as shown in Figure 7-30. The building has a diagonally braced steel framework. Wind-induced vibrations of this building can occur due to turbulence in the wind, vortex shedding (Chapter 3), and unstable galloping (Chapter 4). These phenomena will be considered in this example.

As noted in Sections 7.4 and 8.4.2, a typical density of conventional

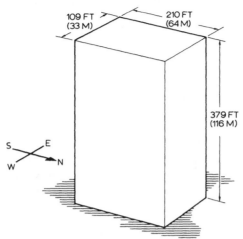

Fig. 7-30 The Alcoa Building.

buildings is $10 \, lb/ft^3$ ($150 \, kg/m^3$) and a typical damping factor of the fundamental mode is $\zeta = 0.01$. The fundamental natural frequency can be estimated from Eq. 7-85. These estimates are compared with the full-scale measurements by Bouwkamp (1974) on the Alcoa Building.

Parameter	Estimate	Measurement
Total mass (kg)	4.0×10^7	3.8×10^7
Natural frequency NS (Hz)	0.77 (Eq. 7-85b)	0.60
Natural frequency EW (Hz)	0.55 (Eq. 7-85b)	0.45
Natural frequency Min. (Hz)	0.37 (Eq. 7-85a)	0.45
Damping NS, ζ	0.01	0.009
Damping EW, ζ	0.01	0.011

The estimates give results within about 20% of the measured parameters.

Two design criteria are ordinarily applied to wind-induced vibration of buildings. First, the building must withstand an extreme wind, such as is expected once every 50 years, without catastrophic failure, and second, for the vast majority of expected wind speeds, the building must not have a dynamic response that disturbs the occupants of the building. Extreme wind data (Fig. 7-18 and Thom, 1968) indicate that for the San Francisco area, the 50-year return wind speed is 70 mph (31 m/sec) and the 2-year return wind speed is 40 mph (18 m/sec). These wind speeds are averaged over approximately 1 minute and they are valid for a height of 10 m (33 ft). A conversion is made first to the gradient wind and then to the height at the top of the building. The power-law profile (Eq. 7-58) with $\alpha = 1/7$ is appropriate for the airport at which the data were taken, and

$\alpha = 0.40$ is appropriate for the urban location of the building. The conversion formula is

$$\bar{U}(L) = \bar{U}_{10}\left(\frac{z_{Gref}}{z_{10}}\right)^{1/7}\left(\frac{L}{z_G}\right)^{0.40} = 0.96\bar{U}_{10},$$

where $z_{Gref} = 300$ m, $z_G = 500$ m (Table 7-5), and $L = 116$ m. The result for the wind velocity at the top of the building is

$$\bar{U}(L) = \begin{cases} 30 \text{ m/sec (67 mph),} & \text{return period} = 50 \text{ yr} \\ 18 \text{ m/sec (40 mph),} & \text{return period} = 2 \text{ yr.} \end{cases}$$

The procedure for calculating the alongwind response is given in Section 7.4.2. The response to the 2 year return wind is given in Table 7-6. The maximum acceleration is expressed in fraction of $g = 9.8$ m^2/sec,

Table 7-6 Wind-induced response of the Alcoa Building

Parameter	Symbol	NS direction	EW direction
Velocity at roof, m/sec	$\bar{U}(L)$	18.0	18.0
Width of section, m	D	33.0	64.0
Reynolds number	$\bar{U}(L)D/\nu$	39.0×10^6	77×10^6
Natural frequency, rad/sec	ω_n	3.8	2.8
Natural frequency, Hz	f_n	0.60	0.45
Breadth/width	B/D	0.51	1.93
Drag coefficient	C_D	2.5	1.6
B.L. exponent	α	0.40	0.40
Mode shape exponent	β	1.0	1.0
Air density, kg/m^3	ρ	1.2	1.2
Damping factor	ζ	0.009	0.011
Vertical correlation coefficient	C_z	10	10
Lateral correlation coefficient	C_y	12	12
Gust length, m	$L_t(L)$	3200	3200
Vertical gust parameter	$f_n L_t(L)/\bar{U}(L)$	107	80
Lateral gust parameter	$C_z L/L_t(L)$	0.36	0.36
Vertical size parameter	$C_z f_n L/\bar{U}(L)$	39	29
Lateral size parameter	$C_y D/C_z L$	0.34	0.66
Gust energy factor	F	0.023	0.027
Background excitation factor	B	0.77	0.70
Surface roughness factor	R	0.58	0.58
Size reduction factor	S	0.016	0.009
Peak-to-rms ratio	$\bar{g}$	3.5	3.5
Gust factor	G	2.8	3.2
Mass/height, kg/m	m	3.3×10^5	3.3×10^5
Mean displacement Eq. 7-67, m	$\bar{X}(L)$	0.0036	0.0082
Mean displacement Eq. 7-84, m	$\bar{X}(L)$	0.0054	0.012
Maximum displacement, m	$G\bar{X}(L)$	0.0126	0.029
Maximum acceleration, Eq. 7-83, % g	—	0.22	0.20
Velocity for vortex resonance, m/sec	$f_n D/S_t$	99	144
Slope of vertical force coefficient, 1/rad	$\partial C_y/\partial\alpha$	3.0	0.0
Onset of gallop, m/sec (Eq. 4-16)	U_{crit}	376.0	∞

the acceleration due to gravity at the surface of the earth. The maximum acceleration at the top of the building for the 2 year return wind is below the 0.5% g human perception threshold. For the 50 year return wind, these accelerations are approximately a factor of $(30/17)^2 = 3.1$ higher, or 0.64% g and these accelerations would be perceptible, although not annoying, to most people.

Onset of a vortex-induced resonance is found to be 99 m/sec using a Strouhal number of $S_t = 0.2$ (Chapter 3). This velocity is well above the expected 50 year wind velocity, so no resonance is expected with vortex shedding. Similarly, the critical velocity of 376 m/sec for onset of galloping of the rectangular section is well above the expected 50 year wind velocity.

Exercises

1. It is proposed to increase the height of the Alcoa Building by an additional 27 stories (380 ft, 116 m). What would be the effect on (a) the velocity at the roof, (b) the alongwind response, (c) the maximum accelerations and their effect on the occupants, (d) the velocity for vortex resonance, and (e) the velocity for onset of galloping?

2. The experimental data given in Chapter 1, Figure 1-3, are for a rectangular section that has proportions very similar to those of the Alcoa Building. Scale these data to predict the amplitude of lateral displacement of the Alcoa Building for a NS wind with velocity from 0 to 700 ft/sec (200 m/sec).

REFERENCES

American Association of State Highway and Transportation Officials (1986) "Standard Specification for Highway Bridges," Washington, D.C.

American Society of Civil Engineers (1980) *Tall Building Criteria and Loading*, Vol. CL, American Society of Civil Engineers, New York.

Armitt, J. (1968) "The Effect of Surface Roughness and Free Stream Turbulence on the Flow Around a Cooling Tower at Critical Reynolds Numbers," *Proceedings Symposium on Wind Effects on Buildings and Structures*, University of Technology, Loughborough, pp. 6.1–6.8.

Axisa, F., et al. (1988) "Random Excitation of Heat Exchanger Tubes by Cross Flow," in *1988 Symposium on Flow-Induced Vibration and Noise*, M. P. Paidoussis, ed., Vol. 2, ASME, New York, pp. 23–46.

Ballentine, J. R., F. F. Rudder, J. T. Mathis, and H. E. Plumblee (1968) "Refinement of Sonic Fatigue Structural Design Criteria," AFFDL-TR-67-156, Wright-Patterson Air Force Base, Ohio.

Basile, D., J. Faure, and E. Ohlmer (1968) "Experimental Study of the Vibrations of Various Fuel Rod Models in Parallel Flow," *Nuclear Engineering and Design*, **7**, 517–534.

Batts, M. E. (1982) "Probabilistic Description of Hurricane Wind Speeds," *ASCE Journal of the Structural Division*, **108**, 1643–1647.

Bendat, J. S., and A. G. Piersol (1986) *Random Data: Analysis and Measurement Procedures*, 2d ed., Wiley-Interscience, New York.

Benjamin, J. R., and C. A. Cornell (1970) *Probability, Statistics, and Decision for Civil Engineers*, McGraw-Hill, New York.

Blakewell, H. P. (1968) "Turbulent Wall Pressure Fluctuations on a Body of Revolution," *Journal of the Acoustical Society of America*, **43**, 1358-1363.

Blevins, R. D. (1979a) *Formulas for Natural Frequency and Mode Shape*, Van Nostrand Reinhold, New York. Reprinted by Robert E. Kreiger, Malabar, Fla., 1984.

―――― (1979b) "Flow-Induced Vibration in Nuclear Reactors: A Review," *Progress in Nuclear Energy*, **4**, 24-49.

―――― (1989) "An Approximate Method for Sonic Analysis of Plate and Shell Structures," *Journal of Sound and Vibration*, **129**, 51-71.

Blevins, R. D., R. J. Gibert, and B. Villard (1981) "Experiments on Vibration of Heat Exchanger Tube Arrays in Cross Flow," *Transactions of the 6th International Conference on Structural Mechanics in Reactor Technology (SMIRT)*, Paper B6/9.

Bolotin, V. V., (1984) *Random Vibrations of Elastic Systems*, Martinus Nijhoff Publishers, The Hague, Netherlands. Translation of the Russian edition.

Bouwkamp, J. G., (1974) "Dynamics of Full Scale Structures," in *Applied Mechanics in Earthquake Engineering*, W. D. Iwan, ed., ASME, New York, pp. 99-133.

Brase, L. O. (1988) "Near Field Exhaust Environment Measurements of a Full Scale Afterbursting Jet Engine with Two-Dimensional Nozzle," AIAA Paper AIAA-88-0182, American Institute of Aeronautics and Astronautics, Washington, D.C.

British Standards Institution (1972) "Wind Loads," *Code of Basic Data for the Design of Buildings*, Chapter V, Part 2, CP3.

Burgreen, D., J. J. Byrnes, and D. M. Benforado (1958) "Vibration of Rods Induced by Water in Parallel Flow," *Transactions of the American Society of Mechanical Engineers*, **80**, 991-1003.

Campbell, C. W. (1986) "Monte Carlo Turbulence Simulation Using Rational Approximations to von Karman Spectra," *AIAA Journal*, **24**, 62-66.

Carlson, T. N., and R. C. Sheets, (1971) "Comparison of Draft Scale Vertical Velocities Computed from Gust Probe and Conventional Data Collected by a DC-6 Aircraft," NOAA Technical Memorandum ERL NHRL-91, National Hurricane Research Laboratory.

Cartwright, D. E., and M. S. Longuet-Higgins (1956) "Statistical Distribution of the Maxima of Random Functions," *Proceedings of the Royal Society Series A*, **237**, 212-232.

Cermak, J. E. (1975) "Application of Fluid Mechanics to Wind Engineering—A Freeman Scholar Lecture," *Journal of Fluids Engineering*, **97**, 9-38.

―――― (1977) "Wind-Tunnel Testing of Structures," *ASCE Journal of the Engineering Mechanics Division*, **103**, 1125-1140.

―――― (1987) "Advances in Physical Modeling for Wind Engineering," *ASCE Journal of Engineering Mechanics*, **113**, 737-755.

Chamberlain, A. C. (1983) "Roughness Length of Sea, Sand and Snow," *Boundary Layer Meteorology*, **25**, 405-409.

Chang, F. K. (1973) "Human Response to Motions in Tall Buildings," *ASCE Journal of the Structural Division*, **98**, 1259-1272.

Changery, M. J., E. J. Dumitriu-Valcea, and E. Simiu (1984) "Directional Extreme Wind Speed Data for the Design of Buildings and Other Structures," NBS Building Science Series 160, U.S. Department of Commerce.

Chen, S. S., and J. A. Jendrzejczyk (1987) "Fluid Excitation Forces Acting on a Square Tube Array," *Journal of Fluids Engineering*, **109**, 415-423.

Chen, Y. N. (1970) "Flow-Induced Vibration in Tube Bundle Heat Exchangers with Cross and Parallel Flow, Part I: Parallel Flow," in *Flow-Induced Vibration in Heat Exchangers*, ASME, New York, pp. 57-66.

Cheng, E. D. H., and A. N. L. Chiu (1985) "Extreme Wind Simulated from Short-Period Records," *ASCE Journal of Structural Engineering*, **111**, 77–94.

Clarkson, B. L., and F. J. Fahy (1968) "Response of Practical Structures to Noise," in *Noise and Acoustic Fatigue in Aeronautics*, Wiley, London, pp. 330–353.

Corcos, G. M. (1964) "The Structure of the Turbulent Pressure Field in Boundary Layer Flow," *Journal of Fluid Mechanics*, **13**, 353–378.

Crandall, S. H., and W. D. Mark (1963) *Random Vibration in Mechanical Systems*, Academic Press, New York.

Dalgliesh, W. A. (1982) "Comparison of Model and Full Scale Tests of the Commerce Court Building in Toronto," in *Wind Tunnel Modeling for Civil Engineering Applications*," T. A. Reinhold, ed. Cambridge University Press, Cambridge.

Davenport, A. G. (1961) "Application of Statistical Concepts to the Wind Loading of Structures," *Proceedings of the Institution of Civil Engineers*, **19**, 449–472.

────── "The Relationship of Wind Structure to Wind Loading," in *Proceedings of a Conference on Buildings and Structures*, National Physical Laboratory, Great Britain, 1963, pp. 54–83.

────── (1964) "Note on the Distribution of the Largest Value of a Random Function with Application to Gust Loading," *Proceedings of the Institution of Civil Engineers*, **28**, 187–197.

────── (1967) "Gust Loading Factors," *ASCE Journal of the Structural Division*, **93**, 11–34.

────── (1970) "On the Statistical Prediction of Structural Performance in a Wind Environment," in *Proceedings of a Seminar on Wind Loads on Structures*, Honolulu, Oct. 19–24, pp. 325–342.

────── (1982) "The Interaction of Wind and Structures," in *Engineering Meteorology*, E. Plate, ed., Elsevier, Amsterdam, pp. 527–572.

Davenport, A. G., and N. Isyumov (1967) "The Application of the Boundary Layer Wind Tunnel to the Prediction of Wind Loading," in *Proceedings of a Seminar on Wind Effects on Structures*, National Research Council of Canada, Ottawa, Sept., pp. 201–230.

Deacon, E. L. (1955) "Gust Variation with Height up to 150 m," *Quarterly Journal of the Royal Meteorological Society*, **81**, 562–573.

Dowell, E. H., et al. (1978) *A Modern Course in Aeroelasticity*, Sijthoff & Noordhoff International Publishers, The Netherlands.

Durst, C. S. (1960) "Wind Speeds Over Short Periods of Time," *Meteorology Magazine*, **89**, 181–186.

Ellis, B. R. (1980) "An Assessment of the Accuracy of Predicting the Fundamental Natural Frequencies of Buildings," *Proceedings of the Institution of Civil Engineers*, **69**, 763–776.

Etkin, B. (1972) *Dynamics of Atmospheric Flight*, Wiley, New York.

Federal Aviation Administration (1988) Title 14, part 25.341, Appendix G to part 25, United States Code of Federal Regulations, Aeronautics and Space, U.S. Government Printing Office, Washington D.C.

Frenkiel, F. N. (1953) "Turbulent Diffusion," *Advances in Applied Mechanics*, Vol. 3, Academic Press, New York, pp. 77–78.

Fung, Y. C. (1969) *An Introduction to the Theory of Aeroelasticity*, Dover, New York.

Geissler, E. D. (1970) *Wind Effects on Launch Vehicles*, AGARDograph 115, The Advisory Group for Aerospace Research and Development of NATO.

Hansen, S. O., and E. G. Sorensen (1985) "A New Boundary Layer Wind Tunnel at the Danish Maritime Institute," *Journal of Wind Engineering and Industrial Acoustics*, **18**, 213–224.

Harris, C. M. (1979) *Handbook of Noise Control*, McGraw-Hill, New York.

Hoerner, S. F. (1965) *Fluid-Dynamic Drag*, published by the author, New Jersey.

Holand, I., et al. (eds.) (1978) *Safety of Structures under Dynamic Loading*, Tapir Publishers, Norway.

Houbolt, J. C. (1973) "Atmospheric Turbulence," *AIAA Journal*, **11**, 421-437.

Houghton, E. L., and N. B. Carruthers (1976) *Wind Forces on Buildings and Structures*, Wiley, New York.

Housner, G. W., and A. G. Brody (1963) "Natural Periods of Vibration of Buildings," *ASCE Journal of the Engineering Mechanics Division*, **89**, 31-65.

Junger, M. C., and D. Feit (1986) *Sound, Structures and Their Interaction*, 2d ed., The MIT Press, Cambridge, Mass.

Justus, C. (1989) "New Height Dependent Magnitudes and Scales for Turbulence Modeling," AIAA Paper 89-0788, AIAA, Washington, D.C.

Kaimal, J. C., et al. (1972) "Spectral Characteristics of Surface-Layer Turbulence," *Journal of the Royal Meteorological Society*, **98**, 563-589.

Kolousek, V., et al. (1984) *Wind Effects on Civil Engineering Structures*, Elsevier, Amsterdam, Prague.

Laganelli, A. L., A. Martellucci, and L. Show (1983) "Prediction of Turbulent Wall Pressure Fluctuations in Attached Boundary Layer Flow," *AIAA Journal*, **21**, 495-502.

Lappe, U. O. (1966) "Low-Altitude Turbulence Model for Estimating Gust Loads on Aircraft," *Journal of Aircraft*, **3**, 41-47.

Lawson, T. V. (1980) *Wind Effects on Buildings*, Applied Science Publishers, London.

Lin, Y. K. (1967) *Probabilistic Theory of Structural Dynamics*, McGraw-Hill, New York.

Liu, H. (June 14, 1979) "Understanding Wind Loads on Plant Buildings," *Plant Engineering*, 187-191.

Loeve, M. M. (1977) *Probability Theory*, 4th ed., Springer-Verlag, New York.

Lowson, M. V. (1968) "Prediction of Boundary Layer Pressure Fluctuations," AFFDL-TR-67-167.

Mehta, K. C. (1984) "Wind Load Provisions ANSI #A58.1-1982," *ASCE Journal of Structural Engineering*, **110**, 769-783.

Mei, C., and C. B. Prasad (1987) "Effects of Nonlinear Damping on Random Response of Beams to Acoustic Loading," *Journal of Sound and Vibration* **117**, 173-186.

Meirovitch, L. (1967) *Analytical Methods in Vibrations*, Macmillan, New York.

Miles, J. W. (1954) "On Structural Fatigue under Random Loading," *Journal of Aeronautical Sciences*, **21**, 753-762.

Nagano, S., M. Naito, and H. Takata (1983) "A Numerical Analyis of Two-Dimensional Flow Past a Rectangular Prism by Discrete Vortex Model," *Computers and Fluids*, **110**, 243-259.

Paidoussis, M. P. (1981) "Fluidelastic Vibration of Cylinder Arrays in Axial and Cross Flow: State of the Art," *Journal of Sound and Vibration*, **76**, 329-360.

——— (1982) "A Review of Flow-Induced Vibration in Reactors and Reactor Components," *Nuclear Engineering and Design*, **74**, 31-60.

Panofsky, H. A., and J. A. Dutton (1984) *Atmospheric Turbulence*, Wiley-Interscience, New York.

Parsons, K. C., and M. J. Griffin (1988) "Whole-Body Vibration Perception Thresholds," *Journal of Sound and Vibration*, **121**, 237-258.

Pettigrew, M. J., and D. J. Gorman (1981) "Vibration of Heat Exchange Components in Liquid and Two-Phase Cross Flow," *Flow-Induced Vibration Guidelines*, P. Y. Chen (ed.), PVP-152, ASME, N.Y., 89-110.

Plate, E. (ed.) (1982) *Engineering Meteorology*, Elsevier, Amsterdam.

Raghunathan, S. (1987) "Pressure Fluctuation Measurements with Passive Shock/Boundary Layer Control," *AIAA Journal*, **25**, 626-628.

Reavis, J. R. (1969) "Vibration Correlation for Maximum Fuel Element Displacement in Parallel Turbulent Flow," *Nuclear Science and Engineering*, **38**, 63-69.

Reed, J. W. (1971) "Wind Induced Motion and Human Discomfort in Tall Buildings,"

Massachusetts Institute of Technology Department of Civil Engineering Research Report R71-42, Structures Publication 310.

Reinhold, T. A. (ed.) (1982) *Wind Tunnel Modeling for Civil Engineering Applications,* Cambridge University Press, Cambridge.

Richards, E. J., and D. J. Meade (1968) *Noise and Acoustic Fatigue in Aeronautics,* Wiley, London.

Rinne, J. E. (1952) "Building Code Provisions for Aseismic Design," *Proceedings of Symposium on Earthquake and Blast Effects on Structures,* Los Angeles, pp. 291–305.

Sachs, P. (1978) *Wind Forces in Engineering,* 2d ed., Pergamon Press, Oxford.

Sandifer, J. B., and R. T. Bailey (1984) "Turbulent Buffeting of Tube Arrays in Liquid Cross Flow," in *ASME Symposium on Flow-Induced Vibrations,* M. P. Paidoussis, et al., eds., ASME Winter Annual Meeting, New Orleans.

Savkar, S. D. (1984) "Buffeting of Cylindrical Arrays in Cross Flow," in *ASME Symposium on Flow-Induced Vibrations,* M. P. Paidoussis, et al., eds., ASME Winter Annual Meeting, New Orleans.

Scruton, C. (1981) "An Introduction to Wind Effects on Structures," *Engineering Design Guide No. 40,* published for the Design Council, British Standards Institution, and the Council of Engineering Institutions, Oxford University Press, Oxford.

Simiu, E. (1980) "Revised Procedure for Estimating Alongwind Response," *ASCE Journal of the Structural Division,* **106,** 1–10.

Simiu, E., and E. M. Hendrickson (1987) "Design Criteria of Glass Cladding Subjected to Wind Loads," *ASCE Journal of Structural Engineering,* **113,** 501–518.

Simiu, E., and D. W. Lozier (1979) "The Buffeting of Structures by Strong Winds—Windload Program," *Computer Program for Estimating Along Wind Response,* National Technical Information Service, NTIS Accession No. PB294757/AS, Springfield Va.

Simiu, E., and R. H. Scanlan (1986) *Wind Effects on Structures,* 2d ed., Wiley-Interscience, New York.

Simiu, E., et al. (1977) "Estimation of Along Wind Building Response," *ASCE Journal of the Structural Division,* **103,** 1325–1338.

Singer, I. A. (1960–61) "A Study of Wind Profile in the Lowest 400 Feet of the Atmosphere," *Brookhaven National Laboratory Progress Reports No. 5 and 9.*

Solari, G. (1982) "Alongwind Response Estimation: Closed Form Solution," *ASCE Journal of the Structural Division* **108,** 225–244.

Taylor, C. E., et al. (1986) "Experimental Determination of Single and Two-Phase Cross Flow-Induced Forces on Tube Rows," in *Flow-Induced Vibration—1986,* S. S. Chen, ed., PVP 104, ASME, New York, pp. 31–40.

Taylor, C. E., et al. (1988) "Vibration of Tube Bundles in Two-Phase Cross Flow: Part 3—Turbulence-Induced Excitation," in *1988 Symposium on Flow-Induced Noise and Vibration,* M. P. Paidoussis, ed., Vol. 2, ASME, New York.

Thom, H. C. S. (1968) "New Distributions of Extreme Winds in the United States," *ASCE Journal of the Structural Division,* **94,** 1783–1801.

Thompson, A. G. R., and R. F. Lambert (1972) "The Estimation of R.M.S. Stresses in Stiffened Skin Panels Subjected to Random Acoustic Loading," Advisory Group for Aerospace Research and Development, North Atlantic Treaty Organization, AGARD-AG-162, Nov. 1972.

Thomson, W. T. (1988) *Theory of Vibration with Applications,* 3d ed., Prentice-Hall, Englewood Cliffs, N.J.

Ungar, E. E., J. F. Wilby, and D. Bliss (1977) "A Guide for the Estimation of Aeroacoustic Loads on Flight Vehicle Surfaces," AFFDL-TR-76-91.

U.S. Atomic Energy Commission (1974) "Design Basis Tornado for Nuclear Power Plants," Regulatory Guide 1.76, Directorate of Regulatory Standards.

Vellozzi, J., and E. Cohen (1968) "Gust Response Factors," *ASCE Journal of the Structural Division,* **97,** 1295–1313.

Vickery, B. J. (1969) "On the Reliability of Gust Loading Factors," in *Proceedings of a Technical Meeting Concerning Wind Loads on Building and Structures, Held in Gaithersburg, Maryland, 1969,* U.S. Government Printing Office SD Catalog No. C13.29/2:30, pp. 93–104.

von Karman, T. (1961) "Progress in the Statistical Theory of Turbulence," in *Turbulence—Classic Papers on Statistical Theory,* S. K. Friedlander, ed., Interscience Publishers, New York, pp. 162–174. Reprint of 1948 paper.

Wambsganss, M. W., and S. S. Chen (1971) "Tentative Design Guide for Calculating the Vibration Response of Flexible Cylindrical Elements in Axial Flow," Argonne National Laboratory Report ANL-ETD-71-07.

Weaver, D. S., and A. Abd-Rabbo (1985) "A Flow Visualization Study of Square Array of Tubes in Water Crossflow," *Journal of Fluids Engineering,* **107,** 354–363.

Weaver, D. S., and H. C. Yeung (1984) "The Effect of Tube Mass on the Flow Induced Response of Various Tube Arrays in Water," *Journal of Sound and Vibration,* **93,** 409–425.

Yang, C. Y. (1986) *Random Vibration of Structures,* Wiley, New York.

Zorumski, W. E. (1987) "Fluctuating Pressure Loads under High Speed Boundary Layers," NASA Technical Memorandum 100517.

Chapter 8

Damping of Structures

This chapter develops analysis for vibration damping of structures in viscous fluids. Experimental data are presented for the damping of piping, buildings and bridges, cables, and spacecraft and aircraft components. Damper design is also discussed.

8.1. ELEMENTS OF DAMPING

Damping is the result of energy dissipation during vibration. Damping is generated by three phenomena: (1) fluid damping due to fluid drag, viscous dissipation, and radiation to the surrounding fluid, (2) internal material damping due to yielding, heating, electromagnetic currents, and internal energy dissipation of materials, and (3) "structural damping" due to friction, impact, scraping, and motion of trapped fluid ("gas pumping," or "squeeze film damping") within a joint. The viscous fluid dissipation of oil within the cylinder of a shock absorber provides most of the damping of an automobile suspension. High-damping viscoelastic materials, such as rubber, can be laminated to structural members to form a highly damped composite (Nashif et al., 1985).

The most widely used and practically useful model for damping forces on structures is the ideal linear viscous damper shown in Figure 8-1. This damper opposes structural motion with a force proportional to velocity,

$$F_d = c \frac{dy}{dt},$$

$$= 2M\zeta\omega_n \frac{dy}{dt}, \tag{8-1}$$

where $y(t)$ is the displacement of the structure, c is a proportionality constant with units of force per unit velocity, and F_d is the damping force. M is the mass of the structure and ω_n is the natural frequency in radians per second. In the second form of this equation, c has been replaced with a quantity that is proportional to the dimensionless damping factor ζ.

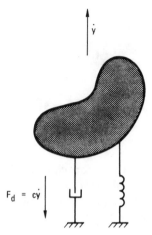

Fig. 8-1 One-dimensional, viscously damped model.

The damping factor ζ is also called the damping ratio, the viscous damping ratio, and the critical damping ratio. Most structures that are subject to vibration are lightly damped and have damping factors of the order of 0.05 or less. For free vibration of these structures, viscous damping induces negligible shifts in natural frequency and damping factor is related to the dynamic response amplification factor Q, the logarithmic decrement δ, and the loss factor η by the following formula:

$$\zeta = \frac{1}{2Q} = \frac{\delta}{2\pi} = \frac{\eta}{2}.$$

These parameters provide avenues for measuring structural damping as discussed in Section 8.3.

The energy expended by damping is the integral of the product of the damping force and the displacement over which the force is applied,

$$\text{Dissipated energy} = \int_{\text{one cycle}} F_d \, dy = \int_t^{t+T} F_d \frac{dy}{dt} \, dt, \qquad (8\text{-}2)$$

where T is the period of vibration. If the vibration is harmonic, $y(t) = A_y \cos \omega t$, then the energy dissipated in one cycle by the viscous damper (Eq. 8-1) is

$$\text{Dissipated energy} = 2\pi M \omega_n \omega \zeta A_y^2. \qquad (8\text{-}3)$$

The total energy of the elastic structure shown in Figure 8-1 is equal to the maximum kinetic (or potential) energy achieved during a cycle. The total energy of the structure of Figure 8-1 during harmonic vibration at frequency ω is

$$\text{Stored energy} = \tfrac{1}{2} M \dot{y}_{\text{max}}^2 = \tfrac{1}{2} M \omega^2 A_y^2. \qquad (8\text{-}4)$$

The ratio of the energy expended per cycle due to viscous damping to the total energy of the structure is proportional to the damping factor:

$$\frac{\text{Dissipated energy}}{\text{Stored energy}} = 4\pi\zeta\,\frac{\omega_n}{\omega} \approx 4\pi\zeta. \qquad (8\text{-}5)$$

The approximation is valid if the structure vibrates at its natural frequency, $\omega = \omega_n$, as is ordinarily the case in flow-induced vibration if the structure does not yield—see Figure 7-10, for example. (See Iwan (1965) and Harris (1988) for discussion of yielding structures.)

Often the damping force will not follow the ideal of the viscous damper. It is possible to define *equivalent viscous damping* as the viscous damping that expends the same energy per cycle as the actual damping force. Equivalent viscous damping factors for Coulomb (friction) damping, hysteretic (displacement dependent) damping, and velocity power-law damping are given in Table 8-1. The equivalent viscous damping factor is a function of vibration amplitude, frequency $[f = \omega/(2\pi)]$, mass, and damping mechanism, that is, design and geometry,

$$\zeta_{\text{e.v.d.}} = \zeta(f, A_y, M, \text{design}).$$

The remainder of this chapter presents analysis and data for the equivalent viscous damping factor as a function of these parameters for a number of practically important structures.

Continuous elastic structures possess multiple modes, and a damping must be assigned to each mode. There are two approaches to this. First, one can model certain physical damping phenomena, such as a viscous fluid or friction between structural elements, and introduce damping into the differential equation of motion through the physical model. The complication here is that unless the damping model is proportional to the mass or stiffness matrices, the modal equations will not decouple (Caughey, 1960). To avoid the mathematical and physical complications of damping-induced cross-modal coupling, uncoupled damping of the

Table 8-1 Equivalent viscous damping[a]

Damper	Damping force	Energy dissipation per cycle	Equivalent viscous damping factor, $\zeta_{\text{equil.}}$
1. Viscous	$c\dot{y}$	$c\pi\omega A_y^2$	$c/(2M\omega)$
2. Coulomb	$F_f\,\text{sgn}\,(\dot{y})$	$4F_f A_y$	$2F_f/(\pi M\omega^2 A_y)$
3. Quadratic	$c_2(\dot{y})^2\,\text{sgn}\,(\dot{y})$	$\frac{8}{3}c_2\omega^2 A_y^3$	$4c_2 A_y/(3\pi M)$
4. nth velocity	$c_n\,|\dot{y}|^n\,\text{sgn}\,(\dot{y})$	$\pi c_n\gamma_n\omega^n A_y^{n+1}$	$c_n\gamma_n\omega^{n-2}A_y^{n-1}/(2M)$
5. Hysteretic	$c_h\dot{y}/\omega$	$\pi c_h A_y^2$	$c_h/(2M\omega^2)$

Source: Adapted in part from Ruzicka and Derby (1971). Also see Nashif et al. (1985).
[a] ω = vibration frequency, radians/sec; sgn $(\;)$ = sign $(\;)$; $\gamma_n = (4/\pi)\int_0^{\pi/2}\cos^{n+1} u\,du$.

form of Eq. 8-1 is generally applied to the individual modes after modal analysis has been completed.

Exercises

1. Verify the last two columns of Table 8-1.
2. Consider the force $F_y = -c(\dot{y})^2$. What damping, if any, does this force provide? Explain.

8.2. FLUID DAMPING

8.2.1. Damping in Still Fluid

Viscous damping. Vibration in a fluid is damped by the surrounding viscous fluid. Fluid damping is the result of viscous shearing of the fluid at the surface of the structure and flow separation. The drag force per unit length on the structure of Figure 8-1 is

$$F_y = F_D = \tfrac{1}{2}\rho \, |U_{\text{rel}}| \, U_{\text{rel}} D C_D, \qquad (8\text{-}6)$$

where ρ is the fluid density, D is a characteristic dimension used in nondimensionalizing the drag force, C_D is the drag coefficient, and U_{rel} is the relative velocity between the structure and the body of fluid; see Chapter 6, Section 6.2. In a reservoir of otherwise still fluid, U_{rel} is the result of motion of the structure alone, $U_{\text{rel}} = -\dot{y}$, and the equation of motion is (Eq. 6-14 with $U = 0$)

$$m\ddot{y} + 2m\zeta_s\omega_n\dot{y} + ky = F_y = -\tfrac{1}{2}\rho \, |\dot{y}| \, \dot{y} D C_D. \qquad (8\text{-}7)$$

The mass per unit length m includes the effect of added mass and ζ_s is the structural damping factor that would be measured if the fluid were absent. This nonlinear equation can be solved for the fluid damping factor.

If the structural motion is harmonic with amplitude A_y, $y(t) = A_y \sin \omega t$, then the nonlinear term on the right-hand side of Eq. 8-7 is expanded in a Fourier series,

$$|\dot{y}| \, \dot{y} = A_y^2 \omega^2 \, |\cos \omega t| \cos \omega t,$$

$$\approx \frac{8}{3\pi} A_y^2 \omega^2 \cos \omega t = \frac{8}{3\pi} \omega A_y \dot{y}. \qquad (8\text{-}8)$$

The factor $8/(3\pi)$ comes from the first term of the Fourier series expansion (Appendix D). Substituting this result back into Eq. 8-7 and rearranging gives

$$m\ddot{y} + 2m\omega_n \left[\zeta_s + \frac{2\rho D C_D A_y \omega}{3\pi m \omega_n} \right] \dot{y} + ky = 0, \qquad (8\text{-}9)$$

which implies the fluid contribution to damping of the structure is

$$\zeta_f = \frac{2}{3\pi} \frac{\rho D^2}{m} \frac{A_y}{D} \frac{\omega}{\omega_n} C_D. \qquad (8\text{-}10)$$

This equation gives the still fluid damping for a known drag coefficient.

For small-amplitude vibration in a viscous fluid, the fluid does not separate and the drag coefficient for a circular cylinder of diameter D at low Reynolds number is given by Eq. 6-40. If we include only the first term in this equation (which is the solution generated by Stokes in 1843—see Batchelor, 1967; Brouwers and Meijssen, 1985; Rosenhead, 1963; Lamb, 1945), then $C_D = (fD/U)(3\pi^3/4)[v/(\pi f D^2)]^{1/2}$, where U is the amplitude of the velocity oscillation and v is the kinematic viscosity of the fluid. Setting the velocity amplitude $U = \omega A_y$ and the oscillation frequency to the natural frequency $f = f_n = \omega_n/(2\pi)$ gives the damping of a cylinder in a viscous fluid,

$$\zeta_{\text{cyl}} = \frac{\pi}{2} \frac{\rho D^2}{m} \left(\frac{v}{\pi f D^2} \right)^{1/2}. \qquad (8\text{-}11)$$

By application of similar theory, the damping of a rigid sphere of radius R in a viscous fluid is (Stephens and Scavullo, 1965)

$$\zeta_{\text{sphere}} = \frac{3\pi}{2} \frac{\rho R^3}{M} \left(\frac{v}{\pi f R^2} \right)^{1/2}. \qquad (8\text{-}12)$$

M is the total mass of the sphere and m is the mass per unit length of the cylinder; both include added mass (Chapter 2). These expressions are in agreement with data taken in air and in water (Stephens and Scavullo, 1965; Ramberg and Griffin, 1977). If a cylinder is centered in a fluid-filled annulus, the damping is a function of the ratio of the diameter of the fixed outer cylinder D_0 to the cylinder diameter D (Chen et al., 1976; Yeh and Chen, 1978). See Figure 2-4 of Chapter 2 with $D = 2R_1$ and $D_0 = 2R_2$. Rogers et al. (1984) developed an approximate expression for the fluid damping of a cylinder in a fluid-filled annulus,

$$\zeta_{\text{cyl.-annulus}} = \frac{\pi}{2} \frac{\rho D^2}{m} \left(\frac{v}{\pi f D^2} \right)^{1/2} \left(\frac{1 + (D/D_0)^3}{[1 - (D/D_0)^2]^2} \right). \qquad (8\text{-}13)$$

This expression also approximates the damping of a tube surrounded by other tubes if D_0 is the diameter of the pattern of surrounding tubes (Pettigrew et al., 1986). The damping and added mass (Table 2-2) increase rapidly as the outer boundary approaches the tube. This is called a *squeeze film* and it has been utilized in the design of bearing dampers and other dampers (Chow and Pinnington, 1987, 1989). Devin (1959) discussed damping of pulsating bubbles in liquids.

Equations 8-11 and 8-12 are valid only for vibration amplitudes much less than one diameter. At amplitudes beyond about 0.3 diameter, the flow will separate and the drag coefficient will be approximately constant (Fig. 6-9). In this case Eq. 8-10 predicts that the damping will increase linearly with amplitude. These phenomena can be seen in Figure 8-2, which shows experimental data of Skop et al. (1976) in comparison with the theory.

Numerical methods have been developed that are capable of predicting the viscous fluid damping of noncircular and coupled structures; see Chilukuri (1987), Blevins (1989), Yang and Moran (1979), Dong (1979), Yeh and Chen (1978), and Pattani and Olson (1988).

Radiation damping. Elastic structures radiate sound to the surrounding fluid as they vibrate. The resultant radiation damping is most important for relatively lightweight plates and shells that have large radiating surfaces, such as the skin panels of automobiles and aircraft. Morse and Ingard (1968), Junger and Feit (1986), and Fahy (1985) review the theory; some results will be presented here for circular and rectangular plates.

Consider a circular piston of radius R set flush in an infinite rigid wall and facing into a reservoir of fluid. The piston vibrates normal to the wall with frequency f, radiating sound energy into the fluid. The force exerted by the fluid on the piston due to acoustic radiation is the sum of a

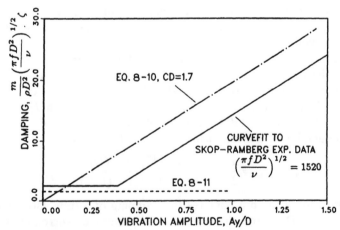

Fig. 8-2 Experimental data and theory for damping of a circular cylinder in a reservoir of viscous fluid.

component in phase with piston velocity and a component in phase with piston acceleration,

$$F = -\rho \pi R^2 c \left(\Theta \dot{y} + \frac{\chi}{\omega} \ddot{y} \right). \tag{8-14}$$

The term Θ is called acoustic resistance and the term χ is called acoustic reactance. The speed of sound in the fluid is c and the raised dots $(\dot{y})$ denote derivation with respect to time. By substituting this force into the equation of motion of an elastically supported piston (Eq. 8-7), we obtain the following equation of motion:

$$\left(M + \frac{\rho \pi R^2 c \chi}{\omega} \right) \ddot{y} + 2M\omega_n \left(\zeta_s + \frac{\rho \pi R^2 c \Theta}{2M\omega_n} \right) \dot{y} + ky = 0, \tag{8-15}$$

which implies that the acoustic radiation damping factor is proportional to the acoustic resistance,

$$\zeta_{\text{acoustic-cir.}} = \frac{1}{4} \frac{\rho R^3}{M} \frac{\lambda}{R} \Theta. \tag{8-16}$$

$M + \rho \pi R^2 c \chi / \omega$ is the total mass of the piston, including the added mass induced by acoustic reactance. $\lambda = 2\pi c / \omega_n$ is the wavelength of the sound wave at the circular frequency ω_n.

The acoustic resistance and acoustic reactance are functions of the ratio of the acoustic wavelength to the radius of the piston (Morse and Ingard, 1968, p. 384),

$$\Theta = 1 - \frac{\lambda}{2\pi R} J_1 \left(\frac{4\pi R}{\lambda} \right) = \begin{cases} 2\pi^2 (R/\lambda)^2, & R/\lambda \ll 0.2, \\ 1, & R/\lambda \gg 0.2, \end{cases}$$

$$\chi = \frac{4}{\pi} \int_0^{\pi/2} \left(\frac{4\pi R}{\lambda} \cos \alpha \right) \sin^2 \alpha \, d\alpha = \begin{cases} (16/3)(R/\lambda), & R/\lambda \ll 0.2, \\ \pi^{-2}(\lambda/R), & R/\lambda \gg 0.2, \end{cases} \tag{8-17}$$

where J_1 is the Bessel function of first order. The radiation damping for a rectangular piston with sides a and b is as follows:

$$\zeta_{\text{acoustic-rect.}} = \frac{1}{4\pi} \frac{\rho a^2 b}{M} \frac{\lambda}{a} \Theta,$$

where

$$\Theta = \begin{cases} (\pi/2)^2 (a^2 + b^2)/\lambda, & a/\lambda \ll 0.2, \\ 1, & a/\lambda \gg 0.2, \end{cases} \tag{8-18}$$

provided that the rectangle is not too far from square. These expressions apply only for small amplitudes and do not include the viscous dissipation or vortex formation, but they can be used to approximate the damping of elastic plates with supported edges. See Junger and Feit (1986) and Stephens and Scavullo (1965).

The mode shape of the fundamental mode of flat plates roughly conforms to a piston (Chapter 7, Section 7.2, Table 7-1). The previous results can be applied to estimate the radiation damping of plates. If the plate has thickness h and is fabricated from material of density ρ_p, then the radiation damping is found from Eqs. 8-14 through 8-18. If the acoustic wavelength greatly exceeds the plate dimensions, $\lambda \gg R, a$,

$$\zeta_{\text{round}} = \frac{\pi}{2} \frac{\rho}{\rho_p} \frac{R^2}{h\lambda}, \qquad \zeta_{\text{rect}} = \frac{\pi}{16} \frac{\rho}{\rho_p} \frac{a^2 + b^2}{h\lambda}. \qquad (8\text{-}19)$$

For example, for an aluminum plate in air, $\rho_{\text{air}}/\rho_p = 0.00044$. If the side length of the square, simply supported plate is 8 in (20 cm) and the plate is 0.050 in (1.3 mm) thick, then its natural frequency is 145 Hz (Table 7-2) and the corresponding wavelength of sound is 1100×12 in/sec/145 = 93 in (2.36 m). The computed radiation damping is $\zeta = 0.0024$, which is an appreciable fraction of the measured structural damping for this configuration (Section 8.4.5).

Exercises

1. The drag of a plate of length L and width b in a flow parallel to the length edge is

$$F_D = \begin{cases} -0.664\rho L^{1/2} b v^{1/2} (\dot{y})^{3/2} \, \text{sgn} \, (\dot{y}) & \text{laminar,} \\ -0.0166\rho L^{6/7} b v^{1/7} (\dot{y})^{13/7} \, \text{sgn} \, (\dot{y}) & \text{turbulent,} \end{cases}$$

where $\dot{y}$ is the flow velocity. Determine the damping factor of an elastically supported plate that oscillates parallel to the edge of length L using the equivalent energy method described in Section 8.1.

2. Derive an expression for the natural frequency of a simply supported, square plate including acoustic reactance. Use Table 7-1 and Eqs. 8-14 and 8-18 as starting points. Evaluate the frequency shift due to acoustic reactance for the plate dimensions given above.

8.2.2. Damping in a Fluid Flow

Cross flow. Consider the elastically supported structure shown in Figure 8-3, which is exposed to a high Reynolds number cross flow. As the structure vibrates, a relative component of flow velocity is induced,

$$U_{\text{rel}}^2 = \dot{y}^2 + (U - \dot{x})^2 \approx U^2 - 2U\dot{x}. \qquad (8\text{-}20)$$

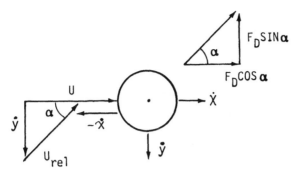

Fig. 8-3 Section in a flow.

The horizontal ($\dot{x}$) and vertical ($\dot{y}$) components of structural velocity are assumed to be small relative to the mean horizontal fluid velocity U. The angle of the relative flow is $\alpha = \tan(\dot{y}/U) \approx \dot{y}/U$. Using these linearizing approximations gives the following net horizontal and vertical forces induced by the relative drag:

$$F_x = F_D \cos\alpha = \tfrac{1}{2}\rho U_{\text{rel}}^2 D C_D \cos\alpha \approx \tfrac{1}{2}\rho U^2 C_D D\left(1 - \frac{2\dot{x}}{U}\right), \quad (8\text{-}21)$$

$$F_y = F_D \sin\alpha = \tfrac{1}{2}\rho U_{\text{rel}}^2 D C_D \sin\alpha \approx -\tfrac{1}{2}\rho U^2 D C_D \frac{\dot{y}}{U}. \quad (8\text{-}22)$$

Substituting these values into an equation of motion of the form of Eq. 8-7 for the horizontal (x) and vertical (y) degrees of freedom, we see that the drag-induced damping due to the cross flow is proportional to flow velocity and inversely proportional to the natural frequency in hertz, f_n, and that the damping is largest for inline motion,

$$\zeta_{x,\text{drag}} = \frac{1}{4\pi f_n D}\frac{U}{D}\frac{\rho D^2}{m}C_D, \qquad \zeta_{y,\text{drag}} = \frac{1}{8\pi f_n D}\frac{U}{D}\frac{\rho D^2}{m}C_D, \quad (8\text{-}23)$$

where m is the mass per unit length, including added mass. These predictions are in rough agreement with experimental data for cylinders at small flow velocities: The measured fluid damping increases with velocity and the inline damping is greater than the transverse damping, but Eq. 8-23 can overpredict the damping. See Figure 5-9, Chen and Jendrzejczyk (1979), and Chen (1981). For reduced velocities greater than about $U/(fD) = 3$, flow-induced vibration associated with vortex shedding and instabilities obscures the fluid damping (Chapters 3 through 5).

Fluid damping in two-phase cross flow is discussed by Hara (1988), Axisa et al. (1988), and Pettigrew et al. (1988).

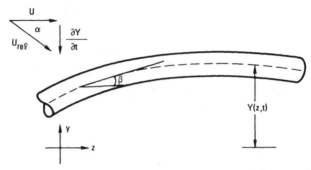

Fig. 8-4 Rod in a parallel flow. U is parallel to the undeformed shape of the rod.

Parallel flow. Consider the rod shown in Figure 8-4 with its undeformed shape parallel to the flow. This flow geometry arises in tubular
heat exchangers and with nuclear reactor fuel pins. Fluid damping is
generated as the rod vibrates normal to the flow. Paidoussis (1966) has
formulated a model for the fluid forces on a circular cylinder in a parallel
flow that has shown qualitative agreement with experimental data. This
model is discussed in Chapter 10, Section 10.2. The fluid forces in
Paidoussis' model that generate fluid damping are proportional to
structural velocity,

$$F_y = 2\rho C_I A U \frac{\partial^2 Y(z, t)}{\partial z \, \partial t} + \frac{1}{2} c_N C_I \frac{\rho A U}{D} \frac{\partial Y(z, T)}{\partial t}. \qquad (8\text{-}24)$$

The first term is the Coriolis force and the second is the result of skin
friction drag. $A = \pi D^2/4$ is the cross-sectional area of the rod, U is the
mean inline flow velocity, C_I is the added mass coefficient (Chapters 2
and 6), and c_N is a friction coefficient.

Using the modal analysis of Appendix A for a single mode,
$Y(z, t) = y(t)\bar{y}(z)$, it can be shown that if the rod vibrates in a single
mode, then the first term (Coriolis force) on the right side of Eq. 8-24
does not contribute to the net force on the rod provided the ends of the
rod do not displace. (If one or both ends do move, such as in a
cantilever, then the Coriolis force term provides additional damping;
Chen, 1981.) If we neglect the first term in Eq. 8-24 and substitute the
second term into either Eq. 8-4 or 8-7, then the damping for one mode of
a circular rod in parallel flow is

$$\zeta_{\text{parallel}} = \frac{1}{8\pi} c_N C_I \frac{U}{f_n D} \frac{\rho D^2}{m}, \qquad (8\text{-}25)$$

where m is the mass per unit length including added mass. The fluid
damping increases with flow velocity.

The added mass coefficient is theoretically $C_I = 1.0$ for an unconfined inviscid fluid and somewhat greater for a viscous fluid or for a confined flow; see Chapters 2 and 6. Three estimates of the friction coefficient are as follows:

$$c_N = \begin{cases} 0.04, & \text{Paidoussis (1966) at } UD/v = 9 \times 10^4; \\ 0.02 \text{ to } 0.1, & \text{Chen (1981);} \\ 1.3(UD/v)^{-0.22} & \text{Connors et al. (1982).} \end{cases} \quad (8\text{-}26)$$

Connors' expression is based on data taken for water flow along a tube in a tube bundle with $P/D = 1.33$. The Chen (1981) value for damping was obtained by fitting a polynomial to experimental data. A reasonable approach to estimating the fluid damping of a circular rod or tube in parallel fluid flow is to estimate the contribution of structural and still fluid damping (Eq. 8-11 and Section 8.4) and then add in the component of fluid dynamic damping,

$$\zeta = \zeta_0 + \frac{1}{8\pi} c_N C_I \frac{U}{f_n D} \frac{\rho D^2}{m}, \quad (8\text{-}27)$$

where c_N is between 0.04 and 0.08 for a Reynolds number, based on diameter, of the order of 10^5. This expression is compared with experimental data in Figure 8-5. Damping increases approximately linearly with increasing velocity from the still-water damping. As discussed in Chapter 10, Section 10.2, relatively low fluid velocities produce damping but a sufficiently high velocity will cause instability.

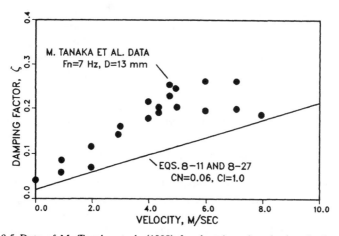

Fig. 8-5 Data of M. Tanaka et al. (1988) for damping of a plastic tube in a parallel water flow, $m/(\rho D^2) = 1.4$, in comparison with theory.

Exercises

1. Consider a uniform tower of diameter D, height L, and mass m per unit length which extends into the boundary layer of wind. The spanwise variation of wind is given by Eq. 7-58. Using Eq. A-23 of Appendix A, determine the equivalent viscous fluid damping for vibration in line with the wind.

2. Stephens and Scavullo (1965) found that the damping of rigid plates oscillating normal to the plane of the plate in a reservoir of still fluid, like a fan moving back and forth, increased linearly with vibration amplitude. What fluid mechanism is the most likely source of this damping?

8.3. STRUCTURAL DAMPING

8.3.1 Techniques for Measurement of Damping

All damping measurement techniques are based on the same idea: The response of the structure is a function of excitation and damping. If a known excitation is applied to a structure, the response can be predicted theoretically as a function of damping. Matching predicted response with measured response determines the damping.

A fluid flow can both excite and damp. For example, at low velocities fluid flow damps tube bundle vibrations in cross flow, but at higher velocity it leads to an instability as shown in Chapter 5, Figure 5.8. A conservative practice with flowing fluid is to measure damping in still fluid. If one wishes to measure only damping generated by material and structural mechanisms, either the measurement must be made in a fluid of sufficiently low density that the fluid damping is negligible or estimated fluid damping must be subtracted from a measurement of total damping.

Free decay method. The equation of motion of the damped single-degree-of-freedom structure shown in Figure 8-1 is

$$M\ddot{y} + 2M\zeta\omega_n\dot{y} + ky = F(t), \tag{8-28}$$

where $F(t)$ is the external excitation, if any, ζ is the damping factor, and M is the mass in motion. If the structure is excited to an amplitude A_y and then the excitation is removed, the vibrations decay slowly in time as shown in Chapter 1, Figure 1-2. If $F(t) = 0$ for $t > 0$, then the solution to Eq. 8-28 for free decay is (Thomson, 1988)

$$y(t) = A_y e^{-\zeta\omega_n t} \sin\left[\omega_n(1 - \zeta^2)^{1/2}t + \phi\right], \tag{8-29}$$

where the phase ϕ is a constant. Note that the frequency is determined by the natural frequency and damping alone. The ratio of the amplitude of any two successive peaks one period $T = 2\pi/[\omega_n(1 - \zeta^2)^{1/2}]$ apart in the decay is

$$\frac{A_i}{A_{i+1}} = e^{2\pi\zeta/(1-\zeta^2)^{1/2}}. \tag{8-30}$$

For lightly damped structures, ζ is much less than 1, so $1 - \zeta^2 \approx 1$ and the previous equation implies $2\pi\zeta = \delta = \log_e (A_i/A_{i+1})$, where δ is called the logarithmic decrement of damping. Since free vibration of lightly damped structures decays slowly, it is easiest to measure the ratio of peak amplitude N cycles apart,

$$2\pi\zeta N = \log_e \left(\frac{A_i}{A_{i+N}}\right). \tag{8-31}$$

If N cycles are required for the amplitude to decay to one-half of the initial amplitude, the damping factor is

$$\zeta = \frac{\log_e 2}{2\pi N} = \frac{0.1103}{N}. \tag{8-32}$$

Damping is measured by the free decay technique using a transducer, such as a strain gauge or accelerometer, an amplifier, and a strip chart recorder. For large, low-frequency structures, the measurement can be done by eye alone. The structure can be mechanically excited by winching up a displacement and cutting the cable, setting off an explosive charge, or tapping the structure with a hammer. Entire buildings can be excited by a single man rocking back and forth on the roof (Czarnecki, 1974). It is possible to map the amplitude dependence of damping from a single decay.

Bandwidth method. The response of a single-degree-of-freedom structure (Eq. 8-28) to steady-state sinusoidal excitation $F(t) = F_0 \sin \omega t$ at frequency $\omega = 2\pi f$ is sinusoidal at the excitation frequency

$$y(t) = A_y \sin (\omega t - \phi). \tag{8-33}$$

The amplitude and phase angle are (Thomson, 1988)

$$\frac{A_y k}{F_0} = \left\{\left[1 - \left(\frac{\omega}{\omega_n}\right)^2\right]^2 + 4\zeta^2\left(\frac{\omega}{\omega_n}\right)^2\right\}^{-1/2}, \tag{8-34a}$$

$$\phi = \tan^{-1}\left(\frac{2\zeta\omega\omega_n}{\omega_n^2 - \omega^2}\right). \tag{8-34b}$$

The response amplitude is shown as a function of frequency in Figure 8-6. The frequency of the peak of the response is found by setting the

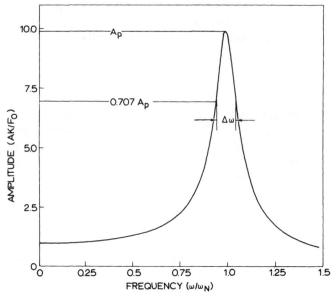

Fig. 8-6 Response of a linear structure to harmonic excitation; $\Delta\omega$ is the bandwidth.

derivative of the amplitude with respect to the excitation frequency to zero. The peak response occurs at approximately the natural frequency, a condition called *resonance*.

$$\frac{\omega}{\omega_n} = (1 - 2\zeta^2)^{1/2} \approx 1. \tag{8-35}$$

The peak resonant response is inversely proportional to damping,

$$\frac{A_p k}{F_0} = \frac{1}{2\zeta(1 - \zeta^2)^{1/2}} \approx \frac{1}{2\zeta}. \tag{8-36}$$

The bandwidth of the response $\Delta\omega$ is defined as the width of the frequency response at $1/\sqrt{2}$ times the peak amplitude, as shown in Figure 8-6. The excitation frequencies (ω_1 and ω_2) that produce these half-power points have response $1/\sqrt{2}$ times the resonant response (Eqs. 8-34a and 8-36),

$$\left\{ \left[1 - \left(\frac{\omega}{\omega_n}\right)^2 \right]^2 + 4\zeta^2 \left(\frac{\omega}{\omega_n}\right)^2 \right\}^{1/2} \approx \left(\frac{2\zeta\omega}{\omega_n}\right) 2^{1/2}. \tag{8-37}$$

The approximations in Eqs. 8-34 through 8-37 are applicable for lightly damped structures, $\zeta \ll 1$, which possess a distinct amplified resonance (Thomson, 1988). Equation 8-37 is solved for the band frequencies by squaring this equation and solving the resultant quadratic,

$$1 - \left(\frac{\omega_1}{\omega_n}\right)^2 = 2\left(\frac{\omega_1}{\omega_n}\right)\zeta, \quad \text{for } \omega_1 < \omega_n,$$

$$1 - \left(\frac{\omega_2}{\omega_n}\right)^2 = -2\left(\frac{\omega_2}{\omega_n}\right)\zeta, \quad \text{for } \omega_2 > \omega_n. \tag{8-38}$$

By subtracting these two equations, the bandwidth is found to be proportional to the damping factor, $\Delta f = (\omega_2 - \omega_1)/(2\pi)$,

$$\zeta = \frac{\Delta f}{2f_n}. \tag{8-39}$$

To measure the damping of a structure using Eq. 8-39, only the resonant frequency f_n and the bandwidth must be obtained. The technique can be generalized by fitting the frequency domain response to a response function of the form of Eq. 8-34a and then determining ζ as a fitted parameter (Brownjohn et al., 1987; Fabunmi et al., 1988).

A related technique, called the *Nyquist* method or *Argand* plot, uses a phase diagram plot to determine damping. The response to the sinusoidal excitation is the sum of components in phase and out of phase with the exciting force $F_0 \sin \omega t$ (Eq. 8-34),

$$y(t) = \frac{A_y k}{F_0} \left\{ \frac{1 - (\omega/\omega_n)^2}{[1 - (\omega/\omega_n)^2]^2 + 4\zeta(\omega/\omega_n)^2} \sin \omega t \right.$$

$$\left. - \frac{2\zeta(\omega/\omega_n)}{[1 - (\omega/\omega_n)^2]^2 + 4\zeta(\omega/\omega_n)^2} \cos \omega t \right\}$$

$$= \text{IN} \sin \omega t + \text{OUT} \cos \omega t. \tag{8-40}$$

IN and *OUT* are the amplitudes of response components in phase and out of phase with the exciting force. Consider a *phase plane* where the IN component is plotted on the horizontal axis and the OUT component is plotted on the vertical axis. This results in a circle that is displaced one radius down from the origin,

$$\left(\frac{\text{IN}}{A_y k/F_0}\right)^2 + \left(\frac{\text{OUT}}{A_y k/F_0} + \frac{\omega_n}{4\zeta\omega}\right)^2 = \left(\frac{\omega_n}{4\zeta\omega}\right)^2. \tag{8-41}$$

The radius of the circle, $\omega_n/(4\zeta\omega)$, is inversely proportional to the damping factor.

The bandwidth and Nyquist techniques can be applied to any structural mode that can be excited. The results are independent of the mode shape and the exciter location. Unbalanced wheel-type exciters and impulse excitation have been used (Fig. 8-7; Bouwkamp, 1974). Fourier transforms (Appendix D) of the input and response signal can be divided

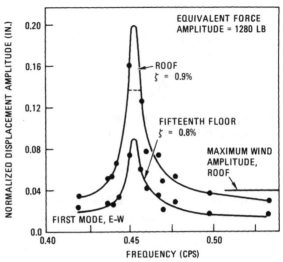

Fig. 8-7 Frequency response of the Alcoa Building, San Francisco (Bouwkamp, 1974).

to form a transfer function and the bandwidth method is applied to this transfer function. Ambient excitation due to wind and traffic has been used to generate a suitable transfer function under the assumption that the exciting spectrum is flat (Brownjohn et al., 1987; Littler and Ellis, 1987).

Magnification factor method. The peak response of a lightly damped structure to steady harmonic excitation is inversely proportional to damping (Eq. 8-36). For lightly damped structures, the damping is

$$\zeta = \frac{F_0}{2kA_p} = \frac{F_0}{2M(\omega_n^2 A_p)} = \frac{1}{2Q}. \tag{8-42}$$

The quantity $(\omega_n^2 A_p)$ is equal to the amplitude of acceleration of the structure at resonance. The static deformation of the structure under a load F_0 is $y_s = F_0/k$. The ratio of the dynamic resonant response to the response to a steady load of the same magnitude is $A_p/y_s = Q = 1/(2\zeta)$, where Q is called the magnification factor. If the amplitude of the resonant response or resonant acceleration can be measured for a known exciting force, or if the magnification factor can be measured, then this equation can be used to determine the damping factor. Its application for nonlinear damping is discussed by Mulcahy and Miskevics (1980). Unlike the free decay or bandwidth methods, the result is dependent on the mode shape, mass distribution, and details of the means used to excite the structures as discussed in the following example.

Exercises

1. Examine Figure 1-2, Chapter 1, and Figure 8-6. Determine the damping factors of these data. Explain the method(s) used and provide a number for the damping factor for each case.

2. Suspend a tennis ball or an apple from 40 in (1 m) of string. Displace the sphere about one-half diameter and let it swing in the pendulum mode. Measure the damping factor and compare with Eq. 8-12. The kinematic viscosity of air at room temperature is $v = 1.6 \times 10^{-4}\, \text{ft}^2/\text{sec}$ $(1.5 \times 10^{-5}\, \text{m}^2/\text{sec})$ and $\rho = 0.075\, \text{lb/ft}^3$ $(1.2\, \text{kg/m}^3)$. A typical apple has a mass of 0.4 lb (0.2 kg).

3. The bandwidth technique can be applied at any two frequencies that have response the same distance below the resonance peak, not just $1/\sqrt{2}$ below the peak. Derive an expression for damping in terms of bandwidth between two points that are $1/\alpha$ below the resonance and show that this reduces to Eq. 8-37 when $\alpha = \sqrt{2}$.

8.3.2. Example: Structural Damping of a Loosely Held Tube

Ordinarily, heat-exchanger tubes are supported at regular intervals by passing through support plates. Clearance is provided, between the tube diameter and the hole, for assembly. As the tube vibrates relative to the plate, energy is dissipated by impact and scraping. This geometry is called a loosely held tube. The results of an experimental study on damping of a loosely held tube are presented here to illustrate some of the practical aspects of damping measurement (Blevins, 1975). Also see Sections 5.3 and 8.4.4.

The test rig is shown in Figure 8-8. It is fabricated of welded steel to achieve natural frequencies well above those of the tubes to be tested. The nickel–iron–chrome alloy tube passes through stainless steel plates bolted to the rig. The tube is excited at midspan by a shaker. Static preloads are placed on the tube by springs that load the tube at the plates. The static loads simulate the steady aerodynamic, gravitational, assembly, and thermal loads on the tube. Tube lengths of 58 and 20 in (1.47 and 0.58 m) were tested. The tube was excited normal and parallel to the preload.

The shaker exciting frequency was uniformly swept through the resonance frequencies at 2.5 Hz/sec with the amplitude of the exciting force held constant. The results were found to be independent of the sweep rate and the direction of the sweep. The midspan acceleration of

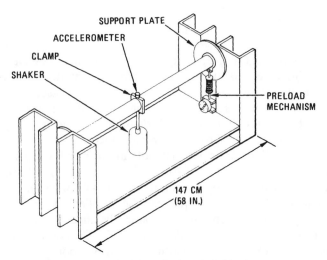

Fig. 8-8 Test rig for measuring damping of a loosely held tube.

the tube and the force amplitude were continuously monitored as shown in Figure 8-9. The force and acceleration signals were passed through a 5 Hz filter centered at the exciting frequency to eliminate noise and the contributions of higher modes. It was felt that the error in data measurement and reduction did not exceed 5%.

A typical sweep is shown in Figure 8-10, where A is the amplitude of the midspan acceleration and F is the amplitude of the sinusoidal exciting force. The uneven appearance of the results, such as the double peak at the third mode, is evidence of nonlinear phenomena such as rattling introduced by the looseness of the end connections. The damping was evaluated using the magnification method, Eq. 8-42. The terms F_0 and M in this equation refer to the generalized force and generalized mass when applied to continuous structures, such as the tube; they are evaluated from modal analysis, which in turn requires an estimate of the mode shape. These modes were estimated to be the modes of a pinned-pinned beam, $\sin(i\pi z/L)$, where i is the mode number and L is the span. For forcing at midspan, F_0 is the applied force and the generalized mass is one-half the total mass of the tube,

$$M = \int_0^L m(z) \sin^2 \frac{i\pi z}{L} \, dz = \frac{mL}{2}.$$

There is considerable scatter in the results for damping. This scatter is generated by the probabilistic nature of the nonlinearities in the system, and this is typical of structural damping. Statistical tools are employed to detect significant trends (Siegel, 1956). In every case tested, the damping

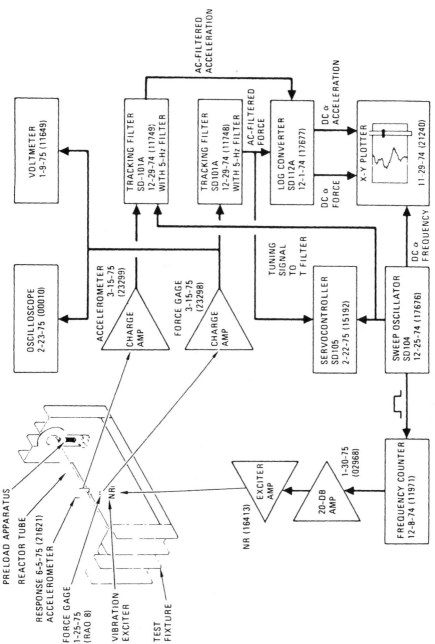

Fig. 8-9 Data circuitry for the damping measurement.

323

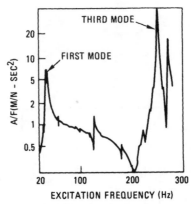

Fig. 8-10 Tube response as a function of excitation frequency.

significantly increased with increased amplitude. The third mode had lower damping than the first mode. Increasing support-plate thickness increased damping, but changing plate roughness or changing direction of excitation relative to preload had no significant influence.

These trends can be seen in the histograms shown in Figures 8-11 and 8-12. The data presented in Figure 8-11 were generated by vibration of the 58 in (147 cm) long tubes with diameters of 1 and 2.16 in (2.54 and 5.49 cm). The support plates were 0.75 and 1.25 in (1.9 and 3.2 cm) thick. It can be seen that the third mode has less damping and a more peaked

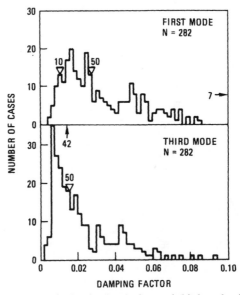

Fig. 8-11 Damping of 58 in (147 cm) tubes in first and third modes (Blevins, 1975).

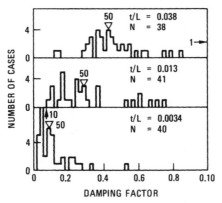

Fig. 8-12 Tube damping as a function of the ratio of plate thickness (*t*) to tube span (*L*) (Blevins, 1975).

distribution than the first mode. The inverted deltas divide the damping values into groups containing 10%/90% and 50%/50% of the sample. *N* is the sample size. Figure 8-12 shows the effect of tube length and plate thickness *t* on damping. All tubes were 1 in in diameter. Increasing the tube support plate thickness to tube length ratio strongly increases damping, probably because the length available for tube scraping increases. These effects have also been observed by Hartlen (1974, quoted by Pettigrew, 1986). Other tests have demonstrated that welding the tube in at a joint, eliminating scraping at joints, reduced the damping factor to values on the order of $\zeta = 0.001$ and less.

A conservative approach to design would be to choose a damping factor such that 90% of the measured values are greater than the chosen damping factor. This procedure and the data of Figure 8-11 give damping factors of 0.009 for first-mode vibration and 0.005 for third-mode vibration. These are similar to damping factors measured in other studies; see Section 8.4.4.

8.4. DAMPING OF BRIDGES, TOWERS, BUILDINGS, PIPING, AND AIRCRAFT STRUCTURES

This section presents the results of damping measurements made on complex practical structures at low amplitudes in order to form a basis for estimating damping and to allow a comparison of these results with new measurements. The results are summarized in Table 8-2. The scatter in measurements of damping requires a statistical approach. It is probably most reasonable to use damping values for design that lie between the first quartile (25% of the data falls below this value) and mean value of

Table 8-2 Summary of damping values[a]

Figure	Sample	Sample size	Standard deviation	Damping factor, ζ					
				Maximum	Average	50% Above	75% Above	90% Above	Minimum
8-14	Suspension bridges	64	0.129	0.0839	0.0117	0.0061	0.0036	0.0024	0.0021
8-15	Steel towers	21	0.0057	0.0286	0.0086	0.0064	0.0048	0.0032	0.0016
8-15	Concrete towers	3	0.0040	0.0191	0.0138	—	—	—	0.0095
8-15	All towers	24	0.0058	0.0286	0.0092	0.0080	0.0051	0.0032	0.0016
8-16	Low excitation, steel buildings	42	0.0105	0.0370	0.0151	0.0130	0.0060	0.0038	0.0029
8-16	Low excitation, concrete buildings	8	0.0070	0.0310	0.0170	0.0140	0.0110	—	0.0100
8-16	Earthquake excitation, steel buildings	24	0.0234	0.1130	0.0510	0.0400	0.0320	0.0200	0.0200
8-16	Earthquake excitation, concrete buildings	34	0.0362	0.1640	0.0685	0.0600	0.0400	0.0200	0.0170
8-16	Low excitation, all buildings	50	0.0100	0.0370	0.0154	0.0130	0.0070	0.0040	0.0029
8-16	Earthquake excitation, all buildings	58	0.0327	0.1640	0.0613	0.0520	0.0350	0.0200	0.0170
8-16	All buildings	108	0.0338	0.1640	0.0400	0.0300	0.0130	0.0060	0.0029
8-17	1- to 10-story steel buildings	54	0.0151	0.0600	0.0257	0.0240	0.0110	0.0060	0.0040
8-17	10- to 20-story steel buildings	52	0.0298	0.2000	0.0253	0.0180	0.0073	0.0060	0.0040
8-17	Over 20-story steel buildings	141	0.0109	0.0500	0.0174	0.0144	0.0092	0.0055	0.0020
8-17	1- to 10-story concrete buildings	116	0.0210	0.1240	0.0266	0.0210	0.0148	0.0100	0.0050

Ref.	Description	N							
8-17	10- to 20-story concrete buildings	69	0.0255	0.1050	0.0319	0.0214	0.0121	0.0096	0.0069
8-17	Over 20-story concrete buildings	81	0.0252	0.1100	0.0257	0.0140	0.0100	0.0080	0.0040
8-17	All 0- to 10-story buildings	170	0.0193	0.1240	0.0263	0.0211	0.0140	0.0085	0.0040
8-17	All 10- to 20-story buildings	121	0.0276	0.200	0.0290	0.0200	0.0110	0.0070	0.0040
8-17	All over 20-story buildings	222	0.0179	0.110	0.0204	0.0141	0.0100	0.0065	0.0020
8-17	All buildings	513	0.0214	0.200	0.0244	0.0180	0.0110	0.0070	0.0020
8-18	Power-plant piping	162	0.0312	0.1770	0.0399	0.0310	0.0190	0.0080	0.0020
8-19	Heat-exchanger tubing, in air	73	0.0145	0.0796	0.0169	0.0120	0.0079	0.0060	0.0020
8-19	Heat-exchanger tubing, in water	84	0.0110	0.0535	0.0196	0.0170	0.0100	0.0073	0.0051
8-19	All heat-exchanger tubing	157	0.0128	0.0796	0.0183	0.0148	0.0092	0.0066	0.0020
8-20	Steam generator	36	0.0123	0.0507	0.0207	0.0194	0.0092	0.0076	0.0066
8-21	Aluminum skin-stringer panels	116	0.0059	0.0380	0.0164	0.0153	0.0130	0.0100	0.0055
8-21	Titanium skin-stringer panels	21	0.0049	0.0275	0.0168	0.0156	0.0123	0.0094	0.0084
8-21	All skin-stringer panels	137	0.0058	0.0380	0.0165	0.0155	0.0120	0.0100	0.0055
8-22	Aluminum honeycomb panels	26	0.0038	0.0270	0.0186	0.0180	0.0150	0.0130	0.0130
8-22	Graphite–epoxy honeycomb panels	42	0.0050	0.0233	0.0111	0.0094	0.0070	0.0060	0.0050
8-22	Kevlar honeycomb panels	7	0.0053	0.0277	0.0193	0.0155	0.0136	—	0.0136
8-22	All honeycomb panels	75	0.0060	0.0277	0.0145	0.0150	0.0083	0.0069	0.0050

[a] See figures for references. 50% Above = Median value, that is, value such that 50% of damping factors exceed this value. 75% Above = Semi-Quartile value, that is, value such that 75% of damping factors exceed this value. The Average, 50% Above, and 75% Above values are recommended for design.

327

the data. The fact that damping generally increases with vibration amplitude provides a degree of conservatism.

8.4.1. Bridges

A bridge is composed of a superstructure that contains the roadway, piers that support the superstructure, and a foundation that supports the piers. Ordinarily, only the superstructure is subject to flow-induced vibration. The histograms of Figure 8-13 show the damping of bridge superstructures in the fundamental mode compiled by Ito et al. (1973). Figure 8-14 presents damping factors for various modes of long-span suspension bridges compiled by Davenport (1981) and Littler and Ellis (1987). See also Brownjohn et al. (1987).

As can be seen from these figures, there is considerable scatter in the data, as is typical for structural damping, and the damping factors of torsion, vertical bending, and horizontal bending modes are comparable. Davenport's data suggest that suspension bridge damping decreases with increasing frequency. Ito's data indicate that short-span (high fundamental frequency) bridges have higher damping than long-span bridges. Ito found no definite relationship between the bridge material (concrete or steel) and damping. The average damping factor of Figure

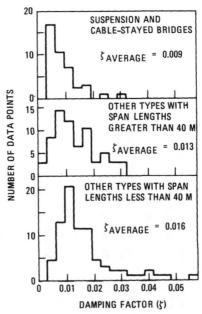

Fig. 8-13 Distribution of damping values of bridge superstructures (Ito et al., 1973).

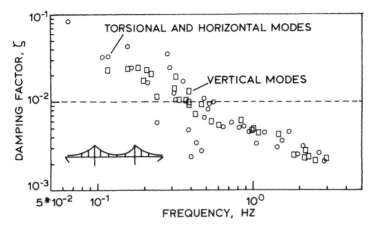

Fig. 8-14 Damping factors for long-span suspension bridges (Davenport, 1981; Littler and Ellis, 1987).

8-14 is $\zeta = 0.0117$, which is similar to the average damping of $\zeta = 0.009$ found by Ito for long-span suspension bridges.

8.4.2 Towers and Stacks

Scruton and Flint's (1964) compilation of damping data for circular stacks and towers is presented in Figure 8-15. The height of these structures in this figure varies from 150 to 710 ft (50 to 240 m) and tip diameter from 2.2 to 24 ft (0.6 to 8 m). All measurements were made at small amplitudes relative to the diameter and height. There is considerable

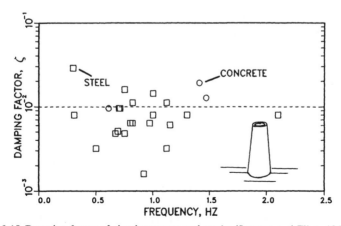

Fig. 8-15 Damping factor of circular towers and stacks (Scruton and Flint, 1964).

scatter. Even among groups of towers having the same material and construction, damping can vary by a factor of 2. Rock foundations produce less damping than do piers in soil. The concrete towers have a higher average damping ($\zeta = 0.0138$) than the welded towers ($\zeta = 0.0086$). The average damping for towers in Figure 8-15 for both materials is $\zeta = 0.0092$.

8.4.3. Buildings

Damping of buildings is shown in Figures 8-16 and 8-17. There are two sets of data in Figure 8-16. The first set is for 12 buildings in the Los Angeles area whose damping was determined by Hart and Vasudevan (1975) by analyzing the response of the buildings to the 1971 San Fernando earthquake. Building heights ranged from 65 to 601 ft (20 to 180 m). The ground accelerations experienced by these buildings were relatively severe, with peak accelerations between 0.10 and 0.27 times the acceleration of gravity. Studies by Rea et al. (1969), Jeary (1988), and Davenport and Hill-Carroll (1986) have shown that damping uniformly increases with vibration amplitude; see also Figure 8-20. The average damping of these buildings for earthquake excitation is $\zeta = 0.0612$.

The second set of data in Figure 8-16 was obtained from forced and ambient vibration tests on eleven buildings (Hart et al., 1975; Bouwkamp, 1974; Czarnecki, 1974). These data are much more

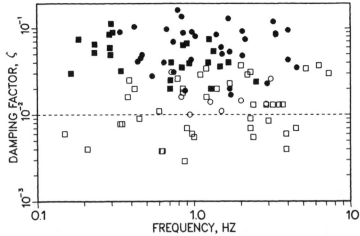

Fig. 8-16 Damping of buildings: ($\bullet$, $\bigcirc$) concrete; ($\blacksquare$, $\square$) steel; ($\blacksquare$, $\bullet$) earthquake excitation; ($\bigcirc$, $\square$) low-level excitation (Bouwkamp, 1974; Hart and Vasudevan, 1975; Hart et al., 1975; Czarnecki, 1974. Additional data from citations given by Hart et al., 1975).

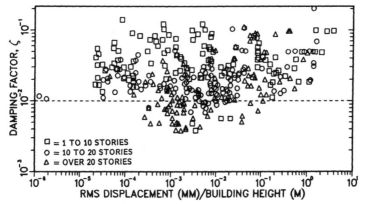

Fig. 8-17 Damping of buildings as a function of excitation amplitude (Davenport and Hill-Carroll, 1986).

representative of the amplitudes likely to be experienced in wind-induced vibration. The buildings ranged in height from a 50 ft (15 m) tall gymnasium to the 1100 ft (335 m) John Hancock building which has first mode dumping $\zeta = 0.006$. The average damping for this data set is $\zeta = 0.0154$, which is one-fourth the average damping for earthquake excitation.

Figure 8-17 shows damping of tall buildings for low-level excitation compiled by Davenport and Hill-Carroll (1986) and plotted as a function of the ratio of excitation amplitude to height. The taller buildings, 20 stories and above, have 60% of the damping of buildings between 5 and 20 stories tall. The concrete buildings of Figure 8-17 have roughly 30% more damping than the steel buildings. The concrete buildings of Figure 8-16 have 15% higher average damping than steel buildings for low-level excitation. Davenport and Hill-Carroll (1986) found that damping increases with vibration amplitude raised to the 1/10 power.

The natural frequency of buildings can be estimated from Eq. 7-85 (Chapter 7).

8.4.4. Pipes and Tubes

Hadjian and Tang (1988a, b) have compiled extensive data on damping of power plant piping for application to seismic analysis. The piping in their studies ranged from 1 to 18 in (25 to 450 mm) in diameter. Their results versus frequency are shown in Figure 8-18. As a result of regression analysis of these data, they recommend the following damping factor for use in the analysis of power plant piping:

$$\zeta = 0.0053 + 0.0024D + 0.0166R + 0.009F - 0.019L, \qquad (8\text{-}43)$$

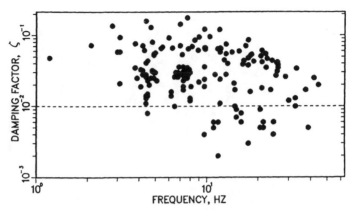

Fig. 8-18 Damping of power plant piping (Hadjian and Tang, 1988).

where D is pipe diameter in inches, R (response level) is 0 if there is no yielding and 1 if the amplitude is sufficient to cause yielding, F (first mode) is 1 in the fundamental mode and 0 in higher modes, and L (online equipment) is 0 if the piping is relatively uniform and 1 if relatively massive valves or other equipment is attached to the piping that adds mass but not damping. For example, for 1 in diameter uniform pipe at low amplitudes, the recommended value of damping ($D = 1$, $R = 0$, $L = 0$) is $\zeta = 0.0167$ in the first mode and $\zeta = 0.0074$ in higher modes. This is similar to but slightly higher than the pipeline damping recommended for heat-exchanger tubing in Section 8.3.3. Hadjian and Tang note that damping is higher if the piping is heavily insulated, such as piping used for liquid-metal nuclear design. Their recommendation for

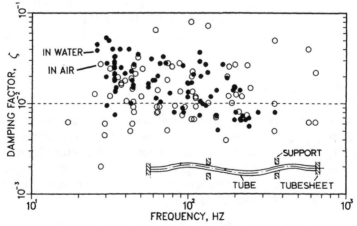

Fig. 8-19 Damping of multi-span heat-exchanger tubing (Pettigrew et al., 1986).

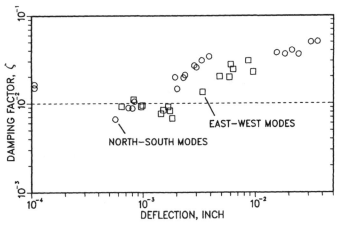

Fig. 8-20 Damping of a nuclear power plant steam generator as a function of amplitude of vibration (Hart and Ibanez, 1973).

heavily insulated piping is

$$\zeta = 0.0924 - 0.0047D - 0.022H + 0.043S, \qquad (8\text{-}44)$$

where H (higher mode) is 0 for the fundamental mode and 1 for higher modes, S (snubbers) is 1 if snubbers are on the pipeline and 0 if there are no snubbers, and D is the diameter of the pipe in inches. As noted in Section 8.3.3, the damping of piping is largely due to friction at joints and insulation. By removing any insulation and welding in the piping, this friction is eliminated and damping factors fall to values on the order of $\zeta = 0.001$ and less.

The damping values compiled by Pettigrew et al. (1986) for multispan heat-exchanger tubing are shown in Figure 8-19. Values in water are generally higher than the values measured in air. They note that damping increases with the width of the support plate, confirming the trend indicated in Figure 8-12. Damping of a nuclear power plant steam generator is presented in Figure 8-20 as a function of amplitude of vibration. Note that the damping increases strongly with vibration amplitude, indicating that local yielding is taking place. The United States Atomic Energy Commission (USAEC, 1973) recommends damping factors for earthquake excitation; also see Hart and Ibanez (1973) and Morrone (1974).

8.4.5. Aircraft and Spacecraft Structures

The exterior surfaces of aircraft are formed from stiffened panels. There are two basic designs: (1) stringer-stiffened panels where sheet metal

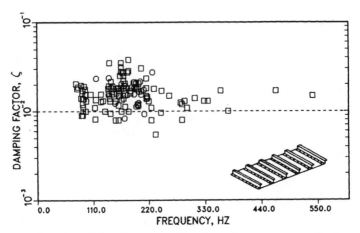

Fig. 8-21 Damping of aircraft stringer-stiffened panels: open squares, aluminum panel; open circles, titanium panel (Schneider, 1974; Ballentine et al., 1968; Rudder, 1972).

stiffeners are riveted to the interior surface of a thin skin and (2) honeycomb sandwich panels where two thin skins (face sheets) are adhesively bonded to opposite sides of a lightweight honeycomb core. Ungar (1973) and Soovere and Drake (1985) have shown that the principal damping mechanisms for panels are (1) friction at joints, (2) "gas pumping," also called squeeze film damping (Chapter 2, Section 2.3), as viscous air is forced across a flexing joint, (3) radiation damping (Section 8.2.1), and (4) material damping. Figure 8-21 shows damping of typical aircraft aluminum and titanium skin stringer panels, and Figure 8-22 shows honeycomb panels fabricated with aluminum, graphite/epoxy,

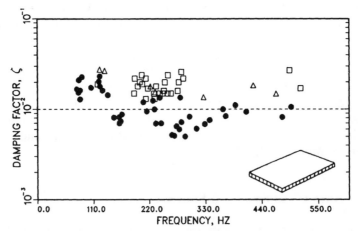

Fig. 8-22 Damping of aircraft honeycomb sandwich panels: open squares, aluminum skins; solid circles, graphite/epoxy skins; open triangles, Kevlar skins (Ballentine et al., 1968; Soovere and Drake, 1985).

and Kevlar face sheets. The majority of these data are for the fundamental mode.

The aluminum skin stringer panels of Figure 8-21 are fabricated from alloy 7075-T6 with riveted Zee, Tee, and Channel section stiffeners. The skins vary in thickness from 0.020 to 0.1 in (0.5 to 3.5 mm) and the stiffeners extend a maximum of 2.90 in (65 mm) off the skin. The average damping of the aluminum panels is $\zeta = 0.0164$. The titanium panels in Figure 8-21 are fabricated from 6 Al-4V alloy titanium. The skins vary in thickness from 0.032 to 0.050 in (0.8 to 1.3 mm). The Zee and Channel section stiffeners extend a maximum of 2.9 in (65 mm) off the skin. The average damping of the titanium panels is $\zeta = 0.0168$. The average damping of all the panels in Figure 8-20 is $\zeta = 0.0165$, which is not significantly different from the value $\zeta = 0.017$ suggested by AGARD (Thomson, 1972) for sonic fatigue analysis of aircraft skin panels. Giavotto et al. (1979) provide additional data on damping of aircraft skin panels.

The average damping of the honeycomb sandwich panels of Figure 8-22 is $\zeta = 0.0145$. The aluminum honeycomb panels have skin thickness between 0.010 and 0.025 in (0.25 and 0.63 mm) with an overall depth of 0.27 to 0.75 in (6.9 to 19 mm). The average damping of the graphite/epoxy honeycomb panels is $\zeta = 0.0111$, which is significantly lower than either the average damping of the Kevlar honeycomb panels, $\zeta = 0.0193$, or the average damping of the aluminum honeycomb panels, $\zeta = 0.0186$.

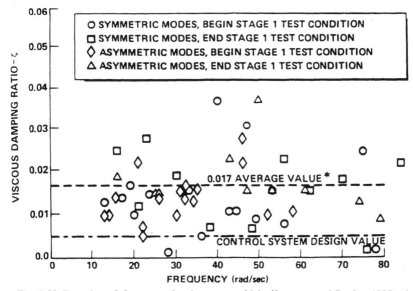

Fig. 8-23 Damping of the space shuttle ascent vehicle (Soovere and Drake, 1985, after Jensen, 1984).

Remarkably, the damping factors of entire missiles and satellites are in the same range as those of the panels (Soovere and Drake, 1985; Rogers, 1988). Figure 8-23 shows the damping of the space shuttle ascent vehicle as measured in vibration tests. The average damping of this large manned missile is $\zeta = 0.017$.

8.5. MATERIAL DAMPING AND DAMPERS

8.5.1. Material Damping

As a material deforms under load, energy is stored in the structure. If the material is perfectly elastic, the strain ϵ in the material is proportional to the applied stress $\sigma = \epsilon/E$, where E is the modulus of elasticity, and the stored energy is equal to the area under the stress strain curve,

$$\text{Stored energy} = \int_{\epsilon=0}^{\epsilon=\epsilon_1} \sigma \, d\epsilon = \frac{\epsilon_1 \sigma_1}{2}. \tag{8-45}$$

As the applied stress σ_1 decreases, the stored energy is released and the deformation returns to zero.

Real materials are not perfectly elastic. Heat is generated by compression and plastic flow. Small tensile stresses produce a slight cooling. Because some energy is lost during loading, a net deformation remains at zero load as shown by the loading path ABC in Figure 8-24. As the loading cycle continues, a hysteresis loop is traced out on the stress–strain plot. The area contained within the loop is the energy expended per cycle in deforming the material.

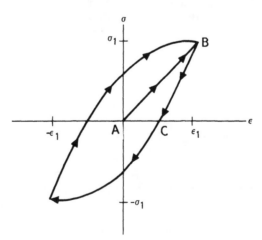

Fig. 8-24 A material hysteresis loop.

The energy expended per cycle is computed from the time-history of stress and strain over one cycle,

$$\text{Dissipated energy} = \oint \sigma \, d\epsilon = \int_0^T \sigma\left(\frac{d\epsilon}{dt}\right) dt. \qquad (8\text{-}46)$$

For an elastic material, stress and strain are in phase and $D = 0$. A viscoelastic material deviates slightly from perfect elasticity because a component of stress lags strain:

$$\sigma = E\left(\epsilon + \frac{\eta}{|\omega|}\frac{d\epsilon}{dt}\right) = E(1 + i\eta). \qquad (8\text{-}47)$$

In the second form of Eq. 8-47, the symbol i denotes the complex constant and complex mathematics is used to highlight the fact that the component of stress proportional to η lags the elastic component $E\epsilon$ by 90 degrees. For the cyclic loading $\epsilon = \epsilon_0 \sin \omega t$ of a viscoelastic material, the dissipated energy per cycle,

$$\text{Dissipated energy} = \pi \eta E \epsilon_0^2, \qquad (8\text{-}48)$$

is proportional to the loss factor η. Taking the ratio of Eqs. 8-48 and 8-45, using $\epsilon_1 = \epsilon_0$, and using $\sigma_1 = E\epsilon_0$,

$$\frac{\text{Dissipated energy}}{\text{Stored energy}} = 2\pi\eta. \qquad (8\text{-}49)$$

Comparing this equation with Eq. 8-5, we see

$$\zeta = \frac{\eta}{2}. \qquad (8\text{-}50)$$

The damping factor (ζ) for homogeneous structures such as Figure 8-25(a) is one-half the material loss factor.

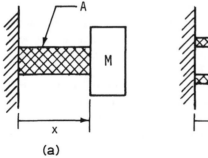 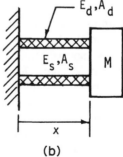

Fig. 8-25 One-dimensional structures with damping material: (a) damping material spring; (b) composite spring.

Table 8-3 Damping of materials

Material	Minimum damping factor	Maximum damping factor	Typical damping factor	Reference
Aluminum, 6063-T6	0.0005	0.005	0.001	Lazan (1968)
Brass	0.002	0.004	0.003	Lazan (1968)
Iron, pure	0.001	0.01	0.005	Lazan (1968)
SAE 1020 steel	0.0004	0.002	0.001	Lazan (1968)
Titanium, pure	0.001	0.05	0.005	Lazan (1968)
Natural rubber	0.01	0.08	0.05	Frye (1988)
Chloroprene rubber	0.03	0.08	0.05	Frye (1988)
Butyl rubber	0.05	0.50	0.2	Frye (1988)
Wood, sitka spruce	—	—	0.006	Zeeuw (1967)
Steel, low alloy	0.00036	0.0015	—	Raggett (1975)
Concrete	0.0018	0.0026	—	Raggett (1975)
Prestressed concrete beams	0.005	0.02	0.01	Raggett (1975)
Gypsum board partitions	0.07	0.40	0.14	Raggett (1975)

Loss factor is a material property that is a function of temperature, frequency, and amplitude. Material damping of several materials is given in Table 8-3 for vibration amplitudes much less than that which produces yield. As the table shows, the damping of the structural materials, such as steel, is small at low amplitudes; much higher damping can be achieved in plastic flow by strains above yield. Rubber and viscoelastic materials have high damping. Figure 8-26 shows the modulus of elasticity and loss factor for a typical viscoelastic damping material as a function of temperature and frequency. Temperature shifts as small as 30°F (15°C) can produce factor of 2 changes in damping, whereas order of magnitude changes in frequency are required to produce the same effect. The American Society for Testing and Materials Standard E756-83 gives standard methods for measurement of vibration properties of damping materials. Nashif et al. (1985), Goodman (1988), Lazan (1968), Bert (1973), Jones (1988), Vacca and Ely (1987), Yildiz and Stevens (1985), and Soovere and Drake (1985) present examples and data for damping of structural and viscoelastic materials. Commercial sources for damping materials can be found in the annual buyer's source issue of *Sound and Vibration,* Acoustical Publications, Bay Village, Ohio.

Two alternatives for reducing vibration of a structure are (1) to increase stiffness by adding structural material and (2) to increase damping. To assess the trade-offs involved in increased damping versus increased structure, consider the single-degree-of-freedom system shown in Figure 8-25(b) whose two-component elastic support consists of a structural material and a damping material.

The structural material, denoted by the subscript s, is assumed to be

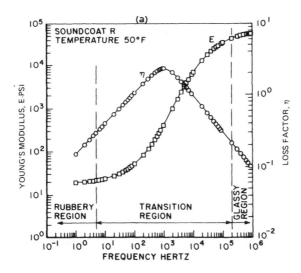

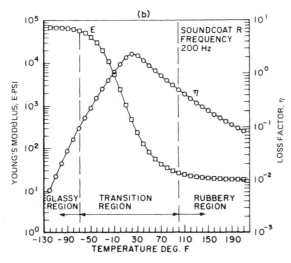

Fig. 8-26 Elastic modulus and loss factor of a viscoelastic damping material (a) versus frequency at constant temperature and (b) versus temperature at constant frequency (Soovere and Drake, 1985).

associated with a structural damping factor ζ_0 due to friction, radiation damping, and losses at joints. The damping material, denoted by the subscript d, has a material loss factor η_d and a modulus E_d. The modulus of the damping material is assumed to be much less than the modulus of the structural material, $E_s \gg E_d$, as is ordinarily the case for structural materials such as steel, aluminum, and concrete, and for rubberlike damping materials. With these assumptions, using Eq. 8-47 with $\epsilon = x/L$, the equation of free vibration of the system of Figure 8-25(b) is

$$M\ddot{x} + \left(\frac{E_d A_d \eta_d}{L\omega_n} + 2M\zeta_0\omega_n\right)\dot{x} + \frac{E_s A_s}{L}x = 0. \tag{8-51}$$

The stiffness is provided by the structural material, which has cross section A_s. The mass of both materials is neglected in comparison with the supported mass M. Comparing this equation with Eq. 8-28 and the previous equations, we obtain the total damping factor,

$$\zeta_t = \zeta_0 + \frac{E_d A_d \eta_d}{2E_s A_s}, \tag{8-52}$$

the sum of structural and material damping.

The response of the structure to external sinusoidal excitation $F_0 \sin \omega t$ is a function of both stiffness and damping (Eqs. 8-33 and 8-34). If the excitation occurs at the natural frequency, then the response is inversely proportional to the product of stiffness and damping (Eq. 8-34 for $\omega = \omega_n$, or Eq. 8-42),

$$x(t) = (F_0/2\zeta_t k) \cos \omega t. \tag{8-53}$$

The response is minimized by maximizing damping times stiffness. Using $k = F_s A_s/L$ and 8-53, the response is minimized by maximizing

$$\zeta_t k = (A_s E_s \zeta_0 + \tfrac{1}{2}A_d E_d \eta_d)/L. \tag{8-54}$$

If $E_s\zeta_0 > E_d\eta_d/2$, it is more efficient to increase the cross section of structural material (A_s) to minimize response; otherwise, it is more efficient to increase the damping material (A_d). The typical loss factor of damping materials (η_d) is two orders of magnitude greater than the typical structural damping factor (ζ_0), but the modulus of elasticity of damping materials (E_d) is typically three orders of magnitude less than that of structural materials (E_s). As a result, adding damping material is an efficient method of reducing resonant response only if the initial structure damping is low, that is, less than $\zeta \approx 0.01$.

Damping material contributes strongly to the overall damping of a structure only if the strain energy in the damping material during deformation is a significant percentage, say 20%, of the total strain energy of the structure. Because the modulus of damping materials is low, this implies that significant amounts of damping material are required. The greatest damping for a given amount of damping material can be obtained by using constrained layer or tuned damper designs as discussed in Section 8.5.2. Adhesively backed damping layers ("damping tape") can be relatively easily added to an existing structure where it is difficult and expensive to increase stiffness by adding structural material.

Exercises

1. Consider the bimaterial oscillator of Figure 8-25(b). Consider aluminum ($E = 10 \times 10^6$ psi, 69×10^9 Pa) to be the structural material and the damping material of Figure 8-26 to be the damping material. It is desired to reduce the maximum response to a sinusoidal force acting at the oscillator natural frequency. At what value of structural damping (ζ_0 without damping material) is it more efficient to add damping material (A_d) rather than add structural material (A_s)? Assume that the mass M is fixed and the exciting frequency equals the natural frequency.

2. Consider broad-band forcing of a structure. The response is proportional to $\zeta^{-1/2}k^{-3/2}$ (see Chapter 7). Is adding damping a more effective method of reducing broad-band response than reducing sinusoidal (narrow-band) resonant response?

8.5.2. Dampers

Damper designs are shown in Figures 8-27 through 8-29. Other dampers are shown in Chapter 4, Figure 4-19. The dampers of Figure 8-27(a) through (g) rely on damping materials. Viscoelastic materials such as polymers, rubber, and rubberlike materials provide the highest energy dissipation per volume, but their application is limited to about 400°F (200°C). At higher temperatures, damping enamel coatings, knitted metal mesh, crushed metal fibers, and loose materials such as sand (Sun et al., 1986) are used. Clever designs incorporating these materials are often very effective in suppressing structural vibration.

The free damping layer, constrained layer damping, and multiple constrained layer dampers of Figure 8-27 are used to damp bending modes of plates and shells. They are most effective if the damping material thickness is comparable to the plate thickness in order to achieve a relatively high strain energy in the damping material. The constrained layer is 2 to 3 times more efficient than the free layer because the constraining layer maximizes shear in the damping material. Multiple shear layers with varying materials broaden the temperature range over which high damping is achieved. Application of the shear damper of Figure 8-27(d) to diagonal bracing of a building is shown in Figure 8-29.

Tuned-mass dampers (Figure 8-27(g) and Figure 8-28) are most effective when located at points of high acceleration. It is theoretically possible to completely suppress a given mode with a tuned damper if the frequency of the tuned damper is carefully adjusted to the frequency of

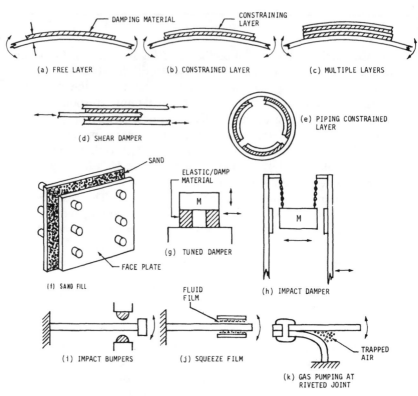

Fig. 8-27 Various damper designs.

the mode to be damped (Den Hartog, 1985; Harris, 1988). Practically speaking, the damping factor produced by a tuned mass damper is roughly equal to the ratio of the damper mass to the structure mass under conditions where tuning can be achieved and is about one-half this ratio with limited tuning (Vickery et al., 1983). Thus, a 5% mass-tuned damper could produce $\zeta \approx 0.025$ in industrial application. Tuned dampers with mass ratios from 1% to 20% have been utilized; 2% to 5% tuned damper mass is typical.

Figures 8-28 and 8-29 show application of dampers to tall buildings. The motivation for these building dampers is to (1) improve the comfort of occupants in the upper stories during high winds (Chapter 7, Section 7.4), (2) protect deformation sensitive details, such as window retention, and (3) ensure against the risk of excessively low damping (note scatter in Figs. 8-15 through 8-18). The tuned-mass damper in the 919 ft (280 m) tall Citicorp Center in New York City consists of a 410 ton concrete block 30 ft (10 m) on a side that slides on oil-film bearings and is positioned

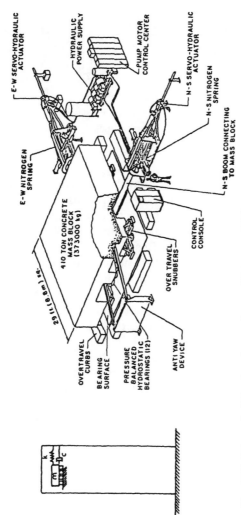

Fig. 8-28 Citicorp building tuned-mass damper (Peterson, 1980).

343

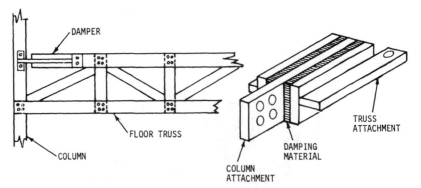

Fig. 8-29 Viscoelastic shear dampers installed in diagonal bracing of building framework.

by pneumatic springs and actively controlled hydraulic actuators (McNamara, 1977; Peterson, 1980). The mass is 2% of the generalized building mass. Its builders project a damping factor equivalent to $\zeta = 0.04$. The 300 ton tuned-mass system installed in the 790 ft (240 m) tall John Hancock Tower in Boston is equal to 1.4% of the building generalized mass in sway and 2.1% in torsion. Tuned-mass dampers have been installed in the CN Tower in Toronto, the Sydney Tower in Sydney, Australia, and the Bronx Whitestone Bridge in New York City. The combination of a 163 ton water tank and a 36 ton tuned-mass damper in the 820 ft (250 m) Sydney Tower increased damping factor by $\Delta \zeta = 0.005$ in the first mode and $\Delta \zeta = 0.011$ in the second mode (Kwok, 1984). Viscoelastic shear dampers applied to building framework structure are shown in Figure 8-29. Hundreds of these dampers are installed in the World Trade Center in New York City and the Columbia Center in Seattle (Keel and Mahmoodi, 1986). Wiesner (1988), Hrovat et al. (1983), Mahmoodi (1969), and McNamara (1977) review building damping systems.

Exercise

1. Consider a steel tube 40 in (102 cm) long with an outside diameter of 0.3 in (7.6 mm) and an inside diameter of 0.28 in (6.6 m). The density of steel is 0.3 lb/in³ (8300 kg/m³). The tube is empty. Its fundamental natural frequency is calculated to be 19.4 Hz if the ends are pinned and 44.1 Hz if the ends are considered to be clamped. Estimate the damping factor of the fundamental mode of the tube for low-amplitude vibration with

pinned ends due to (a) fluid damping of the surrounding air ($\rho = 0.000434\,\text{lb/in}^3$, $1.2\,\text{kg/m}^3$; $\nu = 0.0236\,\text{in}^2/\text{sec}$, 1.5×10^{-5} m^2/sec) using Eq. 8-11, (b) the material damping factor (Table 8-2), and (c) structural damping due to scraping and friction at its ends held in mechanical joints (Section 8.3.3 and Section 8.4.4). What is the total damping? If the ends of the rod are now welded to relatively massive plates, what is the total damping factor?

REFERENCES

Axisa, F., et al. (1988) "Two-Phase Cross-Flow Damping in Tube Arrays," in *Damping— 1988*, PVP—Vol. 133, ASME, New York.

Ballentine, J. R., et al. (1968) "Refinement of Sonic Fatigue Structural Design Criteria," Technical Report AFFDL-TR-67-156, AD83118, Air Force Dynamics Laboratory, Wright-Patterson Air Force Base, Ohio.

Batchelor, G. K. (1967) *An Introduction to Fluid Dynamics*, Cambridge University Press, Cambridge, pp. 356–357.

Bert, C. W. (1973) "Material Damping: An Introductory Review of Mathematical Models, Measures, and Experimental Techniques," *Journal of Sound and Vibration*, **29**, 129–153.

Blevins, R. D. (1975) "Vibration of a Loosely Held Tube," *Journal of Engineering for Industry*, **97**, 1301–1304.

——— (1989) "Application of the Discrete Vortex Method to Fluid Structure Interaction," in *Flow-Induced Vibration—1989*, M. K. Au-Yang, ed., Vol. 154, ASME, New York, pp. 131–140.

Bouwkamp, J. G. (1974) "Dynamics of Full Scale Structures," in *Applied Mechanics in Earthquake Engineering*, W. D. Iwan, ed., ASME, New York, 1974, p. 123.

Brouwers, J. J., and T. E. Meijssen (1985) "Viscous Damping Forces on Oscillating Cylinders," *Applied Ocean Research*, **7**, 18–123.

Brownjohn, M. W., et al. (1987) "Ambient Vibration Measurement of the Humer Suspension Bridge and Comparison with Calculated Characteristics," *Proceedings of the Institution of Civil Engineers, Part 2*, **83**, 561–600.

Caughey, T. K. (1960) "Classical Normal Modes in Damped Linear Systems," *Journal of Applied Mechanics*, **27**, 269–271.

Chen, S. S. (1981) "Fluid Damping for Circular Cylindrical Structures," *Nuclear Engineering and Design*, **63**, 81–100.

Chen, S. S., and J. A. Jendrzejczyk (1979) "Dynamic Response of a Circular Cylinder Subjected to Liquid Cross Flow," *Journal of Pressure Vessel Technology*, **101**, 106–112.

Chen, S. S., M. W. Wambsganss, and J. A. Jendrzejczyk (1976) "Added Mass and Damping of Vibrating Rod in Confined Viscous Fluid," *Journal of Applied Mechanics*, **43**, 325–329.

Chilukuri, R. (1987) "Added Mass and Damping for Cylinder Vibrations Within a Confined Fluid Using Deforming Finite Elements," *Journal of Fluids Engineering*, **109**, 283–288.

Chow, L. C., and R. J. Pinnington (1989) "Practical Industrial Method of Increasing Damping in Machinery, I: Squeeze Film Damping in Air, II: Squeeze Film Damping in Liquids," *Journal of Sound and Vibration*, **118**, 123–139 (1987), **128**, 333–348.

Connors, H. J., S. J. Savorelli, and F. A. Kramer (1982) "Hydrodynamic Damping of Rod Bundles in Axial Flow," in *Flow-Induced Vibration of Circular Cylindrical Structures— 1982*, PVP—Vol. 63, ASME, New York, pp. 109–124.

Czarnecki, R. M. (1974) "Dynamic Testing of Buildings Using Man-Induced Vibration," *Sound and Vibration,* **8(10)**, 18–21.

Davenport, A. G. (1981) "Reliability of Long Span Bridges under Wind Loading," in *Structural Safety and Reliability, Developments in Civil Engineering,* **4,** Proceedings of the ICOSSAR '81 3rd International Conference on Structural Safety and Reliability, Trondheim, Norway, Elsevier Scientific Publishing Company, Amsterdam.

Davenport, A. G., and P. Hill-Carroll (1986) "Damping in Tall Buildings: Its Variability and Treatment in Design," in *Building Motion in Wind,* N. Isyumov et al., eds., Proceeding of ASCE Convention in Seattle, Wash., ASCE, New York, pp. 42–57.

Den Hartog, J. P. (1985) *Mechanical Vibrations,* 4th ed., reprint, Dover, New York.

Devin, C. (1959) "Survey of Thermal, Radiation, and Viscous Damping of Pulsating Air Bubbles in Water," *Journal of the Acoustical Society of America,* **31,** 1654–1667.

Dong, R. G. (1979) "Size Effect in Damping Caused by Water Submersion," *ASCE Journal of the Structural Division,* **105,** 847–857.

Fabunmi, J., P. Chang, and J. Vorwald (1988) "Damping Matrix-Identification using the Spectral Bases Technique," *Journal of Vibration, Acoustics and Reliability in Design,* **110,** 332–337.

Fahy, F. (1985) *Sound and Structural Vibration,* Academic Press, London.

Frye, W. A. (1988) "Properties of Rubber," in *Shock and Vibration Handbook,* 3d ed., C. M. Harris, ed., McGraw-Hill, New York, p. 35–13.

Giavotto, V., et al. (1988) "Damping Problems in Acoustic Fatigue," in *AGARD Conference Proceedings No. 277, Damping Effects in Aerospace Structures,* AGARD-CP-277, AD-A080 451, October.

Goodman, L. E. (1988) "Material Damping and Slip Damping," in *Shock and Vibration Handbook,* 3d ed., C. M. Harris, ed., McGraw-Hill, New York, pp. 36–1–36-28.

Hadjian, A. H., and H. T. Tang (1988a) "Piping System Damping Evaluation," Final Report EPRI NP-6035, Electric Power Research Institute, Palo Alto, Calif.

—— (1986b) "Identification of the Significant Parameters Affecting Damping in Piping Systems," in *Damping—1988,* F. Hara, ed., PVP—Vol. 133, ASME, New York, pp. 107–112.

Hara, F. (1988) "Two-Phase Damping in a Vibrating Circular Structure," in *Damping—1988,* PVP—Vol. 133, ASME, New York, 1988.

Harris, C. M. (1988) *Shock and Vibration Handbook,* 3d ed., McGraw-Hill, New York.

Hart, G. C., and P. Ibanez (1973) "Experimental Determination of Damping in Nuclear Power Plant Structures and Equipment," *Nuclear Engineering and Design,* **25,** 112–125.

Hart, G. C., and R. Vasudevan (1975) "Earthquake Design of Buildings: Damping," *ASCE Journal of the Structural Division,* **101,** 11–30.

Hart, G. C., R. M. DiJulio, and M. Lew (1975) "Torsional Response of High Rise Buildings," *ASCE Journal of the Structural Division,* **101,** 397–416.

Hill-Carroll, P. E. (1985) "The Prediction of Structural Damping Values and Their Coefficients of Variation," thesis for M.S., University of Western Ontario, London, Canada.

Hrovat, D., P. Barak, and M. Rabins (1983) "Semi-Active versus Passive or Active Mass Dampers for Structural Control," *ASME Journal of Engineering Mechanics,* **109,** 691–705.

Ito, M., T. Katayama, and T. Nakazono (1973) "Some Empirical Facts on Damping of Bridges," *Report 7240 Presented at the Symposium on Resistance and Ultimate Deformability of Structures Acted on by Well Defined Loads,* Lisbon, September.

Iwan, W. D. (1965) "A Distributed Element Model for Hysteresis and Its Steady-State Dynamic Response," *Journal of Applied Mechanics,* **32,** 921–925.

Jeary, A. P. (1988) "Damping in Tall Buildings," in *Second Century of the Skyscraper,* L. S. Beedle, ed., Van Nostrand Reinhold, New York, pp. 779–788.

Jensen, D. L. (1984) "Structural Damping of the Space Shuttle Orbiter and Ascent

Vehicle," Technical Report AFWAL-TR-84-3064, Wright-Patterson Air Force Base, Ohio.

Jones, D. I. G. (1988) "Application of Damping Treatments," in *Shock and Vibration Handbook*, 3d ed., C. M. Harris, ed., McGraw-Hill, New York, pp. 37-1–37-34.

Junger, M. C., and D. Feit (1986) *Sound, Structures, and Their Interaction*, 2d ed., MIT Press, pp. 106–149.

Keel, C. J., and P. Mahmoodi (1986) "Design of Viscoelastic Dampers for Columbia Center Building," in *Building Motion in Wind*, N. Isyumov et al., eds., ASME, New York, pp. 66–81.

Kwok, K. C. (1984) "Damping Increase in Building with Tuned Mass Damper," *ASCE Journal of Engineering Mechanics*, **110**, 1645–1648.

Lamb, H. (1945) *Hydrodynamics*, 6th ed., Dover, New York.

Lazan, B. J. (1968) *Damping of Materials and Members in Structural Mechanics*, Pergamon Press, New York.

Littler, J. D., and B. R. Ellis (1987) "Ambient Vibration Measurements of the Humber Bridge," Paper F3, in *Proceedings of the International Conference on Flow-Induced Vibrations*, BHRA, Cranfield, England, pp. 259–266.

McNamara, R. J. (1977) "Tuned Mass Dampers for Buildings," *ASCE Journal of the Structural Division*, **103**, 1785–1798.

Mahmoodi, P. (1969) "Structural Dampers," *ASCE Journal of the Structural Division*, **95**, 1661–1672.

Morrone, A. (1974) "Damping Values of Nuclear Power Plants," *Nuclear Engineering and Design*, **26**, 343–363.

Morse, P. M., and K. U. Ingard (1968) *Theoretical Acoustics*, McGraw-Hill, New York.

Mulcahy, T. M., and A. J. Miskevics (1980) "Determination of Velocity-Squared Fluid Damping by Resonant Structural Testing," *Journal of Sound and Vibration*, **71**, 555–564.

Nashif, A. D., D. I. Jones, and J. P. Henderson (1985) *Vibration Damping*, Wiley-Interscience, New York.

Paidoussis, M. P. (1966) "Dynamics of Flexible Slender Cylinders in Axial Flow, Part 1: Theory, Part 2: Experiments," *Journal of Fluid Mechanics*, **26**, 717–751.

Pattani, P. G., and M. D. Olson (1988) "Forces on Oscillating Bodies in Viscous Fluid," *International Journal for Numerical Methods in Fluids*, **8**, 519–536.

Peterson, N. R. (1980) "Design of Large Scale Tuned Mass Dampers," in *Structural Control*, H. H. E. Leipholz, ed., North-Holland Publishing Company, Netherlands, pp. 581–596.

Pettigrew, M. J., et al. (1986) "Damping of Multi-span Heat Exchanger Tubes," in *Flow-Induced Vibration—1986*, S. S. Chen, ed., PVP-104, ASME, New York, pp. 87–98.

Pettigrew, M. J., et al. (1988) "Vibration of Tube Bundles in Two-Phase Cross Flow: Hydrodynamic Mass and Damping" in *1988 International Symposium on Flow-Induced Vibration and Noise*, M. P. Paidoussis, ed., Vol. 2, American Society of Mechanical Engineers, New York, pp. 79–104.

Raggett, J. D. (1975) "Estimating Damping of Real Structures," *ASCE Journal of the Structural Division*, **101**, 1823–1835.

Ramberg, S. E., and O. M. Griffin (1977) "Free Vibrations of Taut and Slack Marine Cables," *ASCE Journal of the Structural Division*, **103**, 2079–2092.

Rea, D., R. W. Clough, and J. G. Bouwkamp (1969) "Damping Capacity of a Model Steel Structure," *Earthquake Engineering Research Center, University of California, Berkeley, Report EERC 69-14*.

Rogers, L., ed. (1988) *The Role of Damping in Vibration and Noise Control*, DE-Vol. 5, ASME, New York.

Rogers, R. J., C. Taylor, and M. J. Pettigrew (1984) "Fluid Effects on Multi-span Heat Exchanger Tube Vibration," ASME–PVP Conference, San Antonio, Tex.

Rosenhead, L. ed. (1988) *Laminar Boundary Layers,* Oxford University Press, Oxford, 1963, pp. 390–393. Reprinted by Dover, New York.

Rudder, F. F. (1972) "Acoustical Fatigue of Aircraft Structural Component Assemblies," Technical Report AFFDL-TR-71-107, AD893427, Wright-Patterson Air Force Base, Ohio.

Ruzicka, J. E., and T. F. Derby (1971) *Influence of Damping in Vibration Isolation,* U.S. Department of Defense, Shock and Vibration Laboratory, Code 6020, Washington, D.C.

Schneider, C. W. (1974) "Acoustic Fatigue of Aircraft Structures at Elevated Temperatures," Technical Report AFFDL-TR-73-155, AD929282, Air Force Flight Dynamics Laboratory, Wright-Patterson Air Force Base, Ohio.

Scruton, C., and A. R. Flint (1964) "Wind-Excited Oscillations of Structures," *Proceedings of the Institution of Civil Engineers,* **27,** 673–702.

Siegel, S. (1956) *Nonparametric Statistics,* McGraw-Hill, New York.

Skop, R. A., S. E. Ramberg, and K. M. Ferer (1976) "Added Mass and Damping Forces on Circular Cylinders," NRL Report 7970, Naval Research Laboratory, Washington, D.C.

Soovere, J., and M. L. Drake (1985) "Aerospace Structures Technology Damping Design Guide," Report AFWAL-TR-84-3089, Flight Dynamics Laboratory, Wright-Patterson Air Force Base, Ohio.

Stephens, D. G., and M. A. Scavullo (1965) "Investigation of Air Damping of Circular and Rectangular Plates, a Cylinder and a Sphere," NASA TN D-1865, Langley Research Center, Hampton, VA.

Stokes, G. G. (1843) "On Some Cases of Fluid Motion," *Proceedings of the Cambridge Philosophical Society,* **8,** 105–137.

Sun, J. C., et al. (1986) "Prediction of Total Loss Factors of Structures, Part II, Loss Factors of Sand Filled Structure," *Journal of Sound and Vibration,* **104,** 243–257.

Tanaka, M., et al. (1972) "Parallel Flow Induced Damping of PWR Fuel Assembly," in *Damping—1988,* F. Hara, ed., PVP-Vol. 133, ASME, New York, 1988, pp. 121–125.

Thomson, A. G. R. (1972) "Acoustic Fatigue Design Data," Part I, AGARD-AG-162, North Atlantic Treaty Organization, London.

Thomson, W. T. (1988) *Theory of Vibration with Applications,* 3d ed., Prentice-Hall, Englewood Cliffs, N.J.

Ungar, E. E. (1973) "The Status of Engineering Knowledge Concerning the Damping of Built-up Structures," *Journal of Sound and Vibration,* **26,** 141–154.

USAEC Regulatory Guide 1.61 (1973) "Damping Values for Seismic Design of Nuclear Power Plants," October 24.

Vacca, S. N., and R. A. Ely (1987) "Structural Improvement of Operational Aircraft Program," Air Force Wright Aeronautical Laboratory, AFWAL-TR-87-3029, Ohio, prepared by LTV Aircraft Products Group.

Vickery, B. J., N. Isyumov, and A. G. Davenport (1983) "The Role of Damping, Mass and Stiffness in the Reduction of Wind Effects on Structures," *Journal of Wind Engineering and Industrial Aerodynamics,* **11,** 285–294.

Wiesner, K. B. (1988) "The Role of Damping Systems," in *Second Century of the Skyscraper,* L. S. Beedle, ed., Van Nostrand Reinhold, New York, pp. 789–802.

Yang, C. I., and T. J. Moran (1979) "Calculations of Added Mass and Damping Coefficients from Hexagonal Cylinders in a Confined Viscous Fluid," in *Flow-Induced Vibrations,* S. S. Chen, ed., ASME, New York, pp. 97–104.

Yeh, T. T., and S. S. Chen (1978) "The Effect of Fluid Viscosity on Coupled Tube/Fluid Vibrations," *Journal of Sound and Vibration,* **59,** 453–467.

Yildiz, A., and K. Stevens (1985) "Optimum Thickness Distribution of Unconstrained Viscoelastic Damping Layer Treatments for Plates," *Journal of Sound and Vibration,* **103,** 183–199.

Zeeuw, C. H. (1967) "Wood," in *Standard Handbook for Mechanical Engineers,* 7th ed., T. Baumeister, ed., McGraw-Hill, New York, pp. 6–157.

Chapter 9

Sound Induced by Vortex Shedding

Aeroacoustic tones are the sound produced by fluid flow over cylinders and bluff objects. They are heard in the "singing" of powerlines and the sounds of the Aeolian harp and the flute. They often make their unwanted appearance in the compressors of aircraft engines, heat exchangers, and marine hydrophone cables. These sounds have been variously called Aeolian tones, friction tones (Reibungstonen), striking tones (Hiebtonen), edge tones, aeroacoustic tones, and sound produced by vortex shedding.

The sound is associated with periodic vortex shedding. Since the vortices often excite vibration, the fluid dynamics, structural dynamics, and acoustics are intertwined in the generation of sound from flexible structures. This chapter reviews the measurements and theoretical predictions for sound generated by steady low-Mach-number flow over cylinders and tube arrays. The allied problems of structural response, duct acoustics, and sound generated by flow over cavities and noncylindrical structures are also considered. Goldstein (1976), Blake (1986), Muller (1979), and Fahy (1985) review aerocoustics. Vortex shedding is reviewed in Chapter 3.

9.1. SOUND FROM SINGLE CYLINDERS

9.1.1. Experiments

Although Aeolian tones have long been associated with musical instruments and toys (Richardson, 1923), the first quantitative measurements on sound produced by a cylinder in an airstream were reported in 1878 by the Czechoslovakian scientist Strouhal (1878; also see Horak, 1977). His apparatus consisted of wires stretched between radial arms from a rotating shaft. Typically the wires were copper, between 0.07 and 0.33 in (0.18 and 8.5 mm) in diameter and 1.6 ft (0.49 m) in length. His hand-driven apparatus was similar to the motor-driven apparatus of Relf (1921) shown in Figure 9-1.

349

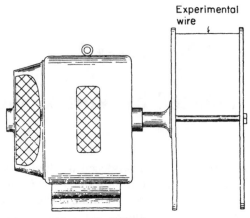

Fig. 9-1 Motor-driven "whirler" of Relf (1921).

When the air velocity U over the wires exceeded about 15 ft/sec (5 m/sec), Strouhal heard audible tones with frequencies of 600 Hz at low velocity and 3000 Hz at high velocity. The frequency of the tone was measured by increasing or decreasing the shaft rotation speed until the pitch matched a reference string (sonometer). Strouhal found that the frequency of the sound was independent of wire tension or length, although the intensity did increase with wire length, and the frequency was approximately predicted by the relationship

$$f = \frac{SU}{D} \quad \text{Hz,} \tag{9-1}$$

where U is the velocity, D is the wire diameter, and S, which we call Strouhal number, varied between 0.156 for the smaller-diameter wires and 0.205 for the larger-diameter wires. Strouhal felt that the tones were the result of friction of wind against the wire, so he called them friction tones (Reibungstonen).

In 1879 Lord Rayleigh observed the vibration and sound of a wire in a chimney draft (see Rayleigh, 1889, 1945). He generally confirmed Strouhal's observations, but he noted that wire motion was predominantly perpendicular rather than parallel to the draft as implied by Strouhal's friction hypothesis. He also postulated that Strouhal number was a function of Reynolds number. The observation of the staggered vortex street by von Karman and Rubach (1912) and Bernard (1908) led von Kruger and Lauth (1914) and Rayleigh (1915) to associate the tones and vibration with periodic vortex shedding. In 1921, Relf confirmed that Strouhal number was a function of Reynolds number using both the motor-driven apparatus in Figure 9-1 and a cylinder in a water channel. Richardson (1923) found that there was a minimum

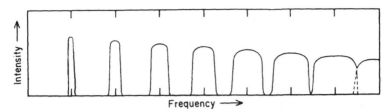

Fig. 9-2 Sound intensity from a wire in air with increasing air velocity observed by Richardson (1923).

Reynolds number for the onset of periodic vortex shedding, which he found to be $UD/v = 33$, and that the tones were most pronounced when the vortex shedding produced wire vibration as shown in Figure 9-2. Also see Chapter 3, Figure 3-29.

Stowell and Deming (1936) used a motor-driven whirler with 0.5 in (1.25 cm) diameter rods to generate sound, which they measured with an electronic microphone. They found that the sound decayed with the inverse square of distance from the whirler and the sound power (W) was proportional to the rod length to the first power and the rod velocity to the 5.5 power,

$$W \sim U^n L, \qquad (9-2)$$

where $n = 5.5$. The sound pressure about the rod falls in a double-lobed pattern, shown in Figure 9-3, with the maximum transverse to the flow. Von Hole (1938) found that n varied over the range $6 \leq n \leq 8$ for velocities over 100 ft/sec (30 m/sec) and 0.04 in (1 mm) and 0.8 in (20 mm) diameter wires. Yudin (1944) found $5.5 \leq n \leq 6$. Inspired by Lighthill's (1952) general theory of aerodynamic sound, Gerrard (1955), Phillips (1956), Etkin et al. (1957), and Keefe (1962) all attempted to

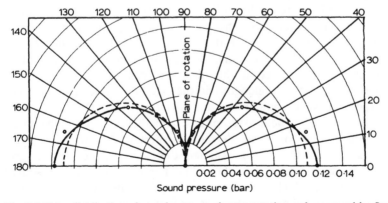

Fig. 9-3 Polar distribution of sound pressure about a rotation rod measured by Stowell and Deming (1936). Solid curve, observed; broken curve, computed.

quantitatively validate the theory using measured sound levels due to vortex shedding from stationary rods. All were frustrated to a degree by the spanwise randomness of vortex shedding at Reynolds numbers beyond 300, by the influence of cylinder vibration, and by test-rig details. Gerrard confirmed the double-lobed pressure pattern (Fig. 9-3). Phillips (1956) incorporated finite spanwise correlation in his theory and showed general agreement with the theoretical value $n = 6$.

Leehey and Hanson (1971) verified the theoretical prediction for aeroacoustic sound due to vortex shedding within 3 dB by measuring the sound radiated by a wire vibrating at low amplitude in a low-turbulence open-jet wind tunnel, simultaneously measuring the spanwise correlation by means of twin hot-wire probes, and deducing the oscillating forces from the wire vibration. They found that small changes in velocity produced substantial variations in lift coefficient and sound, as did Richardson (1923, 1958) and Koopman (1969), which suggests that there is strong fluid–structure coupling with vortex shedding from elastic cylinders. As discussed in Chapter 3, when the vortex shedding frequency approaches the frequency of a natural mode of an elastic cylinder or wire, the vortex shedding locks onto the structural mode and both the spanwise correlation and the magnitude of the vortex forces increase, increasing the tone and the magnitude of vibration.

9.1.2. Theory

Sound for correlated vortex shedding. Consider the circular cylinder in cross flow shown in Figure 9-4. The length is L, diameter D, and flow velocity U. A cylindrical coordinate system (R, θ, ϕ) extends a line of length R from the axis of the cylinder to an observer who records the radiated sound. Vortex-induced fluid forces act on the cylinder in the lift (y) and drag (x) directions. The cylinder can respond to these forces by vibrating. The wavelength of the radiated sound is λ and A_y is the amplitude of cylinder vibration. Theoretical expression for the radiated sound can be developed if the observer is far from the cylinder in comparison with the wavelength and the cylinder diameter and amplitude of vibration are small compared with a wavelength $(\lambda \gg D, \lambda \gg A_y, R \gg \lambda)$. These are called far-field assumptions, and they are employed to avoid the complications of local diffraction of sound about the cylinder contours. In addition, it is assumed that the cylinder length greatly exceeds its diameter, so that the cylinder is effectively a line source of sound, and the Mach number is assumed to be very small, so that convection of traveling sound waves by flow is minimal.

Using these assumptions, the far-field sound pressure radiated by a

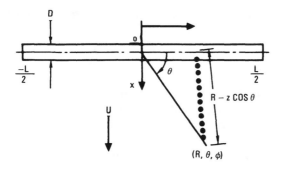

PLAN VIEW

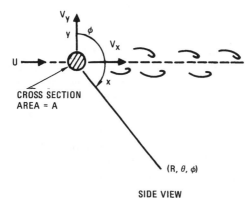

SIDE VIEW

Fig. 9-4 Cylinder in cross flow and coordinate system.

vibrating cylinder is (the derivation is given in Appendix C or by Blake, 1986, or Koopman, 1969)

$$p = \frac{\sin \theta \cos \phi}{4\pi Rc} \int_{-L/2}^{L/2} \left[\rho A \frac{\partial^2 V_y}{\partial t^2} - \frac{\partial F_y}{\partial t} \right] dz$$

$$+ \frac{\sin \theta \sin \phi}{4\pi Rc} \int_{-L/2}^{L/2} \left[\rho A \frac{\partial^2 V_x}{\partial t^2} - \frac{\partial F_x}{\partial t} \right] dz. \quad (9\text{-}3)$$

The y (vertical) coordinate to the observer is $y = R \sin \theta \cos \phi$, the longitudinal coordinate is $x = R \sin \theta \sin \phi$, and z is distance along the cylinder axis. V_y is the vertical component of cylinder velocity, V_x is the cylinder velocity parallel to the free stream, F_y is the fluid force on the cylinder normal to the free stream, and F_x is the fluid force parallel to the free stream. $A = \rho \pi D^2/4$ is the cylinder cross-sectional area. The quantities inside the square brackets are evaluated at retarded time. The retarded time,

$$t' = t - \frac{R}{c} + \frac{z \cos \theta}{c}, \tag{9-4}$$

accounts for the time required for the sound to travel the distance $R - z \cos \theta$, at speed of sound c, from point z on the cylinder to the observer (Fig. 9-4). The sound pressure of Eq. 9-3 is the sum of sound due to the fluctuating fluid force and cylinder vibration normal to the free stream (first term) plus the sound due to fluid force and cylinder vibration parallel to the free stream (second term).

If the vortices are shed in phase along the span of the cylinder and the forces are harmonic at circular frequency $\omega = 2\pi f$, then the fluid forces in lift and drag are

$$F_y = \tfrac{1}{2}\rho U^2 DC_L \sin \omega t, \tag{9-5}$$

$$F_x = \tfrac{1}{2}\rho U^2 DC_D + \tfrac{1}{2}\rho U^2 DC_d \sin (2\omega t + \beta), \tag{9-6}$$

where C_L is the coefficient of oscillating lift (Chapter 3, Fig. 3-16), C_D is the drag coefficient (Chapter 6, Fig. 6-9), and C_d is the coefficient of oscillating drag. Experiments and theory show that the oscillating drag has a frequency twice that of lift. Experimental measurements indicate that C_d is approximately 5% to 10% of the oscillating lift coefficient for both circular and triangular sections (see Chapter 3).

If the cylinder is stationary, $V_x = V_y = 0$, then the radiated sound is produced only by the fluctuating fluid forces. Substituting Eqs. 9-5 and 9-6 into Eq. 9-3 gives an expression for this sound,

$$p(R, \theta, \phi) = -\frac{\sin \theta \cos \phi}{4Rc} \left(\frac{\sin \eta}{\eta}\right) \rho U^3 LC_L S \cos \left[\omega(t - R/c)\right]$$

$$-\frac{\sin \theta \sin \phi}{4Rc} \left(\frac{\sin 2\eta}{2\eta}\right) \rho U^3 LC_d S \cos \left[2\omega(t - R/c) + \beta\right], \tag{9-7}$$

where

$$\eta = \frac{kL}{2} \cos \theta = \frac{\pi L}{\lambda} \cos \theta, \tag{9-8}$$

and the wave number is 2π over the sound wavelength λ,

$$k = \frac{\omega}{c} = \frac{2\pi}{\lambda}, \tag{9-9}$$

and the circular vortex shedding frequency is $\omega = 2\pi SU/D$. The sound is radiated directionally. If the cylinder length (L) is much less than the wavelength of sound ($\lambda = 2\pi c/\omega$), the cylinder radiates sound like a point dipole source. If the cylinder length is much greater than the wavelength of sound, the cylinder radiates sound directionally like a line of dipoles.

A pebble is thrown into a quiet pond causing waves to radiate in ever-widening circles. This is a point source. Now, suppose a long, bare tree branch falls into the pond. Waves radiate parallel to the branch, a line source. While the three-dimensional sound fields radiated by the cylinder are more complex than two-dimensional surface waves, the same principle of increasing directionality with increasing cylinder length applies. In the limit as the ratio of cylinder length to the sound wavelength decreases,

$$\lim_{L/\lambda \to 0} \frac{\sin \eta}{\eta} = \lim_{L/\lambda \to 0} \frac{\sin 2\eta}{2\eta} = 1, \tag{9-10}$$

the radiation pattern approaches that of a single dipole as shown in Figure 9-5.

Figure 9-6 shows a section cut perpendicularly through the cylinder axis with lines of constant sound pressure. Since the component of oscillating drag is much less than the component of oscillating lift, the sound due to oscillating drag is usually neglected when analyzing the sound radiated by oscillating lift. This theoretical pattern is confirmed by the experimental data of Stowell and Deming (1936; Fig. 9-3) and Keefe (1962).

The sound intensity is the average sound power radiated out per unit area into the far field. Considering only the oscillating lift portion of the far-field sound field, using Eq. 9-7, we have

$$I_R = \frac{\overline{p^2}}{\rho c} = \frac{\sin^2 \theta \cos^2 \phi}{32 c^3 R^2} \rho U^6 L^2 C_L^2 S^2 \left(\frac{\sin \eta}{\eta} \right)^2 . \tag{9-11}$$

The overbar ($^-$) denotes averages over many cycles. The sound intensity is proportional to the square of the dynamic pressure times the square of the shedding frequency; hence the sound power is proportional to the free stream velocity raised to the sixth power. Phillips' experiments (1956) have verified the U^6 dependence of sound intensity due to vortex shedding from a cylinder.

The total sound power radiated into the far field is the integral of the sound intensity through a spherical surface with radius much greater than a wavelength:

$$W = \int_0^\pi \int_0^{2\pi} I_R R^2 \sin \theta \, d\phi \, d\theta = \frac{\pi \rho U^6 L^2 C_L^2 S^2}{24 c^3} , \tag{9-12}$$

neglecting sound radiated by drag and assuming $\lambda \gg L$. In comparison, the power extracted from the fluid flow by the average drag on the cylinder is $W_D = F_D U L = \frac{1}{2} \rho U^3 D C_D L$. The ratio of the radiated sound power to the power dissipated in drag is proportional to the cube of Mach

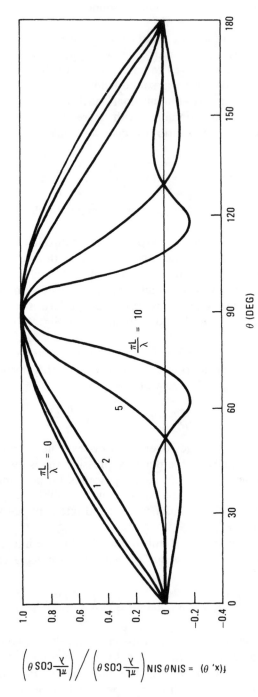

$$f(x, \theta) = \text{SIN } \theta \text{ SIN} \left(\frac{\pi L}{\lambda} \text{ COS } \theta \right) \Bigg/ \left(\frac{\pi L}{\lambda} \text{ COS } \theta \right)$$

Fig. 9-5 Increase in directionality of radiated sound with increasing ratio of cylinder length (L) to wavelength of sound (λ).

356

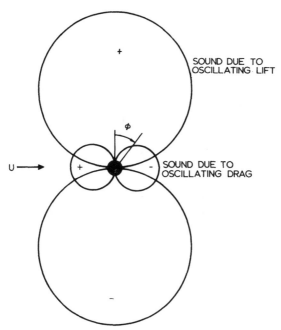

Fig. 9-6 Lines of equal sound intensity for sound radiated from a cylinder, after Keefe (1962).

number, $(U/c)^3$. For low Mach numbers, $U/c \ll 1$, the power radiated as sound is a small fraction of the power dissipated in drag.

Sound for partially correlated vortex shedding. Generally, vortex shedding is not fully correlated along the span of a stationary two-dimensional cylinder. Although vortices are primarily shed at the Strouhal frequency, the phase and frequency of shedding can vary somewhat along the span of a long cylinder; see Chapter 3, Section 3.2 and Figures 3-8 and 3-9. Typically the spanwise correlation length is between two and ten diameters for a stationary circular cylinder in cross flow.

The mean square (i.e., average of the square) of the far-field sound pressure generated by oscillating lift forces on a stationary cylinder is obtained by squaring Eq. 9-3 with $V - x = V_y = F_x = 0$ and averaging the result under the integrals,

$$\overline{p^2}(R, \theta, \phi) = \frac{\sin^2 \theta \cos^2 \phi}{16\pi^2 c^2 R^2} \int_{-L/2}^{L/2} \int_{-L/2}^{L/2} \overline{\frac{\partial F_y}{\partial t}(z_1, t_1') \frac{\partial F_y}{\partial t}(z_2, t_2')} \, dz_1 \, dz_2,$$

(9-13)

where the retarded times for points z_1 and z_2 are

$$t_1' = t - \frac{R}{c} + \frac{z_1 \cos \theta}{c}, \qquad t_2' = t - \frac{R}{c} + \frac{z_2 \cos \theta}{c}. \tag{9-14}$$

The sound at the observer is the integral of sound received from each spanwise point of the cylinder. Because the path length from each point is different, there is some interference between these elements of sound that gives rise to the $(\sin \eta)/\eta$ term in Eq. 9-7. If, in addition, the vortex forces on spanwise points are not correlated, then additional cancellation further reduces the mean square sound at the observer. If we model the vortex shedding process as a narrow-band process that has a randomly varying component of frequency or phase along the cylinder length, $F_y(z, t') = F(z) \sin [(\omega + \Delta\omega)t']$ where $\Delta\omega$ is a small random variable with zero mean, then it can be shown that

$$\overline{\frac{\partial F_y}{\partial t} (z_1, t_1') \frac{\partial F_y}{\partial t} (z_2 \, t_2')} = \overline{\frac{\partial F_y}{\partial t} (z_1, t) \frac{\partial F_y}{\partial t} (z_2, t)} \cos \frac{\omega}{c} (z_1 - z_2) \cos \theta$$

$$\approx \overline{\frac{\partial F_y}{\partial t} (z_1, t) \frac{\partial F_y}{\partial t} (z_2, t)} \qquad \text{if } \lambda \gg L. \tag{9-15}$$

Using this approximation, a spanwise correlation coefficient is defined as follows:

$$r(z_1, z_2) = \overline{\frac{\partial F_y}{\partial t} (z_1, t) \frac{\partial F_y}{\partial t} (z_2, t)} \bigg/ \left[\overline{\left(\frac{\partial F_y}{\partial t} (z_1, t) \right)^2} \right]^{1/2} \left[\overline{\left(\frac{\partial F_y}{\partial t} (z_2, t) \right)^2} \right]^{1/2},$$

$$\tag{9-16}$$

where $[\overline{(\partial F_y/\partial t)^2}]^{1/2} = \frac{1}{2}\rho U^2 D C_{L_{\text{rms}}} \omega^2$ and $C_{L_{\text{rms}}} = (\overline{C_L^2})^{1/2}$ is the root-mean-square lift coefficient. The overbar ($^-$) denotes time-averaged value.

The mean-square sound pressure (Eq. 9-13) for $\lambda \gg L$ is evaluated using Eq. 9-16 with the methodology of Chapter 7, Section 7.3.1. The correlation function is assumed to be only a function of separation between two points $\xi = z_1 - z_2$. The result is

$$I_R = \frac{\overline{p^2}}{\rho c} = \frac{\sin^2 \theta \cos^2 \phi}{16 c^3 R^2} \rho U^6 \overline{C_L^2} S^2 L_c (L - \gamma), \tag{9-17}$$

where the correlation length L_c and the centroid of the correlation area γ are

$$L_c = 2 \int_0^L r(\xi) \, d\xi, \qquad \gamma = \int_0^L \xi r(\xi) \, d\xi \bigg/ \int_0^L r(\xi) \, d\xi. \tag{9-18}$$

γ is always less than the correlation length L_c. If the vortex shedding is fully correlated along the span ($r = 1$), then $L_c = 2L$, $\gamma = L/2$, $\overline{C_L^2} = C_L^2/2$, and this equation reduces to the fully correlated case, Eq.

9-11 with $\lambda \gg L$. If the correlation length is much less than the cylinder length, $L \gg L_c$, then the sound intensity is reduced from the fully correlated case by the ratio of the correlation length to the cylinder length, L_c/L. Leehey and Hanson (1971) experimentally verified Eq. 9-17 by simultaneous measurement of correlation length and sound intensity from a fine wire in a free air jet.

Exercises

1. What is the sound power (Eq. 9-12) radiated by partially correlated vortex shedding?

2. Derive the first line of Eq. 9-15 by assuming $F_y(z, t') = F_0(z) \sin [(\omega + \Delta\omega)t']$, where $\Delta\omega = \Delta\omega(z, t')$ is a function of both space and reduced time. Take the derivative of this expression with respect to time (not retarded time t'), form the product of these derivatives at two points, and average the result, assuming that the average value of $\Delta\omega$ is zero.

9.2. SOUND FROM VIBRATING CYLINDERS

Basic theory. If the frequency of vortex shedding coincides with the natural frequency of the cylinder, the vortex shedding lift forces can induce large-amplitude cylinder vibration normal to the free stream and this vibration contributes to the sound field. The sound radiated from a vibrating cylinder is given by Eq. 9-3 or Appendix C. Following Burton and Blevins (1976), the intensity of the radiated sound,

$$I_R = \frac{\overline{p^2}}{\rho c} = I_{VV} + 2I_{VF} + I_{FF}, \tag{9-19}$$

is the sum of components of sound intensity due to vibration, fluid forces, and their interaction,

$$I_{VV} = \frac{\rho A^2 \sin^2 \theta \cos^2 \phi}{16\pi^2 R^2 c^3} \int \int_{-L/2}^{L/2} \overline{\frac{\partial^2 V_1}{\partial t^2} \frac{\partial^2 V_2}{\partial t^2}} \, dz_1 \, dz_2, \tag{9-20}$$

$$I_{VF} = \frac{A \sin^2 \theta \cos^2 \phi}{16\pi^2 R^2 c^3} \int \int_{-L/2}^{L/2} \overline{\frac{\partial^2 V_1}{\partial t^2} \frac{\partial F_2}{\partial t}} \, dz_1 \, dz_2, \tag{9-21}$$

$$I_{FF} = \frac{\sin^2 \theta \cos^2 \phi}{16\pi^2 R^2 \rho c^3} \int \int_{-L/2}^{L/2} \overline{\frac{\partial F_1}{\partial t} \frac{\partial F_2}{\partial t}} \, dz_1 \, dz_2, \tag{9-22}$$

where F is the fluid force normal to the free stream and V is the velocity of cylinder response normal to the free stream. The subscripts 1 and 2 denote that the quantities are evaluated at spanwise locations z_1 and z_2 and at retarded times given by Eq. 9-14.

The integrals are considerably simplified if we restrict attention to the far field $(R \gg \lambda)$ and cylinders whose length is short compared with the wavelength, $\lambda = c/f = 2\pi c/\omega$, $\lambda \gg L$, and the fluid forces and cylinder vibration are assumed to act predominately at a circular frequency ω. Under these assumptions, the sound from different parts of the cylinder arrive at the observer at the same instant so the retarded time effects do influence the integration and the differentiation in Eqs. 9-20 through 9-22 can be considerably simplified,

$$I_{VV} = \frac{\rho A^2 \sin^2 \theta \cos^2 \phi \omega^4}{16\pi^2 R^2 c^3} \int \int_{-L/2}^{L/2} \overline{V_1 V_2} \, dz_1 \, dz_2, \tag{9-23}$$

$$I_{VF} = \frac{A \sin^2 \theta \cos^2 \phi \omega^3}{16\pi^2 R^2 c^3} \int \int_{-L/2}^{L/2} \overline{V_1 F_2^+} \, dz_1 \, dz_2, \tag{9-24}$$

$$I_{FF} = \frac{\sin^2 \theta \cos^2 \phi \omega^2}{16\pi^2 R^2 \rho c^3} \int \int_{-L/2}^{L/2} \overline{F_1 F_2} \, dz_1 \, dz_2. \tag{9-25}$$

In these expressions the subscripts 1 and 2 denote spanwise locations 1 and 2 and at time t, not retarded time. F_2^+ denotes that this term is advanced in phase by 90 degrees. I_{VV} is the intensity of sound radiated by cylinder vibration. The cross term I_{VF} is the sound intensity due to the interaction of fluid forces in phase with cylinder acceleration and cylinder motion.

I_{FF}, Eq. 9-22 or 9-25, is the sound intensity due to fluid forces on the cylinder. It is identical to that for the stationary cylinder case (Eq. 9-11 or 9-17) except that here the fluid force F includes both the oscillating lift force and the fluid force imposed on the cylinder by the added mass (Chapter 2, Section 2.2) of fluid entrained by the vibrating cylinder. The added mass force acts in phase with cylinder acceleration and out of phase with cylinder velocity. The added mass force is approximately equal to the mass of fluid displaced by the cylinder times cylinder acceleration $F_{am} = -\rho A \omega \, \partial V / \partial t$. Substituting this into Eq. 9-22 or 9-25, we see that the sound intensity due to added mass force is equal to the sound intensity due to cylinder vibration (I_{VV}) and the cross-term intensity is also equal to the sound intensity due to cylinder vibration ($I_{VF} = I_{VV}$). Thus,

$$I_R = I_{VV} + 2I_{VF} + I_{FF} = I_{VV} + 2I_{VV} + (I_{ff} + I_{VV}) = 4I_{VV} + I_{ff}, \tag{9-26}$$

where

$$I_{ff} = \frac{\sin^2 \theta \cos^2 \phi \omega^2}{16\pi^2 R^2 \rho c^3} \int \int_{-L/2}^{L/2} \overline{F_L(z_1, t) F_L(z_2, t)} \, dz_1 \, dz_2,$$

$$I_{VV} = \frac{\rho A^2 \sin^2 \theta \cos^2 \phi \omega^4}{16\pi^2 R^2 c^3} \left[\int_{-L/2}^{L/2} \overline{V(z, t)} \, dz \right]^2. \tag{9-27}$$

The radiated sound intensity is the sum of a component due to cylinder motion (I_{VV}) and a component due to the vortex lift force (I_{ff}). I_{ff} is identical to the sound intensity previously found for a stationary cylinder (Eqs. 9-11, 9-17) if the lift coefficients are the same in both cases.

These expressions are easily evaluated for fully correlated vortex shedding from a spring-mounted circular cylinder that moves as a rigid body in sinusoidal motion, $A = \pi D^2/4$, $\overline{V(z, t)^2} = \omega^2 y_{rms}^2$, $\overline{F_L^2} = [\frac{1}{2}\rho U^2 D C_{L_{rms}}]^2$,

$$I_{VV} = \frac{\sin^2 \theta \cos^2 \phi \rho A^2 \omega^6 y_{rms}^2 L^2}{16\pi^2 R^2 c^3}, \qquad I_{ff} = \frac{\sin^2 \theta \cos^2 \phi \rho U^6 C_{L_{rms}}^2 S^2 L^2}{16 R^2 c^3}.$$

$$(9\text{-}28)$$

The ratio of sound produced by cylinder motion to that produced by lift forces is

$$\frac{I_{VV}}{I_{ff}} = \frac{16\pi^6 S^4}{C_{L_{rms}}^2} \left(\frac{y_{rms}}{D}\right)^2.$$

$$(9\text{-}29)$$

Setting this ratio to 1 allows us to solve for the root-mean-square cylinder displacement at which the sound generated by motion equals the sound generated by lift forces,

$$\frac{y_{rms}}{D} = \frac{C_{L_{rms}}}{4\pi^3 S^2} \approx 0.07$$

$$(9\text{-}30)$$

for $C_{L_{rms}} = 0.35$ and $S = 0.2$.

In general, the sound radiated as a result of cylinder motion dominates the sound field if the cylinder amplitude exceeds about 5% to 10% of the cylinder diameter. The sound produced by the oscillating lift forces dominates for amplitudes less than 5% of diameter. This result holds even if the wavelength of the radiated sound is of the same order as the cylinder length.

Sound for sinusoidal mode shapes. Often a single mode resonates with vortex shedding. This case arises with marine tow cables, slender beams, and towers. The majority of the vibration is perpendicular to the flow. The spanwise distribution of displacement and velocity is approximately sinusoidal,

$$V(z, t) = v(t)\bar{y}(z), \qquad \text{where } \bar{y}(z) = \begin{cases} \cos(n\pi z/L), & n = 1, 3, 5, \ldots, \\ \sin(n\pi z/L), & n = 2, 4, 6, \ldots, \end{cases}$$

$$(9\text{-}31)$$

where z varies over the span of the cylinder $-L/2 \le z \le L/2$ (Fig. 9-4) and n is the modal index. The sinusoidal mode shape is appropriate for a tightly stretched cable and pinned–pinned beams, and it also approximates the higher modes of beams.

Using the techniques of the previous paragraphs, it can be shown that the total radiated sound intensity is the sum of a component due to the lift forces and a component due to cylinder motion,

$$I = 4I_{VV} + I_{ff}. \tag{9-32}$$

This expression is applicable to sound wavelengths comparable to or larger than the cylinder length (Burton and Blevins, 1976). The sound intensity due to cylinder motion is evaluated from Eq. 9-20,

$$
\begin{aligned}
I_{VV} &= \frac{\rho A^2 \sin^2\theta \cos^2\phi \omega k^2 \overline{v^2}}{16\pi^2 R^2 c^3} \left[\int_{-L/2}^{L/2} \cos\frac{n\pi z}{L} \cos(kz\cos\theta)\,dz \right]^2 \\
&= \frac{\rho A^2 \sin^2\theta \cos^2\phi \omega k^3 \overline{v^2} L^2}{4\pi^2 R^2} \left(\frac{n\pi \cos(b/2)}{n^2\pi^2 - b^2} \right)^2; \qquad n = 1, 3, 5, \ldots,
\end{aligned}
$$

$$\tag{9-33}$$

where $b = kL\cos\theta$, and the wave number is $k = \omega/c = 2\pi/\lambda$. This expression is appropriate when n is an odd number. For n even, $\cos(n\pi z/L)$, $\cos(kz\cos\theta)$, and $\cos(b/2)$ are replaced by $\sin(n\pi z/L)$, $\sin(kz\cos\theta)$, and $\sin(b/2)$.

The radiated sound power (W) is obtained by integrating $I_{ff} + 4I_{VV}$ over a sphere of radius R that is much greater than a wavelength (Eq. 9-12):

$$W = 4W_V + W_f = \frac{\rho U^6 L^2 S^2}{c^3}(4W_V' + W_f'), \tag{9-34}$$

where the sound power due to cylinder motion is

$$W_V = \int_0^{2\pi} \int_0^\pi I_{VV} R^2 \sin\theta\, d\theta\, d\phi = \rho\frac{1}{2\pi} ck^4 A^2 L^2 J^2 \overline{v^2}, \tag{9-35}$$

and the mode shape–wavelength–cylinder length coupling factor $J(kL, n)$ is

$$J^2(kL, n) = \frac{1}{kL}\int_0^{kL}\left[1 - \left(\frac{b}{kL}\right)^2\right]\frac{n^2\pi^2\cos^2(b/2)}{n^2\pi^2 - b^2}\,db, \qquad n = 1, 3, 5, \ldots. \tag{9-36}$$

When n is even, $\sin^2(b/2)$ replaces $\cos^2(b/2)$ in this expression. $J(kL, n)$ is plotted in Figure 9-7 for the first six sinusoidal modes, $n = 1$ through $n = 6$. The asymptotic behavior for very low wave numbers ($\lambda \gg L$) is shown at the left side of this figure. At these wave numbers, the cylinder is a compact source of sound. For odd n, the source has the angular dependence of a dipole ($\sin^2\theta\cos^2\phi$) and $\sim k^4 J^2 \sim (1/\lambda)^4$. For even n, $\int \sin(n\pi z/L)\,dz = 0$ and cancellation occurs because sound radiated from one part of the mode shape is canceled by sound radiated from other spanwise portions that have opposite sign. The low-wave-number limit of Eq. 9-36, for even n, has angular dependence of $\sin^2\theta\cos^2\theta\cos^2\phi$, that

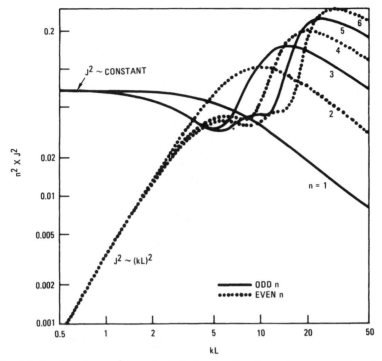

Fig. 9-7 Coupling factor J^2 for velocity-induced radiated sound power as a function of wave number ($kL = 2\pi L/\lambda$) (Burton and Blevins, 1976).

of a lateral quadrapole, and the radiated sound power diminishes as $(1/\lambda)^6$. At high wave numbers, interference effects give the curves in Figure 9-7 a complicated appearance.

The nondimensionalized components of radiated sound power (Eq. 9-34) are plotted in Figure 9-8 as a function of the mean-square cylinder velocity for $n = 1$, $L/D = 50$, $S = 0.2$, $kL = 1$, and $U/c = 0.016$. At high amplitudes $(\overline{v^2})^{1/2}/U > 0.2$, the sound power is dominated by the contribution due to cylinder motion. The solid portions of the curves are the results of analysis (Burton and Blevins, 1976); the dashed portions are estimated.

Exercises

1. Considering the results of this section, explain the physical process that led to Figure 9-4 and Richardson's observations on it.

2. By using the partially correlated model for sound produced by vortex shedding, determine the ratio of sound power (W) due to cylinder motion to that produced by the vortex lift force for partially correlated vortex shedding.

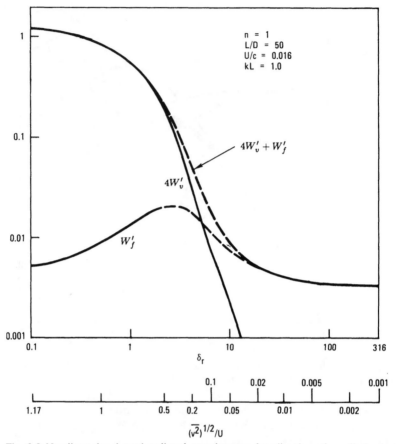

Fig. 9-8 Nondimensional total radiated sound power for vibration of a cylinder in the fundamental sinusoidal mode as the sum of components due to fluid lift force (W_f') and due to cylinder motion ($4W_v'$). The reduced damping is defined in Chapter 3 (Burton and Blevins, 1976).

9.3. SOUND FROM MULTIPLE TUBES AND HEAT EXCHANGERS

9.3.1. Experiments

Intense sound and vibration can be produced by gas flow through a heat-exchanger array of tubes. Baird (1954) reported that noise and vibration in the heat exchanger of the Etiwanda Steam Power Station occurred when power exceeded 86 MW while burning oil and 105 MW while burning natural gas. "The severity of the vibration and its destructive effect on both the metal and refractory material was greater than any [previously] observed . . . The vibration was accompanied by intense sound which could easily be heard in the concrete control room some distance away." He found the oscillating pressure corresponded to a transverse standing wave across the heat exchanger duct, as shown in Figure 9-9, at 40–50 Hz.

Baird attributed the sound to flow-induced 'fluid pulsation' in sympathy with transverse acoustic modes of the heat-exchanger duct. In 1956, Grotz and Arnold found that the sound changed mode as the air velocity was increased. They found that the excited frequencies corresponded to the transverse duct mode natural frequencies and that the vibration could be reduced by the installation of baffles. Putnam (1965) found that the resonance frequencies corresponded to vortex-shedding Strouhal numbers (Eq. 9-1) ranging from $S = 0.2$ to $S = 0.46$.

Acoustic resonance has occurred in inline tube arrays, staggered tube arrays, single rows of tubes, rectangular ducts, cylindrical ducts, helically coiled tubes, finned tubes, chemical process exchangers, air heaters, power-generation boilers, marine boilers, conventional power plants, nuclear power plants, heat-recovery heat exchangers, turbojet engine compressors, turning vanes of wind tunnels, plates in a wind tunnel, and the combustion chambers of rocket engines (Mathias et al., 1988; Blevins, 1984; Flandro, 1986; Brown, 1985; Parker, 1972). It has been observed with air, flue gas, steam, hydrocarbons, and two phase gas–vapor flow. Acoustic resonance is always associated with an acoustic mode that is transverse to both the flow and the tube or plate axes. Typical sound pressure levels in a tube array during resonance are (Eq. 9-38) 160–176 dB within a tube array and approximately 20–40 dB lower outside the heat-exchanger shell. These sound levels can fatigue the heat-exchanger ducting and they are very disturbing to nearby humans.

Sound propagating through an array of heat-exchanger tubes is slowed slightly by the presence of the tubes. The rate of decrease of sound speed

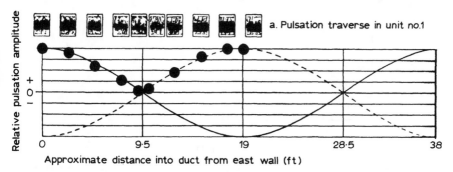

Fig. 9-9 Gas pulsation waves in power-plant duct measured by Baird (1954).

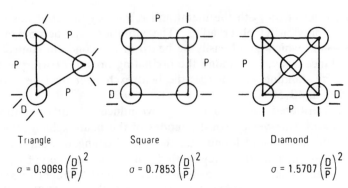

Fig. 9-10 Tube patterns and solidity. Flow is right to left.

is dependent on the fraction of the volume occupied by tubes (Parker, 1979; Blevins, 1985):

$$\frac{c}{c_0} = \frac{1}{(1+\sigma)^{1/2}}. \tag{9-37}$$

The speed of sound in free space is c_0 and c is the speed at which sound propagates perpendicular to the tubes' axes through an array of cylinders

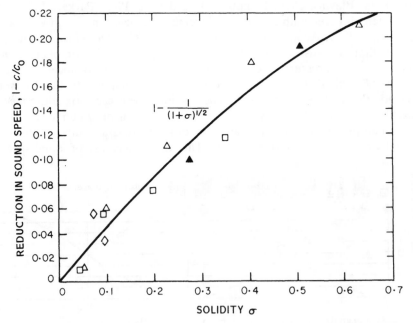

Fig. 9-11 Theory and experiment for sound speed through an array of tubes (Blevins, 1986).

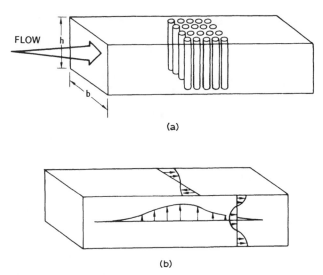

Fig. 9-12 (a) Tube array in a duct (tube size exaggerated for clarity) and (b) illustration of a typical cut-off acoustic mode shape.

or tubes. The solidity σ is the fraction of volume occupied by tubes in a regular array, Figure 9-10. Equation 9-37 agrees well with the data shown in Figure 9-11.

The sound waves reflect off duct walls to create a pattern of standing waves, called *acoustic modes,* at the natural acoustic frequencies of the duct. Because the sound speed within the tube array differs from that in the adjacent ducting, certain acoustic modes decay exponentially with distance from the array. These modes are called *cut-off* or *bound modes* because they do not propagate energy away from the tube array (Tyler and Sofrin, 1962; Grotz and Arnold, 1956; Parker, 1979; Blevins, 1984). A cut-off duct mode of a tube array is shown in Figure 9-12. When acoustic resonance excites a tube-array bound mode, the high sound levels are confined to within one or two duct widths of the tube array.

Blevins (1985) found that sound levels in excess of 140 dB can entrain vortex shedding from cylinders, shift its frequency, and increase spanwise correlation (Chapter 3, Figure 3-9). Ffowcs Williams and Zao (1989) showed that, with appropriate feedback from the near wake, sound could either enhance or suppress vortex shedding from a cylinder. Welsh et al. (1984) observed effects of sound on vortex shedding from blunt-ended plates (Figure 9-13). Kim and Durbin (1988) and Farrell (1980) showed that sound influences vortex shedding from cavities, cylinder pairs, and rectangular sections.

It is interesting to note that it is not the sound pressure that affects vortex shedding but rather the velocity induced by the sound (Blevins,

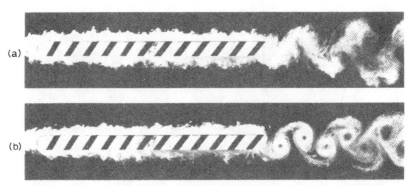

Fig. 9-13 Vortex shedding from a 0.48 in (12.1 mm) thick plate: (a) without sound, $S = 0.213$, $U = 81$ ft/sec (24.6 m/sec); (b) with resonant sound, $SPL = 145.5$ dB, $S = 0.224$, $U = 95$ ft/sec (29 m/sec) (Welsh et al., 1984).

1985). Thus, for example, it is easier to entrain shedding from a tube or plate located in the center of a duct where acoustic velocities are maximum than at the edge of a duct where acoustic pressures are maximum but acoustic velocities are minimum. The sound field behaves much like an oscillating velocity field as far as an individual tube is concerned, and oscillating flows are well known to entrain vortex shedding; see Chapter 6, Section 6.4 and Chapter 3, Section 3.3. The greater the transverse velocity (or sound-pressure-induced velocity), the greater the entrainment.

Vortex shedding persists in arrays of tubes and plates (Figure 9-14; Weaver and Abd-Rabbo, 1985; Abd-Rabbo and Weaver, 1986; Stoneman et al., 1988). The sound generated by vortex shedding from the

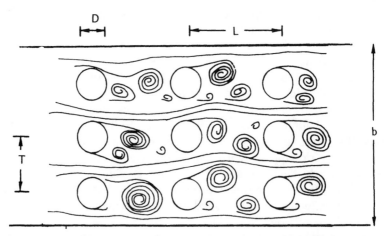

Fig. 9-14 Flow pattern according to a photograph taken by Wallis (1939) for flow in a tube array (After Chen, 1973).

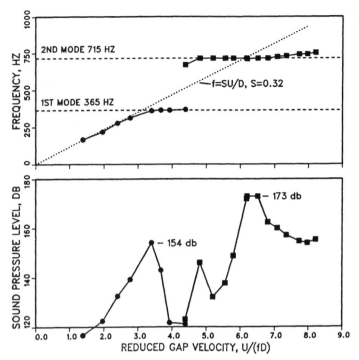

Fig. 9-15 Sound produced by air flow through a duct with a tube array with pitch-to-diameter ratio of 3 and diamond pattern (Fig. 9-10). The reduced velocity is based on average flow velocity through minimum gap and the frequency of the fundamental transverse acoustic mode, 365 Hz (Blevins and Bressler, 1987).

tubes and plates (Section 9.1) feeds back via the duct acoustic modes to influence the vortex-shedding process. Figure 9-15 shows sound generated by flow through a tube array and the dominant frequency of that sound. The shedding frequency (Eq. 9-1) increases with velocity. When the shedding frequency approaches one of the duct transverse acoustic natural frequencies, the resonant sound increases above about 140 dB, and the shedding locks onto the duct acoustic natural frequency, creating a loud persistent resonance. As velocity is further increased, the resonance is broken until the next higher acoustic mode comes into resonance.

The sound pressure level (SPL) in Figure 9-15 is defined as follows:

$$SPL = 20 \log_{10} (p_{rms}/0.00002 \text{ Pa}). \tag{9-38}$$

Compare Figure 9-15 with Figure 3-14 in Chapter 3.

9.3.2.   Prediction and Suppression of Resonance

Prediction. Methods of prediction of acoustic resonance are based on calculating the acoustic and vortex shedding frequencies, determining whether there is a match, and then estimating the magnitude of the resonance from data.

Step 1. The natural frequencies in hertz of transverse acoustic modes of closed rectangular and cylindrical volumes with rigid walls are (Blevins, 1979):

$$f_{a,j} = \begin{cases} \dfrac{c}{2}\dfrac{j}{b}, & j = 1, 2, 3, \ldots, & \text{rectangular volume,} \\[2mm] \dfrac{c}{2\pi}\dfrac{\lambda_j}{R}, & j = 1, 2, 3, \ldots, & \text{cylindrical volume,} \end{cases} \qquad (9\text{-}39)$$

where b is the width of the rectangular duct transverse to the flow and tube axis (Fig. 9-9) and R is the radius of a cylindrical duct; c is the speed of sound; $\lambda_1 = 1.841$ is the dimensionless frequency parameter associated with the fundamental diametrical acoustic mode of a cylinder; $\lambda_2 = 3.054$ is the second diametrical mode. A rectangular duct and a cylindrical duct heat-exchanger design are shown in Chapter 5, Figure 5-11. Techniques for estimating three-dimensional duct modes are discussed by Parker (1979) and Blevins (1986). The effect of mean flow is discussed by Quinn and Howe (1984).

As noted earlier, the speed of sound through tube arrays is showed by the presence of tubes (Eq. 9-37). However, in general tubes do not fill the entire volume of the ducting and the acoustic modes spill out into the ducting forward and aft of the tube array. The natural frequencies of these modes lies between that calculated with the speed of sound reduced for the presence of tubes (Eq. 9-37) and that obtained by neglecting the influence of tubes (Blevins and Bressler, 1987).

Step 2. The dominant frequency of the acoustic sound generated by flow over tubes is the shedding frequency,

$$f = \frac{SU}{D} \quad \text{Hz.} \qquad (9\text{-}40)$$

Here U is the velocity averaged across the minimum gap between tubes and D is the tube diameter. The dimensionless Strouhal number S is nearly independent of Reynolds number for Reynolds numbers greater than 1000, but it is dependent on tube-to-tube spacing. The Fitzhugh correlation for Strouhal numbers in tube arrays is given in Chapter 3, Figures 3-4 and 3-5, Weaver et al. (1987) suggest $S = 0.5$ for inline arrays based on a square grid and $S = 0.58$ for arrays based on an equilateral triangular grid with the base of the triangle normal to the flow. There is considerable uncertainty in these correlations for closely spaced arrays because the close spacing broadens the shedding frequency and a distinct, unique shedding frequency becomes difficult or impossible to identify (Fitzpatrick et al., 1988; Donaldson and McKnight, 1979; Weaver et al., 1987).

Vortex shedding also occurs from finned heat-exchanger tubes. Mair et al. (1975) found that the following effective diameter collapses the Strouhal number for finned tubes onto the bare tube data:

$$D_e = \frac{1}{s}[(s - t)D_r + tD_f], \qquad (9\text{-}41)$$

where t is the fin thickness, s is the spanwise spacing between the fin centers, D_f is the diameter of the fins, and D_r is the root (i.e., bare) diameter of the tube.

Since the resonance can shift the natural shedding frequency up or down, a band must be placed on the excitation frequency (Eq. 9-40) when comparing it with the acoustic natural frequencies $f_{a,j}$ (Eq. 9-39) to determine whether a potential resonance exists,

$$(1 - \alpha)\frac{SU}{D} < f_{a,j} < (1 + \beta)\frac{SU}{D}, \qquad \text{for resonance.} \qquad (9\text{-}42)$$

Barrington (1973) and Rogers and Penterson (1977) recommend $\alpha = \beta = 0.2$. Blevins and Bressler (1987) measured typical values of $\alpha = 0.19$ and $\beta = 0.29$, but maximum values were $\alpha = 0.4$ and $\beta = 0.48$. If Eq. 9-42 holds for any transverse acoustic mode, a resonance is predicted with that mode and high sound levels may result.

Step 3. In roughly 30% to 40% of practically important cases, a predicted resonance does not emerge. One reason for this is that the resonant sound amplitude is a function of the pitch and pattern of the tube arrays specified by the tube array parameters T/D and L/D for closely spaced tube arrays (Figs. 9-10, 9-14). In the limit as $L = D$, the one tube touches its downstream neighbor, forming a longitudinal wall that blocks transverse acoustic modes. Small pitch-to-diameter ratios, less than 1.6, also tend to suppress vortex shedding (Price and Paidoussis, 1989). Figure 9-16 shows the maximum resonant amplitude measured in tests on a large number of tube arrays for resonance in the first acoustic mode. For pitch-to-diameter ratios less than 1.6, small changes in spacing can make very large changes in sound level.

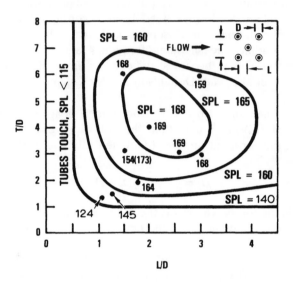

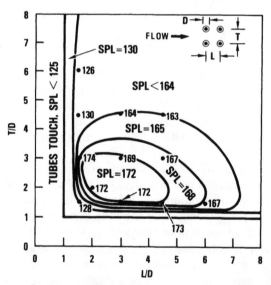

Fig. 9-16 Peak resonant sound pressure level for resonance in a tube array as a function of tube-to-tube spacing; Mach number = 0.2, Reynolds number = 80 000. Measurements by the author and M. M. Bressler.

The sound pressure radiated by vortex shedding increases with Mach number (U/c, Eq. 9-7). The energy that creates the sound comes from the loss in static pressure (Δp_{drop}) as gas flows over the tubes. Blevins and Bressler (1992-section 10.4) have found the following expression predicts maximum resonant sound in terms of pressure drop and Mach number,

$$p_{rms} = 12\Delta p_{drop} \frac{U}{c} \qquad (9\text{-}43)$$

As a consequence, higher acoustic modes of tube arrays are more prone to acoustic resonance than the fundamental mode because higher modes are excited at higher Mach numbers (Grotz and Arnold, 1956; Funakawa and Umakoshi, 1970; Ziada et al., 1988). The Mach number of the sound for Figure 9-16 is 0.2. The pressure drop is 20 inches of water. This figure is probably conservative for lower Mach numbers and nonconservative (i.e., a lower bound on the magnitude of sound) for higher Mach numbers and higher pressure drops.

Suppression. Acoustic resonance in heat-exchanger tube arrays has been suppressed by (1) detuning the resonance with baffles that shift the acoustic natural frequency upwards, (2) increasing acoustic damping with Helmholtz resonators, (3) altering the tube surface, and (4) removing tubes.

Practically speaking, resonance does not occur for $U/(f_a D) < 2$, where U is the average velocity in the minimum reference gap between tubes, f_a is the fundamental acoustic frequency, and D is the tube diameter. See Figure 9-15. Increasing the fundamental acoustic natural frequency (Eq. 9-39) above $f_a > U/(2D)$ brings the acoustic natural frequency above the range of shedding frequencies, bringing the array out of resonance. This is accomplished by installing baffles parallel to both the flow and the tube axis to decrease the effective transverse width b of the section. Baird (1954), Grotz and Arnold (1956), Cohan and Deane (1965), and many others have successfully employed baffles to suppress resonance. Baffle installations are shown in Figure 9-17.

Figure 9-18 shows the effect of a single and multiple baffles on the sound generated by an inline tube array. A single baffle reduces the sound in the fundamental mode but has no effect on the seond acoustic mode. Multiple baffles suppress higher modes. A central baffle is more effective than upstream or downstream baffles and a solid baffle is more effective than a perforated baffle, although perforated baffles have proved adequate in some cases (Blevins and Bressler, 1987; Byrne, 1983). Relatively thin, 1/8 in (3.2 mm), sheet-metal baffles have proved effective. Baffle effectiveness requires that its mass be much greater, say a factor of 10, than the mass of gas in the exchanger volume, rather than that its natural frequency be above the acoustic frequency.

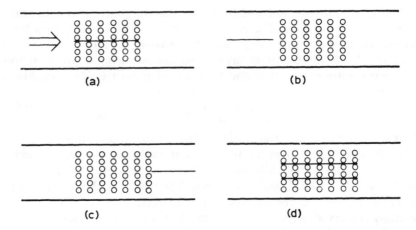

Fig. 9-17 Baffle positioning in array. Tube size is exaggerated for clarity. (a) Central baffle, (b) upstream baffle, (c) downstream baffle, (d) two central baffles at 1/3 and 2/3 transverse locations.

A Helmholtz resonator is a bottle-shaped chamber that is attached to the side of the heat-exchanger shell, communicating with the shell side gas through a narrow neck that damps acoustic modes by viscous losses as the air is forced into and out of the neck. Blevins and Bressler (1987) found that a carefully tuned resonator with volume equal to 3.2% of the volume of the heat-exchanger duct produced a 13 decibel reduction in resonance of the fundamental acoustic mode and that a 1.5% volume resonator had negligible effect. The advantage of resonators over baffles

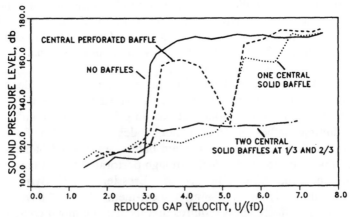

Fig. 9-18 Sound levels in an inline, square tube array with $P/D = 2.0$ with various baffles; $f = 365\,Hz$ (fundamental acoustic frequency) and $D = 0.75\,in$ (19 mm) is tube diameter (Blevins and Bressler, 1987).

is that the intrusion into the heat-exchanger cavity is minimal. The disadvantages are that they must be tuned to the frequency of the acoustic mode to be effective and the resonators themselves are sizable vessels; see Baylac and Gregorie (1975).

Two less robust methods for suppressing resonance are tube removal and modifying the tube surface by soot accumulation. Walker and Reising (1968), Barrington (1973), and Zdravkovich and Nuttall (1974) have found that by removing about 3% of "judiciously placed" tubes, usually near the middle of the array, resonance was suppressed in some cases. In other case, with strong resonances, tube removal is ineffective (Blevins and Bressler, 1987). The build-up of dirt and soot in a tube bank has been associated with the reduction of resonant sound (Rogers and Penterson, 1977; Baird, 1959). The author has also observed this effect in a large process heat exchanger. However, in laboratory tests, substantially roughening the surface of tubes with sand paper had no effect on a loud and persistent resonance. Probably tube removal and soot accumulation are most effective on marginal resonances where small alterations could trip the array out of resonance.

Exercise

1. Consider the heat exchanger described in Chapter 5, Section 5.5. If air ($c = 1100$ ft/sec, 343 m/sec), instead of water, is pumped through the shell side, what are the transverse acoustic natural frequencies of the first two modes? At what volume do flows create resonances between vortex shedding and these modes? What is the resultant sound pressure? How would your answer change if the pitch-to-diameter ratio were 2.0 instead of 1.25?

9.4. SOUND FROM FLOW OVER CAVITIES

Flow over and into cavities arises in aerodynamic and hydrodynamic applications including hydraulic flumes and gates, aircraft air scoops and bays, pipe branches and valves, bellows corrugations, rocket engines, orifices, and musical instruments (Flatau and Van Moorhem, 1988; Harris et al., 1988; Jungowski et al., 1987; Rockwell and Naudascher, 1978; Fletcher, 1979). The cavity flow is associated with a series of acoustic tones at the acoustic natural frequencies of the cavity. These tones are excited by the unstable shear layer across the cavity mouth. They have been observed in both subsonic and supersonic flow over cavities whose openings are parallel to the flow and cavities with openings normal to the flow.

Consider flow over a cavity in a wall shown in Figure 9-19. The boundary layer at the upstream edge of the cavity has thickness δ. It flows over the upstream edge and forms a free shear layer that trails aft over the cavity opening. The free shear layer is inherently unstable; upstream disturbances develop waves that roll into vortices at discrete frequencies. The shear layer vortices impinge on the downstream edge of the cavity, creating alternate inward and outward flow to the cavity that feeds back upstream as cavity pressure oscillations. The impinging shear layer instability is the mechanism for sound generation by a flute (Fletcher, 1979). It can also cause intense unwanted noise in aircraft and fluid power systems.

The somewhat complex details of the shear layer impingement and feedback are reviewed by Blake (1986), Rockwell and Naudascher (1978), Rockwell and Schachenmann (1982), Ronneberger (1980), Gharib and Roshko (1987), Bhattacharjee et al. (1986), and Harris et al. (1988). Here we will attempt the less heroic task of describing the frequencies of the oscillations, onset of acoustic resonance, and methods of suppression.

The waves that dominate shear wave instability have a preferential wavelength associated with the most unstable frequencies. When these wavelengths are integer multiples of the length of the cavity opening (L) in the direction of flow, the feedback from the impingement on the downstream edge enhances the shear wave instability. Experiments have shown that the associated frequencies in hertz are

$$f_n = \begin{cases} 0.33(n - 1/4)U/L, & \text{turbulent boundary layer (Franke and Carr,} \\ & \text{1975)}, n = 1, 2, 3, \ldots, \\ 0.52nU/L, & \text{laminar boundary layer (Ethembabaoglu,} \\ & \text{1978)}, n = 1, 2, 3, \ldots. \end{cases}$$

$$(9\text{-}44)$$

These relationships are shown in Figure 9-20: n is the shear wave mode number, the number of shear wavelengths across the opening, and U is the free stream velocity. The frequencies are also influenced by boundary layer thickness and Mach number (Fig. 9-20; Rockwell, 1977).

As the resonance condition is approached, the vortex-induced pressure oscillations in the cavity increase in magnitude, synchronizing the vortex formation and increasing the strength of the shed vortices until a resonance is formed at the cavity acoustic natural frequency. Cavity shear wave oscillations of this type are not limited to two-dimensional laboratory flows. They occur with round, square, two-dimensional, and axisymmetric openings and both shallow and deep cavities as well as from bellows corrugations, air scoops, and cavities that face into the flow. The

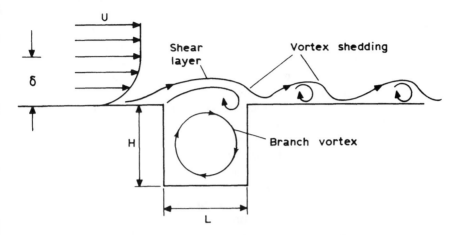

Fig. 9-19 Flow over a cavity, after Jungowski et al. (1987).

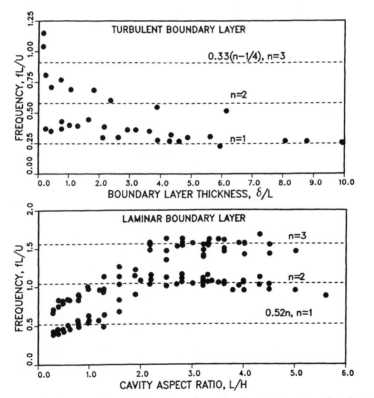

Fig. 9-20 Experimental data for shear wave frequency for turbulent boundary layer flow over a cavity: (a) turbulent boundary layer (DeMetz and Farabee, 1977); (b) laminar boundary layer (Ethembabaoglu, 1978).

377

oscillations occur in both subsonic and supersonic flow. They can increase drag and in certain ranges reduce drag (Gharib and Roshko, 1987). Cavity resonances can create sound pressure levels in excess of 175 dB.

For resonances between the shear wave excitation frequencies and the cavity acoustic frequencies, the cavity must be sufficiently long or deep so that the acoustic natural frequencies coincide with the preferred shear wave instability frequencies (Eq. 9-44). For relatively shallow cavities, $L/H < 1$, this is the case if the acoustic wavelength is comparable to or smaller than the length of the opening, $\lambda \leq L$, and the resonance appears as a longitudinal standing wave. For deep cavities, $L/H > 1$, the acoustic mode tends to occur in the depth and it is called an organ pipe mode. If the cavity opening leads to an expanding but closed cavity, then the acoustic mode is said to be a Helmholtz mode and even lower frequencies are possible. The natural frequencies of these cavity acoustic modes in hertz are (Blevins, 1979)

$$
f_a = \begin{cases}
(c/2)\sqrt{(i/L)^2 + (j/2H)^2}, & \text{rectangular cavity, } i = 1, 3, 5, \ldots, \\
& j = 0, 1, 2, 3, \ldots, \\
jc/(4H), & \text{deep cavity with narrow opening,} \\
& j = 1, 3, 5, \ldots, \\
(c/2\pi)\sqrt{A/Vd}, & \text{Helmholtz resonator.}
\end{cases}
$$

$$(9\text{-}45)$$

As shown in Figure 9-19, L is the length of a rectangular cavity, H is its depth, and c is the speed of sound. The second form of the equation applies to a narrow, deep cavity of any cross section. For the three-dimensional Helmholz cavity, A is the cross-sectional area of the opening, d is the depth of the opening plus a correction factor equal to approximately 1.6 times the radius of the opening to account for an effective depth, and V is the volume within the resonator. The volume of the Helmholtz type cavity coupled with a relatively small opening is capable of producing much lower-frequency sounds than a constant cross section cavity of the same depth.

Five methods that have proven effective in reducing acoustic resonance in open cavities are (1) rounding or ramping the trailing edge, (2) installing a fence or turning vane at the upstream edge to deflect the flow over the cavity, (3) breaking up the opening with an "egg crate" into a large number of smaller openings, (4) introducing mass flow into or out of the cavity, and (5) damping the acoustic mode. Figure 9-21 shows that a double ramp can achieve a 20 dB drop in the sound level. Other experiments show that the effect is also achieved by tapering or rounding the trailing edge, rather than both the leading and trailing edges (Rockwell and Naudascher, 1978). Bernstein and Bloomfield (1989) by rounding the

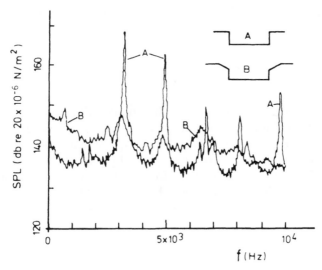

Fig. 9-21 Effect of double ramps on the attenuation of shallow-cavity acoustic oscillations (Franke and Carr, 1975).

edge to a radius equal to 30% of the cavity diameter reduced oscillating pressure by a factor of 200 in a safety valve. The reason for the effectiveness of sounding the trailing edge is believed to be that it prevents mass flow back into the cavity, interrupting the feedback. It also diffuses the vortex impact into a series of smaller and less-coherent events. Another method of disrupting the vortex impingement is to install a deflector along the upstream edge so that the free shear layer no longer impacts the downstream edge of the cavity (Willmarth et al., 1978; Rockwell and Naudascher, 1978). Deflection fences are widely employed on the sun roofs of automobiles.

It is possible to suppress cavity acoustic resonance by injecting mass into the cavity or bleeding mass from the cavity (Sarohia and Massier, 1976; Marquart and Grubb, 1987). Cavity bleed and injection apparently stabilize the vortex motion in the opening and suppress the feedback that leads to the resonance. The author has observed this effect in an air scoop. When a downstream valve, which controlled flow from the air scoop, was closed, sound pressures to 175 dB resulted at the frequency of the air scoop duct-to-valve cavity. When the valve was opened to permit mass flow, the oscillations were reduced by 20 dB. Perforating the valve with holes whose area equalled or exceeded 7% of the duct area also suppressed the resonance.

The magnitude of the resonance is an inverse function of the damping of the acoustic cavity. Lightly damped acoustic modes lead to sharp resonances such as shown in Figure 9-21. By increasing the damping of the cavity acoustic mode, such as with an auxiliary tuned Helmholtz resonator, it is possible to reduce the magnitude of the resonance by 20 dB.

Exercise

1. Obtain a soft drink or similar bottle. Empty it. Predict its Helmholtz natural acoustic mode using Eq. 9-45 and a speed of sound in air of 1100 ft/sec (343 m/sec). Estimate the velocity at which you can blow air by blowing small pieces of paper along the room. Estimate the shear wave frequencies from Eq. 9-44 for the bottle. Excite the acoustic mode. Do theory and experiment agree?

REFERENCES

Abd-Rabbo, A., and D. S. Weaver (1986) "Flow Visualization Study of Flow Development in a Staggered Tube Array," *Journal of Sound and Vibration,* **106,** 241–256.

Baird, R. C. (1954) "Pulsation-Induced Vibration in Utility Steam Generation Units," *Combustion,* **25**(10), 38–44.

——— (1959) discussion of Putnam (1959), *Journal of Engineering for Power,* **81,** 420.

Barrington, E. A. (1973) "Acoustic Vibration in Tubular Heat Exchangers," *Chemical Engineering Process,* **69**(7), 62–68.

Baylac, G., and J. P. Gregorie (1975) "Acoustic Phenomena in a Steam Generating Unit," *Journal of Sound and Vibration,* **42,** 31–48.

Bernard, H. (1908) "Formation des Centres de Giration a l'arriere d'un Obstacle in Mouvement," *Comptes Rendus Hebdomadaire des Seances de l'Academie des Sciences, Paris,* **147,** 839–842.

Bernstein, M. D., and W. J. Bloomfield (1989) "Malfunction of Safety Valves due to Flow-Induced Vibration," in *Flow-Induced Vibration—1989,* M. K. Au-Yang (ed.), Vol. 154, American Society of Mechanical Engineers, New York, pp. 155–164.

Bhattacharjee, S., et al. (1986) "Modification of Vortex Interactions in a Reattaching Separated Flow," *AIAA Journal,* **4,** 623–629.

Blake, W. K. (1986) *Mechanics of Flow-Induced Sound and Vibration,* Academic Press, New York.

Blevins, R. D. (1979) *Formulas for Natural Frequency and Mode Shape,* Van Nostrand Reinhold, New York. Reprinted by Krieger, Malabar, Fla., 19984.

——— (1984) "Review of Sound Induced by Vortex Shedding from Cylinders," *Journal of Sound and Vibration,* **92,** 455–470.

——— (1985) "The Effect of Sound on Vortex Shedding from Circular Cylinders," *Journal of Fluid Mechanics,* **161,** 217–237.

——— (1986) "Acoustic Modes of Heat Exchanger Tube Bundles," *Journal of Sound and Vibration,* **109,** 19–31.

Blevins, R. D., and M. M. Bressler (1987) "Acoustic Resonance in Heat Exchanger Tube Bundles—Part I: Physical Nature of the Phenomena, Part II: Prediction and Suppression of Resonance," *Journal of Pressure Vessel Technology,* **109,** 275–288.

Brown, R. S., et al. (1985) "Vortex Shedding Studies in a Simulated Coaxial Dump Combustor," *AIAA Journal of Propulsion and Power,* **1,** 413–415.

Burton, T. E., and R. D. Blevins (1976) "Vortex Shedding Noise for Oscillating Cylinders," *Journal of the Acoustical Society of America*, **60**, 599–606.

Byrne, K. P. (1983) "The Use of Porous Baffles to Control Acoustic Vibrations in Cross Flow Tubular Heat Exchangers," *Journal of Heat Transfer*, **105**, 751–758.

Chen, Y. N. (1973) "Karman Vortex Streets and Flow-Induced Vibration in Tube Banks," *Journal of Engineering for Industry*, **95**, 410–412.

Cohan, L. J., and W. J. Deane (1965) "Elimination of Destructive Vibrations in Large, Gas and Oil-Fired Units," *Journal of Engineering for Power*, **87**, 223–228.

DeMetz, F. C., and T. M. Farabee (1977) "Laminar and Turbulent Shear Flow Induced Cavity Resonances," AIAA Paper 77-1293, AIAA, New York.

Donaldson, I. S., and W. McKnight (1979) "Turbulence and Acoustic Signals in a Cross-Flow Heat Exchanger Model," in *Flow-Induced Vibrations*, ASME, New York, pp. 123–128.

Ethembabaoglu, S. (1978) "On the Fluctuating Flow Characteristics in the Vicinity of Gate Slots," Division of Hydraulic Engineering, University of Trondheim, Norwegian Institute of Technology, 1973. Quoted by Rockwell.

Etkin, B., et al. (1957) "Acoustic Radiation from a Stationary Cylinder in a Fluid Stream (Aeolian Tones)," *Journal of the Acoustical Society of America*, **29**, 30–36.

Fahy, F. (1985) *Sound and Structural Vibration*, Academic Press, London.

Farrell, C. (1980) "Uniform Flow Around Circular Cylinders: A Review," in *Advancements in Aerodynamics, Fluid Mechanics, and Hydraulics*, R. E. Arndt (ed.), American Society of Civil Engineers, New York, pp. 301–313.

Ffowcs Williams, J. E., and B. C. Zhao (1989) "The Active Control of Vortex Shedding," *Journal of Fluids and Structures*, **3**, 115–122.

Fitzpatrick, J. A., et al. (1988) "Strouhal Numbers for Flows in Deep Tube Array Models," *Journal of Fluids and Structures*, **2**, 145–160.

Flandro, G. A. (1986) "Vortex Driving Mechanism in Oscillatory Rocket Flows," *AIAA Journal of Propulsion and Power*, **2**, 206–214.

Flatau, A., and W. K. Van Moorhem (1988) "Flow, Acoustic and Structural Resonance Interaction in a Cylindrical Cavity," in *1988 International Symposium on Flow-Induced Vibration and Noise*, W. L. Keith (ed.), Vol. 6, American Society of Mechanical Engineers, New York, pp. 1–12.

Fletcher, N. H. (1979) "Air Flow and Sound Generation in Musical Wind Instruments," *Annual Review of Fluid Mechanics*, **11**, 123–146.

Franke, M. E., and D. L. Carr (1975) "Effect of Geometry on Cavity Flow-Induced Pressure Oscillations," AIAA Paper 75-492, AIAA, New York.

Funakawa, M., and R. Umakoshi (1970) "The Acoustic Resonance in a Tube Bank," *Journal of Japanese Society of Mechanical Engineers*, **13**, 348–355.

Gerrard, J. H. (1955) "Measurements of Sound from Circular Cylinders in an Air Stream," *Proceedings of the Physical Society of London, Section B*, **68**, 453–461.

Gharib, M., and A. Roshko (1987) "The Effect of Flow Oscillations on Cavity Drag," *Journal of Fluid Mechanics*, **177**, 501–530.

Goldstein, M. E. (1976) *Aeroacoustics*, McGraw-Hill, New York.

Grotz, B. J., and F. R. Arnold (1956) "Flow-Induced Vibrations in Heat Exchangers," Department of Mechanical Engineering, Stanford University, Stanford, Calif., Technical Report No. 31, DTIC Number 104568.

Harris, R. E., D. S. Weaver, and M. A. Dokainish (1988) "Unstable Shear Layer Oscillation Past a Cavity in Air and Water Flows," in *1988 International Symposium on Flow-Induced Vibration and Noise*, W. L. Keith (ed.), Vol. 6, American Society of Mechanical Engineers, New York, pp. 13–24.

Horak, Z. (1977) "In Appreciation of Cenek Strouhal," *Journal of Industrial Aerodynamics*, **2**, 185–188.

Jungowski, W. M., et al. (1987) "Tone Generation by Flow Past Confined, Deep Cylindrical Cavities, AIAA Paper AIAA-87-2666, AIAA, New York.

Keefe, R. T. (1962) "An Investigation of the Fluctuating Forces Acting on Stationary Circular Cylinder in a Subsonic Stream, and of the Associated Sound Field," *Journal of the Acoustical Society of America,* **34,** 1711–1714.

Kim, H. J., and P. A. Durbin (1988) "Investigation of the Flow Between a Pair of Circular Cylinders in the Flopping Regime," *Journal of Fluid Mechanics,* **196,** 431–448.

Kinsler, L. E., et al. (1982) *Fundamentals of Acoustics,* 3d ed., Wiley, New York.

Koopman, G. H. (1969) "Wind Induced Vibrations and Their Associated Sound Fields," Dissertation, Catholic University of America.

Leehey, P., and C. E. Hanson (1971) "Aeolian Tones Associated with Resonant Vibration," *Journal of Sound and Vibration,* **13,** 456–483.

Lighthill, M. J. (1952) "On Sound Generated Aerodynamically, I. General Theory," *Proceedings of the Royal Society of London, Series A,* **211,** 564–587.

Mair, W. A., P. D. F. Jones, and R. K. W. Palmer (1975) "Vortex Shedding from Finned Tubes," *Journal of Sound and Vibration,* **39,** 293–296.

Marguart, E. J., and J. P. Grubb (1987) "Bow Shock Dynamics of a Forward-Facing Nose Cavity," AIAA Paper 87-2709, Presented at AIAA 11th Aeroacoustics Conference, Oct. 19–21, 1987, Sunnyvale, Calif.

Mathias, M., et al. (1988) "Low Level Flow-Induced Acoustic Resonances in Ducts," *Fluid Dynamics Research,* **3,** 353–356.

Muller, E. A. (ed.) (1979) *Mechanics of Sound Generation in Flows,* Springer-Verlag, New York.

Parker, R. (1972) "The Effect of the Acoustic Properties of the Environment on Vibration of a Flat Plate Subject to Direct Excitation and to Excitation by Vortex Shedding in an Airstream," *Journal of Sound and Vibration,* **20,** 93–112.

―――― (1979) "Acoustic Resonances in Passages Containing Banks of Heat Exchanger Tubes," *Journal of Sound and Vibration,* **57,** 245–260.

Phillips, O. M. (1956) "The Intensity of Aeolian Tones," *Journal of Fluid Mechanics,* **1,** 607–624.

Price, S. J., and M. P. Paidoussis (1989) "The Flow-Induced Response of a Single Flexible Cylinder in an In-line Array of Rigid Cylinders," *Journal of Fluids and Structure,* **3,** 61–82.

Putnam, A. A. (1965) "Flow-Induced Noise in Heat Exchangers," *Journal of Engineering for Power,* **81,** 417–422.

Quinn, M. C., and M. S. Howe (1984) "The Influence of Mean Flow on the Acoustic Properties of a Tube Bank," *Proceedings of the Royal Society of London, Series A,* **396,** 383–403.

Rayleigh, J. W. S. (1889) "Acoustical Observations II," *Philosophical Magazine,* **7,** 149–162. Contained in *Scientific Papers,* Dover, New York, 1964.

―――― (1894) *Theory of Sound,* Reprinted by Dover, New York, 1945.

―――― (1915) "Aeolian Tones," *Philosophical Magazine,* **29,** 434–444. Contained in *Scientific Papers,* Dover, New York, 1964.

Relf, E. F. (1921) "On the Sound Emitted by Wires of Circular Section when Exposed to an Air Current," *Philosophical Magazine,* **42,** 173–176.

Richardson, E. G. (1923–1924) "Aeolian Tones," *Proceedings of the Physical Society of London,* **36,** 153–157.

―――― (1958) "The Flow and Sound Field Near a Cylinder Towed through Water," *Applied Science Research,* **A7,** 341–350.

Rockwell, D. (1977) "Prediction of Oscillation Frequencies for Unstable Flow Past Cavities," *Journal of Fluids Engineering,* **99,** 294–299.

Rockwell, D., and E. Naudascher (1978) "Review—Self-Sustaining Oscillation of Flow Past Cavities," *Journal of Fluids Engineering,* **100,** 152–165.

Rockwell, D., and A. Schachenmann (1982) "The Organized Shear Layer due to Oscillation for a Turbulent Jet through an Axisymmetric Cavity," *Journal of Sound and Vibration*, **85**, 371–382.

Rogers, J. D., and C. A. Penterson (1977) "Predicting Sonic Vibration in Cross Flow Heat Exchangers—Experience and Model Testing," ASME Paper 77-WA/DE-28.

Ronneberger, D. (1980) "The Dynamics of Shearing Flow over Cavity," *Journal of Fluid Mechanics*, **71**, 565–581.

Sarohia, V., and P. F. Massier (1978) "Control of Cavity Noise," AIAA Paper 75-528, Presented at Third AIAA Aeroacoustics Conference, Palo Alto, Calif., July 20–23.

Stoneman, S. A. T., et al. (1988) "Resonant Sound Caused by Flow Past Two Plates in Tandem in a Duct," *Journal of Fluid Mechanics*, **192**, 455–484.

Stowell, E. Z., and A. F. Deming (1936) "Vortex Noise from Rotating Cylindrical Rods," *Journal of the Acoustical Society of America*, **7**, 190–198.

Strouhal, V. (1878) "Ueber eine Besondere Art Der Tonenegung," *Annalen der Physik and Chemie (Leiipzig)*, Series 5, **5**, 216–251.

Tyler, J. M., and T. G. Sofrin (1962) "Axial Compressor Noise Studies," *Society of Automotive Engineers Transactions*, **70**, 309–332.

von Hole, W. (1938) "Frequenz und Schallstarkemessungen an Hiebtonen," *Akustische Zeitschrift*, **3**, 321–331.

von Karman, T., and H. Rubach (1912) "Uber den Mechanismus des Flussigkeits- und Luftwiderstandes," *Physikalische Zeitschrift*, **13**, 49–59.

von Kruger, F., and A. Lauth (1914) "Theorie der Hiebtone," *Annalen der Physik (Leipzig)*, **44**, 801–812.

Walker, E. M., and G. F. S. Reising (1968) "Flow-Induced Vibrations in Cross Flow Heat Exchangers," *Chemical Process Engineering*, **49**, 95–103.

Wallis, R. P. (1939) "Photographic Study of Fluid Flow Between Banks of Tubes," *Engineering* **148**, 423–425.

Weaver, D. S., and A. Abd-Rabbo (1985) "A Flow Visualization Study of a Square Array of Tubes in Water Cross Flow," *Journal of Fluids Engineering*, **107**, 354–363.

Weaver, D. S., J. A. Fitzpatrick, and M. ElKashlan (1987) "Strouhal Numbers for Heat Exchanger Tube Arrays in Cross Flow," *Journal of Pressure Vessel Technology*, **109**, 219–223.

Welsh, M. C., A. N. Stokes, and R. Parker (1984) "Flow-Resonant Sound Interaction in a Duct Containing a Plate," *Journal of Sound and Vibration*, **95**, 305–323.

Willmarth, W. W., et al. (1978) "Management of Free, Turbulent Shear Layers Associated with Isolated Regions of Separated Flow," *Journal of Aircraft*, **15**, 385–386.

Yudin, E. Y. (1944) "On the Vortex Sound from Rotating Rods," *Zhurnal Teknicheskoi Fizik*, **14**, 561. Translated into English as *NACA TM* 1136, 1947.

Zdravkovich, M. M., and J. A. Nuttall (1974) "On the Elimination of Aerodynamic Noise in a Staggered Tube Bank," *Journal of Sound and Vibration*, **34**, 173–177.

Ziada, S., et al. (1988) "Acoustical Resonance in Tube Arrays," in *Flow-Induced Vibration and Noise in Cylinder Arrays*, M. P. Paidoussis et al., eds., Vol. 3 of 1988 International Symposium on Flow-Induced Vibration and Noise, ASME, New York, pp. 219–254.

Chapter 10

Vibrations of a Pipe Containing a Fluid Flow

A fluid flow through a pipe can impose pressures on the pipe walls that deflect the pipe. Water hammer is deflection of a pipe produced by accelerating fluid. The most familiar form of water hammer is the rumbling of household plumbing heard on quiet mornings when a faucet is turned on. Valve chatter is associated with the opening and closing of valves in response to these pressures. Pipe whip is the dynamic response of a pipe line to an instantaneous rupture. A steady flow can also deflect a pipe. A high-velocity flow through a thin-walled pipe can cause the pipe to buckle or vibrate at large amplitude. These instabilities and instabilities caused by leakage and external axial flow are reviewed in this chapter.

10.1. INSTABILITY OF FLUID-CONVEYING PIPES

10.1.1. Equations of Motion and Solution for Pinned Boundaries

Ashley and Haviland (1950) studied the vibration of a fluid-conveying pipe in conjunction with the flow-induced vibrations of the trans-Arabian pipeline. Housner (1952) was the first to correctly derive the governing equation of motion and predict instability. The type of instability depends on the end conditions on the pipe. Pipes supported at both ends bow out and buckle when the flow velocity exceeds the critical velocity (Housner, 1952; Dodds and Runyan, 1965; Holmes, 1978). Cantilever pipes flail about with large amplitude once the flow velocity exceeds a critical velocity (Gregory and Paidoussis, 1966; Paidoussis, 1970). The most familiar form of this instability is the flailing about of an unrestrained garden hose.

Solutions have been found for articulated pipes (Benjamin, 1961), pipes with unsteady flow (Ginsberg, 1973; Paidoussis, 1987), pipes with lumped masses (Hill and Swanson, 1970; Chen and Jendrzejczyk, 1985), pipes with circumferential (i.e., shell) modes (Shayo and Ellen, 1978;

384

Weaver and Paidoussis, 1977), short pipes (Paidoussis et al., 1986; Matsuzaki and Fung, 1977), viscous fluids and coaxial pipes (Chebair et al., 1988), curved pipes (Unny et al., 1970; Chen, 1973; Misra et al., 1988), and actively controlled pipes (Doki and Aso, 1989). Paidoussis and Issid (1974) and Paidoussis (1987) provide excellent general reviews of dynamics of fluid-conveying pipes. In this section the equations of motion of a straight fluid-conveying pipe are developed and solved following Niordson (1953) and Gregory and Paidoussis (1966).

Figure 10-1 shows a pipe span that has transverse deflection $Y(x, t)$ from its equilibrium position. A fluid of density ρ flows at pressure p and constant velocity v through the internal area A. The length of the pipe is L, the modulus of elasticity of the pipe is E, and its area moment of inertia is I. Consider the small elements cut from the pipe in Figure 10-2. The fluid element in Figure 10-2(a) has been extracted from the pipe element in Figure 10-2(b) for clarity. As the fluid flows through the deflecting pipe, it experiences centrifugal acceleration because of the changing curvature of the deforming pipe. The accelerations are opposed by the vertical component of fluid pressure applied to the fluid element and the pressure force F per unit length applied on the fluid element by the pipe walls. A balance of forces on the fluid element in the y direction for small deformations gives

$$F - pA\frac{\partial^2 Y}{\partial x^2} = \rho A\left(\frac{\partial}{\partial t} + v\frac{\partial}{\partial x}\right)^2 Y. \tag{10-1}$$

The pressure gradient in the fluid along the length of the pipe is opposed by the shear stress of fluid friction against the pipe walls. For constant flow velocity, summing forces parallel to the pipe axis in Figure 10-2(a) gives

$$A\frac{\partial p}{\partial x} + \tau S = 0, \tag{10-2}$$

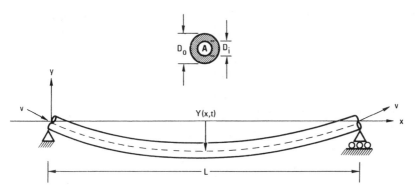

Fig. 10-1 A fluid-conveying pipe with pinned ends.

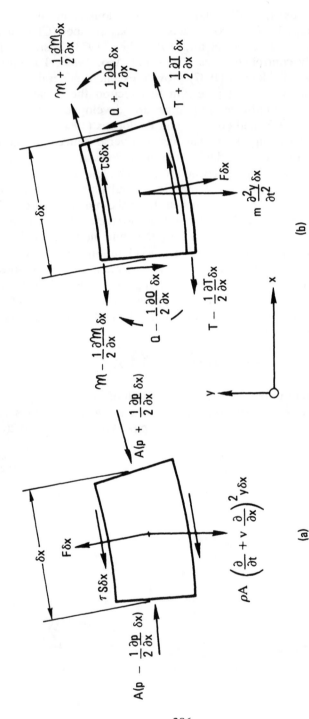

Fig. 10-2 Forces and moments acting on elements of (a) the fluid and (b) the pipe.

where S is the inner perimeter of the pipe and τ is the shear stress on the inner surface of the pipe. The equations of motion of the pipe element are derived from Figure 10-2(b). Summing forces parallel to the pipe axis gives

$$\frac{\partial T}{\partial x} + \tau S - Q \frac{\partial^2 Y}{\partial x^2} = 0, \tag{10-3}$$

where T is the longitudinal tension in the pipe and Q is the transverse shear load carried by the pipe. The forces on the element of the pipe that act normal to the pipe axis accelerate the pipe element in the y direction. For small deformations,

$$\frac{\partial Q}{\partial x} + T \frac{\partial^2 Y}{\partial x^2} - F = m \frac{\partial^2 Y}{\partial t^2}, \tag{10-4}$$

where m is the mass per unit length of the empty pipe.

The transverse shear force Q in the pipe is related to the bending moment M_b in the pipe and the pipe deformation by

$$Q = -\frac{\partial M_b}{\partial x} = -EI \frac{\partial^3 Y}{\partial x^3}. \tag{10-5}$$

Since Q is proportional to $\partial Y^3/\partial x^3$, the third term on the left side of Eq. 10-3 is of the order of Y^2 and is neglected for linear, small-deformation analysis. Equations 10-1, 10-4, and 10-5 are combined to eliminate F and Q:

$$EI \frac{\partial^4 Y}{\partial x^4} + (pA - T) \frac{\partial^2 Y}{\partial x^2} + pA \left(\frac{\partial}{\partial t} + v \frac{\partial}{\partial x} \right)^2 Y + m \frac{\partial^2 Y}{\partial t^2} = 0. \tag{10-6}$$

The shear stress τ is eliminated from Eqs. 10-2 and 10-3 to give

$$\frac{\partial(pA + T)}{\partial x} = 0. \tag{10-7}$$

This equation implies that $pA - T$ is independent of position along the span of the pipe. If we assume that at the end of the pipe the tension in the pipe is zero and the fluid pressure is equal to the ambient pressure, $p = T = 0$ at $x = L$, then Eq. 10-7 implies that $pA - T = 0$ for all x. If the pipe is fitted with a convergent nozzle, consideration of momentum gives $pA - T = \rho A v(v_j - v)$, where v_j is the throat velocity (Gregory and Paidoussis, 1966).

Substituting $pA - T = 0$ into Eq. 10-6 gives the equation of motion for free transverse vibration of a straight, tension-free fluid-conveying pipe:

$$EI \frac{\partial^4 Y}{\partial x^4} + \rho A v^2 \frac{\partial^2 Y}{\partial x^2} + 2\rho A v \frac{\partial^2 Y}{\partial x \partial t} + M \frac{\partial^2 Y}{\partial t^2} = 0, \tag{10-8}$$

where $M = m + \rho A$ is the mass per unit length of the pipe and the fluid in the pipe. The boundary conditions associated with a pinned-pinned span of Figure 10-1 are that the displacement and moment are zero at the ends,

$$Y(0, t) = Y(L, t) = 0, \qquad \frac{\partial^2 Y}{\partial x^2}(0, t) = \frac{\partial^2 Y}{\partial x^2}(L, t) = 0. \qquad (10\text{-}9)$$

For a cantilever pipe, the displacement and slope at the clamped end are zero,

$$Y(0, t) = \frac{\partial Y}{\partial x}(0, t) = 0, \qquad \frac{\partial^3 Y}{\partial x^3}(L, t) = \frac{\partial^2 Y}{\partial x^2}(L, t) = 0, \qquad (10\text{-}10)$$

and there is no moment or shear at the free end. Derivations of these equations of motion and the boundary conditions using an energy approach are given by Housner (1952), Crandall (1968), and Paidoussis and Issid (1974).

The first and last terms in Eq. 10-8 are pipe stiffness and inertia terms that are independent of flow. The second term from the left is centrifugal force due to acceleration of fluid through the curvature of the deformed pipe. This term is identical in form to an axial compression term (compare with the second term from the left in Eq. 10-6) and it produces a reduction in natural frequency and, eventually, buckling. The third term from the left in Eq. 10-8 is the force required to rotate fluid elements with local pipe rotation. It is called a *Coriolis force*. Its mixed derivative causes an asymmetric distortion of classical mode shapes and it leads to a flutterlike instability. It also makes Eq. 10-8 difficult to solve.

Equation 10-8 does not possess classical normal modes (Appendix A). Its solution cannot be separated simply into time and space components. For example, if a trial solution of the form

$$Y(x, t) = \bar{y}(x) \sin \omega t \qquad (10\text{-}11)$$

is substituted into Eq. 10-8, it can be seen that the Coriolis force term varies as $\cos \omega t$, while the remainder of the terms have a $\sin \omega t$ time dependence. This suggests that solutions should be written as

$$Y(x, t) = a_1 \bar{y}(x) \sin \omega t + a_2 \bar{y}(x) \cos \omega t, \qquad (10\text{-}12)$$

where a_1 and a_2 are interdependent.

The pinned–pinned boundary conditions of the pipe span shown in Figure 10-1 and given by Eq. 10-9 are satisfied by the set of sinusoidal mode shapes,

$$\bar{y}(x) = \sin \frac{n \pi x}{L}, \qquad n = 1, 2, 3, \ldots. \qquad (10\text{-}13)$$

These mode shapes pass through the first, second, and fourth terms in Eq. 10-8 unaltered, but the mixed-derivative Coriolis force generates spatially asymmetric terms for a symmetric mode shape ($n = 1, 3, 5, \ldots$) and spatially symmetric terms for an asymmetric mode shape ($n = 2, 4, 6, \ldots$). These considerations imply that the solution of the equation of motion of a fluid-conveying pipe with pinned end conditions is the sum of symmetric and asymmetric spatial modes with sine and cosine time components (Housner, 1952),

$$Y_j(x, t) = \sum_{n=1,3,5,\ldots} a_n \sin\frac{n\pi x}{L} \sin\omega_j t + \sum_{n=2,4,6,\ldots} a_n \sin\frac{n\pi x}{L} \cos\omega_j t,$$

$$j = 1, 2, 3, \ldots, \quad (10\text{-}14)$$

where ω_j is the natural frequency of the j vibration mode. This trial solution is substituted into Eq. 10-8. The Coriolis force produces terms containing $\cos(n\pi x/L)$. These cosine terms can be expanded in a Fourier half-range series of sine functions over the pipe span $0 \leq x \leq L$:

$$\cos\frac{n\pi x}{L} = \sum_{p=1,2,3,\ldots} b_{np} \sin\frac{p\pi x}{L}, \quad n = 1, 2, 3, \ldots, \quad (10\text{-}15)$$

where

$$b_{np} = \begin{cases} 0, & n + p = \text{even}, \\ 4p/[\pi(p^2 - n^2)] & n + p = \text{odd}. \end{cases} \quad (10\text{-}16)$$

The Fourier series converges relatively slowly over the span of the pipe and not at all at the ends, but it does allow the spatial dependence to be factored out of the solution. With these substitutions, the terms in Eq. 10-8 can be grouped according to whether they contain $\sin\omega t$ or $\cos\omega t$. The coefficients of each group are set to zero to give the following equations:

$$a_n\left[EI\left(\frac{n\pi}{L}\right)^4 - \rho A v^2\left(\frac{n\pi}{L}\right)^2 - M\omega_j^2\right] = \frac{8\rho A v \omega_j}{L} \sum_{p=2,4,6,\ldots} a_p \frac{pn}{n^2 - p^2},$$

$$n = 1, 3, 5, \ldots, \quad (10\text{-}17)$$

$$a_n\left[EI\left(\frac{n\pi}{L}\right)^4 - \rho A v^2\left(\frac{n\pi}{L}\right)^2 - M\omega_j^2\right] = -\frac{8\rho A v \omega_j}{L} \sum_{p=1,3,5,\ldots} a_p \frac{pn}{n^2 - p^2},$$

$$n = 2, 4, 6, \ldots. \quad (10\text{-}18)$$

These equations can be put in matrix form:

$$|[K] - \omega_j^2 M[I]|\{\bar{a}\} = 0, \quad (10\text{-}19)$$

where $\bar{a}$ is an N-by-1 column vector containing $a_1, a_2, \ldots, [I]$ is the identity matrix with values of one on the diagonal and all other entries equal to zero, and $[K]$ is the stiffness matrix with the following entries:

$$
k_{np} = \begin{cases} EI(\pi n/L)^4 - \rho A v^2(\pi n/L)^2, & n = p, \\ -(8\rho A v \omega_j/L)[np/(n^2 - p^2)], & n = \text{odd}, n + p = \text{odd}, \\ (8\rho A v \omega_j/L)[np/(n^2 - p^2)], & n = \text{even}, n + p = \text{odd}, \\ 0, & n \neq p, n + p = \text{even}. \end{cases} \quad (10\text{-}20)
$$

Nontrivial solutions to Eq. 10-19 are sought by setting the determinant of the coefficient matrix to zero:

$$
\|[K] - \omega_j^2 M[I]\| = 0. \quad (10\text{-}21)
$$

Because the system has an infinite number of natural modes, practical solution of Eq. 10-21 requires that only the first few modes be considered. It only the first two modes are included in an appropriate analysis, then a_3, a_4, . . . are set equal to zero and Eq. 10-21 becomes

$$
\left[1 - \left(\frac{v}{v_c}\right)^2 - \left(\frac{\omega_j}{\omega_N}\right)^2\right]\left[16 - 4\left(\frac{v}{v_c}\right)^2 - \left(\frac{\omega_j}{\omega_N}\right)^2\right]
$$
$$
- \frac{256}{9\pi^2}\left(\frac{v}{v_c}\right)^2\left(\frac{\rho A}{M}\right)\left(\frac{\omega_j}{\omega_N}\right)^2 = 0, \quad (10\text{-}22)
$$

where ω_N is the circular fundamental natural frequency of the pipe in the absence of fluid flow,

$$
\omega_N = 2\pi f_N = \frac{\pi^2}{L^2}\left(\frac{EI}{M}\right)^{1/2}, \quad (10\text{-}23)
$$

and the critical velocity of flow for buckling of the pipe is

$$
v_c = \frac{\pi}{L}\left(\frac{EI}{\rho A}\right)^{1/2}. \quad (10\text{-}24)
$$

The exact solution of Eq. 10-22 determines the first two natural frequencies of the pipe:

$$
\left(\frac{\omega_j}{\omega_N}\right)^2 = \alpha \pm \left\{\alpha^2 - 4\left[1 - \left(\frac{v}{v_c}\right)^2\right]\left[4 - \left(\frac{v}{v_c}\right)^2\right]\right\}^{1/2}, \quad j = 1, 2, \quad (10\text{-}25)
$$

where

$$
\alpha = \frac{17}{2} - \left(\frac{v}{v_c}\right)^2\left[\frac{5}{2} - \left(\frac{128}{9\pi^2}\right)\left(\frac{\rho A}{M}\right)\right]. \quad (10\text{-}26)
$$

ω_1 and ω_2 are real for all $v/v_c \leq 1$. ω_1 is only a weak function of mass ratio. The approximation to Eq. 10-25,

$$
\frac{\omega_1}{\omega_N} = \left[1 - \left(\frac{v}{v_c}\right)^2\right]^{1/2}, \quad (10\text{-}27)
$$

is accurate to within 2.6% for all $v \leq v_c$ and $\rho A/M \leq 0.5$ and is within 12.8% for all $v \leq v_c$ and $\rho A/M \leq 1$.

At zero velocity the first two natural frequencies of the pipe are ω_N and $4\omega_N$, which are the classical natural frequencies (Appendix A). The natural frequencies decrease with increasing velocity of fluid flow. When the velocity of flow through the pipe equals the critical velocity, the lowest natural frequency of the pipe goes to zero (Eq. 10-25 or Eq. 10-27),

$$\lim_{v \to v_c} \omega_1 = 0. \tag{10-28}$$

When $v = v_c$, the pipe bows out and buckles, because the force required to make the fluid conform to the pipe curvature resulting from small deformation exceeds the stiffness of the pipe. In academe, this instability is called a *static cusp bifurcation*.

Mathematically, the instability of fluid-conveying pipes with both ends supported is due to the centrifugal force term $\rho A v^2 \, \partial^2 Y/\partial x^2$ in Eq. 10-8 (see Exercise 2 below). As noted earlier, this term is identical in form to the term $-T \, \partial^2 Y/\partial x^2$ associated with tension in the pipe. This suggests that the critical velocity for instability of pipes supported at both ends can be estimated by equating the coefficients of these terms at the compression required for buckling,

$$v_c \approx \left[\frac{|T_b|}{\rho A} \right]^{1/2}, \tag{10-29}$$

where T_b is the load required to buckle the pipe. This procedure is exact for the simply supported pipe and compares well with the numerical results of Naguleswaran and Williams (1968) for clamped–clamped spans.

Equation 10-27 is compared with experimental results in Figure 10-3. The experiments were made on a 10.5 ft (3.2 m) long span aluminum pipe with a 1 in (2.54 cm) diameter and a 0.065 in (1.65 mm) thick wall. Water was supplied to the pipe from a reservoir. The remaining experimental parameters were $\rho A = 0.008$ slug/ft (0.384 kg/m), $E = 10 \times 10^6$ psi (68.9×10^9 Pa), $I = 1.0 \times 10^{-6}$ ft^4 (8.64×10^{-9} m^4), $m = 0.00712$ slug/ft (0.342 kg/m). The velocity required for instability of this long-walled pipe is 129 ft/sec (39.4 m/sec) (Eq. 10-24).

Equation 10-19 can be solved for the ratio a_1 to a_2, which determines the mode shape. For the fundamental mode,

$$\frac{a_2}{a_1} = -\frac{8}{3\pi^2} \left(\frac{\omega_1 L}{v_c} \right) \left(\frac{v}{v_c} \right) \left[16 - 4 \left(\frac{v}{v_c} \right)^2 - \left(\frac{\omega_1}{\omega_N} \right)^2 \right]^{-1}, \tag{10-30}$$

where ω_1 is the lowest natural frequency solution to Eq. 10-25. For all cases with v less than v_c, it can be easily shown that the ratio $|a_2/a_1|$ is less than 0.094. Thus, the first sinusoidal bending mode will dominate the response.

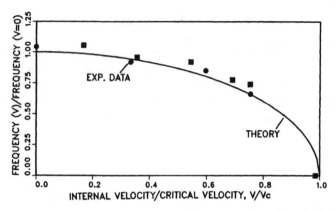

Fig. 10-3 Comparison of theoretical (Eq. 10-27) and experimental results for reduction of fundamental frequency of a pinned-pinned pipe with increasing flow velocity (Dodds and Runyan, 1965).

Exercises

1. Starting with Eq. 10-6, it is easy to show that the equation of motion of a fluid-conveying pipe with internal tension and pressure is

$$EI\frac{\partial^4 Y}{\partial x^4} + (pA - T)\frac{\partial^2 Y}{\partial x^2} + \rho A v^2 \frac{\partial^2 Y}{\partial x^2} + 2\rho A v \frac{\partial^2 Y}{\partial x\, \partial t} + M\frac{\partial^2 Y}{\partial t^2} = 0,$$

where $pA - T$ is constant along the span. If $p = 0$, what are the solutions corresponding to Eqs. 10-25 through 10-27 with the inclusion of a steady tension T in the pipe? Does tension increase or decrease pipe stability?

2. Because the buckling instability of a fluid-conveying pipe is a static rather than a dynamic instability, the time-independent inertia and Coriolis terms in Eq. 10-8 ($\partial^2 Y/\partial x\, \partial t$ and $\partial^2 Y/\partial t^2$) can be omitted in determining a critical velocity. Use this approach and determine the critical velocity for (a) a pinned–pinned pipe and (b) a clamped–clamped pipe whose mode shape is given by Eq. 10-36 but with $L\lambda_1 = 4.73$ and $\sigma = 0.9825$.

3. Consider a pinned–pinned tension-free pipe ($T = 0$) that has internal pressure p. Is the pressure stabilizing or destabilizing? Does this depend on whether there is axial constraint at the downstream end? Under what pressure, or suction, will the pipe buckle? Can internal pressurization increase the natural frequencies of a hollow helicoper rotor blade? Naguleswaran and Williams (1968) and Paidoussis (1987) discuss the solution.

10.1.2. Cantilever and Curved Pipes

The free vibrations of the straight fluid-conveying cantiliver shown in Figure 10-4 are described by the solution of the equation of motion (Eq. 10-8) with the boundary conditions of Eq. 10-10. Several methods of solution are given by Gregory and Paidoussis (1966). The approximate solution presented in this section is analogous to the modal expansion used in the pinned–pinned beam case, and it can be used to generate solutions of any order.

The deflection of the cantilever is assumed to be of the form

$$Y(x, t) = \text{Real}\,[\Psi(x/L)e^{i\omega t}], \tag{10-31}$$

where Real [ ] denotes the real part and i is the imaginary constant $\sqrt{-1}$. Thus, if ω is real,

$$e^{i\omega t} = \cos \omega t + i \sin \omega t. \tag{10-32}$$

This equation describes steady harmonic vibration at frequency ω. If ω is imaginary, $\omega = i\omega_R$, where ω_R is a real number and $e^{i\omega t} = e^{-\omega_R t}$. This describes either exponential decay of vibration with time if $\omega_R > 0$ or exponential growth if $\omega_R < 0$. In general, ω has real and imaginary parts, so the vibrations of the cantilever are contained by an exponentially growing or decaying envelope.

If the trial solution of Eq. 10-31 is substituted into the equation of motion (Eq. 10-8), the result is

$$\Psi'''' + V^2\Psi'' + 2i\beta^{1/2}V\Omega\Psi' - \Omega^2\Psi = 0, \tag{10-33}$$

where the primes denote the derivatives with respect to x/L. The

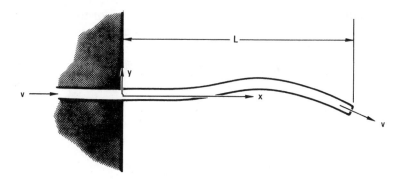

Fig. 10-4 Fluid-conveying cantilever pipe.

dimensionless constants β, Ω, and V are

$$\beta = \frac{\rho A}{M}, \quad \Omega = \omega L^2 \left(\frac{M}{EI}\right)^{1/2}, \quad V = vL\left(\frac{\rho A}{EI}\right)^{1/2}, \quad M = \rho A + m. \quad (10\text{-}34)$$

M equals the mass of the empty pipe plus the mass of internally contained fluid per unit length.

The mode shapes of the fluid-conveying cantilever are approximated by a series comprising the cantilever mode shapes that would be found in the absence of flow:

$$\Psi\left(\frac{x}{L}\right) = \sum_{r=1}^{\infty} a_r \bar{y}\left(\frac{x}{L}\right), \quad (10\text{-}35)$$

where

$$\bar{y}\left(\frac{x}{L}\right) = \cosh\left(L\lambda_r x/L\right) - \cos\left(L\lambda_r x/L\right)$$
$$- \sigma_r[\sinh\left(L\lambda_r x/L\right) - \sin\left(L\lambda_r x/L\right)]. \quad (10\text{-}36)$$

For the first three modes, values of $L\lambda_r$ and σ_r are $L\lambda_1 = 1.875$, $L\lambda_2 = 4.694$, $L\lambda_3 = 7.855$, $\sigma_1 = 0.734099$, $\sigma_2 = 1.018466$, $\sigma_3 = 0.999225$ (Blevins, 1979). These modes satisfy the boundary conditions of Eq. 10-10 and are orthogonal over the span of the cantilever,

$$\int_0^1 \bar{y}_r\left(\frac{x}{L}\right)\bar{y}_s\left(\frac{x}{L}\right) d\left(\frac{x}{L}\right) = \begin{cases} 1, & r = s, \\ 0, & r \neq s. \end{cases} \quad (10\text{-}37)$$

If the series of Eq. 10-35 is substituted into Eq. 10-33, the equation of motion of the fluid-conveying cantilever becomes

$$\sum_{r=1}^{\infty} [\bar{y}_r'''' - \Omega^2 \bar{y}_r + V^2 \bar{y}_r'' + 2i\beta^{1/2}V\Omega\bar{y}']a_r = 0. \quad (10\text{-}38)$$

This set of equations determines the natural frequencies and mode shapes of the fluid-conveying cantilever. Solutions can be sought by expressing the derivatives of the mode shapes as series in terms of the mode shape as was done in the previous section for pinned boundary conditions:

$$\bar{y}_r' = \sum_{s=1}^{\infty} b_{rs}\bar{y}_s, \quad \bar{y}_r'' = \sum_{s=1}^{\infty} c_{rs}\bar{y}_s, \quad \bar{y}_r'''' = \lambda_r^4 \bar{y}_r, \quad (10\text{-}39)$$

where

$$b_{rs} = \frac{4}{(\lambda_s/\lambda_r)^2 + (-1)^{r+s}},$$

$$c_{rs} = \begin{cases} \dfrac{4(\lambda_r\sigma_r - \lambda_s\sigma_s)}{(-1)^{r+s} - (\lambda_s/\lambda_r)^2}, & r \neq s, \\ \lambda_r\sigma_r(2 - \lambda_r\sigma_r) & r = s. \end{cases} \quad (10\text{-}40)$$

Substituting these series into Eq. 10-38 gives

$$\sum_{r=1}^{\infty} \left[(\lambda_r^4 - \Omega^2)\bar{y}_r + V^2 \sum_{s=1}^{\infty} c_{rs}\bar{y}_s + 2i\beta^{1/2}V\Omega \sum_{s=1}^{\infty} b_{rs}\bar{y}_s \right] a_r = 0. \quad (10\text{-}41)$$

If this equation is multiplied through by $\bar{y}_s$ and the resultant equation is integrated over the span, then by using the orthogonality conditions of Eq. 10-37, the equations can be written in matrix form:

$$|[K] - \Omega^2[I]|\{\bar{a}\} = 0, \quad (10\text{-}42)$$

where the entries of the stiffness matrix $[K]$ are

$$k_{rs} = \begin{cases} \lambda_r^4 + V^2 c_{sr} + 2i\beta^{1/2}V\Omega b_{sr}, & r = s, \\ V^2 c_{sr} + 2i\beta^{1/2}V\Omega b_{sr}, & r \neq s. \end{cases} \quad (10\text{-}43)$$

Nontrivial solutions of the matrix equations (Eq. 10-42) exist only if the determinant of the coefficient matrix is zero:

$$|[K] - \Omega^2[I]| = 0. \quad (10\text{-}44)$$

The solution of the equation determines the dimensionless natural frequencies Ω_j as functions of the dimensionless mass ratio, β, and the dimensionless velocity, V. If we set a value of Ω, Eq. 10-44 fixes a relationship between these two parameters.

The dimensionless frequency (Eq. 10-34) can have real and imaginary parts,

$$\Omega = \Omega_R + i\Omega_I, \quad (10\text{-}45)$$

but all the remaining entries in Eq. 10-44 are real. Ω_R produces vibration within an exponentially decaying envelope if $\Omega_I > 0$ or an expanding envelope if $\Omega_I < 0$. Then condition $\Omega_I = 0$ defines neutral stability.

A stability map, determined by $\Omega_I = 0$ with multimodal analysis, is shown in Figure 10-5. With mass ratios in the neighborhood of $\rho A/M = 0.295, 0.67$, and 0.88, increases in flow velocity cause the pipe to lose stability, then briefly regain stability. The real component of frequency is shown in Figure 10-6 for neutral stability, $\Omega_I = 0$. Gregory and Paidoussis (1966) found that at least three modes must be included in the approximate stability analysis to reproduce most of the features of the "exact" numerical results. Their theory is shown in comparison with experimental data in Figure 10-7. Other experiments on fluid-conveying pipes are summarized by Jendrzejczyk and Chen (1985).

The real part of frequency does not go to zero at the onset of instability for a cantilever pipe, as it did for buckling of the pinned–pinned pipe. The fluid-conveying cantilever does not bow out and buckle when flow velocity exceeds a critical velocity like the pinned–pinned pipe. Instead, it flails about with finite frequency like an unrestrained garden hose.

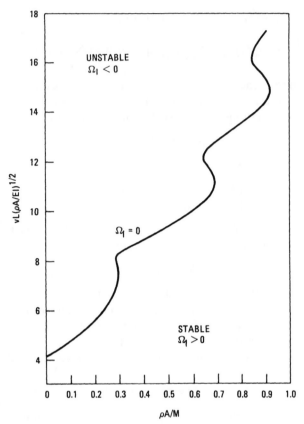

Fig. 10-5 Critical flow velocity for onset of instability in a cantilever pipe as a function of mass ratio (Gregory and Paidoussis, 1966).

The instability of a cantilever pipe is associated with the Coriolis force, which is a follower force and creates a nonconservative system (Nemat–Nassar, 1966). Following Chen (1981), we can show that the Coriolis force does net work on a fluid-conveying pipe only if at least one end of the pipe is free to move laterally:

$$\Delta W_c = 2\rho A v \int_0^Y \int_0^L \frac{\partial^2 Y}{\partial x\, \partial t}\, dy\, dx = \rho A v \int_0^Y \left(\frac{\partial Y}{\partial t}\right)^2 \Big|_0^L dy.$$

If there is movement at the ends, $\partial Y/\partial t \neq 0$; hence $\Delta W_c > 0$ and the Coriolis force provides an avenue for instability. Mathematically, the cantilever pipe instability is called a *Hopf bifurcation* (see Holmes, 1978; Bajaj et al., 1980; Hagedorn, 1988) and the resultant vibrations can be chaotic (Paidoussis and Moon, 1988; Tang and Dowell, 1988).

Table 10-1 summarizes the stability of fluid-conveying straight and

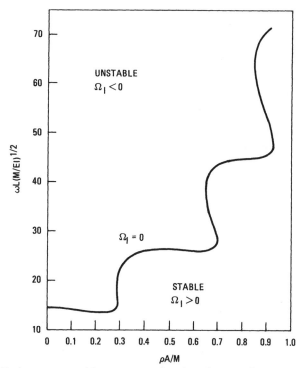

Fig. 10-6 Real component of frequency as a function of mass ratio at onset of instability of a cantilever pipe (Gregory and Paidoussis, 1966).

curved pipes. For curved pipes, the curvature couples lateral and axial (i.e., extensional) motions. A curved pipe must bear the steady inplane load associated with internal pressure and change in angular momentum of the flowing fluid (see Eq. 10-58), and this steady extensional load stabilizes a curved pipe as long as both ends are fixed (Chen, 1973; Hill and Davis, 1974; Svetlitsky, 1977). If one or both ends can move axially, then the curved pipe can experience buckling and flutterlike instabilities (Misra et al., 1988; Holmes, 1978).

Exercise

1. Use a power series $\Psi(x) = \sum a_r x^r$ to express the spatial dependence of the solution of Eq. 10-8 for a cantilever pipe. Develop expressions for the entries of the stiffness matrix if only two terms corresponding to $r = 1$ and $r = 2$ are retained in the analysis.

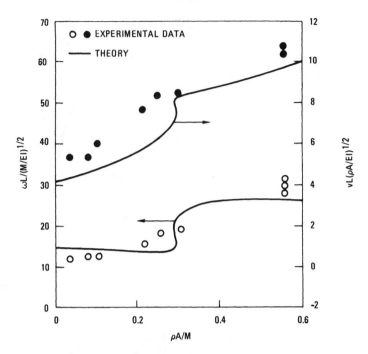

Fig. 10-7 Comparison of theory and experimental data for onset of instability of a fluid-conveying cantilever pipe (Gregory and Paidoussis, 1966).

Table 10-1 Stability of fluid-conveying pipes

Pipe geometry	Boundary conditions[a]	In-plane or out-of-plane	Instability mode[b]
Straight	Clamped–clamped	Both	Buckling
Straight	Clamped–pinned	Both	Buckling
Straight	Pinned–pinned	Both	Buckling (Eq. 10-24)
Straight	Clamped–free	Both	Flutter (Figure 10-5)
Curved	Clamped–clamped	Both	Stable[c]
Curved	Clamped–pinned	Both	Stable[c]
Curved	Pinned–pinned	Both	Stable[c]
Curved	Clamped–free	In-plane	Buckling
Curved	Clamped–free	Out-of-plane	Flutter
Curved	Clamped–sliding	In-plane	Flutter

Source: Adapted from Misra et al. (1988).
[a] Clamped = prevents transverse displacement, transverse slope, and axial motion; pinned = prevents transverse displacement and axial motion; sliding = prevents transverse displacement and slope but permits axial motion.
[b] First instability mode with increasing flow velocity.
[c] Inplane steady extensional load provides stability.

10.1.3. Example

Consider the pinned–pinned pipe span shown in Figure 10-1 and described by the parameters in text between Eqs. 10-29 and 10-30. The mass ratio and critical velocity (Eq. 10-24) for internal flow are

$$\frac{\rho A}{M} = 0.537, \qquad v_c = 129 \text{ ft/sec } (39 \text{ m/sec}).$$

If two possible flow velocities, 75 ft/sec and 150 ft/sec (23 m/sec and 46 m/sec) are proposed, then the higher flow velocity can immediately be rejected for use because it exceeds the critical velocity that would cause the pipe to buckle. For a flow velocity of 75 ft/sec (23 m/sec), Eq. 10-24 gives

$$\frac{v}{v_c} = 0.581.$$

The reduction in the fundamental frequency due to the flow velocity is given by Eq. 10-27:

$$\frac{\omega_1}{\omega_N} = 0.814.$$

The natural frequency in the absence of flow, ω_N, can be calculated from Eq. 10-23 to be $\omega_N = 28.4$ radians/sec (4.52 Hz); thus, the fundamental natural frequency of the fluid-conveying cantilever is 3.68 Hz.

If this same pipe is clamped at one end and the constraint is released at the other end, the pipe becomes a cantilever. Figure 10-5 predicts that the onset of instability of the cantilever arises at a flow velocity of

$$v = \frac{9.4(EI/\rho A)^{1/2}}{L} = 387 \text{ ft/sec } (118 \text{ m/sec}).$$

Note that the flow velocity required for instability of a cantilever is higher than that for an equivalent pinned–pinned span. This implies that the addition of a constraint to the free end of the cantilever, such as that imposed by grasping the free end, can cause a stable cantilever to buckle immediately, since the additional constraint produces a geometry that approaches the pinned–pinned case.

10.2. EXTERNAL AXIAL FLOW

External axial flow over pipes and rods occurs in parallel-flow heat exchangers, in nuclear reactor fuel pins, in towed marine systems, and even in the swimming of certain microorganisms (Taylor, 1952). Here the stability of circular rods and cylinders in an axial flow will be explored. Paidoussis (1987) reviews this field.

The responses of an elastic cylinder to external axial flow and to internal flow (Sections 10.1.1 and 10.1.2) are very similar in terms of the fluid forces and the dynamic response. The geometry of flow is shown in Figure 7-8, Section 7.3.3 (this section discusses the response of stable cylinders to turbulence in the parallel flow). The equation of motion derived by Paidoussis and Ostoja-Starzewski (1981) for axial fluid flow over an elastic cylinder is

$$EI\frac{\partial^4 Y}{\partial x^4} + \rho A U^2 \frac{\partial^2 Y}{\partial x^2} + 2\rho A U \frac{\partial^2 Y}{\partial x \, \partial t} - \tfrac{1}{2}\rho U^2 DC_f\Big(1 + \frac{D}{D_h}\Big)(L - x)\frac{\partial^2 Y}{\partial x^2}$$

$$+ \tfrac{1}{2}\rho DU C_f \frac{\partial Y}{\partial t} + \tfrac{1}{2}\rho U^2 DC_f\Big(1 + \frac{D}{D_h}\Big)\frac{\partial Y}{\partial x} + M\frac{\partial^2 Y}{\partial t^2} = 0. \quad (10\text{-}46)$$

The fluid forces on each element of the cylinder are assumed to be identical to those that would be exerted on an infinitely long rigid cylinder of the same velocity and orientation to the flow. Much of the notation is identical to that of the previous sections; that is, $Y(x, t)$ is the lateral deformation of the pipe, E is the modulus of elasticity, I is the area moment of inertia, ρ is the density of the fluid, x is axial distance along the span, U is the axial fluid velocity over the cylinder, and $M = m + \rho A$ is the mass per unit length of the cylinder, m, plus the external added mass of fluid, ρA, where $A = \pi D^2/4$ is the cross-sectional area of the cylinder of diameter D in a large reservoir; see Chapter 2, Section 2.2. C_f is the friction coefficient for fluid flow over the cylinder such that the mean axial force on the cylinder per unit length is $\tfrac{1}{2}\rho U^2 DC_f$. C_f is determined by boundary layer skin frictions and roughness of the surface; $C_f = 0.02$ is a typical value.

(The derivations of Eqs. 10-46 and 10-8 are complex because of the difficulty in formulating a consistent model for the fluid force terms. This has led to errors. See Housner's 1952 revision of Ashley and Haviland's 1950 derivation for internal flow, Dowling's 1988 revision of Paidoussis's 1966 derivation for external flow, and Ginsberg's 1973 revision of Chen's 1971 derivation for unsteady flow. The author has continued this tradition with a sign error in the first edition, which was resolved by Mr. C. S. Lin. The corrections have been made.)

Equation 10-46 for external flow over a cylinder is identical in form to Eq. 10-8, the equation of motion of a pipe with internal flow, if the skin friction coefficient C_f is set to zero and if v is substituted for U. Thus, the solutions and the observed phenomena are very similar for internal and external axial flow. As with internal flow, the external axial flow over a cylinder can excite both buckling and flutterlike instabilities at velocities above a critical velocity defined approximately by Eq. 10-24 or 10-29 (Paidoussis, 1966). Similarly, axial tension stabilizes the cylinder against instability. The axial fluid friction terms in Eq. 10-46 can generate this stabilizing tension if the downstream end is free to move axially, such as with towed cylinders. Lee (1981) and Triantafyllou and Chryssostomidis (1985) predict that stability of towed cylinders is assured if either (a) the tension at the downstream end exceeds $\rho A U^2$ or (b) the overall length-to-diameter ratio exceeds $\pi/(2C_f)$. Long, towed, neutrally buoyant hydrophone strings are always stable as a result of the friction-induced tension (Dowling, 1988). Short cylinders can be stabilized by a drogue or sea anchor at the downstream end.

The trailing end shape has a large influence on the stability and turbulence-induced vibration of cantilever cylinders in external axial flow because the end shape determines drag-induced tension and because cantilevers are very responsive to forcing at the free end. The experimental results of Wambsganss and Jendrzejczyk (1979) are in agreement with Paidoussis (1976) in finding that a squared-off end cap minimizes excitation of cantilever cylinders in a parallel flow. Wambsganss and Jendrzejczyk's 1979 compilations of trailing edge geometry and the relative response of cantilevers to axial flow excitation are given in Table 10-2. See Chapter 7, Section 7.3.3 for a discussion of turbulence-induced vibration of cylinders in axial flow.

Unfortunately, the predictions of instability of cylinders in axial flow generally agree with experiments to no better than a factor of 2. Transition between various instability mechanisms and turbulence excitation is not well understood. Scaled tests are required if quantitatively accurate predictions are needed for a particular system at velocities that can produce instability.

Riley et al. (1988) review the related field of the response of compliant coatings to flow.

10.3. PIPE WHIP

All pipes have flaws and most pipes have some corrosion. If a flaw in a pipe grows to a critical length or corrosion erodes enough material from the pipe wall, the pipe can quickly rupture through the cross section. If the pipe is part of a high-pressure system, fluid will blow down through the rupture into the air. The rush of fluid and the unrestrained fluid pressure place an impulsive reaction on the pipe which can cause the pipe to whip about and threaten personnel and structures. It is a common practice for designers of power plants to postulate pipe ruptures and then perform analysis to determine what restraints or armor is required to prevent secondary failures (U.S. Atomic Energy Commission, 1973).

Table 10-2 Effect of trailing edge geometry on response of cantilevers in axial flow

Trailing end geometry[a]	Displacement response		Damping	
	$(y/d)_{rms}$	Normalized[b]	ζ (%Cr)	Normalized[b]
	0.024	0.80	0.20	0.77
	0.024	0.80	0.24	0.92
	0.028	0.93	0.14	0.54
	0.030	1.0	0.26	1.00
	0.042	1.4	0.29	1.12
	0.046	1.5	0.38	1.46
	0.046	1.5	0.40	1.54
	0.050	1.7	0.36	1.38
	0.074	2.5	0.43	1.65
	0.084	2.8	0.26	1.00
	0.112	3.7	0.23	0.88

Source: Wambsganss and Jendrzejczyk (1979).
[a] Ordered relative to effectiveness in attenuating vibration.
[b] Normalized to the value for the square-end geometry (A).

The break is usually postulated to occur just past a junction or bend in the pipe, as shown in Figure 10-8. The fluid spews out at right angles to the local pipe axis and the fluid reaction force tends to bend the pipe directly rather than promote an instability. The reaction force on the pipe can be evaluated from the fluid momentum equation using a stationary control volume shown in Figure 10-8 (Blevins, 1984),

$$\mathbf{F} = -\int_S p\mathbf{n}\,ds - \frac{d}{dt}\int_{vol}\rho\mathbf{V}\,d(\text{vol}) - \int_S \rho\mathbf{V}(\mathbf{V}\cdot\mathbf{n})\,dS, \qquad (10\text{-}47)$$

where $\mathbf{F}$ is the vector fluid reaction force on the pipe, $\mathbf{n}$ is the unit normal vector outward from the control surface S, which is fixed in space and contains volume vol, and $\mathbf{V}$ is the vector fluid velocity relative to a fixed frame. Evaluation of Eq. 10-47 is not trivial because it requires a

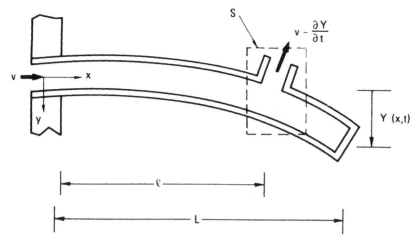

Fig. 10-8 A ruptured cantilever pipe.

transient model for the fluid velocity and pressure as well as evaluation of the pipe motion. In a high-pressure air or steam system, the flow chokes at the break and the fluid exiting the pipe is sonic, relative to the pipe, until the system blows down over perhaps a second.

If we neglect the transient term in Eq. 10-47 and assume that the fluid exits the break with velocity v relative to the pipe, the transverse force on the pipe is

$$F_y = \rho A \left[v - \frac{\partial Y(l, t)}{\partial t} \right]^2 + (p - p_s)A. \qquad (10\text{-}48)$$

This force is essentially identical to the force on a right-angle pipe bend in a steady flow (Blevins, 1979). There are two force components: First, a component required to turn the fluid 90 degrees as it exits the break at absolute velocity $v - \partial Y/\partial t$; and second, pressure-induced force due the internal pressure relative to the ambient atmospheric pressure p_s. For a highly pressurized system $(p \gg p_s)$ at the instant of rupture, the force is due to pressure alone, $v = \partial Y/\partial t = 0$,

$$F_y \approx pA. \qquad (10\text{-}49)$$

This approximation is widely used even though it neglects fluid dynamic terms that are important as the system blows down.

From Eq. 10-49, the equation of motion of a slender uniform pipe that bends in response to the instantaneous fluid force at the rupture is

$$EI \frac{\partial^4 Y(x, t)}{\partial x^4} + M \frac{\partial^2 Y(x, t)}{\partial t^2} = \begin{cases} 0, & t < 0 \text{ or } x \neq l, \\ pA, & x = l, t \geq 0, \end{cases} \qquad (10\text{-}50)$$

where E is the modulus of elasticity of the pipe material, I is the area moment of inertia of the pipe, and M is the mass per unit length of the pipe and the fluid it contains. Solutions are sought by a modal expansion

$$Y(x, t) = \sum_{j=1}^{N} \bar{y}_j(x) y_j(t), \tag{10-51}$$

where $\bar{y}_j(x)$ are the mode shapes associated with free vibration of the ruptured pipe. Equation 10-51 is substituted into Eq. 10-50, the result is multiplied through by $y_k(x)$ and integrated over the pipe span, and the orthogonality conditions of Eq. 10-37 are employed to give a series of linear ordinary differential equations that describe the response of each mode,

$$\ddot{y}_j(t) + \omega_j^2 y_j(t) = pA\bar{y}_j\left(\frac{l}{L}\right)\left[ML \int_0^L \bar{y}_j^2\left(\frac{x}{L}\right) d\left(\frac{x}{L}\right)\right]^{-1}$$

$$\text{if } t \geq 0, \text{ otherwise } 0, j = 1, 2, 3, \ldots \tag{10-52}$$

ω_j are the circular natural frequencies of the pipe. The pipe is assumed to be motionless before the rupture so the initial conditions for Eq. 10-52 are $y_j(0) = \dot{y}_j(0) = 0$.

Consider the steady (i.e., static) solution obtained by neglecting the term $\ddot{y}_j$ in Eq. 10-52. The static deflection for each mode, $Y_j(x)$, is equal to the generalized force on each mode divided by the modal stiffness,

$$Y_j(x) = \frac{pA}{\omega_j^2 ML} \frac{\bar{y}(l/L)}{\int_0^L \bar{y}_j^2(x/L) \, d(x/L)} \bar{y}_j(x), \tag{10-53}$$

where $\omega_j^2 ML$ is the stiffness of the pipe in the j mode. (Consider that the natural frequency of a spring–mass system is $\omega^2 = k/m$. Thus, the stiffness is $k = \omega^2 m$.) pA times mode shape terms forms the generalized force. The response is ordinarily dominated by the first few modes. For example, if the rupture occurs at the tip of the cantilever shown in Figure 10-8, and the response in the first mode is 1.0, then the amplitudes of the second and third modes will be 0.0255 and 0.00324, respectively.

The exact transient deflection is obtained by Duhamel's integral (Thomson, 1988). The result for the jth mode, which can be verified by back-substitution into Eq. 10-50, is

$$Y_j(x, t) = Y_j(x)(1 - \cos \omega_j t), \qquad j = 1, 2, \ldots . \tag{10-54}$$

The transient deflection oscillates about the steady deflection (Eq. 10-53), reaching a maximum value of twice the steady solution. The oscillations persist because damping has been neglected. It is possible to include damping in the Duhamel's integral solution by adding the term $2\zeta_j\omega_j\dot{y}_j(t)$ to the left side of Eq. 10-50. The transient solutions with damping are

$$Y_j(x, t) = Y_j(x)[1 - e^{-\zeta_j\omega_j t}(1 - \zeta_j^2)^{-1/2} \cos (\sqrt{1 - \zeta_j^2} \, \omega_j t - \phi)], \tag{10-55}$$

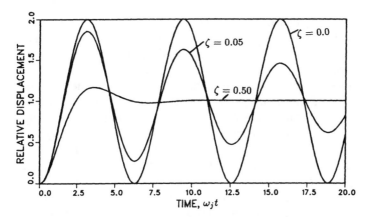

Fig. 10-9 Pipe displacement as a function of time and damping.

where the phase angle is defined by $\tan \phi = \zeta_j/(1 - \zeta_j^2)^{1/2}$. These damped transient solutions are shown in Figure 10-9. Note that damping causes the dynamic solution to decay to the steady response but does not significantly reduce the dynamic response for short times, $t \ll 1/(\zeta_j \omega_j)$.

As the pipe bends, a moment is induced in the pipe,

$$M_b = EI \frac{\partial^2 Y}{\partial x^2} = EI \sum_{j=1}^{N} y_j(t) \frac{\partial^2 \bar{y}_j(x)}{\partial x^2}. \tag{10-56}$$

In many cases, the bending moment exceeds the moment that can be elastically borne by the pipe. The pipe yields plastically. The dynamic analysis of yielded pipes can be made by using plastic hinges connecting intermediate rigid segments of pipe (Salmon and Verma, 1976) or a finite-element, elastic–plastic model (Anderson and Singh, 1976).

Exercises

1. What is the maximum response of the damped pipe (Eq. 10-55)?

2. Use Duhamel's integral or a numerical method to determine the response of an elastic pipe to a transient force $F_y = pAe^{-\alpha t}$, where α is constant.

10.4. ACOUSTICAL FORCING AND LEAKAGE-INDUCED VIBRATION

10.4.1. Pipe Acoustical Forcing

Fluid flow through valves, bends, and orifices generates turbulence within the flow and radiates acoustical energy upstream and downstream (Blake,

1986; Reethoff, 1978). The acoustic waves reflect at changes in area such as at valves, tanks, and constrictions to form a series of standing acoustic waves which are the acoustical modes of the piping. The acoustic waves exert forces on bends and changes in area that result in pipe vibration. If the acoustical source possesses sufficient energy and the acoustic natural frequencies of the piping coincide with structural natural frequencies of the piping, then large acoustically driven pipe motions can result that can be very disturbing to plant operators and may eventually fatigue the piping. These acoustically forced piping vibrations have occurred in steam and water lines of power plants and in hydrocarbon lines in the process industry (Gibert, 1977; Hartlen and Jaster, 1980).

Consider the piping system shown in Figure 10-10. Fluid flows from a tank reservoir to a bend, through an area change, and through another bend to a flow-regulating valve. The pipe run between the reservoir and the valve can be considered a one-dimensional acoustic system. To a first approximation, the boundary condition is zero acoustic velocity at the valve (i.e., closed) and a pressure-relieving boundary (i.e., open, zero acoustic pressure) at the tank. If we neglect the area change, the natural frequencies in hertz of these systems are (Blevins, 1979)

$$f_i = \begin{cases} ic/(4L), & \text{closed–open system, } i = 1, 3, 5, \dots, \\ ic/(2L), & \text{closed–closed or open–open system, } i = 1, 2, 3, \dots, \end{cases}$$

$$(10\text{-}57)$$

where L is the axial length along the pipe centerline between boundaries and c is the speed of sound in the fluid. This sound speed is slightly reduced by the elasticity of the pipe walls (Wylie and Streeter, 1978). In multiphase fluids, or fluids near critical state, the sound speed can be

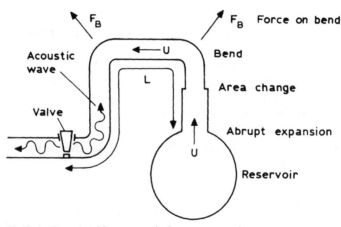

Fig. 10-10 A pipe run with an acoustical source at a valve.

evaluated thermodynamically from the expression $c = \partial p / \partial \rho$ at constant entropy. Bradshaw (1976), Siikonen (1983), and To (1984) have developed computer codes for numerical evaluation of the acoustic modes of piping systems. In large power plant and petrochemical process lines, the long spans of piping produce low acoustic natural frequencies. Hartlen and Jaster (1980) report 7.5 Hz as the dominant frequency in a steam line of a conventional power plant. The author has observed a 5 Hz oscillation in a 42 inch (1.1 m) diameter hydrocarbon line of a geothermal power plant.

Chadha et al. (1980) found that a valve acoustic source is broad-band in nature and has an overall level approximately equal to 1% of the steady pressure drop across the valve. The broad-band frequency content drops off sharply beyond a maximum frequency of approximately $f_{cutoff} \approx 0.05 U/d$, where d is the diameter of the valve throat and U is the downstream velocity. The components of this broad-band pressure at or near the acoustic natural frequencies (Eq. 10-57) will be amplified. It is the author's experience that maximum acoustic pressures in the fundamental acoustic mode can be expected to reach 1% to 2% of the steady pressure loss across the contiguous valve or orifice. That is, if a valve produces a steady pressure drop of 300 psi (2 MPa), then peak resonant fluctuating acoustic pressures in the contiguous pipe on the order of 3 to 6 psi (20 to 40 kPa) could be expected. Figure 10-11 shows the magnitude of pressure fluctuations in a power-plant steam line.

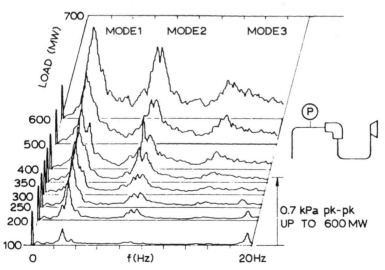

Fig. 10-11 Spectra of pressure in power-plant piping as a function of plant power (Hartlen and Jaster, 1980).

Note that the energy is centered at the acoustic modes and increases with plant power as the pressure drop across the regulating valve increases. Blevins and Bressler (1992) have proposed an equation (Eq. 9-43) for the magnitude of sound radiated by an internal acoustic resonance in terms of the pressure drop. The agreement with data is shown in Fig. 10-12.

The oscillating acoustic pressure imposes forces on bends and changes in areas. The fluid force on a right-angle bend is

$$\mathbf{F} = [(p - p_s) + \rho U^2]A\mathbf{i} - [(p - p_s) + \rho U^2]A\mathbf{j}, \qquad (10\text{-}58)$$

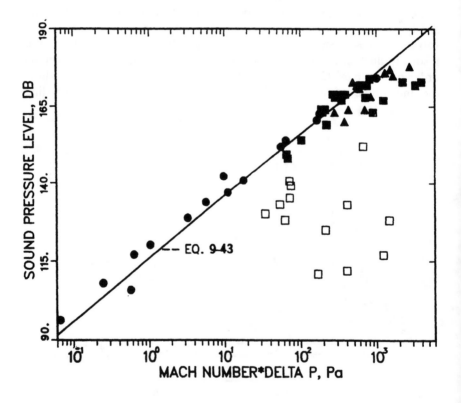

Fig. 10-12 Maximum resonant sound levels from experiment in comparison with Eq. 9-43. (Blevins and Bressler, 1992).

where **F** is the inplane force that the fluid applies to the bend, **i** is the unit vector in the direction of the incoming flow, and **j** is the unit vector in the direction of the exiting flow. U is the average velocity of fluid of density ρ in the pipe, which has pressure p and internal cross-sectional area A; p_s is the pressure in the surrounding atmosphere. The force on a change in area is $(p - p_s)(A_1 - A_2)$, where $A_1 - A_2$ is the change in flow area. The force acts inline with the pipe axis.

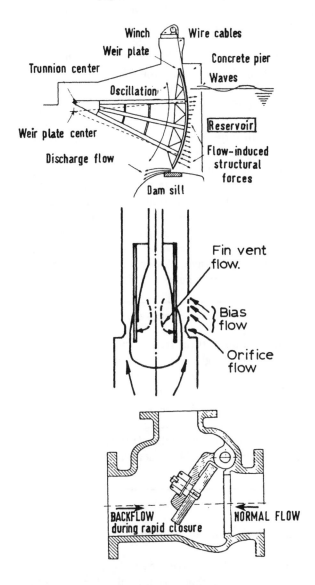

Fig. 10-13 Leakage flows that have produced instability of the valve elements (Mulcahy, 1984; Naudascher and Rockwell, 1980; Weaver et al., 1978; Parkin et al., 1983).

Both the pressure p and the fluid velocity U will have steady and oscillating components,

$$U = U_0 + u_a \sin \omega t, \qquad p = p_0 + p_a \sin \omega t, \qquad (10\text{-}59)$$

where U_0 is the mean flow velocity and p_0 is the mean pressure. The acoustically induced velocity is $u_a = p_a/(\rho c)$. The oscillating acoustic pressures of amplitude p_a produce oscillating pressure forces on a bend at the acoustic natural frequencies. The velocity-induced forces produce components of force at both the acoustic frequencies and their harmonics because of the squared term in Eq. 10-58.

The largest pipe response will occur when the acoustic natural frequencies, or their harmonics, coincide with a structural natural frequency of the piping. The author has observed this coincidence to produce 2-ft (0.6-m) deflections of 42 in (1.1 m) diameter pipe. There are three approaches to reducing these coincident vibrations: first, reducing the acoustic source through operational changes that lead to a reduced pressure drop across the valve or orifice; second, using low-noise valves or an inline muffler to reduce the exciting acoustic pressure (Beranek, 1980); third, providing stiff supports at bends and changes in area. By supporting bends and changes in area with a stiff external frame, the acoustic loads are reacted to ground without inducing pipe vibration.

Water hammer is the extreme example of pipe acoustical forcing. With water hammer a sudden break or valve opening or closing generates an intense transient acoustical wave. Water hammer analysis is discussed by Wylie and Streeter (1978) and Tullis (1988).

Exercise

1. The fundamental natural frequency of a pinned–pinned span of thin-walled pipe, neglecting the mass of internal fluids, is $f = [D/(8\pi L^2)](2E/\rho_m)^{1/2}$ Hz. ρ_m is pipe material density and E is elastic modulus. At what pipe span will this frequency coincide with the fundamental acoustic frequency of a closed–open acoustic system? Show this result by plotting the two frequencies for 12 in (0.3 m) and 24 in (0.6 m) diameter (D) steel pipe carrying compressed air ($c = 1100$ ft/sec, 343 m/sec) as a function of the unsupported span L.

10.4.2. Leakage Flow-Induced Vibration

Often flows are regulated by valves, gates, gags, or seals that consist of elastic or pivoted elements that constrict the flow as shown in Figure 10-13. The local high-velocity leakage flow loads the valve element, causing it to deform, altering the local flow field and the force on the element. The result can be an instability that produces wear and flow oscillations. Such phenomena are called leakage flow-induced vibration.

Leakage flow instability has occurred in water control valves (Weaver, et al. 1978; D'Netto and Weaver, 1987), in gates of hydraulic systems (Thang and Naudascher, 1986; Jongeling, 1988; Naudascher and Rockwell, 1970), with fluid rod segments in gas-cooled reactors (France and Rowney, 1988), in the center bodies of annular diffusers (Hobson and Jebwab, 1988), in flow-regulating devices of gas systems (Parkin et al., 1983), and in weirs (Equchi and Tanaka, 1989). There appear to be few general rules for analysis of these systems. The instability is a function of the pressure drop and the shape of the upstream constriction and downstream diffuser. The instability can be influenced by downstream vortex formation or jet switching and it can couple to the acoustic modes of the piping, but this is not always so.

Mulcahy (1986) reports that an upstream constriction is generally less stable than a downstream constriction; Weaver et al. (1978) reports that a gentle, rather than abrupt, close-off of flow by a valve element is stabilizing; and Parkin et al. (1983) similarly found that machining longitudinal grooves into the center body (valve element) increases stability. However, each case of leakage flow-induced vibration must currently be treated experimentally on a case-by-case basis (Mulcahy, 1988).

REFERENCES

Anderson, J. C., and A. K. Singh (1976) "Inelastic Response of Nuclear Piping Subject to Rupture Forces," *Journal of Pressure Vessel Technology*, **98**, 98–104.

Ashley, H., and G. Haviland (1950) "Bending Vibrations of a Pipe Line Containing Flowing Fluid," *Journal of Applied Mechanics*, **17**, 229–232.

Bajaj, A. K., et al. (1980) "Hopf Bifurcation Phenomena in Tubes Carrying Fluid," *SIAM Journal of Applied Mathematics*, **39**, 213–230.

Benjamin, T. B. (1961) "Dynamics of System of Articulated Pipes Conveying Fluid, Parts 1 and 2," *Proceedings of the Royal Society of London, Series A*, **261**, 457–499.

Beranek, L. L. (1980) *Noise Reduction*, Robert E. Kreiger Publishing, Malabar, Fla. Reprint of 1960 edition.

Blake, W. K. (1986) *Mechanics of Flow-Induced Sound and Vibration*, Academic Press, New York.

Blevins, R. D. (1979) *Formulas for Natural Frequency and Mode Shape*, Van Nostrand Reinhold, New York. Reprinted by Robert E. Kreiger, Malabar, Fla., 1984.

———— (1984) *Applied Fluid Dynamics Handbook*, Van Nostrand Reinhold, New York, p. 70.

Blevins, R. D. and M. M. Bressler, "Experiments on Acoustic Resonance in Heat Exchanger Tube Bending" ASME PVP-Vol. 243, volume 4, 59-80, 1992.

Bradshaw, R. T. (1976) *WAVENET: Waves in Fluid Networks: User Guide*, Waltham, Mass.

Chadha, J. A. et al. (1980) "Acoustic Source Properties of Governor Valves," in

Flow-Induced Vibration of Power Plant Components, PVP-41, ASME, New York, pp. 125–138.

Chebair, A. E., et al. (1988) "Theoretical Study of the Effect of Unsteady Viscous Forces on Inner- and Annular-Flow Induced Instabilities of Cylindrical Shells," in *1988 International Symposium on Flow-Induced Vibration and Noise*, ASME, New York.

Chen, S. S. (1971) "Dynamic Stability of a Tube Conveying Fluid," *ASCE Journal of the Engineering Mechanics Division*, **97**, 1469–1485.

―――― (1973) "Out-of-Plane Vibration and Stability of Curved Tubes Conveying Fluid," *Journal of Applied Mechanics*, **40**, 362–368.

―――― (1981) "Fluid Damping of Circular Cylindrical Structures," *Nuclear Engineering and Design*, **63**, 81–100.

Chen, S. S., and J. A. Jendrzejczyk (1985) "General Characteristics, Transition, and Control of Instability of Tubes Conveying Fluid," *Journal of the Acoustical Society of America*, **77**, 887–895.

Crandall, S. H. (1968) *Dynamics of Mechanical and Electromechanical Systems*, McGraw-Hill, New York, pp. 300–395.

D'Netto, W., and D. S. Weaver (1987) "Divergence and Limit Cycle Oscillations in Valves Operating at Small Openings," *Journal of Fluids and Structures*, **1**, 3–18.

Dodds, H. L., and H. Runyan (1965) "Effect of High-Velocity Fluid Flow in the Bending Vibrations and Static Divergence of a Simply Supported Pipe," National Aeronautics and Space Administration Report NASA TN D-2870, June 1965.

Doki, H., and K. Aso (1989) "Dynamic Stability and Active Control of Cantilever Pipes Conveying Fluid," in *Flow-Induced Vibration—1989*, Vol. 154, M. K. Au-Yang (ed.), ASME, New York, pp. 25–30.

Dowling, A. P. (1988) "The Dynamics of Towed Flexible Cylinder, Parts 1 and 2," *Journal of Fluid Mechanics*, **187**, 507–571.

Equchi, Y., and N. Tanaka (1989) "Fluid–Elastic Vibration of Flexible Overflow Weir," in *Flow-Induced Vibration—1989*, Vol. 154, M. K. Au-Yang (ed.), ASME, New York, pp. 43–50.

France, E. R., and B. A. Rowney (1988) "Flow-Induced Vibration of Control Rods in an Advanced Gas Cooled Reactor," in *1988 International Symposium on Flow-Induced Vibration and Noise*, Vol. 4, ASME, New York, pp. 147–164.

Gibert, R. J. (1977) "Pressure Fluctuations Induced by Fluid Flow in Singular Points of Industrial Circuits," CEA-N-1925, Paper B 3/5, Structural Mechanics in Reactor Technology Conference.

Ginsberg, J. H. (1973) "The Dynamical Stability of a Pipe Conveying a Pulsatile Flow," *International Journal of Engineering Science*, **11**, 1013–1024.

Gregory, R. W., and M. P. Paidoussis (1966) "Unstable Oscillations of Tubular Cantilevers Conveying Fluid, Parts 1 and 2," *Proceedings of the Royal Society of London, Series A*, **293**, 512–542.

Hagedorn, P. (1988) *Non-Linear Oscillations*, Oxford Science Publications, Clarendon Press, Oxford.

Hartlen, R. T., and W. Jaster (1980) "Main Stream Vibration Driven Flow-Acoustic Excitation," in *Practical Experiences with Flow-Induced Vibrations*, Symposium in Karlsruhe, Germany, September 3–6, 1979, Springer-Verlag, New York, pp. 144–152.

Hill, J. L., and C. G. Davis (1974) "Effect of Initial Forces on the Hydroelastic Vibration and Stability of Planar Curved Tubes," *Journal of Applied Mechanics*, **41**, 355–359.

Hill, J. L., and C. P. Swanson (1970) "Effects of Lumped Masses on the Stability of Fluid-Conveying Tubes," *Journal of Applied Mechanics*, **37**, 494–497.

Hobson, D. E., and M. Jebwab (1988) "Investigation of the Effect of Eccentricity on the Unsteady Fluid Forces on the Centrebody of an Annular Diffuser," in *1988 International Symposium on Flow-Induced Vibration and Noise*, Vol. 4, ASME, New York, pp. 111–124.

Holmes, P. J. (1978) "Pipes Supported at Both Ends Cannot Flutter," *Journal of Applied Mechanics*, **45**, 619–622.

Housner, G. W. (1952) "Bending Vibrations of a Pipe Line Containing Flowing Fluid," *Journal of Applied Mechanics*, **19**, 205–208.

Jendrzejczyk, J. A., and S. S. Chen (1985) "Experiments on Tubes Conveying Fluid", *Journal of Thin-Walled Structures*, **3**, 109–134.

Jongeling, T. H. G. (1988) "Flow-Induced Self-Excited In-Flow Vibration of Gate Plates," *Journal of Fluids and Structures*, **2**, 541–566.

Lee, T. S. (1981) "Stability Analysis of the Ortloff–Ives Equation," *Journal of Fluid Mechanics*, **110**, 293–295.

Matsuzaki, Y., and Y. C. Fung (1977) "Unsteady Fluid Dynamic Forces on a Simply-Supported Circular Cylinder of Finite Length Conveying a Flow," *Journal of Sound and Vibration*, **54**, 317–330.

Misra, A. K., et al. (1988) "On the Dynamics of Curved Pipes Transporting Fluid, Parts I and II," *Journal of Fluids and Structures*, **2**, 221–244.

Mulcahy, T. M. (1984) "Avoiding Leakage Flow-Induced Vibration by a Tube-in-Tube Slip Joint," Argonne National Laboratory Report ANL-84-82, Argonne, Ill.

—— (1986) "Leakage Flow-Induced Vibration for Variations of a Tube-in-Tube Slip Joint," Argonne National Laboratory Report ANL-86-11, Argonne, Ill.

—— (1988) "One-Dimensional Leakage-Flow Vibration Instabilities," *Journal of Fluids and Structures*, **2**, 383–403.

Naguleswaran, S., and C. J. H. Williams (1968) "Lateral Vibrations of a Pipe Conveying a Fluid," *Journal of Mechanical Engineering Science*, **10**, 228–238.

Naudascher, E., and D. Rockwell (eds.) *Practical Experiences with Flow-Induced Vibrations*, Springer-Verlag, New York.

Nemat-Nassar, S., et al. (1966) "Destabilizing Effects on Velocity Dependent Forces in Non-conservative Systems," *AIAA Journal*, **4**, 1276–1280.

Niordson, F. I. N. (1953) "Vibrations of a Cylindrical Tube Containing Flowing Fluid," *Transations of the Royal Institute of Technology, Stockholm*, **73**.

Paidoussis, M. P. (1966) "Dynamics of Flexible Cylinders in Axial Flow, Parts 1 and 2," *Journal of Fluid Mechanics*, **26**, 717–751.

—— (1970) "Dynamics of Tubular Cantilevers Conveying Fluid," *Journal of Mechanical Engineering Science*, **2**, 85–103.

—— (1976) "Elastohydrodynamics of Towed Slender Bodies: The Effect of Nose and Tail Shapes on Stability," *Journal of Hydronautics*, **10**, 127–134.

—— (1987) "Flow-Induced Instabilities of Cylindrical Structures," *Applied Mechanics Reviews*, **40**, 163–175.

Paidoussis, M. P., and N. T. Issid (1974) "Dynamic Stability of Pipes Conveying Fluid," *Journal of Sound and Vibration*, **33**, 267–294.

Paidoussis, M. P., and F. C. Moon (1988) "Nonlinear and Chaotic Fluidelastic Vibrations of a Flexible Pipe Conveying Fluid," *Journal of Fluids and Structures*, **2**, 567–591.

Paidoussis, M. P., and M. Ostoja-Starzewski (1981) "Dynamics of a Flexible Cylinder in Subsonic Axial Flow," *AIAA Journal*, **19**, 1467–1475.

Paidoussis, M. P., et al. (1986) "Dynamics of Finite-Length Tubular Beams Conveying Fluid," *Journal of Sound and Vibration*, **106**, 311–331.

Parkin, M. W., E. R. France, and W. E. Boley (1983) "Flow Instability Due to a Diameter Reduction of Limited Length in a Long Annular Passage," *Journal of Vibration, Acoustics, Stress, and Reliability in Design*, **105**, 353–360.

Reethoff, G. (1978) "Turbulence-Generated Noise in Piping," *Annual Review of Applied Mechanics*, **10**, 333–367.

Riley, J. J., M. Gad-el-Hak, and R. W. Metcalfe (1988) "Compliant Coating," *Annual Review of Fluid Mechanics*, **20**, 393–420.

Salmon, M. A., and V. Verma (1976) "Rigid Plastic Beam Model for Pipe Whip Analysis,"
    ASCE Journal of the Engineering Mechanics Division, **102,** 415–430.
Shayo, L. K., and C. H. Ellen (1978) "Theoretical Studies of Internal Flow-Induced
    Instabilities of Cantilever Pipes," *Journal of Sound and Vibration*, **56,** 463–474.
Siikonen, T. (1983) "Computational Method for the Analysis of Valve Transients," *Journal
    of Pressure Vessel Technology*, **105,** 227–233.
Svetlitsky, V. A. (1977) "Vibrations of Tubes Conveying Fluids," *Journal of the Acoustical
    Society of America*, **62,** 595–600.
Tang, D. M., and E. H. Dowell (1988) "Chaotic Oscillations of a Cantilevered Pipe
    Conveying Fluid," *Journal of Fluids and Structures*, **2,** 263–283.
Taylor, G. I. (1952) "Analysis of the Swimming of Long and Narrow Animals,"
    Proceedings of the Royal Society of London, Series A, **214,** 158–183.
Thang, N. D., and E. Naudascher (1986) "Self-Excited Vibrations of Vertical Lift Gates,"
    Journal of Hydraulics Research, **24,** 391–404.
Thomson, W. T. (1988) *Theory of Vibrations with Applications*, 3d ed., Prentice-Hall,
    Englewood Cliffs, N.J.
To, C. W. S. (1984) "The Acoustic Simulation and Analysis of Complicated Reciprocating
    Compressor Piping Systems, Part II, Program Structure and Applications," *Journal of
    Sound and Vibration*, **96,** 195–205.
Triantafyllou, G. S., and C. Chryssostomidis (1985) "Stability of a String in Axial Flow,"
    Journal of Energy Resources Technology, **107,** 421–425.
Tullis, J. P. (1988) *Hydraulics of Pipelines, Pumps, Valves, Cavitation and Transients*,
    Wiley-Interscience, New York.
Unny, T. E., E. L. Martin, and R. N. Dubey (1970) "Hydroelastic Instability of Uniformly
    Curved Pipe-Fluid Systems," *Journal of Applied Mechanics*, **37,** 617–622.
U.S. Atomic Energy Commission (1973) "Protection Against Pipe Whip Inside Contain-
    ment," *Regulatory Guide 1.46.*
Wambsganss, M. W., and J. A. Jendrzejczyk (1979) "The Effect of Trailing End Geometry
    on the Vibration of a Circular Cantilevered Rod in Nominally Axial Flow," *Journal of
    Sound and Vibration*, **65,** 251–258.
Weaver, D. S., and M. P. Paidoussis (1977) "On Collapse and Flutter Phenomena in Thin
    Tubes Conveying Fluid," *Journal of Sound and Vibration*, **50,** 117–132.
Weaver, D. S., et al. (1978) "Flow-Induced Vibrations of a Hydraulic Valve and Their
    Elimination," *Journal of Fluids Engineering*, **100,** 239–245.
Wylie, E. B., and V. L. Streeter (1978) *Fluid Transients*, McGraw-Hill, New York.
    Reprinted by FEB Press, Ann Arbor, Mich., 1983.

Appendix A

Modal Analysis

The objective of modal analysis is to reduce the complex partial differential equations describing the motion of continuous structures to sets of much simpler, ordinary differential equations that describe the motion of equivalent one-dimensional structures. A theoretical treatment of modal analysis is given in Meirovitch (1967). In this appendix, examples of modal analysis are given for galloping and fluid elastic instability of a slender beam.

The partial differential equation describing the motion of a slender beam is

$$\frac{\partial^2}{\partial z^2}\left(EI\frac{\partial^2 Y(z, t)}{\partial z^2}\right) + m\frac{\partial^2 Y}{\partial t^2} = F(z, t), \qquad (A-1)$$

where Y is the displacement normal to the beam axis in some direction, z is the distance along the beam axis, m is the mass per unit length of the beam, F is the external force per unit length applied normal to the beam axis in the direction of Y, and I is the area moment of inertia for bending:

$$I = \int_A \xi^2 \, dA, \qquad (A-2)$$

where A is the cross-sectional area, ξ is the distance from the shear center of the beam in the direction Y, and μ is the mass per unit volume of the section. In general, the moment of inertia, the elastic modulus, and the mass per unit length vary along the span of the beam; however, in this example they are considered to be uniform over the span.

Solutions to Eq. A-1 are sought in terms of the associated equation for the free vibrations of the beam:

$$EI\frac{\partial^4 Y}{\partial z^4} + m\frac{\partial^2 Y}{\partial t^2} = 0. \qquad (A-3)$$

There are two kinds of boundary conditions on the beam:

1. Geometric boundary conditions. These arise from geometric constraints on the beam. For example, if the end of a beam at $z = 0$ is held by a pin, then $Y(0, t) = 0$. If the end of a beam at $z = 0$ is clamped, then $\partial Y(0, t)/\partial z = 0$.

2. Kinetic boundary conditions. These arise from the forces and moments on a beam. For example, the moment on the end of a pinned beam must be zero. Since the moment in a beam is $EI\,\partial^2 Y/\partial z^2$, a pinned end at $z = 0$ implies that $\partial^2 Y(0, t)/\partial z^2 = 0$.

For a pinned–pinned beam with a length of L, the appropriate boundary conditions are

$$Y(0, t) = Y(L, t) = 0,$$

$$\frac{\partial^2 Y(0, t)}{\partial z^2} = \frac{\partial^2 Y(L, t)}{\partial z^2} = 0. \qquad \text{(A-4)}$$

Solutions to Eqs. A-3 and A-4 that can be separated into components in space and time are sought:

$$Y(z, t) = \bar{y}(z)y(t). \qquad \text{(A-5)}$$

Substituting Eq. A-5 into Eq. A-3 and rearranging, Eq. A-3 becomes

$$\frac{1}{\bar{y}(z)} \frac{d^4\bar{y}(z)}{dz^4} = -\frac{m}{EI} \frac{1}{y(t)} \frac{d^2y(t)}{dt^2} = \text{constant}. \qquad \text{(A-6)}$$

The solution to Eq. A-6, using the boundary conditions of Eq. A-4,

$$\bar{y}(0) = \bar{y}(L) = 0,$$

$$\bar{y}''(0) = \bar{y}''(L) = 0, \qquad \text{(A-7)}$$

is

$$y(t) = A \sin \omega_n t + B \cos \omega_n t, \qquad n = 1, 2, 3, \ldots,$$

$$\text{where } \omega_n = \frac{n^2\pi^2}{L^2} \left(\frac{EI}{m}\right)^{1/2}, \qquad \text{(A-8)}$$

and

$$\bar{y}_n(z) = \sin(n\pi z/L), \qquad n = 1, 2, 3, \ldots. \qquad \text{(A-9)}$$

$\bar{y}_n$ are the natural modes of the structure, and ω_n are the circular natural frequencies of the structure. The lowest frequency mode, which is given by $n = 1$, is called the fundamental mode. The natural mode shapes and frequencies for beams with a variety of boundary conditions are given by Blevins (1984), Meirovitch (1967), Thomson (1988), and Timoshenko et al. (1974). The complete solution to Eqs. A-6 and A-7 is

$$Y(z, t) = \sum_{n=1}^{\infty} (A \sin \omega_n t + B \cos \omega_n t) \sin(n\pi z/L), \qquad \text{(A-10)}$$

where A and B are constants.

Note that the natural modes are orthogonal over the span of the beam:

$$\int_0^L \bar{y}_i \bar{y}_j \, dz = \begin{cases} 0, & i \neq j, \\ L/2, & i = j. \end{cases} \qquad \text{(A-11)}$$

The natural modes of a structure are guaranteed to be orthogonal if the system equations are self-adjoint (Meirovitch, 1967). (Some authors prefer to introduce a constant into the mode shape, so that $\int_0^L \bar{y}_j^2 \, dz = L$ for all j. This has not been done in this book.)

Solutions to the original forced equation of motion (A-1) are sought in terms of an expansion of the natural modes:

$$Y(z, t) = \sum_{n=1}^{\infty} y_n(t)\bar{y}_n(z). \qquad \text{(A-12)}$$

If this expression is substituted into Eq. A-1, then

$$\sum_{n=1}^{\infty} \left[EI\left(\frac{\pi n}{L}\right)^4 y_n(t) + m\ddot{y}_n(t) \right] \sin (n\pi z/L) = F(z, t). \qquad \text{(A-13)}$$

Multiplying this equation through by $\sin (j\pi z/L)$, integrating over the length of the beam, and using the orthogonality property gives

$$\ddot{y}_n + \omega_n^2 y_n = \frac{\displaystyle\int_0^L F(z, t) \sin (n\pi z/L)\, dz}{\displaystyle\int_0^L m \sin^2 (n\pi z/L)\, dz}, \qquad n = 1, 2, 3, \ldots. \qquad \text{(A-14)}$$

In general, if a beam possesses orthogonal modes (Eq. A-11), as is often the case,

$$\ddot{y}_n + \omega_n^2 y_n = \frac{\displaystyle\int_0^L F(z, t)\bar{y}_n(z)\, dz}{\displaystyle\int_0^L m\bar{y}_n^2(z)\, dz}, \qquad n = 1, 2, 3, \ldots. \qquad \text{(A-15)}$$

These ordinary differential equations describe the motion of a set of one-dimensional, spring-supported structures, responding to the generalized forces given by the right-hand side of Eq. A-15. The sum of the responses of the equivalent structures gives the response of the continuous structure (Eq. A-12). Thus, if the structure possesses orthogonal modes, the partial differential equation (A-1) can be reduced to a set of equivalent ordinary differential equations (A-15). Even in cases in which the modes of the structure are not orthogonal, Eq. A-15 often gives a close approximation to the response of the structure in a given mode.

If the mass of the structure varies along its span, then the beam may not possess orthogonal modes. However, if the variable mass does not significantly affect the mode shapes, then an equivalent mass per unit length can be defined as

$$m = \frac{\displaystyle\int_0^L m(z)\bar{y}_n^2(z)\, dz}{\displaystyle\int_0^L \bar{y}_n^2(z)\, dz}. \qquad \text{(A-16)}$$

The equivalent mass m is a function of the mode shape. However, if the mass per unit length of the structure is constant, m is always equal to the mass per unit length of the structure. Ordinarily, the concept of equivalent mass is used only when there is good reason to believe the structure vibrates primarily in a single mode of known shape. Using Eq. A-16, Eq. A-15 becomes

$$\ddot{y}_n + \omega_n^2 y_n = \frac{\displaystyle\int_0^L F(z, t)\bar{y}_n(z)\, dz}{m \displaystyle\int_0^L \bar{y}_n^2(z)\, dz}, \qquad n = 1, 2, 3, \ldots. \qquad \text{(A-17)}$$

The right-hand side of A-17 is $1/m$ times the generalized force of the nth mode, in the y direction. The force $F(z, t)$ is the sum of damping and exciting components:

$$F = F^d + F^e.$$ (A-18)

The component that damps the structure, F^d, is approximated by a viscous damper, which retards motion with a force proportional to velocity:

$$F^d = -c \frac{\partial Y}{\partial t}.$$ (A-19)

Substituting Eqs. A-18 and A-19 into Eq. A-17 gives

$$\ddot{y}_n + 2\zeta_n\omega_n\dot{y} + \omega_n^2 y = \frac{\displaystyle\int_0^L F^e(z, t)\bar{y}_n(z)\,dz}{m\displaystyle\int_0^L \bar{y}_n^2(z)\,dz}, \qquad n = 1, 2, 3, \ldots,$$ (A-20)

where the equivalent viscous damping factor for each mode has been defined as

$$\zeta_n = \frac{c}{2m\omega_n}.$$ (A-21)

A.1. GALLOPING OF A BEAM

If an exciting force is produced by aerodynamic galloping, then, as shown in Chapter 4, the exciting force per unit length at some spanwise point z, where the flow velocity is U, is

$$F_y^e = \tfrac{1}{2}\rho U^2 D \sum_{i=1}^{\infty} a_i \left[\frac{(\partial Y/\partial t)}{U} \right]^i,$$ (A-22)

where a_i is a constant determined by the lift and drag coefficients of the structure. Two cases can be easily analyzed. First, a stability analysis is made by neglecting all but the term corresponding to a_1. Second, an analysis is made of finite-amplitude vibrations of the fundamental mode.

For a stability analysis of the zero solution $[Y(z, t) = 0]$, only the linear term ($i = 1$) in Eq. A-22 must be considered. If the flow velocity U varies over the span of the structure, then substituting the linear term of Eq. A-22 into Eq. A-20 and neglecting all but a single mode, $Y(z, t) = y(t)\bar{y}(z)$, gives

$$\ddot{y} + 2\zeta\omega_n\dot{y} + \omega_n^2 y = \tfrac{1}{2}\rho D a_1 \frac{\dot{y}\displaystyle\int_0^L U(z)\bar{y}^2(z)\,dz}{m\displaystyle\int_0^L \bar{y}^2(z)\,dz}, \qquad n = 1, 2, 3, \ldots,$$ (A-23)

where $U(z)$ is the flow velocity at each spanwise point. An equivalent flow velocity for the onset of galloping is defined:

$$U = \int_0^L U(z)\bar{y}^2(z)\,dz \Big/ \int_0^L \bar{y}^2(z)\,dz. \tag{A-24}$$

This equivalent velocity is generally a function of mode shape. The onset of galloping occurs when the coefficient of the $\dot{y}$ terms is zero,

$$2\zeta_n\omega_n = \frac{\rho D a_1 U}{2m}. \tag{A-25}$$

The equivalent flow velocity required for the onset of instability is

$$\frac{U}{f_n D} = \frac{4m(2\pi\zeta_n)}{\rho D^2 a_1}. \tag{A-26}$$

This is identical to the result (Eq. 4-16) derived in Chapter 4, for the spring-supported structure, with the substitution of equivalent flow velocity (Eq. A-24) and equivalent mass (Eq. A-16) for the uniform flow velocity and mass used in Chapter 4.

Since Eq. A-26 predicts that the flow velocity required for the onset of instability of each mode increases with the frequency of the mode, the fundamental mode is most susceptible to galloping. Ordinarily, galloping analysis is restricted to the fundamental mode. If Eq. A-22 is substituted into Eq. A-20, and the analysis is limited to a single mode ($\bar{y}_n = \bar{y}$ for $n = 1$, $\bar{y}_n = 0$ for $n \neq 1$), then

$$\ddot{y} + 2\zeta\omega\dot{y} + \omega^2 y = \frac{\rho D}{2m}\sum_{i=1}^{\infty}\beta_i a_i \dot{y}^i, \tag{A-27}$$

where

$$\beta_i = \int_0^L U^{2-i}(z)\bar{y}^{i+1}(z)\,dz \int_0^L \bar{y}^2(z)\,dz. \tag{A-28}$$

Equation A-27 describes the nonlinear, finite-amplitude galloping vibrations of a beam that vibrates only in a single mode. If the flow velocity and mass are constant over the span of the structure, Eqs. A-27 and A-28 reduce to Eqs. 4-26 of Chapter 4.

A.2. FLUID ELASTIC INSTABILITY OF A TUBE ARRAY

The aerodynamic force per unit length, normal to a free stream with velocity U at the spanwise point z on the j tube in a closely spaced array, is shown in Chapter 5 (see Fig. 5-7) to be

$$F_{yj}^e = \frac{\rho U^2}{4} K_y(X_{j+1} - X_{j-1}), \tag{A-29}$$

where X is displacement parallel to the free stream. The tubes are assumed to vibrate only in a single mode. Suppose that the mode shape of the tubes in the fundamental modes is given by $\bar{y}(z)$ for displacement in the x and y directions:

$$\begin{aligned}X_i(z, t) &= x_i(t)\bar{y}(z),\\Y_j(z, t) &= y_j(t)\bar{y}(z).\end{aligned} \tag{A-30}$$

If Eqs. A-29 and A-30 are substituted into Eq. A-20, the equation of motion for the j tube in the y direction is

$$\ddot{y}_j + 2\zeta_y^j \omega_y^j \dot{y}_j + \omega_y^{j2} y_j = \frac{\rho}{4m} \frac{\displaystyle\int_0^L U^2(z)\bar{y}^2(z)\,dz}{\displaystyle\int_0^L \bar{y}^2(z)\,dz} K_y(x_{j+1} - x_{j-1}). \qquad (A\text{-}31)$$

If the flow velocity varies along the span, an equivalent flow velocity for whirling is defined as

$$U^2 = \frac{\displaystyle\int_0^L U^2(z)\bar{y}^2(z)\,dz}{\displaystyle\int_0^L \bar{y}^2(z)\,dz}, \qquad (A\text{-}32)$$

where $U(z)$ is the flow velocity at each point on the span. Then the equation of motion becomes

$$\ddot{y}_j + 2\zeta_y^j \omega_y^j \dot{y}_j + \omega_y^{j2} = \frac{\rho U^2}{4} K_y(x_{j+1} - x_{j-1}). \qquad (A\text{-}33)$$

This is identical to the equation of motion derived for spring-supported tubes (Eq. 5-9 with $C_y = 0$), except for the substitution of an equivalent flow velocity (Eq. A-32) for the uniform velocity considered in Chapter 5.

REFERENCES

Blevins, R.D., Formulas for Natural Frequency and Mode Shape, Krieger, Melbourne, Fla, 1984.

Meirovitch, L. (1967) *Analytical Methods in Vibrations*, Macmillan, New York, pp. 429–432.

Thomson, W. T. (1988) *Vibration Theory and Applications*, 3d ed., Prentice-Hall, Englewood Cliffs, N.J.

Timoshenko, S., D. H. Young, and W. Weaver, Jr. (1974) *Vibration Problems in Engineering*, 4th ed., Wiley, New York.

Appendix B

Principal Coordinates

The purpose of the principal-coordinate approach is to generate a set of coordinates describing the displacement of a structure that is inertially coupled in its natural coordinates but uncoupled in its principal coordinates. In this appendix, a principal-coordinate analysis is made of the two-degree-of-freedom oscillator shown in Figure B-1; this analysis can easily be extended to the three-degree-of-freedom case. Continuous structures can be placed in principal coordinates once the partial differential equations describing the structure are reduced to second-order, ordinary differential equations using the modal analysis described in Appendix A.

The following analysis assumes that bending-torsion galloping may be represented by a two-dimensional model in which the structural resistance to vertical bending and torsion is represented by springs and dampers, as shown in Figure B-1. Two-dimensional flow is assumed to hold. This is strictly applicable to large-aspect-ratio structures but may hold for relatively small aspect ratios if the end effects are minimal.

For a small θ, the absolute position (X, Y) of each point (η, ξ) on the cross section of Figure B-1, in terms of the relative displacements (y, θ) are

$$X = \eta\theta + \xi, \qquad Y = y + \xi\theta - \eta. \tag{B-1}$$

The corresponding velocities are

$$\dot{X} = \eta\dot{\theta}, \qquad \dot{Y} = \dot{y} + \xi\dot{\theta}, \tag{B-2}$$

where X and Y are the absolute positions, y is the displacement of the shear center normal to the free stream, θ is the rotation about the shear center, and the ξ, η coordinate system is fixed in the body (Fig. B-1).

The kinetic energy of the section is

$$T = \frac{1}{2}\int_A (\dot{X}^2 + \dot{Y}^2)\mu\, d\xi\, d\eta = \frac{1}{2}m\dot{y}^2 + \frac{1}{2}J_\theta\dot{\theta}^2 + S_x\dot{\theta}\dot{y}, \tag{B-3}$$

where

$$J_\theta = \int_A (\xi^2 + \eta^2)\mu\, d\xi\, d\eta,$$

$$m = \int_A \mu\, d\xi\, d\eta, \tag{B-4}$$

$$S_x = \int_A \xi\mu\, d\xi\, d\eta,$$

421

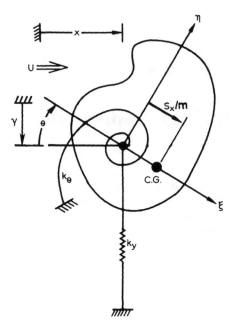

Fig. B-1 Two-degree-of-freedom structural model (not showing dampers parallel to springs).

μ is the density per unit volume of the cross section A, and J_θ is the mass polar moment of inertia. The potential energy of the structure per unit length is

$$V = \tfrac{1}{2}k_y y^2 + \tfrac{1}{2}k_\theta \theta^2, \tag{B-5}$$

where k_y and k_θ are the spring constants per unit length. The equations of motion are derived by Lagrange's equations (Meirovitch, 1967),

$$L = T - V, \tag{B-6}$$

$$Q_i = \frac{d}{dt}\left(\frac{\partial L}{\partial \dot{q}_i}\right) - \frac{\partial L}{\partial q_i}, \tag{B-7}$$

where Q_i is the generalized force with respect to the generalized coordinate q_i and is obtained from virtual work considerations:

$$\delta W = \sum_{i=1}^{2} Q_i\, \delta q_i. \tag{B-8}$$

Thus,

$$Q_y = F_y, \qquad Q_\theta = F_M, \tag{B-9}$$

where F_y and F_M are the sums of aerodynamic and damping forces along the y axes and in torsion, respectively. Applying Lagrange's equations (Eq. B-7) to Eqs. B-3, B-5, and B-6 gives

$$m\ddot{y} - S_x\ddot{\theta} + k_y y = F_y, \tag{B-10}$$

$$J_\theta \ddot{\theta} - S_x\ddot{y} + k_\theta \theta = F_M. \tag{B-11}$$

These equations are identical to Eqs. 4-39, 4-40 and 4-45, 4-46. The y and θ displacements are inertially coupled by S_x. It is convenient to place these equations in matrix form:

$$[M]\{\ddot{x}\} + [K]\{x\} = \{F\}, \tag{B-12}$$

where

$$[M] = \begin{bmatrix} m & S_x \\ S_x & J_\theta \end{bmatrix}, \qquad [K] = \begin{bmatrix} k_y & 0 \\ 0 & k_\theta \end{bmatrix},$$

$$\{x\} = \begin{Bmatrix} y \\ \theta \end{Bmatrix}, \qquad \{F\} = \begin{Bmatrix} F_y \\ F_M \end{Bmatrix}, \tag{B-13}$$

and $[M]$ and $[K]$ are the mass and stiffness matrices, respectively. Principal coordinates are sought such that

$$\{x\} = \{\bar{P}_j\} \cos \omega_j t \tag{B-14}$$

by formulating the eigenvalue problem

$$[-\omega_j^2[M] + [K]]\{\bar{P}_j\} = 0, \tag{B-15}$$

which is equivalent to

$$\begin{bmatrix} k_y - m\omega^2 & -S_x\omega^2 \\ -S_x\omega^2 & k_\theta - J_\theta\omega^2 \end{bmatrix} \{\bar{P}\} = 0. \tag{B-16}$$

The determinant of the coefficient matrix must be zero for nontrivial solutions. Setting the determinant to zero gives the characteristic frequencies

$$\omega_{1,2}^2 = \frac{\omega_y^2 + \omega_\theta^2 \pm \{(\omega_y^2 + \omega_\theta^2)^2 - 4\omega_y^2\omega_\theta^2(1 - S_x^2/J_\theta m)\}^{1/2}}{2(1 - S_x^2/J_\theta m)}. \tag{B-17}$$

The characteristic vectors are

$$\{\bar{P}_1\} = \begin{Bmatrix} \alpha_1 \\ 1 \end{Bmatrix}, \qquad \{\bar{P}_2\} = \begin{Bmatrix} 1 \\ \alpha_2 \end{Bmatrix}, \tag{B-18}$$

where

$$\alpha_1 = \frac{S_x}{m} \frac{\omega_1^2}{\omega_y^2 - \omega_1^2}, \qquad \alpha_2 = \frac{S_x}{J_\theta} \frac{\omega_2^2}{\omega_\theta^2 - \omega_2^2}. \tag{B-19}$$

Equation B-15 implies that

$$-\omega_1^2[M]\{\bar{P}_1\} + [K]\{\bar{P}_1\} = 0, \tag{B-20}$$

$$-\omega_2^2[M]\{\bar{P}_2\} + [K]\{\bar{P}_2\} = 0. \tag{B-21}$$

If the transpose (j row becomes the j column) of these equations is taken, then, since $[M]$ and $[K]$ are symmetric,

$$-\omega_1^2\{\bar{P}_1\}^T[M] + \{\bar{P}_1\}^T[K] = 0, \tag{B-22}$$

$$-\omega_2^2\{\bar{P}_2\}^T[M] + \{\bar{P}_2\}^T[K] = 0. \tag{B-23}$$

The superscript T denotes transpose. If Eq. B-20 is premultiplied by $\{\bar{P}_2\}^T$, Eq.

B-23 is postmultiplied by $\{\bar{P}_1\}$, and the resultant equations are subtracted, then

$$(\omega_2^2 - \omega_1^2)\{\bar{P}_2\}^T[M]\{\bar{P}_1\} = 0. \tag{B-24}$$

Thus,

$$\{\bar{P}_2\}^T[M]\{\bar{P}_1\} = 0 \tag{B-25}$$

if $\omega_1 \neq \omega_2$. Similarly,

$$\{\bar{P}_2\}^T[K]\{\bar{P}_1\} = 0 \tag{B-26}$$

if $\omega_1 \neq \omega_2$.

Equations B-25 and B-26 imply that the matrix composed of the characteristic vectors will diagonalize the mass and stiffness matrices:

$$[\{\bar{P}_1\}\{\bar{P}_2\}]^T[M][\{\bar{P}_1\}\{\bar{P}_2\}] = \begin{bmatrix} * & 0 \\ 0 & * \end{bmatrix}, \tag{B-27}$$

$$[\{\bar{P}_1\}\{\bar{P}_2\}]^T[K][\{\bar{P}_1\}\{\bar{P}_2\}] = \begin{bmatrix} * & 0 \\ 0 & * \end{bmatrix}, \tag{B-28}$$

provided that ω_1 and ω_2 are distinct, nonidentical frequencies. An asterisk (*) denotes a nonzero entry in the matrix.

If the principal-coordinate transform is defined as

$$\{x\} = [\{\bar{P}_1\}\{\bar{P}_2\}]\{p\}, \tag{B-29}$$

where the principal coordinates p are

$$\{p\} = \begin{Bmatrix} p_1 \\ p_2 \end{Bmatrix}, \tag{B-30}$$

the principal-coordinate transform is substituted into the system equations (Eq. B-12), and the resultant equation is premultiplied through by $[\{\bar{P}_1\}\{\bar{P}_2\}]^T$, the result is a system of uncoupled differential equations for the principal coordinates:

$$\ddot{p}_i + \omega_i^2 p_i = f_i, \qquad i = 1, 2, \tag{B-31}$$

where

$$\begin{Bmatrix} f_1 \\ f_2 \end{Bmatrix} = ([\{\bar{P}_1\}\{\bar{P}_2\}]^T[M][\{\bar{P}_1\}\{\bar{P}_2\}])^{-1}[\{\bar{P}_1\}\{\bar{P}_2\}]^T \begin{Bmatrix} F_1 \\ F_2 \end{Bmatrix}, \tag{B-32}$$

if the system has distinct, nonidentical, characteristic frequencies. The inversion of the matrix in parentheses in Eq. B-32 is a simple matter since this matrix is diagonal (Eq. B-27).

The onset of galloping instability of the principal coordinates is found by the techniques used for uncoupled oscillators in Chapter 4. The forces f_1 and f_2 are expanded in a Taylor series in p_1, p_2, $\dot{p}_1$, and $\dot{p}_2$. The onset of galloping is determined by setting the coefficients of the terms, whose coefficients are $\dot{p}_1$ in f_1 and $\dot{p}_2$ in f_2, to zero. This implies that the onset of galloping is found from

$$k_1 \frac{\partial F_y}{\partial \dot{p}_1} + \frac{\partial F_m}{\partial \dot{p}_1} = 0, \tag{B-33}$$

$$\frac{\partial F_y}{\partial \dot{p}_2} + k_2 \frac{\partial F_m}{\partial \dot{p}_2} = 0, \tag{B-34}$$

where F_y and F_m are the aerodynamic forces defined in Chapter 4. The derivatives are evaluated at $p_1 = p_2 = \dot{p}_1 = \dot{p}_2 = 0$. These two equations reduce to Eq. 4-43. Of course, if $\omega_1 \approx \omega_2$, there is no guarantee that the principal coordinates uncouple the system, and the onset of galloping must be found using the other techniques described in Chapter 4.

REFERENCE

Meirovitch, L. (1967) *Analytical Methods in Vibrations*, Macmillan, New York, pp. 47–50.

Appendix C

Aerodynamic Sources of Sound

In this appendix, the equations for the aerodynamic sound produced in an unbounded fluid volume by forces on the fluid are derived. The derivation follows the work of Koopman (1969), with some corrections, and incorporates the basic results of Curle (1955) and Lighthill (1952). Blake (1986) considers the derivation in considerable detail.

Lighthill's treatment of the generation of aerodynamic sound can be developed by first considering the exact equations of continuity of mass and momentum of a fluid:

$$\frac{\partial \rho}{\partial t} + \frac{\partial}{\partial y_i}(\rho u_i) = 0, \tag{C-1}$$

$$\frac{\partial}{\partial t}(\rho u_i) + \frac{\partial}{\partial y_j}(\rho u_i u_j - \tau_{ij}) = 0, \tag{C-2}$$

where $\rho =$ mass density, $t =$ time, $u_i =$ components of fluid velocity vector ($i = 1, 2, 3$), $\rho u_i =$ momentum density, $y_i =$ coordinates of source point ($i = 1, 2, 3$), and $\tau_{ij} =$ Stokes' stress tensor,

$$\tau_{ij} = -p\,\delta_{ij} - \tau'_{ij}, \tag{C-3}$$

where p is pressure and τ'_{ij} is the stress due to viscous fluid stresses.

The summation convention is used in all the equations in this appendix. All terms in which the same subscript appears twice are summed over that subscript. It is convenient to introduce the stress tensor T_{ij}:

$$T_{ij} = \rho u_i u_j + (p - c^2 \rho)\delta_{ij} + \tau'_{ij}. \tag{C-4}$$

T_{ij} is the difference between the effective stress in the flow field and the stresses in the uniform acoustic medium at rest. The equations of motion can then be written in the form

$$\frac{\partial}{\partial t}(\rho u_i) + c^2 \frac{\partial \rho}{\partial y_i} = -\frac{\partial T_{ij}}{\partial y_j}, \tag{C-5}$$

$$\frac{\partial \rho}{\partial t} + \frac{\partial}{\partial y_i}(\rho u_i) = 0. \tag{C-6}$$

Physically, this states that a fluctuating flow, occurring within a uniform acoustic medium at rest, generates the same density fluctuations as would be produced in a stationary acoustic medium that is acted upon by a system of externally applied

426

stresses T_{ij}. Eliminating the term ρu_i from Eqs. C-4 and C-5 gives

$$\frac{\partial^2 \rho}{\partial t^2} - c^2 \frac{\partial^2 \rho}{\partial y_i^2} = \frac{\partial^2 T_{ij}}{\partial y_i \partial y_j}. \tag{C-7}$$

T_{ij} can be simplified for subsonic flow by neglecting compressible effects in this term and further neglecting viscous fluid stresses; see Howe (1975). With these approximations

$$T_{ij} = \rho u_i u_j.$$

The tensor $\rho u_i u_j$ is called the Reynolds stress. Further, using these approximations it can be shown that Eq. C-7 reduces to (Howe, 1975; Kambe and Minota, 1981; Blake, 1986)

$$\frac{\partial^2 \rho}{\partial t^2} - c^2 \frac{\partial^2 \rho}{\partial y_i^2} = \rho \nabla \cdot (\omega \times U), \tag{C-8}$$

where ω is the vorticity vector and U is the vector fluid velocity. The low Mach number sources of sound in the fluid are thus associated with stretching of vortex filaments in the fluid velocity.

Sound can be produced by interaction with a boundary as well as within the fluid. Curle (1955) considered the general solution of this equation, including the presence of boundaries at a surface S, using the standard Kirchhoff solution of the nonhomogeneous wave equation in the infinite region outside a fixed internal surface, and obtained a solution of the form

$$c^2(\rho - \rho_0) = \frac{\partial^2}{\partial x_i \partial x_j} \int_V \left[\frac{T_{ij}}{4\pi r} \right] dV(y) + \frac{1}{4\pi} \int_S \frac{l_i}{r} \left[\frac{\partial}{\partial y_j} (\rho u_i u_j + p_{ij}) \right] dS(y)$$

$$+ \frac{1}{4\pi} \frac{\partial}{\partial x_i} \int_S \frac{l_j}{r} [\rho u_i u_j + p_{ij}] \, dS(y), \tag{C-9}$$

where $\rho - \rho_0$ = field density fluctuation, x_i = coordinates of the field point of the observer ($i = 1, 2, 3$), $r = |x_i - y_i|$, and l_i = direction cosines taken in the outward normal direction from the fluid volume. The brackets denote retarded time $t - r/c$.

The turbulent effects of the fluid are assumed to occur within a volume V, in which an object having a closed surface S is located. The sound field can be interpreted as that which would be generated in a uniform acoustic medium at rest acted upon by (1) a volume distribution of quadrupoles with strength T_{ij} distributed throughout the flow; (2) sources distributed around the internal boundary S with a strength equal to the rate of variation of local mass outflow from S; and (3) dipoles distributed around S with a strength equal to the local rate of momentum output from S. It can be shown that, for low Mach number flows, the contribution of the quadrupole source is much less than that of the dipole source. Therefore, the quadrupole source is neglected by setting $T_{ij} = 0$.

One limitation of Eq. C-9 is that the surface S is fixed in space. This implies that solutions to Eq. C-9 apply only to structures that are at rest or undergo only small vibrations. However, Frost and Harper (1975) have shown that, for an oscillating sphere, surface motions introduce terms that are smaller than the solution to Eq. C-9 by the order of $(D/\lambda)^2$, where D is the cross-section dimension and λ is the wavelength of sound. Thus, it is reasonable in far-field analysis $(D/\lambda \ll 1)$ to apply Eq. C-9 directly, even to cases where the vibration amplitude is of the same order as the cross-section dimension.

The second term in Eq. C-9 can be simplified by substituting Eq. C-2 into Eq. C-9. If the second-order effects of fluid compressibility are neglected, which is reasonable for a low Mach number flow, Eq. C-9 becomes

$$\rho - \rho_0 = -\frac{\rho_0}{4\pi c^2} \int_S \frac{l_i}{r} \left[\frac{\partial u_i}{\partial t} \right] dS(y) + \frac{1}{4\pi c^2} \frac{\partial}{\partial x_i} \int_S \frac{l_j}{r} [\rho_0 u_i u_j + p_{ij}] dS(y). \quad \text{(C-10)}$$

Equation C-10 can be expressed in a more workable form by applying Gaussian theory to the first integral to yield

$$\int_S \frac{l_i}{r} \left[\frac{\partial u_i}{\partial t} \right] dS(y) = -\int_V \frac{\partial}{\partial y_i} \frac{[\partial u_i/\partial t]}{r} dV(y),$$

$$= -\int_V \left\{ \frac{1}{r} \left[\frac{\partial^2 u_i}{\partial t \partial y_i} + \frac{1}{c} \frac{\partial^2 u_i}{\partial t^2} \frac{(x_i - y_i)}{r} \right] + \left[\frac{\partial u_i}{\partial t} \right] \frac{(x_i - y_i)}{r^3} \right\} dV(y). \quad \text{(C-11)}$$

The form of Eq. C-11 is due to the fact that $\partial u_i/\partial t$ is evaluated at retarded time; thus,

$$\frac{\partial}{\partial y_i} \left[\frac{1}{r} f(y_i, t - r/c) \right] = \frac{1}{r} \frac{\partial f}{\partial y_i} - \left(\frac{1}{r^2} f + \frac{1}{cr} \frac{\partial f}{\partial (t - r/c)} \right) \frac{\partial r}{\partial y_i}. \quad \text{(C-12)}$$

The sign of Eq. C-11 is due to definition of the direction of the outward vectors l_i. If the surface S contains a volume $V(y)$ which is incompressible ($\rho = $ constant in Eq. C-1), then the first term in the integral on the right-hand side of Eq. C-11 is zero.

Operating on the second surface integral in Eq. C-10 with $\partial/\partial x_i$ gives

$$\frac{\partial}{\partial x_i} \int_S \frac{l_j}{r} [\rho_0 u_i u_j + p_{ij}] dS(y)$$

$$= -\int_S l_j \left\{ \frac{(x_i - y_i)}{cr^2} \left[\frac{\partial}{\partial t} (\rho_0 u_i u_j + p_{ij}) \right] + \frac{(x_i - y_i)}{r^3} [\rho_0 u_i u_j + p_{ij}] \right\} dS(y). \quad \text{(C-13)}$$

If the typical dimension of the cross section of the body D, contained by the surface S, is such that $D \ll c/\omega$, where ω is a typical sound frequency, the retarded-time term $r(x, y)/c$ may be written $r(x)/c$ for evaluating the integral. If

$$t' = t - \frac{r(x)}{c}$$

and

$$p_i = -l_j p_{ij},$$

and the distance to the field point is assumed to be large compared with the size of the body (i.e., $x_i \gg y_i$), then, using Eqs. C-11 through C-13, Eq. C-10 becomes

$$\rho - \rho_0 = \frac{\rho_0 x_i}{4\pi c^3 r^2} \int_V \frac{\partial^2 u_i}{\partial t^2} (y, t') dV(y) + \frac{\rho_0 x_i}{4\pi c^2 r^3} \int_V \frac{\partial u_i}{\partial t} (y, t') dV(y)$$

$$- \frac{x_i}{4\pi c^3 r^2} \int_S \frac{\partial p_i}{\partial t} (y, t') dS(y) - \frac{x_i}{4\pi c^2 r^3} \int_S p_i(y, t') dS(y)$$

$$- \frac{\rho_0 x_i}{4\pi c^3 r^2} \int_S l_j \frac{\partial u_i u_j}{\partial t} (y, t') dS(y) - \frac{\rho_0 x_i}{4\pi c^2 r^3} \int_S l_j u_i u_j(y, t') dS(y). \quad \text{(C-14)}$$

Since the body contained within S is assumed to be incompressible, it can be shown that the last two terms in Eq. C-14 vanish.

If the cross section of the body is small compared with the wavelength of sound at the frequency of interest, the integrals in Eq. C-14 can be evaluated without reference to the shape of the cross section. If the body has a central axis extending from $z = 0$ to $z = L$, then u_i takes on the values on the axis.

The force per unit length exerted by the body on the fluid is

$$F_i = -\int_S p_i(y, t') \, ds, \qquad (C\text{-}15)$$

where s denotes a line of integration about the axis of the region. For small acoustic disturbances, the density fluctuations are proportional to the pressure fluctuations:

$$\rho - \rho_0 = \frac{p - p_0}{c^2}, \qquad (C\text{-}16)$$

and Eq. C-14 becomes

$$p - p_0 = \frac{x_i}{4\pi r^2} \int_0^L \left[\frac{\rho_0 A}{c} \frac{\partial^2 V_i}{\partial t^2} + \frac{1}{c} \frac{\partial F_i}{\partial t} + \frac{1}{r} \left(\rho_0 A \frac{\partial V_i}{\partial t} + F_i \right) \right] dz, \qquad (C\text{-}17)$$

where V_i is the velocity of the axis of the body, and V_i and F_i are evaluated at retarded time. This is the fundamental equation for the aerodynamic sound pressure (p) produced by a moving body which exerts a force on an infinite fluid volume.

REFERENCES

Blake, W. (1986) *Mechanics of Flow-Induced Sound and Vibration*, Academic Press, New York.

Curle, N. (1955) "The Influence of Solid Boundaries Upon Aerodynamic Sound," *Proceedings of the Royal Society of London, Series A*, **231**, 505–514.

Frost, P. A., and E. Y. Harper (1975) "Acoustic Radiation from Surfaces Oscillating at Large Amplitude and Small Mach Number," *Journal of the Acoustical Society of America*, **58**, 318–325.

Howe, M. S. (1975) "Contributions to the Theory of Aerodynamic Sound, with Application to Excess Jet Noise and the Theory of the Flute," *Journal of Fluid Mechanics*, **71**, 625–673.

Kambe, T., and T. Minota (1981) "Sound Radiation from Vortex Systems," *Journal of Sound and Vibration*, **74**, 61–72.

Koopman, G. H. (1969) "Wind Induced Vibrations and Their Associated Sound Fields," Dissertation, Catholic University of America.

Lighthill, M. J. (1952) "On Sound Generated Aerodynamically. I. General Theory," *Proceedings of the Royal Society of London, Series A*, **211**, 564–587.

Digital Spectral and Fourier Analysis

If a continuous function of time $y(t)$ is defined over an interval $0 \le t \le T$, it can be represented by a Fourier series over the same interval,

$$y(t) = a_0 + \sum_{k=1}^{\infty} [a_k \cos (2\pi k/T)t + b_k \sin (2\pi k/T)t], \qquad 0 \le t \le T, \qquad \text{(D-1)}$$

where the Fourier coefficients are obtained by integration,

$$a_0 = \frac{1}{T} \int_0^T y(t) \, dt, \qquad a_k = \frac{2}{T} \int_0^T y(t) \cos (2\pi k/T)t \, dt,$$

$$b_k = \frac{2}{T} \int_0^T y(t) \sin (2\pi k/T)t \, dt. \qquad \text{(D-2)}$$

The Fourier series converges to $y(t)$ as the number of terms included in the series increases (Bracewell, 1978). Inspecting this result, we see that the Fourier coefficients a_k and b_k are the magnitudes of components of $y(t)$ at frequencies $f_k = k/T$, $k = 1, 2, 3, \ldots, \infty$. The Fourier coefficients can be plotted against their associated frequencies to display the frequency content of the function. Such a plot is called a *spectrum*.

In analogy to the Fourier coefficients of Eq. D-2, the finite Fourier transform is defined as a transformation of a function over the interval $0 \le t \le T$ (Bendat and Piersol, 1986),

$$H(f_k) = \int_0^T y(t) e^{-i2\pi f_k t} \, dt, \qquad \text{(D-3)}$$

where i is the imaginary constant such that

$$i = \sqrt{-1}, \qquad e^{-i\theta} = \cos \theta - i \sin \theta. \qquad \text{(D-4)}$$

The Fourier transform $H(f_k)$ is a complex function. The real part of $H(f_k)$ is equal to $(T/2)a_k$ of the Fourier series coefficients and the imaginary part is equal to $-(T/2)b_k$ (Eqs. D-1 and D-2).

The finite Fourier transform can be computed digitally by discretizing the time interval $0 \le t \le T - \Delta T$ into $N - 1$ time intervals of width $\Delta T = T/N$ and N points numbered $n = 0, 1, 2, 3, \ldots, N - 1$, as shown in Figure D-1,

$$t_k = 0, \Delta T, 2 \Delta T, \ldots, (N-1) \Delta T. \qquad \text{(D-5)}$$

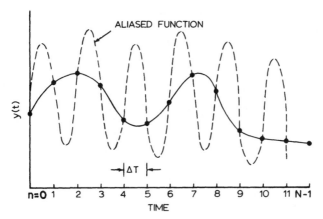

Fig. D-1 Digitization of a time-history.

The finite Fourier transform with discretized time-history is

$$H(f_k) = \Delta T \sum_{n=0}^{N-1} y(n\,\Delta T)e^{-i2\pi f_k n\,\Delta T}, \qquad k = 0, 1, 2, 3, \ldots, (N-1). \quad \text{(D-6)}$$

The frequencies in the finite Fourier transform,

$$f_k = \frac{k}{T} = \frac{k}{(N\,\Delta T)}, \qquad k = 0, 1, 2, 3, \ldots, (N-1), \quad \text{(D-7)}$$

span the range from 0 to $(N-1)/T$. This has important consequences for data sampling. For example, if we wish to characterize ocean waves that have periods between 10 and 20 sec, the sampling interval should exceed 20 sec, say $T = 100$ sec. If the sample size is 1024 points, then data will be taken every $\Delta T = 100/1024 = 0.0976$ sec. The resolution of the spectral information is $\Delta f = f_{k+1} - f_k = 1/T = 1/100$ sec $= 0.01$ Hz. The lowest nonzero frequency is $1/T = 0.01$ Hz. The highest frequency included in the analysis is $(N-1)/T = 10.23$ Hz but useful information is only obtained up to the Nyquist frequency, $N/(2T) = 5.12$ Hz.

Substituting Eq. D-7 into Eq. D-6 completes the discrete finite Fourier transform computation,

$$H(k/T) = \Delta T \sum_{n=0}^{N-1} y(n\,\Delta T)e^{-i2\pi nk/N}, \qquad k = 0, 1, 2, 3, \ldots, (N-1). \quad \text{(D-8)}$$

The index k is the frequency counter and the index n is the integration counter. Numerical evaluation of Eq. D-8 is straightforward. N^2 evaluations of the quantity inside the summation are required. The author calls this the slow Fourier transform (SFT). In 1965 Cooley and Tukey invented a numerical algorithm called the Fast Fourier Transform (FFT) that reduces the number of computations to $N \log_2 N$, provided N is a multiple of 2 ($N = 2^p$, where $p =$ integer), making it practical to calculate finite Fourier transforms on small computers. A related transform called the Hartley transform has recently been developed that offers even faster computing times (Bracewell, 1986, 1989).

The FFT algorithms given by Brigham (1974) in FORTRAN and ALGOL and by Rameriz (1985) in BASIC contain approximately 50 lines of coding. These FFT algorithms calculate only the summation in Eq. D-8,

$$FFT(k/T)(=) \sum_{n=0}^{N-1} y(n \, \Delta T)e^{-i2\pi nk/N}, \qquad k = 0, 1, 2, 3, \ldots, (N-1). \qquad \text{(D-9)}$$

The result is a real and an imaginary component of the transform for each frequency. With the exception of the zero frequency, the real part of the discrete Fourier transform is symmetric about the center frequency $f_{N/2} = N/(2T)$, also called the *Nyquist frequency*. The imaginary part is antisymmetric, as shown in Figure D-2 (Brigham, 1974, p. 143). This symmetry is associated with a phenomenon called *aliasing*.

Aliasing is the result of discretization of the time-history. As shown in Figure D-1, the discretized points could have been drawn from either a low-frequency function (solid line) or a very much higher-frequency function (dashed line). As a consequence, for each low-frequency component in the discrete Fourier transform, there is a symmetrical high-frequency component. If we have sampled at a sufficient rate or filtered out frequencies above the frequency of interest before the time-history is digitized, then the high-frequency components that appear in the Fourier transform above the Nyquist frequency $f_k = N/(2T)$ have no physical significance. Aliasing of high frequencies in a time-history into low frequencies can be avoided by (1) filtering out components above the frequency of interest before the time-history is digitized and (2) digitizing at a rate ΔT equal to one half the period of the highest-frequency component of interest (Brigham, 1974). The former is accomplished experimentally with so-called anti-aliasing filters inserted before the digitizer.

A single-sided spectrum that truncates the high-frequency portions of the discrete Fourier transform is defined as follows:

$$G_y(k/T) = \begin{cases} H(0), & k = 0; \\ 2H(k/T), & 1 \le k < N/2 - 1; \\ 0, & N/2 \le k \le N - 1. \end{cases} \qquad \text{(D-10)}$$

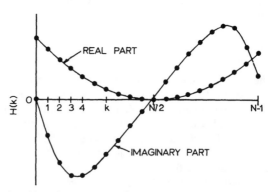

Fig. D-2 Real and imaginary parts of the discrete Fourier transform.

$G_y(k/T)$, like $H(k/T)$, is a complex function. The *auto spectral density*, also called *power spectral density*, is a real function equal to the mean square of each frequency component divided by the width of the frequency interval Δf,

$$S_y(0) = \frac{1}{T}[G_y(0)]^2,$$

$$S_y(k/T) = \frac{1}{2T}|G_y(k/T)|^2 = \frac{1}{2T}G_y^*(k/T)G_y(k/T) = (a_k^2 + b_k^2)\frac{T}{2}, \qquad \text{(D-11)}$$

$$k = 1, 2, 3, \ldots, (N/2 - 1),$$

where * denotes complex conjugate. $S_y(k/T)$ has the units of y^2/Hz, that is, $(\text{meter})^2/\text{Hz}$ for displacement or $(\text{psi})^2/\text{Hz}$ for pressure. The *auto spectrum* is defined as the mean square in each frequency interval,

$$S_{yy}(k/T) = \Delta f\, S_y(k/T) = \frac{1}{T}S_y(k/T). \qquad \text{(D-12)}$$

$\Delta f = f_{k+1} - f_k = 1/T$ is the frequency resolution. The magnitude of the auto spectrum is the root mean square of each auto spectrum component,

$$\text{Mag}\{S_{yy}(k/T)\} = \sqrt{S_{yy}(k/T)}, \qquad k = 0, 1, 2, 3, \ldots, (N/2 - 1). \quad \text{(D-13)}$$

The magnitude of the auto spectrum has the same units as $y(t)$. The Fourier coefficients (Eq. D-1) are

$$a_k = \text{Real}\{G_y(k/T)\}/T, \; b_k = -\text{Im}\{G_y(k/T)\}/T, \; k = 0, 1, \ldots, (N/2 - 1).$$
$$\text{(D-14)}$$

These algorithms have been incorporated into experimental analysis equipment (Bruel & Kjaer, 1985; Harris, 1981).

Practically speaking, the number of sampling points N is fixed by hardware and the sampling interval T is chosen so that the frequency range of interest lies below the Nyquist frequency. T is not generally an integer multiple of the period of the time-history. Yet the finite Fourier transform (Eqs. D-1, D-6) produces a series that is periodic of period T (Eq. D-1). As a result, this finite Fourier series representation will generally be discontinuous at time T since $y(0) \neq y(T)$, and this discontinuity produces extraneous components in the Fourier components that appear as side lobes about the spectral peaks. This phenomenon is called *leakage*. The most direct method of suppressing leakage in the spectra is to force the time-history to be periodic at period T. This is done by *windowing* the time-history by premultiplying the time-history by a window function that is periodic with period T.

Four widely used windows are (Brigham, 1974; Harris, 1978, 1981; Gade and Herlufsen, 1987):

Rectangular $W(t) = 1.0$
Triangular $W(t) = 1 - |(2t/T) - 1|$
Hanning $W(t) = \frac{1}{2} - \frac{1}{2}\cos(2\pi t/T)$
Kaiser–Bessel $W(t) = 1 - 1.24\cos(2\pi t/T) + 0.244\cos(4\pi t/T)$
$\qquad\qquad - 0.00305\cos(6\pi t/T).$

These windows apply over the sampling interval $0 \le t \le T$. They are plotted in

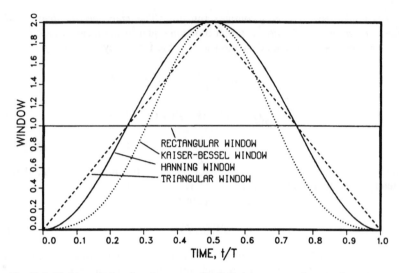

Fig. D-3 Various window functions normalized to the same maximum value.

Figure D-3. Gade and Herlufsen (1987) and Harris (1981, 1978) have found that the Kaiser–Bessel window is superior for resolving differences between two coherent frequencies and that the Hanning window is superior for broad-band signals. As can be seen in Figure D-3, these two windows are very similar when normalized to the same maximum amplitude.

Because the windows affect the average amplitude of the time-history signal, they also influence the magnitude of the Fourier components and an additional factor is required to ensure that the mean-square value is maintained. Bendat and Piersol (1986) suggest that the Fourier components $H(k/T)$ should be increased by the factor $\sqrt{8/3}$ if the Hanning window is used in order to regain the proper magnitude of the spectrum. It is the author's experience that this factor varies somewhat with frequency.

Exercises

1. Compute the finite Fourier series of a square wave over a 1 sec period; $y(t) = -1/2$, $0 < t < 0.5$ sec, $y(t) = 1/2$, $0.5 \leq t \leq 1$ sec.

2. What is the relationship between the Fourier components (Eqs. D-1, D-14) and the auto spectrum, magnitude of auto spectrum, and power spectral density?

3. An analyzer is limited to 2048, that is, 2^{11}, sample points. It is desired to characterize the vibration of ducting using an accelerometer. The maximum duct frequency is felt to be about 200 Hz. What digitization rate ΔT should be used? What is the sampling duration T? What is the Nyquist frequency? At what frequency should the anti-aliasing filter be set? At what discrete frequencies will results be obtained? What is the frequency resolution? What will be the units of the finite discrete Fourier transform,

the power spectral density, the auto spectrum, and the magnitude of the auto spectrum? What would be the benefit if 4096 points could be used? Is there any way to obtain the benefits of using 4096 points if we are hardware-limited to 2048 points at a time?

REFERENCES

Bendat, J. S., and A. G. Piersol (1986) *Random Data: Analysis and Measurement Procedures,* 2d ed., Wiley, New York.

Bracewell, R. N. (1978) *The Fourier Transformation and Its Application,* 2d ed., McGraw-Hill, New York.

Bracewell, R. N. (1986) *The Hartley Transform,* Oxford University Press, Oxford.

Bracewell, R. N. (1989) "The Fourier Transform," *Scientific American,* **260**(6), 86–95.

Brigham, E. O. (1974) *The Fast Fourier Transform,* Prentice-Hall, Englewood Cliffs, N.J.

Bruel & Kjaer (1985) *Instruction Manual for the Dual Channel Signal Analyzer Type 2032,* Vol. 2, Naerum, Denmark.

Gade, S. H., and H. Herlufsen (1987) "Use of Weighting Function in DFT/FFT Analysis," *Bruel & Kjaer Technical Review,* Nos. 3 and 4.

Harris, F. J. (1978) "On the Use of Windows for Harmonic Analysis with Discrete Fourier Transforms," *Proceedings of the IEEE,* **66**(1), 51.

Harris, F. J. (1981) "Trigonometric Transforms," Scientific-Atlanta, Spectral Dynamics Division Publication DSP-005 (8-81), San Diego, Calif.

Ramirez, R. W. (1985) *The FFT Fundamentals and Concepts,* Prentice-Hall, Englewood Cliffs, N.J.

Section III, Division 1, Appendix N of 1992 ASME Boiler and Pressure Vessel Code

N-1300 FLOW-INDUCED VIBRATION OF TUBES AND TUBE BANKS

N-1310 INTRODUCTION AND SCOPE

The flow-induced vibration (FIV) potential of structures has been known for a long time (Refs. 79 through 84). FIV analyses are required to determine the adequacy of a design, or in areas of uncertainty, to be aware of the need for experimental verification (Refs. 85 and 86) if high reliability of the component is a necessity. FIV may be due to any one of several excitation mechanisms because power systems include many types of flexible components subject to a variety of fluid flows, such as pipe, channel, and jet flows followed by mixing in plenums and heat exchangers. Since a single component is often subjected to different turbulent flows from several directions because of the influence of adjacent structures and boundaries, FIV analyses for more than one excitation mechanism is not unusual.

The quantitative data and correlations available to perform FIV analyses are unique to the flow geometry created by each component. More quantitative information and design methods are available for some components than others. In particular, the circular cylinder has been studied most. N-1320 through N-1340 of this Appendix are presented to illustrate one or more acceptable steps for the FIV analysis of arrays of cylinders subject to the three most significant excitation mechanisms. The general methods employed are applicable to other types of components, but the data are specifically for single cylinders and cylindrical arrays. Because of the large number of FIV mechanisms, the methodology of analysis is referenced, but enough information is given to understand a mechanism and make design calculations. Because of the developing nature of the subject, more than one set of design data or methods may be recommended with the implication to the designer to use either the more appropriate or the more conservative predictions.

Semiempirical correlations based on experimental data, but guided by the equations of motion, often form the basis of a design method. The state-of-the-art regarding description of the FIV mechanisms is that many mathematical models have been proposed for fluid-structure coupling forces, but general agreement on the physics of many of the phenomena has not been attained, although models simulating the behavior may be available.

N-1311 Definitions

In this section some commonly used terminologies in flow-induced vibration analysis are defined and briefly described.

(a) Fluid forces can be defined into two broad categories to describe FIV excitation mechanisms (Refs. 79 through 83). Fluid excitation forces are created by the incident flow on a structure, and they would occur, in some form, even without structural motion. Fluid-structure coupling forces are induced by structural motion, and they occur in both flowing and nonflowing fluids.

(b) Added mass and added damping have been successfully used to characterize the fluid-structure coupling forces created by the motion of a structure in a nonflowing fluid (Refs. 87 through 93). Added mass and added damping increase the effective mass and damping of a structure vibrating in a fluid. In addition, the presence of a dense fluid between otherwise unconnected, adjacent structures can couple their vibrations and result in significantly different natural frequencies, mode shapes, and damping from those obtained in a vacuum. For a low density fluid (e.g., air); the added mass is often negligible.

(c) In a *weakly coupled fluid-structure system*, the FIV excitation mechanism causes small structural motion and the fluid forces induced by the structural motion can be linearly superimposed onto the fluid excitation forces which are largely independent of the structural motion. The fluid-structure coupling forces can be expressed to a first order of approximation in terms of added mass, stiffness, and damping matrices. The fluid excitation forces can be determined separately from the coupling forces either by analysis or by model tests with only the hydraulics simulated.

Examples of FIV excitation mechanisms producing weakly coupled fluid-structure systems are incident flow turbulence and turbulent boundary layers over rods, plates, and shells (Refs. 81 and 82); some wake flows produced by flow across bluff bodies; and many sources of acoustic noise (Refs. 80 and 95). In these cases, the fluid excitation energy is generated at some point in the fluid circuit and the structure is the recipient of the energy. The forces due to flow turbulence and attached boundary layers typically are broadband random, while separated wake flows that roll into periodically shed vortices can produce very discrete frequency forces (Refs. 82, 87, and 97).

(d) In a *strongly coupled fluid-structure system*, the FIV excitation mechanism causes the structural motion to become large enough to change the flow field; some of the fluid forces amplify, rather than inhibit, the structural motion that produced them. Clearly distinguishing between fluid-structure coupling forces and fluid excitation forces is difficult in strongly coupled fluid-structure systems. In general, the coupling forces are highly nonlinear functions of structural motion and flow velocity.

(e) Fluid-elastic instability of closely packed heat exchanger tube bundles (Refs. 80, 81, 82, and 88) is an example of a strongly coupled fluid-structure system. The motion of each tube affects the fluid forces and the motion of the other tubes to produce self-excitation. The occurrence of the instability has been interpreted as due to adverse changes in the structural mass, damping, and fluid-structure coupling force (Ref. 88). However, most of the expressions for predicting the onset of instability are based on compilations of direct measurements of the critical velocities at the onset of instability.

(f) Cross flow is a flow perpendicular to the structural longitudinal axis. Cross flow

is one example where an FIV mechanism is produced that can create either a weakly or a strongly coupled fluid-structure system. Vortex shedding in the wake of a tube in cross flow produces both fluid excitation forces and fluid-structure coupling forces that amplify structural motion. For ideal cross flow, where a long, smooth surface tube is isolated in uniform (2-D) cross flow with little or no turbulence in the approaching flow stream, very periodic, two-dimensional vortices are shed. These vortices produce alternating lift forces normal to the tube axis and flow and are nearly as large as the steady, flow direction drag forces, if the Reynolds number, based on the tube diameter, is below 2×10^5 (Refs. 82, 87, and 89). If the vortex shedding frequency is sufficiently different from the structural natural frequencies, the alternating lift forces act as fluid excitation forces only. However, if the vortex shedding frequency and one of the structural natural frequencies are sufficiently close to each other and the fluid excitation forces can produce large enough motion, then coupled fluid-structure forces occur, which apparently further amplify the motion. Enough experimental data are available to bound the fluid excitation forces, but the representation of the coupled fluid-structure forces is still being researched. Most of the representations are based on highly phenomenological models that stimulate, to various degrees, a small amount of data covering only a narrow range of idealized conditions.

(g) The *joint acceptance* is a measure of the probability that a structure vibrating in one mode will remain in the same mode when excited by a random force; the *cross acceptance* is a measure of the probability that a structure vibrating in one mode will change to another mode when excited by a random force. For many applications only the joint acceptance is assumed to be important. When mode shapes are normalized to unity, the sum of the joint acceptances is equal to 1 (see N-1342.1). Therefore, the assumption that the joint acceptance is equal to 1 gives conservative estimates of structural responses.

N-1312   Nomenclature

$C_n =$ reduced damping in nth mode
$C_L =$ lift coefficient
$D =$ cylinder diameter
$E =$ Young's modulus
$f_n =$ natural frequency of nth vibration mode, hertz
$f_s =$ frequency of periodic vortex shedding, hertz
$F =$ force
$G_f =$ single-sided power spectral density of the forcing function, in (force/length)2 per Hz
$G_f^i = G_f$ spectrum for the ith span of a multi-span tube
$G_y =$ single-sided power spectral density of response
$H_j =$ transfer function of jth vibration mode
$I =$ area moment of inertia
$J^2 =$ joint acceptance
$J_{jk}^2 =$ cross acceptance for the jth and kth vibration modes
$(J_{jk}^i)^2 =$ acceptance for the ith span
$\ell_c =$ axial correlation length

$\quad = 2 \int_0^L r(x')\, dx'$ where $R(x')$ is the correlation function and x' is the separation distance

$l_c^i =$ coffelation length in the ith span

$L_e =$ cylinder length subject to vortex shedding

$L_i =$ span length

$m =$ mass per unit length

$m_A =$ added fluid mass per unit length

$m_c =$ contained fluid mass per unit length

$m_f =$ cylinder displaced fluid mass/length

$m_s =$ structural mass per unit length

$m_t =$ total mass per unit length of tube

$\quad = m_A + m_c + m_s$

$M_j =$ modal mass

$M_n =$ effective modal mass/length for nth vibration mode

$n =$ vibration mode, $n = 1$ is fundamental mode

$p =$ pressure

$P =$ tube pitch

$\quad =$ distance between tube centers

$q =$ dynaniic pressure,

$Re =$ Reynolds number, VD/ν

$R_p =$ Cross correlation of the pressure field

$S =$ Strouhal number, $f_s D/V$

$S_f =$ cross spectral density of the forcing function on a cylinder, (force/length)^2 per Hertz

$S_{fo} =$ power spectral density of the forcing function

$S_p =$ cross spectral density of the pressure field

$S_y =$ power spectral density of cylinder response

$t =$ time

$U_c =$ convection velocity

$V =$ mean velocity

$x =$ axial distance

$y^*_n =$ maximum displacement in nth vibration mode

$y^2 =$ mean square response of a cylinder

$\alpha_n =$ amplification factor in nth vibration mode

$\gamma_n =$ mode shape factor in nth vibration mode

$\Gamma =$ coherence of forcing function on a cylinder

$\Gamma^i =$ coherence for ith span

$\delta_m =$ mass-damping parameter, $2\pi\xi_n m_t/\rho D^2$

$\delta_n =$ log decrement for nth vibration mode

$\quad = 2\pi\xi_n$

$\xi_n =$ fraction of critical damping for nth mode

$\rho =$ fluid mass density

$\phi_n =$ nth vibration mode shape

$\phi^*_n =$ maximum value of ϕ_n

$\theta =$ angle between direction of flow and normal to tube axis

$\nu =$ kinematic viscosity

$\omega =$ frequency, radians/sec

FIG. N-1321-1   VORTICES SHED FROM A CIRCULAR CYLINDER

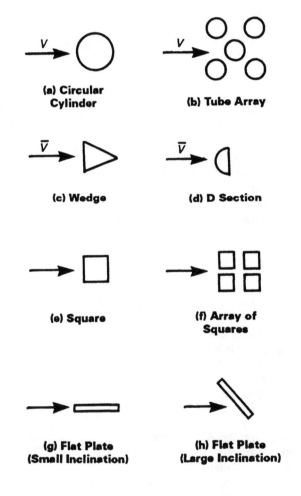

Fig. N–1321–2 Some Typical cross sections of Bluff Bodies that can experience vortex shedding.

N-1320   VORTEX SHEDDING

N-1321   Vortex Shedding From a Fixed Bluff Body

For a bluff body in uniform cross flow, the wake behind the body is no longer regular, but contains distinct vortices of the pattern shown in Fig. N-1321-1 for a circular cylinder. The vortices are shed alternatively from each side of the body in a regular manner and give rise to an alternating lift force. Experimental studies of this vortex shedding process have shown (Refs. 94 and 95) that the frequency in hertz of the alternating lift force can be expressed as:

$$f_s = SV/D \tag{62}$$

Some common types of bodies or structures for which vortex shedding occurs are shown in Fig. N-1321-2. The following discussions are based on the circular cylinder; however, the concepts apply equally well to other bluff bodies.

The oscillating lift force produced on an isolated single cylinder of diameter D and length L by uniform cross flow can be expressed as (Refs. 96 and 97):

$$F = C_L JqDL[\sin{(2\pi f_s t)}] \tag{63}$$

where C_L, f_s, and J are functions of the Reynolds number Re and must be determined experimentally. In uniform cross flow, the energy of vortex shedding occurs over a very narrow frequency band with a center frequency f_s, except over a transition band of Reynolds number (2×10^5 to 3×10^6) where the character of the frequency content may vary from almost periodic to completely random. The measured Strouhal number is $S \approx 0.2$ for $10^3 < Re < 2 \times 10^5$; for larger Re, experimental values of S and C_L show considerable scatter.

The alternating vortex fluid forces are not generally correlated over the entire cylinder length L. As a consequence, two limiting cases of the joint acceptance exist for a uniform rigid-body-mode (Ref. 97).

$$\begin{aligned} J^2 &= \ell_c L \quad &&\text{if } \ell_c \ll L \\ &= 1 \quad &&\text{if fully correlated} \end{aligned} \tag{64}$$

The correlation length in the lift direction for stationary cylinders has been found to be approximately 3 to 7 diameters ($3D < \ell_c < 7D$), for $10^3 < Re < 2 \times 10^5$ (Ref. 87). For larger Reynolds numbers, the correlation lengths for stationary cylinders can be expected to be even smaller because the attached boundary layer becomes fully turbulent. J^2 is usually much less than 1 for long stationary cylinders. Motion of the cylinder at the frequency of vortex shedding substantially increases the correlation length (Refs. 82 and 87) as discussed in N-1323 and N-1324.

Vortex shedding also induces a force in the streamwise or drag direction. The drag force occurs at twice the vortex shedding frequency for single cylinders (Ref. 87). However, the magnitude of the oscillating drag force is typically an order of magnitude smaller than the oscillating lift force.

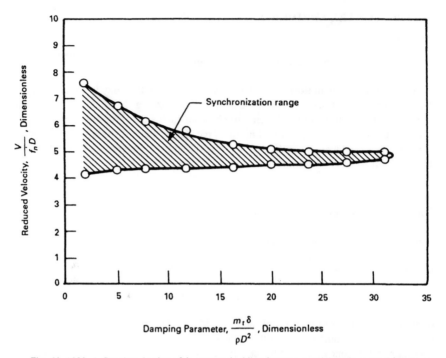

Fig. N–1323–1 Synchronization of the vortex shedding frequency and the tube natural frequency for a single, flexibly-mounted circular cylinder. Synchronization occurs within the shaded region. (ref. 106)

N-1322   Practical Cross Flow

The case of ideal cross flow is rarely found except in the laboratory. Many practical conditions reduce the effectiveness and strength of vortex shedding as an excitation mechanism:

(a) If the body is located in a turbulent flow, or if the tube surface is rough, the turbulence tends to widen the band of shedding frequencies and decrease the energy at the dominant shedding frequency (Ref. 129).

(b) If the cylinder is inclined to the flow, the shedding frequency can be adequately predicted by employing the component of flow velocity normal to the cylinder axis

$$f_s = (SV/D) \cos \theta \tag{65}$$

where θ is the angle between the direction of flow and the normal to the cylinder axis. Inclined flow tends to reduce the magnitude of the vortex shedding forces (Ref. 98).

(c) Spanwise variations in flow velocity imply that the vortex shedding frequency also varies in the spanwise direction. This effect will generally reduce the magnitude of the net vortex shedding excitation.

(d) There is some evidence (Refs. 99, 100) that vortex shedding does not occur in two-phase flow and that vortex shedding is only a concern in single-phase flows.

(e) While the vortex shedding characteristics discussed above have general appl-
icability, the effects of adjacent bodies have not been specifically included. Studies
of two (Ref. 101) or more circular cylinders show that vortex shedding does occur,
but its character is very sensitive to the relative location and spacing of the cylinders.
For the important case of tube arrays, the values of S, J, and C_L to be employed in
Eq. (63) of N-1321 are much more uncertain than for single cylinders, as is evident
by the considerable scatter in the experimental data (Refs. 100, 102, 103, 104, 105,
139, and 140).

N-1323   Flexible Cylinders

When the vortex shedding frequency f_s is sufficiently different from the structural
natural frequencies, a condition called off-resonance, the representation of the vortex
shedding lift force by F given in Eq. (63) of N-1321 is valid, and is conservative if
$C_L = 1$ and $J = 1$ is chosen. This conservative representation of the force can be
extended to nonuniform loading, and modal response can be employed to simplify the
analysis of cylinders where many modes are active. Normally, off-resonance response
is small. However, as resonance is approached, large motions are encountered.

In the case of a single flexible or resiliently supported tube, once vibration begins
the shedding frequency and the tube natural frequency can become synchronized if the
two are sufficiently close. For a spring-supported cylinder in an air stream, it was
shown (Ref. 106) that the velocity range over which synchronization persists depends
upon the damping parameter, $m_t\delta_n/\rho D^2$. In Fig. N-1323-1, the shaded area is the region of synchronization. The ordinate, V/f_nD, is a reduced velocity, where f_n is the
natural frequency of the spring-mounted cylinder. Note, in particular, that with increasing $m_t\delta_n/\rho D^2$ the reduced velocity range over which synchronization persists decreases, and no synchronization occurs for $m_t\delta_n/\rho D^2 > 32$. Outside the shaded area,
the cylinder experiences an alternating lift force at the vortex shedding frequency for
a stationary cylinder, as given previously in Eq. (63) of N-1321.

The consequences of synchronization are many. As the flow velocity is either increased or decreased so that the vortex shedding frequency approaches the structure
frequency, the following will occur.

(a) The vortex shedding frequency shifts to the structural natural frequency, i.e.,
it synchronizes with or "locks-in" to the structural frequency even if the flow velocity
or the structural frequency is varied within the range of sychronization as indicated in
Fig. N-1323-1.

(b) The spanwise correlation of the vortex shedding forcing function increases rapidly as structural response increases.

(c) The lift force becomes a function of structural amplitude.

(d) The drag force on the structure increases.

(e) The strength of the shed vortices increases.

Within the synchronization band, substantial resonant vibration in lightly damped
structures often occurs. Vibration amplitudes up to three diameters peak-to-peak have
been observed in dense fluids, such as water, over cables and tubing. The vibrations
are predominantly transverse to the flow and they are self-limiting (Refs. 82, 87, and
94).

Large-amplitude, synchronized vibrations in the drag direction have been observed for a single cylinder in water. These oscillations initiate at relatively low flow velocities corresponding to subharmonic frequencies of vortex shedding, i.e., at 1/4, 1/3, or 1/2 the flow velocity required for synchronization according to Eq. (62) of N-1321 (Refs. 107 and 108). However, the synchronization in the drag direction is not as strong as in the lift direction, and usually occurs only for lightly damped structures in dense fluids (Refs. 87 and 110). Lock-in has not been observed in two-phase flow or deep (more than a few rows) inside a closely spaced tube bundle.

N-1324   Design Procedures for a Circular Cylinder

Whenever possible, lock-in operating conditions should be avoided, but complex designs often make this impossible. Thus, criteria are given for which lock-in can be avoided, and off-resonance structural dynamic analysis can be employed, as well as design procedures to calculate the response during lock-in.

N-1324.1 Avoiding Lock-In Synchronization. Lock-in for a single cylinder can be avoided by one of the following four methods (Refs. 82, 106, 108, and 109). For tube arrays, only (a), (b), and (c) are applicable methods, and V must be the flow velocity in the minimum gap $(P - D)$.

(a) If the reduced velocity for the fundamental vibration mode ($n = 1$) satisfies:

$$V/f_1 D < 1 \tag{66}$$

then both lift and drag direction lock-in are avoided.

(b) If for a given vibration mode the reduced damping is large enough

$$C_n > 64 \tag{67}$$

then lock-in will be suppressed in that vibration mode.

(c) If for a given vibration mode

$$V/f_n D < 3.3 \tag{68}$$

and

$$c_n > 1.2 \tag{69}$$

then lift direction lock-in is avoided and drag direction lock-in is suppressed.

(d) If the structural natural frequency falls in the ranges $f_n < 0.7f_s$ or $f_n > 1.3f_s$, then lock-in in the lift direction is avoided in the nth mode.

The reduced damping C_n is calculated according to

$$C_n = \frac{4\pi\xi_n M_n}{\rho D^2 \displaystyle\int_{L_e} \phi_n^2(x)\, dx} \tag{70}$$

where $\xi_n = \delta_n/2\pi$ is the fraction of critical damping measured in air and M_n is the generalized mass

$$M_n = \int_0^L m(t)\phi_n^2(x)\, dx \tag{71}$$

with ϕ_n the nth mode shape function and $m_t(x)$ is the cylinder mass per unit length. The range L_e in the denominator implies that the integration is over only the region of the cylinder length subject to lock-in cross flow. Note that m_t is calculated according to:

$$m_t(x) = m_s(x) + m_c(x) + m_A(x) \tag{72}$$

For an isolated cylinder, m_A is the displaced fluid mass. If sections of the cylinder are close to other bodies, then the possibility of increased added mass and fluid damping must be taken into account (Refs. 81, 82, 90 to 93).

N-1324.2 Vortex-Induced Response. Off resonance, the response can be calculated using standard methods (Ref. 96) of forced-vibration analysis and Eq. (63) for the forcing function (Refs. 82 and 89). The resultant response is ordinarily small. If operating conditions are such that lock-in cannot be avoided or suppressed, then the resonant vortex induced response must be calculated. Three approaches for calculating the response are recommended for three classes of structures and flows: single uniform cylinder in uniform flow, tube arrays, and nonuniform cylinders in nonuniform flow.

(a) Uniform Structure and Flow. If a uniform cylinder is subject to uniform cross flow over its span, then both the vortex shedding frequency and the vortex force are constant over the span of the cylinder. The periodic vortex induced lift force is given by Eq. (63) of N-1321. At lock-in, the vortex shedding frequency equals the natural frequency of the nth vibration mode, $f_s = f_n$, and the cylinder response is given by (Refs. 82 and 89)

$$\frac{y^*_n}{D} = \frac{C_L J \phi^*_n}{16\pi^2 S l^2 [m_t \xi_n / (\rho D^2)]} \tag{73}$$

This equation provides a conservative upper bound estimate to the amplitude of periodic vortex-induced vibration if the lift coefficient is taken as unity, $C_L = 1$, and the vortex shedding is fully correlated along the span of the cylinder, $J = 1$. Other values of C_L and J may be used in circumstances where experimental data are available. However, Eq. (73) with $C_L = 1$ and $J = 1$ has been found to give overly conservative predictions owing to the tendency of the actual lift coefficient to decrease at vibration amplitudes exceeding 0.5 diameter and the lack of perfect spanwise correlation at lower amplitudes. To obtain less conservative predictions, three semiempirical nonlinear methods are given in Table N-1324.2(a)-1. The mode shape factor γ generally varies between 1.0 and 1.3 (Ref. 82) and C_n is determined according to Eq. (70) of N-1324.1 using ξ_n determined in air.

(b) Within Tube Arrays. Coherent vortex shedding has been found to exist only in the first few rows in arrays of cylinders with center-to-center spacing less than 2 diameters, and the design procedures for a single cylinder are applicable using the velocity in the minimum gap $(P - D)$. Within the array vortex shedding exists over a broad range of frequencies rather than at a single distinct frequency. The response within the array is generally less than that of a comparable single cylinder. The techniques that have been developed to predict vibration within the array are based on the theory of random vibration and are given in N-1340.

TABLE N-1324.2(a)-1
SEMIEMPIRICAL CORRELATIONS FOR PREDICTING
RESONANT VORTEX-INDUCED VIBRATION
AMPLITUDE

Reference	Predicted Resonant Amplitude
111	$$\frac{y^*_n}{D} = \frac{1.29\gamma}{[1 + 0.43(2\pi S^2 C_n)^{3.35}}$$
82	$$\frac{y^*_n}{D} = \frac{0.07\gamma}{(C_n + 1.9)S^2}\left[0.3 + \frac{0.2}{(C_n + 1.9)S}\right]^{1/2}$$
110	$$\frac{y^*_n}{D} = \frac{0.32}{[0.06 + (2\pi S^2 C_n)^2]^{1/2}}$$

(c) Nonuniform Structures and Flow. Many cylindrical structures have nonuniform distribution of mass and stiffness, and they are exposed to flow velocities that vary over the span. In this case, only a part of the span of structure will resonate with vortex shedding and contribute to the excitation.

One method for treating nonuniform structures in nonuniform flows is:

(1) determine the natural frequencies and mode shapes of the structure;

(2) determine the spanwise distribution of the flow;

(3) identify portions of the structure that can resonate with vortex shedding for each mode. This can be done by calculating the spanwise distribution of the vortex shedding frequency and estimating the potential for resonance by a band of plus or minus 30% from this frequency.

(4) Apply a lift force given by Eq. (63) of N-1321 with $f_n = f_s$ and $C_L = 1$ to those segments of the span that are resonant.

Procedures (1) through (4) are illustrated in Refs. 89 and 112. For a uniform cylinder in uniform cross flow, assumption (4) of complete correlation and $C_L = 1$ gives overly conservative predictions. Other values for C_L may be used where experimental data are available.

N-1330   FLUID-ELASTIC INSTABILITY

Many FIV mechanisms exist wherein as energy supplied to the system is increased, usually as increased flow velocity, a critical value is attained at which a large increase in response occurs. Continued increases in the supplied energy results in continued static or dynamic divergences (rapid increases) of the response. In general, fluid-elastic instability is a result of strong coupling between the structure and the fluid.

N-1331   Instability of Tube Arrays in Cross Flow

Fluid flow across an array of elastic tubes can induce a dynamic instability that can result in very large amplitude vibrations once a critical cross flow velocity is exceeded. Often, motion is limited only by tube-to-tube impacting. The flow of fluid over the tubes results in both fluid excitation and fluid-structure coupling forces on the tubes. The fluid-structure coupling excitation forces fall into several groups:

(a) forces that vary approximately linearly with displacement of a tube from its equilibrium position (displacement mechanisms) (Ref. 113);

(b) fluctuations in the net drag forces induced by the oscillating tube's relative velocity with respect to the mean flow (fluid damping mechanism) (Ref. 88); and

(c) combinations of the above forces that exhibit step changes as a certain amplitude is exceeded because of the abrupt shift in the point of flow separation (jet switch mechanism) (Ref. 114). Instability may result from any or all of these fluid forces which are functions of the tube motion.

The general characteristics of tube vibration during instability are as follows.

(d) Tube Vibration Amplitude. Once a critical cross flow velocity is exceeded, vibration amplitude increases very rapidly with flow velocity V, usually as V^n where $n = 4$ or more, compared with an exponent in the range $1.5 < n < 2.5$ below the instability threshold. This can be seen in Fig. N-1331-1, which shows the response of an array of metallic tubes to water flow. The initial hump is attributable to vortex shedding that tends to produce larger amplitudes in water flow than air flows.

(e) Vibration Behavior With Time. Often the large amplitude vibrations are not steady in time, but rather beat with amplitudes rising and falling about a mean value in a pseudorandom fashion (Ref. 115).

(f) Synchronization Between Tubes. Most often the tubes do not move as individuals, but rather move with neighboring tubes in somewhat synchronized orbits, as shown in Fig. N-1331-2. This behavior has been observed in tests both in water and air (Refs. 113, 115, 116, and 117), with orbit shapes ranging from near circles to near straight lines. As the tubes whirl in their oval orbits they extract energy from the fluid. The stiffness mechanism requires motion of the adjacent tubes, but the damping mechanism does not.

(g) Influence of Structural Variations. Restricting the motion or introducing frequency differences between one or more tubes often increases the critical velocity for instability (Refs. 115, 116, and 118). Such increases are generally no greater than about 40%. Often the onset of instability is more gradual in a tube bank with tube-to-tube frequency differences than in a bank with identical tubes which are free to vibrate.

N-1331.1 Prediction of the Critical Velocity. Dimensional analysis considerations imply that the onset of instability is governed by the following dimensionless groups: the mass ratio $m_t/\rho D^2$; the reduced velocity V/fD; the damping ratio ξ_n, measured in the fluid; the pitch to diameter ratio P/D; the array geometry (see Fig. N-1331-3), and the Reynolds number VD/ν. In this section, V is the flow velocity in the gaps between the tubes, and is determined by the product of $P/(P - D)$ and the (approach) flow velocity that would occur if the tubes were not present. Note the added mass part of m_t may be much larger than the displaced fluid mass because of the confining effect of adjacent tubes (Refs. 81, 90, and 92). Also, for most cases, the flow is fully turbulent ($VD/\nu > 2000$) and the Reynolds number is not expected to play a major role in the instability. In such cases, the reduced critical velocity for the onset of instability can be expressed as a function of the remaining nondimensional parameters.

The relationship between the parameters can be investigated theoretically or experimentally. One general form that has been used to fit experimental data is:

$$V_c/f_n D = C(m_t/\rho D^2)^a (2\pi \xi_n)^b \tag{74}$$

where C and the indices a and b are functions of the tube array geometry. Experimental data suggest that a and b fall in the range $0.0 < a, b < 1.0$ (Refs. 115, 116, 119, 138, and 139).

N-1331.2 Recommended Formula. Mean values for the onset of instability can be established by fitting semiempirical correlations to experimental data. The correlation form chosen is

$$V_c/f_n D = C[m_t(2\pi\xi_n)/\rho D^2]^a \tag{75}$$

where,

   U_c = critical cross flow velocity

   f_n = natural frequencies of the immersed tube. For uniform cross flow, the tubes will be stable if the representative cross flow velocity V is less than the critical velocity V_c. If the flow is nonuniform over the tube lengths, an equivalent uniform cross flow gap velocity can be defined as either the maximum cross flow velocity, or the modal weighted velocity:

$$V_e^2 = \int_0^L V^2(x)\phi_n^2(x)\, dx \bigg/ \int_0^L \phi_n^2(x)\, dx \tag{76}$$

where $V(x)$ is the cross flow velocity at each axial location of the tube. The tubes will be stable if $V_e < V_c$ for all modes.

The available 170 data points for onset of instability (Ref. 120) are shown in Fig. N-1331-4. In the range $m_t(2\pi\xi_n)/\rho D^2 > 0.7$, there are sufficient data to permit fitting of Eq. (75) to data for each array type. The mean values of C are

	Triangle	Rotated Triangle	Rotated Square	Square	All
C_{mean}	4.5	4.0	5.8	3.4	4.0

Based on theory for the displacement mechanism (Ref. 113), which is active in this parameter range, $a = 0.5$ was chosen in these fits. For $m_t(2\pi\xi_n)/\rho D^2 < 0.7$, where the fluid damping mechanism is primarily active, neither the theory nor data are sufficient to establish values of C and a in Eq. (75). Conservative estimates of the mean values of $V_c/f_n D$ for $m_t(2\pi\xi_n)/\rho D^2 < 0.7$ can be obtained using Eq. (75) with $a = 0.5$ and the mean C given in the table above. The use of Eq. (75) with $a = 0.5$ and $C = 3.3$ has been recommended (Refs. 80, 100) for the entire mass damping parameter range of Fig. N-1331-4.

N-1331.3 Suggested Inputs. Accurately predicting the critical velocity requires scale model testing to determine the value of C and the damping ratio in each application, because practical flow and structural geometries contain features nonexistent in the simpler, controlled laboratory tests used to establish the data base of Fig. N-1331-4 (Ref. 120). Usually, industrial tube arrays (bundles) involve multiple spans

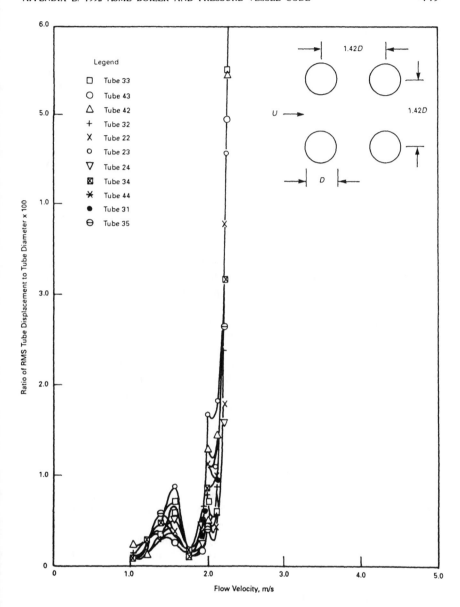

Fig. N–1331–1 Response of a tube bank to cross flow. (ref. 115)

with intermediate supports provided by plates with holes slightly larger than the tube diameter. Also, flow may pass around the edge of the bundle and does not have the pure cross flow direction shown in Fig. N-1331-3, even within the bundle. Furthermore, when the vibration amplitude is small, such as that experienced during sub-critical vibration, not all support plates are active. Damping ratios in this vibration

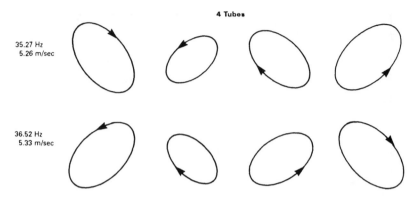

Fig. N-1331-2 Tube vibration patterns at fluidelastic instability for a four-tube row. (ref. 118)

mode are typically small, from 0.1% in gas to about 1% in steam or water. When the vibration amplitude is large, as characterized by the onset of instability, support plate-to-tube interaction greatly increases the damping ratio which can reach 5% or more.

All the practical features discussed above tend to raise the critical flow velocity. Thus, the data base of Fig. N-1331-4 can be used to determine a conservative criterion for avoiding fluid-elastic instabilities of tube arrays: if the design equivalent uniform cross flow gap velocity [Eq. (76)] is less than the critical velocity [Eq. (75)] computed with the suggested design values defined by the solid line ($C = 2.4$, $a = 0.5$) in Fig. N-1331-4 and with a damping ratio of 0.5% in gas, or 1.5% in "wet" steam or liquid, then instability is almost certainly not a problem, and scale model testing will not be necessary. Otherwise, more accurate values of C and the immersed tube's damping ratios, or the critical velocity itself, must be determined by either model testing or from operational experience.

N-1340   TURBULENCE

In general, the coolant flow paths and flow rates promote and maintain turbulent flows that are optimal for purposes of heat transfer, but provide sources for structural excitation. Also, turbulence in the flow can affect the existence and strength of other excitation mechanisms associated with separating boundary layer flows (wakes), as discussed in N-1320 on vortex shedding. This section will concentrate on turbulence as a source of fluid excitation forces.

N-1341   Random Excitation

Where turbulent flow comes into contact with the surface of a structure some of the momentum in the flow is converted into fluctuating pressures. In addition to any forces produced by the mean flow component, random surface pressure fluctuations are produced by the turbulent velocity component. The time history of the surface pressure fluctuations, like the flow turbulence, is complex and amenable to descrip-

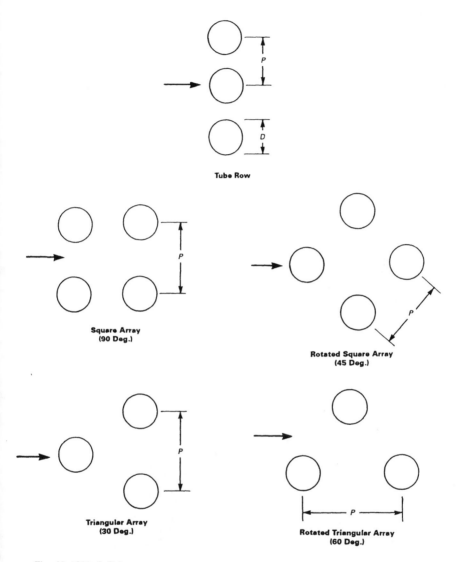

Fig. N-1331-3 Tube arrangements.

tion only on a statistical basis. However, the fluctuating pressure and the resulting flow-induced response usually can be regarded as ergodic and analyzed with a finite-time record not dependent upon the time origin.

For purposes of structural analysis and design, most useful information on the fluctuating pressures becomes available once the spatial spectral densities of the pressure

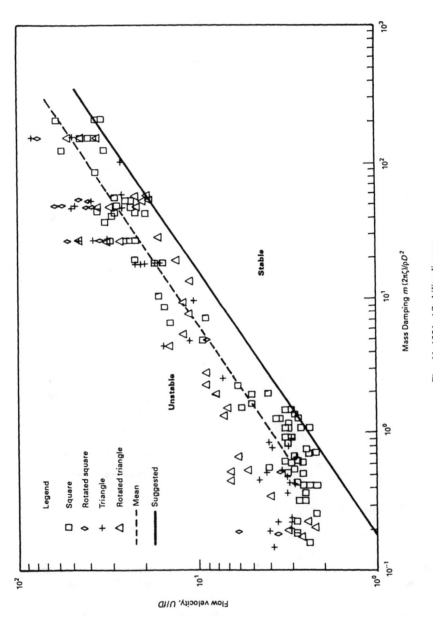

Fig. N–1331–4 Stability diagram.

452

field, $S_p(x_1, x_2, \omega)$, are determined. The spectral density is the Fourier transform of the cross correlation of the pressure field

$$S_p(x_1, x_2, \omega) = \frac{1}{2\pi} \int_{-\infty}^{\infty} \left\{ \lim_{t_0 \to \infty} \frac{1}{2t_0} \int_{-t_0}^{t_0} p(x_1, t) \right.$$
$$\left. \times \; p(x_2, t + \tau) \, d\tau \right\} e^{-i\omega t} \, dt \tag{77}$$

and provides information about the average products of components of the pressure $p(x, t)$ as a function of the circular frequency ω in radian/sec for every possible pairing of structural points, x_1 and x_2, including the same point. S_p has units of (pressure)^2 (second). The frequency content of the spectra are band limited, from zero to a maximum frequency determined by the turbulence source. The magnitude of the power spectra increases when the energy of the turbulence at frequency ω increases. The size of a region of the structure over which the pressures at different points are coherent, or has some cause-effect relationship, is interpreted as a correlation length of the pressure field, or the size of the associated turbulent eddy (Ref. 82).

Only select parts of the surface pressures are effective in exciting dynamic structural response: those parts with frequency content in narrow bands centered on the structural natural frequencies and with correlation lengths similar in size to the spatial wavelength of the associated vibration mode (Ref. 122). The resulting structural response occurs in the narrow frequency bands with random amplitudes, and the widths of the frequency bands are determined by the system damping. Knowledge of the surface pressure statistics enables prediction of the associated structural response statistics utilizing the probabilistic theory of structural dynamics.

N-1342   Structural Response of Tubes and Beams

N-1342.1 Response to Homogeneous Turbulence Excitation. The assumption of a linear structure is justifiable for the small vibrations associated with the turbulence excitation of weakly coupled fluid-structure systems, and the linear structural dynamic analysis theory for arbitrary random loading of beams is highly developed. Since the energy dissipation mechanism of turbulent flow rapidly smooths disturbances caused by the structural boundaries of and in the flow channel, the statistical character of turbulent cross flow often varies gradually over the total length of select spans of single tubes and tube bundles, especially within the bundles. Thus, in many applications, the assumption of a uniform mean velocity and homogeneous turbulence is reasonable.

Assuming a homogeneous and ergodic pressure field, the equations of motion can be uncoupled to allow solution by modal analysis (see N-1222). The expression for the power spectral density of the cylinder response is (Ref. 123)

$$S_y(x_1, x_2, \omega = S_{f_0} L_j, \sum_{k=1}^{\infty} \sum \phi_j \phi_k J_{jk}^2 H_j H^*_k \tag{78}$$

The mode shapes $\phi_j(x)$ satisfy the orthogonality relation

$$\int_0^L m_t(x)\phi_i(x)\phi_j(x) \, dx = M_j \delta_{ij} \tag{79}$$

where M_j is the generalized mass, defined here to have the same dimensions as m_t. Thus, if m_t is constant, $M_j = m_t$ and the orthogonality condition reduces to

$$\int_0^L \phi_i(x)\phi_j(x)\, d = \delta_{ij} \tag{80}$$

The transfer function for the jth mode is

$$H_j(\omega) = [M_j(\omega_j^2 - \omega^2 + 2i\xi_j\omega\omega_j)]^{-1} \tag{81}$$

where ω_j and ξ_j are the modal natural frequency and damping, respectively. The acceptance integral is

$$J_{jk}^2(\omega) = L^{-1} \int_0^L \int_0^L \phi_j(x_1)\Gamma(x_1, x_2, \omega)\phi_k(x_2)\, dx_1\, dx_2 \tag{82}$$

where the complex coherence function is

$$\Gamma(x_1, x_2, \omega) = S_f(x_1, x_2, \omega)/S_{fo}(\omega) \tag{83}$$

and S_f is the cross spectral density of the turbulent forcing function per unit length between two different points on the cylinder's length, $x = x_1$ and $x = x_2$. When $x_1 = x_2 = x$, $S_f(x, x, \omega) = S_{fo}(\omega)$ is the power spectral density (or autospectrum) that is independent of location for a homogeneous pressure field. The joint acceptance $J_{jj}(\omega)$ reflects the relative effectiveness of the forcing function to excite the jth vibration mode while the cross acceptance $J_{jk}(\omega)$, $j \neq k$, reflects contributions due to coupling between different modes. In general, the responses in two different modes are dependent upon each other.

The mean square response $y^2(x)$ is the most useful measure for the amplitude or stress and strain design, and is found by integration of the power spectral density of the response, $S_y(x, \omega) = S_y(x, x, \omega)$ over the frequency band, or

$$\overline{y^2}(x) = \int_{-\infty}^{\infty} S_y(x, \omega)\, d\omega \tag{84}$$

The distribution of the positive and negative peaks in displacement has been found for the fundamental mode of a rod in parallel flow to be Gaussian (Refs. 81, 82, and 124). Assuming a Gaussian fluctuating pressure distribution, a Rayleigh distribution is expected for the absolute amplitude of response (Refs. 82 and 125).

Based on physical reasoning and experimental data, the complex coherence function of the homogeneous turbulence pressure field for the tubes in cross flow has been characterized as (Refs. 122 and 126)

$$\Gamma(x_1, x_2, \omega) = \exp[-2|x_1 - x_2|/l_c]$$
$$\times \exp[-i\omega(x_1 - x_2)\sin\theta/U_c] \tag{85}$$

where $l_c \ll L$ is the correlation length, which is a measure of the coherence range of the turbulent pressure field; U_c is the convection velocity, or the velocity at which the turbulent eddies move with the flow; and θ is the angle between the direction of the flow and the normal to the axis of the tube. Note that $U_c/\sin\theta$ is the phase velocity of the pressure signal along the tube.

In the case of lightly damped structures with well-separated modes, cross modal

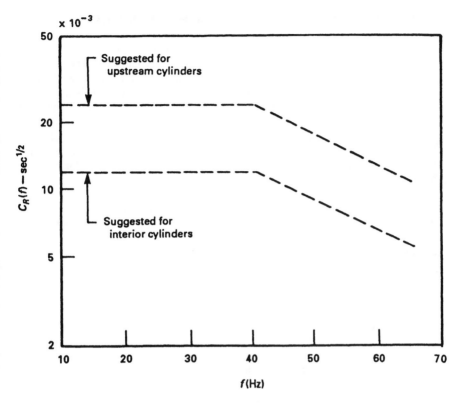

Fig. N–1343–1 Random excitation coefficient for arrays in cross flow. (ref. 100)

contribution to the response can be ignored, and Eq. (84) can be analytically evaluated to be (Refs. 112 and 125):

$$\overline{y^2}(x) = \sum_j 2\pi\xi_j\omega_j S_y(\omega_j) = \sum_j \pi\xi_j f_j G_y(f_j) \tag{86}$$

In the second equality, the response is expressed in terms of the more commonly used engineering variables frequency f (in Hz) and the single-sided power spectral density G as a function of f, where

$$G(f) = 4\pi S(\omega) \quad \text{for} \quad f = \omega/2 > 0 = 0 \quad \text{for} \quad f < 0 . \tag{87}$$

Under the same assumption of light structural damping

$$\overline{y^2}(x) = \sum_j \frac{L G_f(f_j)\phi^2_j(x)}{64\pi^3 M_j^2 f^3_j \xi_j} J_{jj}^2 \tag{88}$$

where $G_f(f_j)$ is the single-sided power spectral density in (force/length)^2/Hz generated by the turbulent pressure field at the natural frequency f_j of the jth vibration mode.

N-1343   Design Procedures for Tubes and Beams in Turbulent Cross Flow

In most situations, the component of turbulent flow normal to the axis of a cylinder is a more dominant excitation mechanism than the parallel component. The exception occurs when the flow direction is parallel or barely inclined to the cylinder axis. Thus, cross flow analysis using the component of flow normal to the cylinder should always be made, supplemented by the parallel flow analyses of N-1345 at small angles of inclination.

The theory in the subsections that follow can be applied to one-dimensional structures in general, but the specific information on the statistics of the pressure field must be limited to tubes (circular cylinders) until adequate information is available for other beam cross sections.

N-1343.1 Uniform Cross Flow. In the simplest case of uniform cross flow over the entire length of a lightly damped, rigid cylinder with an evenly distributed total mass and spring supports at both ends, $\phi = L^{-1/2}$, $\theta = 0$ in Eq. (85), and the joint acceptance integral of Eq. (82), with $i = j = 1$, reduces to (Refs. 112 and 127)

$$J_{11}^2 \approx l_c/L, \qquad l_c << L \tag{89}$$

The correlation length l_c in most cross flows over a tube (circular cylinder) is no more than three diameters (see N-1322). Also, although Eqs. (89) were derived for the fundamental mode of the transverse vibrations of a rigid, spring-supported cylinder, they can be used to estimate the joint acceptance of the fundamental mode of cylinders which are simply supported or clamped at both ends. Of course, in determining the RMS response with Eq. (88), the mode shapes corresponding to the actual boundary conditions and normalized according to Eq. (80) are used. For other boundary conditions and higher modes, the joint acceptance integral will have to be evaluated either numerically (most cases) or in closed form from Eq. (85). Since $J_{11}^2 \leqslant 1.0$, (Ref. 127), an upper bound response estimate can be found by setting all the $J_{jj} = 1.0$ in Eq. (88).

The random characteristics of the forces exerted on the tubes by the turbulent flow must be obtained from tests. Two expressions for the power spectral density of the turbulent force per unit length on tubes in a tube array are:

$$G(f) = [C_R(f)\rho V_g^2 D/2]^2 \qquad \text{(Ref. 100)} \tag{90}$$

and

$$G(f) = [C_L(f)\rho V_g^2 D/2]^2 (D/V_g) \qquad \text{(Ref. 128)} \tag{91}$$

where the gap velocity V_g is related to the velocity upstream of the tubes, V_∞, by

$$V_g/V_\infty = P/(P - D) \tag{92}$$

The coefficient $C_R(f)$ in Eq. (90) has units of sec$^{-1/2}$ and is given in Fig. N-1343-1 as a function of the frequency f. Therefore, the application of Eq. (90) should be limited to the parameter range for which the data were taken, namely, high turbulent water flow (1 to 2 m/sec) entering closely spaced heat exchanger tubes of 12 to 19 mm in diameter. The decrease in $C_R(f)$ with penetration into the bundle is attributed to the highly turbulent inlet flow and possible vortex excitation observed in the first few tube rows. Data for the nondimensional lift coefficient $C_L(f)$ of Eq. (91) have not been obtained for as many tube array configurations (Refs. 141 and 142), but use of this alternative expression may predict less conservative responses (Ref. 142).

For an isolated tube in cross flow, which is not subject to conditions of lock-in vortex shedding (see N-1323), the power spectral density $G_f(f)$ and the correlation length are strong functions of the turbulence created in the incident flow stream by the upstream structures. Relatively small amounts of turbulence can cause significant reductions in the effectiveness of vortex shedding as an excitation mechanism, and all periodicity can be eliminated with sufficiently strong incident turbulence. For given turbulence intensities and scale lengths of the incident flow, $G_f(f)$ is available (Ref. 129) and the correlation length l_c may be approximated by the scale length of the incident flow. In the absence of specific information about the incident flow, the random turbulence coefficient for the upstream tube in Fig. N-1343-1(a) can be used to estimate $G_f(f)$ for most isolated tubes in cross flow because of the wide variety of incident flow conditions contained in the data base. In the latter case, the velocity used in Eqs. (90) and (91) should be the free stream velocity of the flow.

If upstream structures produce well defined vortices, strong excitation mechanisms may be created on isolated cylinders more than twenty diameters downstream (Ref. 130). Such configurations should be avoided.

N-1343.2 Multiple Spans of Uniform Cross Flow. In many applications, a cylinder is subject to one or more partial spans of uniform, but different, velocity and density cross flows that are uncorrelated with each other or with the flow over the remainder of the cylinder's length. These conditions often exist when baffles are used to channel different density flows in the interior of the pressure vessels (heat exchangers, reactors, etc.). The mean square response for such conditions can be determined by simple generalizations of the results given in N-1342.2 for homogeneous turbulence excitation.

Since the uniform cross flow over the span of length L_i is uncorrelated with the uniform cross flows over the other spans, $G_f J_{jj}^2$ in Eq. (88) can be calculated (Ref. 127) by summing the products of the locally defined spectra G_f^i and joint acceptances $(J_{jj}^i)^2$ over all the spans i over which there is significant cross flow. Thus, the mean square response becomes

$$\overline{y^2}(x) \approx \sum_j \sum_i \frac{L_i G'_f(f_j)\phi_j^2(x)}{64\pi^3 M_j^2 f_j^3 \xi_j} (J_{jj}^i)^2 \tag{93}$$

where

$$G'_f(f_j) = G_f(f_j) \int_0^{L_i} \phi_j^2(x)\, dx \tag{94}$$

$(J_{jj}^i)^2$ are determined with Eq. (82) using that part of ϕ_j, denoted by ϕ_j^i, that is active over the ith span with length L_i. As discussed in N-1343.1, if ϕ_j^i is similar to the fundamental mode shape of a one-span beam with simple or clamped supports at each end, then $(J_{jj}^i)^2 \approx l_c^i/L_i$. The correlation lengths inside a tube bundle are smaller than that for an isolated tube, being about 1–2 tube diameters. After specifying $G_f^i(f)$ and using Eqs. (90) and (91) for instance, the mean square response can be determined.

N-1343.3 Nonuniform Cross Flow. In industrial heat exchangers, the cross flow velocities are seldom uniform over the entire length, or even one span of the tubes.

While an average cross flow velocity can be used to estimate the force spectra in Eqs. (90), (91), and (93), when the velocity distribution is available, better estimates can be obtained by using mode shape-weighted power spectral densities similar to the generalized forces used in deterministic analysis (Ref. 127):

$$G_f(f_j) = (D/2)^2 C_R^2(f_j) \int_0^L [\rho U^2(x)]^2 \phi_j^2(x)\, dx \qquad (95)$$

for a single-span tube of uniform mass density and

$$G_f^i(f_j) = (D/2)^2 C_R^2(f_j) \int_0^{L_i} [\rho_i U^2(x)]^2 \phi_j^2(x)\, dx \qquad (96)$$

for a multi-span of spanwise uniform mass density. These estimates are not rigorously derivable, but they will lead to more accurate estimates of response, especially when the peaks in the velocity distributions are close to the antinodes of the vibration modes.

N-1344 Vortex-Induced Vibrations in a Tube Bundle

The existence of vortex shedding deep in a tube bundle is much less clearly defined than for a single cylinder. Experimental measurements involving tube bundles showed that even if a resonance peak exists in the dynamic pressure power spectral density, it is much broader and not as well defined as in the case of a single tube. Furthermore, these peaks are bounded by the pressure power spectral density given by Eq. (90). However, if lock-in vortex-induced vibration occurs in a particular span, the forcing function and the tube mode shape will be fully correlated and in-phase for that span. This means that the span joint acceptance, $J_{jj}^i = 1.0$. To be conservative, a lock-in vortex-induced vibration amplitude deep in a multi-span tube bundle can be calculated by substituting $J_{jj}^i = 1.0$ into Eq. (93), for all spans and all modes where lock-in cannot be avoided or suppressed according to N-1324.1.

Classical vortex shedding does occur in the boundary tubes. For the first two to three rows of tubes in a tube bundle, vortex-induced vibration analysis following the procedure outlined in N-1320 is recommended.

N-1345 Cylinders in Axial Flow

Turbulence generally is a much weaker excitation mechanism in axial flow compared with cross flow, where the flow separates from the vibrating body. Also, axial flow is a source of flow damping which increases with flow rate (Refs. 81, 82, and 131). As a result, RMS vibration amplitudes of tubes in axial flow are typically only a few percent of the tube diameter.

The surface pressure fluctuations that excite a tube in axial flow are due to many sources: local turbulence created by the shear flow in the developing boundary layer; free stream turbulence created by upstream disturbances (grid supports, abrupt changes in channel size, elbows, valves, etc.) that quickly attenuate downstream of the disturbance; localized acoustic noise (waves); and system acoustic noise that can propagate long distances (Ref. 136). For pipes and single rods in annuli subject only to fully developed flow, relatively homo general experimental characterizations of the homogeneous pressure fields are possible (Ref. 131), because they depend only on the local channel geometry and the flow rates. However, general characterizations have not been developed that account for upstream disturbances and adjacent bodies, although many specific systems have been studied (Refs. 131 through 137). Evidently, accurate predictions can be made when the pressure field is characterized in the same system as the response is measured, but the predictions from system to system may vary by an order of magnitude for the same axial flow velocity. Because response is usually much easier to measure than the pressure fluctuations necessary to characterize a pressure field, especially a nonhomogeneous one, empirical correlations of response have been developed for important component geometries (Refs. 99 and 137). The component and prototype tests upon which the correlations are based include all component geometries and excitation sources. Of course, the use of these correlations must be limited to the type of components and parameter variations for which they were developed.

N-1345.1 Recommended Design Procedures

(a) When the characterization of the pressure field is available, then the response of the structure can be predicted by the general method outlined in N-1342. But, unlike cross flows, the convection velocity U_c is important and must be known in axial flow before the acceptance integral, Eqs. (82) and (85), can be evaluated. In axial flows, it is not generally true that the larger the correlation length l_c, the larger the response as in cross flows. Rather, the response is governed by the matching of the structural mode shape and the phase-coherence of the pressure field (Refs. 112 and 122), in addition to its power spectral density.

(b) Regardless of whether a pressure field characterization is available, the maximum amplitude of motion can be estimated to within an order of magnitude using the empirical correlation (Ref. 99)

$$\frac{y^*}{D} = \left[\frac{5 \times 10^{-4} K_N}{\alpha^4} \right] \left[\frac{u^{1.6} \epsilon^{1.8} Re^{0.25}}{1 + u^2} \right]$$
$$\times \left[\frac{D_h}{D} \right]^{0.4} \left[\frac{\beta^{2/3}}{1 + 4\beta} \right] \tag{97}$$

if the cylinder parameters are within the ranges covered by the correlation:

$2.1 \times 10^{-3} \leqslant u^2 = m_A V^2 L^2 / (EI) \leqslant 8 \times 10^{-1}$
$26.8 \leqslant \epsilon = L/D \leqslant 58.7$
$2.6 \times 10^4 \leqslant Re = VD/\nu \leqslant 7 \times 10^5$
$4.9 \times 10^{-4} \leqslant \beta = m_A/m_t \leqslant 6.2 \times 10^{-1}$
$2.10 \leqslant \alpha^2 = \omega_1 [m_t L^4 / EI]^{1/2} \leqslant 20.8$
$1 \leqslant K_n \leqslant 5$

where K_n is a noise factor representing a departure from quiet, steady axial flows of $K_n = 1$. Commercial systems are expected to be bounded by $K_n = 5$. E is the modulus of elasticity, I is the beam area moment of inertia, and ν is the fluid kinematic viscosity.

REFERENCES

(79) Chen, P. Y. (ed.) *Flow-Induced Vibration Design Guidelines*, ASME, Vol. pp. 1–52, New York, 1981.

(80) Paidoussis, M. P. A Review of Flow-Induced Vibrations in Reactors and Reactor Components. *Nuclear Science and Engineering:* 74, pp. 31–60, 1983.

(81) Chen, S. S. *Flow-Induced Vibration of Circular Cylindrical Structures*, Hemisphere Publishing Corporation, Washington, DC, 1987.

(82) Blevins, R. D. *Flow-Induced Vibration*, 2nd Ed., Van Nostrand Reinhold, New York, 1990. Krieger, 1993.

(83) Mulcahy, T. M., and Wambsganss, M. W. Flow-Induced Vibration of Nuclear Reactor System Components, *Shock Vib. Dig.* 8(7), pp. 33–45 1976.

(84) Naudascher, E., and Rockwell, D. (eds.) *Practical Experiences with Flow Induced Vibrations*, Springer-Verlag, New York, 1980.

(85) Mulcahy, T. M. Flow-Induced Vibration Testing Scale Modeling Relations. *Flow-Induced Vibration Design Guidelines*, pp. 111–126.

(86) Bohm, G. J., and Tagart, S. W., Jr. Flow-Induced Vibration in the Design of Nuclear Components. *Flow-Induced Vibration Design Guidelines*, pp. 1–10.

(87) Sarpkaya, T. Vortex-Induced Oscillations—A Selective Review. *Journal of Applied Mechanics*, 6, pp. 241–258 1979.

(88) Chen, S. S. Vibration of a Group of Circular Cylinders Subjected to a Fluid Flow. *Flow-Induced Vibration Design Guidelines*, pp. 75–88.

(89) Connors, H. J., Jr. Vortex Shedding Excitation and the Vibration of Circular Cylinders. *Flow-Induced Vibration Design Guidelines*, pp. 47–74.

(90) Chen, S. S. Fluid Damping for Circular Cylindrical Structures. *Nucl. Eng. Des.* 63(1), pp. 81–109, 1981.

(91) Mulcahy, T. M. Fluid Forces on Rods Vibrating in Finite Length Annular Regions. *Journal of Applied Mechanics* 102(2), pp. 234–240, 1980.

(92) Blevins, R. D. *Formulas for Natural Frequency and Mode Shape*, Van Nostrand Reinhold Company, New York, 1979. Reprinted Robert E. Krieger Publishing Co., Malabar, FL.

(93) Au-Yang, M. K. Generalized Hydrodynamic Mass for Beam Mode Vibration of Cylinders Coupled by Fluid Gap. *Journal of Applied Mechanics* 44, pp. 172–173, 1977.

(94) King, R. A Review of Vortex Shedding Research and Its Application. *Ocean Engineering* 4, pp. 141–171, 1977.

(95) Blevins, R. D. Review of Sound Induced by Vortex Shedding from Cylinders. *J. Sound Vib.* 92, pp. 455–470, 1984.

(96) Den Hartog, J. P. *Mechanical Vibrations*, 4th Ed., McGraw-Hill, New York, p. 305, 1956.

(97) Keefe, R. T. An Investigation of the Fluctuating Forces Acting on a Stationary Circular Cylinder in a Subsonic Stream and of the Associated Sound Field. University of Toronto, UTIA Report No. 76, 112, 1961.

(98) Ramberg, S. E. The Influence of Yaw Angle Upon the Vortex Wakes of Stationary and Vibrating Cylinders. Naval Research Laboratory Memorandum Report 3822, Washington, DC, 1978.

(99) Paidoussis, M. P. Fluid-elastic Vibration of Cylinder Arrays in Axial and Cross-Flow State of the Art. *Flow-Induced Vibration Design Guidelines*, pp. 11–46.

(100) Pettigrew, M. J., and Gorman, D. J. Vibration of Heat Exchanger Tube Bundles in Liquid and Two-Phase Cross-Flow. *Flow-Induced Vibration Design Guidelines*, pp. 89–110.

(101) Zdravkovich, M. M. Review of Flow Interference between Two Circular Cylinders in Various Arrangements. *Journal of Fluids Engineering* 99, pp. 618–633, 1977.

(102) Owen, P. R. Buffeting Excitation of Boiler Tube Vibration. *J. Mech. Eng. Sci.* 7, p. 437, 1965.

(103) Chen, Y. N. Flow-Induced Vibration and Noise in Tube Bank Heat Exchangers Due to von Karman Streets. *Journal of Engineering for Industry* 90(1), pp. 135–146, 1968.

(104) Fitz-Hugh, J. S. Flow-Induced Vibration in Heat Exchangers. *International Symposium on Vibration Problems in Industry*, Keswick, England, Paper 427, 1973.

(105) Chen, Y. N. Fluctuating Lift Forces of the Karman Vortex Streets on Single Circular Cylinders and in Tube Bundles, Part 3—Lift Forces in Tube Bundles. *Journal Engineering for Industry* 94, pp. 603–628, 1972.

(106) Scruton, C. On the Wind Excited Oscillations of Stacks, Towers, and Masts. *National Physical Laboratory Symposium on Wind Effects on Buildings and Structures*, Paper 16, pp. 798–832, 1963.

(107) Bishop, R. E. D., and Hassan, Y. A. The Lift and Drag Forces on a Circular Cylinder in a Flowing Field. *Proc. Royal Soc.*, London, A 227, pp. 51–75, 1964.

(108) King, R. On Vortex Excitation of Model Piles in Water. *J. Sound Vib.* 29(2), pp. 169–188, 1973.

(109) Mulcahy, T. M. Avoidance of the Lock-in Phenomenon in Partial Cross Flow. *J. Sound Vib.* 112(3), pp. 570–574, 1987.

(110) Sarpkaya, T. Fluid Forces on Oscillating Cylinders. *J. Water Way Port Coastal and Ocean Division* 104, pp. 275–290, ASCE, 1978.

(111) Griffin, O. M., Skop, R. A., and Ramberg, E. The Resonant, Vortex-Excited Vibrations of Structures and Cable Systems. *Offshore Technology Conference*, Houston, TX, Paper No. OTC-2319, 1975.

(112) Au-Yang, M. K. Flow-Induced Vibration—Guidelines for Design, Diagnosis, and Trouble Shooting of Common Power Plant Components. *Joint Flow-Induced Vibration Symposium*, ASME Winter Annual Meeting, New Orleans, LA, 1984; and *Journal of Pressure Vessel Technology* 107, pp. 326–334, 1985.

(113) Connors, H. J. Fluid-elastic Vibration of Tube Arrays Excited by Cross Flow. *Symposium on Flow-Induced Vibration in Heat Exchangers*, ASME Winter Annual Meeting, Dec. 1970.

(114) Roberts, B. W. Low Frequency, Aero-elastic Vibrations in a Cascade of Circular Cylinders. Mechanical Engineering Science Monograph No. 4, 1966.

(115) Chen, S. S., Jendrzejczyk, J. A., and Lin, W. H. Experiments on Fluid-elastic Instability in Tube Banks Subject to Liquid Cross Flow, Part 1: Rectangular Arrays. Argonne National Laboratory Report ANL-CT-78-44, July 1978.

(116) Weaver, D. S., and Grover, L. K. Cross Flow Induced Vibrations in a Tube Bank. *Journal of Pressure Vessel Technology* 101, 1979.

(117) Guerrero, H. N., et al. Flow Induced Vibrations of a PWR Upper Guide Structure Tube Bank Model. Paper presented at *Topical Meeting on Nuclear Reactor Thermal Hydraulics*, Saratoga, NY, Oct. 1980; Combustion Engineering Technical Paper, Windsor, TIS-6297.

(118) Southworth, D. J., and Zdravkovich, M. M. Cross Flow Induced Vibrations of Finite Tube Banks with In-Line Arrangements. *J. Mech. Eng. Sci.* 17, pp. 190–198, 1975.

(119) Paidoussis, M. P. Flow-Induced Vibrations in Nuclear Reactors and Heat Exchangers. In *Practical Experience with Flow Induced Vibrations* (E. Naudascher and D. Rockwell, eds.), Springer-Verlag, New York, pp. 1–81, 1980.

(120) Chen, S. S. Guidelines for the Instability Flow Velocity of Tube Arrays in Cross Flow. *J. Sound Vib.* 93(1), pp. 439–455, 1984.

(121) Blevins, R. D. Discussion of Guidelines for the Instability Flow Velocity of Tube Arrays in Cross Flow. *J. Sound Vib.* 97, pp. 641–644, 1984.

(122) Au-Yang, M. K., and Connelly, W. H. A Computerized Method for Flow-induced Random Vibration Analysis of Nuclear Reactor Internals. *Nucl. Eng. Des.* 42, pp. 257–263, 1977.

(123) Lin, Y. K. *Probabilistic Theory of Structural Dyanmics*, McGraw Hill, New York, 1967.

(124) Wambsganss, M. W., and Boers, B. L. Parallel-Flow-Induced Vibration of a Cylindrical Rod. ASME Paper No. 68-WA/NE-15, Dec. 1968.

(125) Crandall, S. H., and Marks, W. D. *Random Vibration in Mechanical Systems*, Academic Press, New York, 1963.

(126) Corcos, G. M. The Structure of the Turbulent Pressure Field in Boundary Layer Flow. *J. Fluid Mech.* 13, 1964.

(127) Au-Yang, M. K. Turbulent Buffeting of a Multi-Span Tube Bundle. *Journal of Vibration, Stress and Reliability in Design,* 108, pp. 150–154 1986.

(128) Blevins, R. D., Gibert, R. J., and Villard, B. Experiments on Vibration of Heat Exchanger Tubes in Cross Flow. *Sixth International Conference on Structural Mechanics in Reactor Technology,* Paris, France, Paper B6/9, 1981.

(129) Mulcahy, T. M. Fluid Forces on a Rigid Cylinder in Turbulent Cross Flow. *Symposium on Flow-Induced Vibrations, Vol. 1—Excitation and Vibration of Bluff Bodies in Cross Flow,* pp. 5–28, ASME, New York, 1984.

(130) Chen, S. S. A Review of Flow-Induced Vibration of Two-Circular Cylinders in Cross Flow. *Journal of Pressure Vessel Technology* 108, pp. 382–393, 1986.

(131) Chen, S. S., and Wambsganss, M. W. Parallel-Flow-Induced Vibration of Fuel Rods. *Nucl. Eng. Des.* 18, pp. 253–278, 1972.

(132) Mulcahy, T. M., Wambsganss, M. W.. Lin, W. H., Yeh, T. T., and Lawrence, W. P. Measurements of Wall Pressure Fluctuations on a Cylinder in Annular Water Flow with Upstream Disturbances. *Sixth International Conference on Structural Mechanics in Reactor Technology,* Paris, France, Paper B6/5*, 1981.

(133) Gibert, R. S. Etude des fluctuations of pression dans les circuits para courus par des fluides— Sources de fluctuations engendrees par les singularites d'Scoulement. Note CEA-N 1925, 1976.

(134) Mulcahy, T. M., Yeh, T. T., and Miskevics, A. J. Turbulence and Rod Vibrations in an Annular Region with Upstream Disturbances. *J. Sound Vib.* 69(1), pp. 59–69, 1980.

(135) Lin, W. H., Wambsganss, M. W., and Jendrzejczyk, J. A. Wall Pressure Fluctuations Within a Seven Rod Array. General Electric Report GEAP-24375 (DOE/ET/34209-20), San Jose, CA, Nov. 1981.

(136) Kadlec, J., and Ohlmer, E. On the Reproducibility of the Parallel-Flow Induced Vibration of Fuel Pins. *Nucl. Eng. Des.* 17, pp. 355–360, 1971.

(137) Wambsganss, M. W., and Mulcahy, T. M. Flow-Induced Vibration of Nuclear Reactor Fuel. *Shock Vib. Dig.* 11(11), pp. 11–22, and 11(12), pp. 11–13, 1979.

(138) Weaver, D. S., and Yeung, H. C. The Effect of Tube Mass on the Flow Induced Response of Various Tube Arrays in Water. *J. Sound Vib.* 93(3), pp. 409–425, 1984.

(139) Weaver, D. S., and Fitzpatrick, J. A. A Review of Flow Induced Vibrations in Heat Exchangers. *International Conference on Flow Induced Vibrations,* Bowness-on-Windermere, England, Paper A1, pp. 1–17, May 1987.

(140) Weaver, D. S., Fitzpatrick, J. A., and ElKashlan, M. Strouhal Numbers for Heat Exchanger Tube Arrays in Cross Flow. *Journal of Pressure Vessel Technology* 109, pp. 219–223, 1987.

(141) Chen, S. S., and Jendrzejczyk, J. A. Fluid Excitation Forces Acting on a Square Tube Array. *Journal of Fluids Engineering* 109, pp. 415–423, 1987.

(142) Axisa, F., Antunes, J., Villard, B., and Wullschleger, M. Random Excitation of Heat Exchanger Tubes by Cross Flow. *1988 International Symposium on Flow-Induced Vibration and Noise,* ASME, Chicago, IL, Vol. 2—*Flow-Induced Vibration of Cylinder Arrays in Cross Flow,* pp. 23–47, 1988.

Author Index

Abbott, I. H., 132
Abd-Rabba, A., 368
Abel, J. M., 104
Achenbach, E., 45, 48, 50, 75
Ackerman, N. L., 24
Advisory Board on the Investigation of Suspension Bridges, 84
Ainsworth, P., 53
Airey, R. C., 71
American Association of State Highway and Transportation Officials, 275
American Society of Civil Engineers, 108, 127, 129
Anand, N. M., 48, 57, 58
Anderson, J. C., 405
Angrilli, F., 219
Apelt, C. J., 80
Arbhabhirama, A., 24
Aref, H., 35, 38, 47
Arie, M., 176
Armitt, J., 290
Arnold, F. R., 365, 367, 373
Ashley, H., 131, 384, 400
Aso, K., 385
Au-Yang, M. K., 31, 174
Axisa, F., 267, 268, 313

Bailey, R. T., 287
Baird, R. C., 364, 365, 373, 375
Bajaj, A. K., 396
Ballentine, J. R., 241, 250, 256, 334
Balsa, T. F., 163
Bardowicks, H., 110
Barnett, K. M., 47, 48
Barrington, E. A., 371, 375
Bascom, W., 227
Basile, D., 241, 250, 256
Basista, H., 173
Basu, R. I., 63, 72
Batchelor, G. K., 210, 309
Batts, M. E., 274

Baylac, G., 375
Bearman, P. W., 36, 45, 47, 48, 58, 82, 105, 118, 125, 127, 210, 217, 218, 219, 220
Bellman, R., 30, 136, 166
Benard, H., 43
Bendat, J. S., 240, 241, 430
Benjamin, J. R., 275
Benjamin, T. B., 384
Beranek, L. L., 470
Bernard, H., 43
Bernstein, M. D., 378
Bert, C. W., 338
Bhattacharjee, S., 376
Bishop, J. R., 196, 213, 214, 215
Bishop, R. E. D., 54, 55, 63, 67, 225
Bisplinghoff, R. L., 104, 131
Blagoveschensky, S. N., 228, 229
Blake, W., 427
Blake, W. K., 249, 352, 376, 405
Blakewell, H. P., 263
Blessman, J., 45, 47
Blevins, R. D., 17, 24, 39, 52, 63, 67, 73, 74, 91, 104, 144, 153, 158, 169, 178, 188, 199, 210, 245, 268, 271, 280, 310, 325, 359, 362, 370, 375, 394, 406
Bloomfield, W. J., 368
Bokaian, A., 58, 104, 127, 185
Bolotin, V. V., 239
Borthwick, A. G. L., 206, 209, 215, 219
Bouwkamp, J. G., 297, 319, 320, 330
Boyer, R. C., 174
Bracewell, R. N., 431
Bradshaw, R. T., 407
Bressler, M. M., 369, 370, 375
Brigham, E. O., 432
British Standards Institute, 271
Brody, A. G., 288
Brooks, I. H., 81
Brooks, N. P. H., 104, 109, 113, 116, 119
Brouwers, J. J., 309
Brown, R. S., 365
Brown, S. J., 31
Brownjohn, M. W., 319, 320, 328

Subject Index